ENERGY STORAGE IN THE EMERGING ERA OF SMART GRIDS

Edited by **Rosario Carbone**

Energy Storage in the Emerging Era of Smart Grids
Edited by Rosario Carbone

Published by InTech
Janeza Trdine 9, 51000 Rijeka, Croatia

Publishing Process Manager Sandra Bakic

Technical Editor Teodora Smiljanic

Cover Designer InTech Design Team

Additional hard copies can be obtained from orders@intechopen.com

Energy Storage in the Emerging Era of Smart Grids, Edited by Rosario Carbone
p. cm.
ISBN 978-953-307-269-2

For EU product safety concerns:
IN TECH d.o.o., Prolaz Marije Krucifikse Kozulić 3, 51000 Rijeka, Croatia,
info@intechopen.com or visit our website at intechopen.com.

Meet the editor

Dr. Rosario Carbone was born in Italy in 1965. In 1990., he received his degree on Electrical Engineering, from University of Calabria, Italy. In 1995., he received his Ph.D. on Electrical Engineering, from University "Federico II" of Naples, Italy. In 1995., he became a researcher on electrical power systems, at the University of Neaples, Italy. At the moment, he is an Associate Professor on electrical power systems, at the University "Mediterranea" of Reggio Calabria – Italy. His educational activities mainly concern electrical power systems, power electronics and electrical safety. His research activities deal with: (I) power electronic apparatus with high performances; (II) analysis of electrical power systems in presence of power electronic apparatus; (III) high performance distributed generation plants from renewables. Rosario Carbone is author of about 70 papers; he won a "best paper award" at the IEEE International Conference "I.C.H.Q.P. 2000", Orlando, USA .

Contents

Preface

Traditional electrical power systems were essentially based on centralized and fuel consuming power generation plants, where end-users were supplied via unidirectional transmission and distribution grids.

The increasing demand for electrical energy and, at the same time, the need for reducing CO_2 emissions are now changing these strongholds, and power systems are more and more integrated by "distributed generation" (DG), that is to say small and medium size generator plants managed by end-users (now called "prosumers", to underline that they are both consumers and producers) and essentially based on renewables.

In order not to compromise reliability and quality of the supply, modern power systems now have to become "smarter", for properly managing power, received both from centralized and distributed sources; of course, this could be accomplished by means of sophisticated control and communication technologies but, in our opinion, energy-storage systems can also have a central role.

In fact, electricity generated from renewables by distributed plants, unlike to that generated by fuel consuming centralized plants, is highly "intermittent" and this worsens the problem of optimally matching electricity availability with electricity demand of end-users.

Without solving this problem, reliability, quality and stability of modern power system are seriously compromised.

Reliable, high-efficient and cost-effective energy storage systems - undoubtedly – can play a crucial role for a large-scale integration on power systems of DG and for enabling the starting and the consolidation of the new era of so called smart-grids.

A non exhaustive list of benefits of the energy storage properly located on modern power systems with DG could be as follows: it can increase voltage control, frequency control and stability of power systems, it can reduce outages, it can allow the reduction of spinning reserves to meet peak power demands, it can reduce congestion on the transmission and distributions grids, it can release the stored energy when energy is most needed and expensive, it can improve power quality or service reliability for customers with high value processes or critical operations and so on.

At this moment, a large number of energy storage technologies and systems are available and effectively viable; nevertheless, existing storage technologies can be complimented with innovative researches in order to find new, more reliable and cost-effective solutions.

The main goal of the book is to give a date overview on: (i) basic and well proven energy storage systems, (ii) recent advances on technologies for improving the effectiveness of energy storage devices, (iii) practical applications of energy storage, in the emerging era of smart grids.

The book is organized into three sections.

In the first section (chapters from 1 to 7), the basic and well proven technologies for making up an energy storage system are reviewed.

In Chapter 1, electrochemical energy storage technologies are reviewed, also showing how batteries, electrochemical flow cell systems, hydrogen based systems and capacitors can be effectively used on modern distribution grids for integrating renewable energy sources (RES), in order to improve their availability, reliability and power quality.

In Chapter 2, specific reference is made to supercapacitors. It is shown that they can be used as an effective alternative to traditional secondary batteries, especially in applications where batteries have to be cycled with shallow depth of discharges, in order to achieve long cycle lives. High-efficiency power electronic converters suitable for overcoming some specific problems related to practical utilization of supercapacitors (voltage imbalance in series connections and terminal voltage variations during charging/discharging process) are also introduced and discussed.

In Chapter 3, the state of art of high-speed flywheels is overviewed. Particular attention is dedicated to the problem of the optimization of the rotor design process and well proven approaches for solving it are introduced and discussed, with specific reference to a hybrid composite flywheel rotor.

In Chapter 4, the operation planning of hydrothermal systems is analysed, with the main aim to optimize the reservoirs of the hydroelectric systems so that thermoelectric generation can be profitably replaced, whenever possible, by hydroelectric generation. The specification of reservoir operation rules by means of Genetic Fuzzy Systems is investigated and the Mamdani fuzzy inference systems is efficiently used to estimate the operating volume of each hydroelectric plant based on the value of the energy stored in the hydroelectric system.

In Chapter 5, a lab scale system is experimented to demonstrate the real possibility to capture the energy from lightning return strokes, as that can also be considered a clean energy sources. The capacitor is used as energy storage device while a high-frequency switching is used to isolate it (and its stored energy), once it has been supplied by means of a lab-generated lightning impulse voltage.

In Chapter 6, the storage of thermal energy is considered. Phase change materials (PCMs) are specifically introduced and analysed due to their well proven capability to store thermal energy as latent heat, thanks to a constant-temperature phase-change process. In order to obtain a high heat transfer rate, micro encapsulated PCMs are proposed to be used; their fabrication process and their characterization are discussed in depth.

Finally, in Chapter 7, the energy storage process that characterises living organisms is introduced. Energy transduction in animal living cells takes place on the mitochondrion and the energy is stored in the body in the form of high-energy molecules such as Adesonine Three-Phosphate (ATP), maintaining the body at a constant temperature of about 37°C; the investigation of this process could be helpful in searching for an alternative energy storage systems, if studied from a thermodynamic point of view. In this chapter, a thermodynamic model for ATP synthesis is proposed and a quantitative comparison between the rate of energy loss and efficiency of energetic and thermogenic mitochondria is operated. Quantitative evaluation of different mitochondria leads to a better understanding of their thermodynamic functions.

In the second section (chapters from 8 to 13), recent advances on technologies for improving energy storage systems are introduced and analysed.

In Chapter 8, bidirectional dc-dc converters are considered. The most common and economical energy storage devices in medium-power range are batteries and super-capacitors, and bidirectional dc-dc converters could be a key element because they can allow energy exchange between storage devices and the rest of system. High efficiency, lightweight, compact size and high reliability are, of course, some important requirements for bidirectional dc-dc converters. In this chapter, they are reviewed and classified; isolated bidirectional dc-dc converters, employing soft-switching techniques, and they are investigated with particular emphasis.

Chapter 9 includes a detailed analysis of bidirectional dc-dc converters, as well as the characterization of their performances for power buffer utilizations. Furthermore, the dynamic behaviour of an electrical energy generating set (like that for military applications), with a power buffer based on supercapacitors, is specifically studied and discussed.

In Chapter 10, a new possibility for improving electrochemical performances of lithium rechargeable batteries (probably, the most leading candidates for large scale energy storage devices) is investigated. In particular, it is shown that nano-structured electrodes, based on the bio-material templates, possess superior electrochemical performances, such as specific capacity, rate capability and cyclability, due to the improved Li-ions and electrons supply, and strain accommodation upon cycling; that is to say, bio-inspired synthesis can be considered as a promising way for fabricating Lithium rechargeable batteries with improved performances.

In the same direction, Chapter 11 introduces and discusses the possibility of using thioether polymers for fabricating novel cathode active materials for lithium batteries.

Thioether polymers show: high discharge specific capacity (up to 800 $mAhg^{-1}$), discharge voltage above 2V and good cycling stability.

Chapter 12 deals with the improvement of performances of electrochemical capacitors. Future generations of electrochemical capacitors are expected to come close to current Li-ion batteries in energy density, maintaining their high power density; this may be achieved by using ionic liquids with a voltage window more than 4V, by discovering new materials that combine double-layer capacitance and pseudo-capacitance, and by developing hybrid devices. Concerning the materials issues, MnO_2 is one of the most promising: it has a very high theoretical capacitance of ~1380 F/g but suffers from poor conductivity. To further improve performances of MnO_2-based capacitors, it is necessary to design MnO_2 materials into nano-architectures with desirable physic-chemical features or composites with other materials, such as porous carbon or conductive polymer.

Finally, in Chapter 13 fuel cells are considered. Fuel cells based on polymer electrolyte membranes are considered to be one of the most promising alternative energy conversion device and can have an important role also as energy storage systems. An improvement of this kind of fuel cell can be obtained by incrementing their operating temperature (at the moment it is limited at about 80 °C); high temperature avoids the existence of two phase flow in the flow field so enhancing stability and reliability of the system, it also reduces the power loss caused by the electrochemical polarization of cathode and it is also beneficial to effectively making use of the exhaust heat and to enhance the CO endurance of the anode. In this chapter, the issue of developing a new type of proton exchange membrane that can be endurable to high temperatures (over 100 °C), still maintaining a high proton conduction, is introduced and discussed; an effective solution is introduced and tested.

In the third section (chapters from 14 to 21), a lot of practical applications of energy storage are considered and their effectiveness is evidenced.

In Chapter 14, the role of an energy storage system in balancing a local distribution network area is introduced and discussed. In particular, after concerning with the technical challenges that arise from intentional islanding of micro-grids that include micro-generation sources, a combination of an energy storage system and a backup generator is proposed as an effective solution for intentional islanding. A micro-grid model is defined and studied by means of a simulation software. A methodology for calculating the requirements of the energy storage system is introduced and utilized referring to a case-study and, then, it is shown as the combined use of a backup generator with an energy storage system can be profitably used for supporting the islanding operation mode of a distribution network area.

In Chapter 15, having in mind a future scenario where a power system could be made with 100% of renewable resources, the crucial role of an energy storage system is evidenced and analysed referring, as a case-study, to the Spanish electric power system.

Chapter 16 analyses problems and benefits of utilizing and managing a mix of short-term and long-term energy storage devices in a power system with intermittent renewable resources. The chapter evidences the fundamental role of modern power electronic converter designing and utilization together with that of a proper selection for a complementary control algorithm.

Chapter 17 gives an overview of practical utilization of different kind of energy storage devices in industrial applications.

In Chapter 18 it is shown that flywheel energy storage devices can be profitably used in practice. Also in this case, the fundamental role of power electronic converters is amply evidenced and a matrix converter is introduced and analysed together with a proper control technique, to control the power transfer process between the flywheel and the distribution grid. The reported results show that flywheel energy storage devices with "predictive optimal" matrix converter control can be used as a voltage restorer to excel in the mitigation of voltage sags and swells as well, as voltage distortion at critical loads.

Chapter 19 shows how a three-phase matrix converter can make ease the interaction of power electronic converters with grids, for controlling active and reactive power flows, also without an energy storage system. Their benefits in replacing the classical topologies with associations of back-to-back converters are evidenced by means of simulation results and experiments referring to some case-studies.

The ocean wave energy short term variability is the main topic of Chapter 20. In this context, the short term energy storage is considered as a possible element in the amelioration of this fluctuating resource. With reference to a case-study based on an oscillating water column type wave energy device, it is shown as a judicious combination of mechanical and electrical energy storage can reduce power fluctuations to the grid, reducing peak-to-average ratios.

In the field of electric trains, Chapter 21 deals with the possibility to use an on-board energy storage system for the rolling stock. The proposed energy storage system is based on both rechargeable batteries and electric double layer capacitors. In this chapter, it is shown how an on-board energy storage system can effectively enable energy savings and, at the same time, it is a promising tool to prevent regenerative energy failure for rolling stock. In practice, the energy storage devices and their charge/discharge converters are proposed to be shunt connected to main DC power source in a typical configuration. By means of simulation case studies and experiments, it is shown that the regeneration power can be effectively absorbed and saved and that braking can be realized without additional energy consumptions.

Rosario Carbone
University "Mediterranea" of Reggio Calabria
Italy

Part 1

Energy Storage Systems

1

Electrochemical Energy Storage

Pier Luigi Antonucci and Vincenzo Antonucci
Mediterranea University of Reggio Calabria,
CNR Institute for Advanced Energy Technologies,
Italy

1. Introduction

The problems related to the differed time between production and use of electrical energy produced by renewable sources makes storage systems an integral part of Renewable Energy Sources (RES), especially for stand-alone systems. Furthermore, for grid-connected systems, the stability of the electric system and the quality and stability of the delivered voltage will result in a high quality system in the presence of intermediate storage. Storage systems are particularly onerous for RES and, as a consequence, their cost and life-time significantly affect the total cost of the whole system.

The relation between technical and economical characteristics in terms of dimension and technology to be employed has to be singled-out and optimized depending on the circumstances and, for different applications, in terms of release time of the stored energy and outbound power.

The alternatives to obtain a continuous, reliable and cost-effective availability of electricity include:

- diffusion of RES, production from clean carbon technologies and others sources
- technical evolution of the transmission grid for the connection of these sources to the loads
- programs to meet power/energy demands and management of the loads.

The diffusion of storage technologies in the public grid include several benefits:

- the optimization of the grid for the fulfilment of base-load
- the way to facilitate the power trade-off in RES systems fluctuating over time or available only during day-time
- easy integration in grid of the energy requirements for hybrid vehicles
- the chance of investments differentiation in the distribution grid to be adapted to temporary peak-loads.
- resources for the provision of auxiliary services directly to the electricity provider.

The different storage technologies can be classified on the basis of the different methodologies utilized:

- mechanical (compressed air energy storage, flywheels)
- electrochemical (lead-, nickel-, high temperature salts-, redox-batteries, hydrogen.
- electrical (capacitors, supercapacitors).

Although some storage technologies could work for several applications, the most part of the different options is not economically applicable to different functional categories. Their assessment must be done on the basis of several parameters which establish their applicability:

- power level (nominal, pulsed)
- energy storage level (at different charge and discharge rates)
- memory effect
- power density
- energy density
- overall cycle efficiency
- life-time (number of cycles and performance)
- operative characteristics
- environmental impact (LCA)
- recycle opportunity and costs
- investment costs
- maintaining costs

Technology	**State of the art**	**Short term application**	**Medium term application**
Lead acid, NiCd	Commercially available	Off grid demonstration	Niche markets
HT batteries	Commercially available	Demonstration	Probable, potentially with PV
Redox batteries	Lab prototypes, dem	Dev & demonstration	Probable
Lithium batteries	Lab prototypes, dem	Dev & demonstration	Probable
Hydrogen	Lab prototypes & dem	Demonstration	Possible, niche applications
Capacitors	Near to the market (small size)	Required development	Possible attainment of commercial targets

Table 1. Technological and commercial maturity of the different technologies is summarized

Technology	**Commercial maturity**	**Costs**
Lead acid	◆	◆
Ni-Cd	▲	▲
Ni-Mh	▲	▲
Na/S	■	■
ZEBRA	■	▲
Zn/Br	■	▲
Lithium ion	■	▲
Redox (V)	■	▲
Supercap	▲	▲
RFC	●	●

Table 2. State of the art of electrochemical technologies

Symbol	Commercial maturity	Costs
◆	Mature product, several units sold	Price list available
▲	Commercial product	Prices on request
■	Prototypes under construction tests in progress	Defined per single project
●	Only projects available	Estimated

Table 2. State of the art of electrochemical technologies (continuation)

This chapter deals with the analysis of electrochemical technologies for the storage of electricity in stationary applications able to meet present and future challenges for the three following goals:

- Power quality: stored energy to be delivered for seconds in order to guarantee the continuity of stabilized electricity supply
- Bridging power: energy furnished for minutes to guarantee the continuity of the service during the transition from an energy source to another one.
- Energy management, for the time-lag between the production and the utilization of the produced energy. A typical utilization is the optimization of the load level when the energy cost is low and utilization when the cost from the grid is high.
- (Antonucci P.L., 2010, Antonucci V., 2011, Strbac G. & Black M. 2004, Stuntz. L, 2004., Makansi J. & Abboud J, 2002).

Application	Power	Energy
Domestic	1 kW	5 kWh
Commercial	10-100 kW	25 kWh
Distribution grid	10-100 MW	10-100 MWh

Table 3. Typical intervals and parameters of the different applications

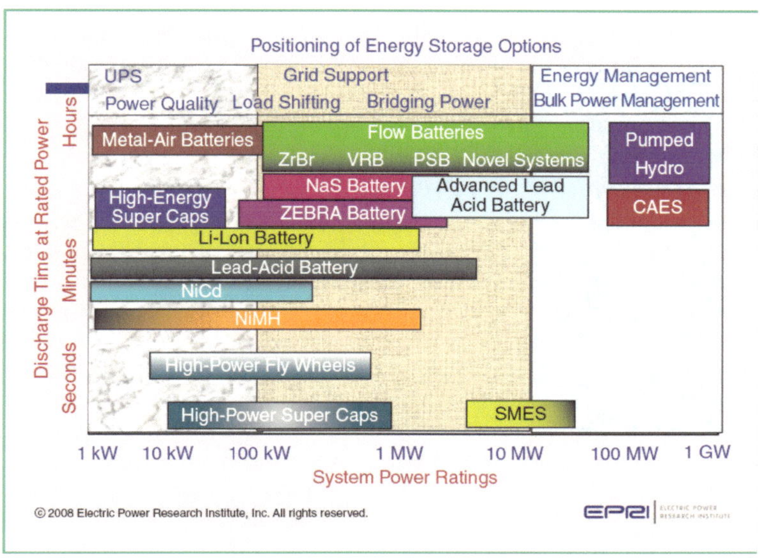

Fig. 1. The state of the art of storage technologies (source: EPRI)

At present, the most common electrochemical storage technology is represented by lead-acid batteries. In USA the current market of lead-acid batteries for commercial, industrial and automotive applications is about 3 billion dollars per year, with an annual rate of growth of 8.5%.
For what concerns the most recent applications (distributed generation, peak shawing, power quality), the use of lead-acid batteries has been widely demonstrated, but costs and life-cycle characteristics are not satisfactory for these applications, characterized by high number of cycles.
On the other hand, the use of Li ions batteries for mobile applications is rapidly increasing, with an annual growth rate of 50-60%. For stationary applications (in particular, for mitigation of the variability of renewable sources), their potential has not yet been thoroughly explored, as well as their cost-effectiveness, except for some auxiliary storage systems.
Others electrochemical technologies for backup systems include sodium-sulfide, zinc-bromide and redox vanadium batteries. The sodium sulphur and zinc bromide technologies are rapidly expanding in Japan and USA. Nickel cadmium and Nickel MH are used in power backup systems.
The pressing demand of better storage technologies have produced incentives to the R&D sector and financing through venture capital. In this regard, R&D is addressed to a base-electrochemistry level in order to single-out the combinations of chemicals having the maximum potential in energy storage systems.

2. Batteries

Batteries are devices that convert the chemical energy contained in an electrochemically active material directly into electrical energy by means of a redox reaction. For a rechargeable system, the battery allows to store a defined amount of chemical energy and can be re-charged when the electrochemically active material has been transformed.
They are the most established way of storing electricity. Several types of rechargeable systems exist, from the mature lead acid to different newer technologies at various developmental stages.
Recently, new demands of portable and transport applications, as well as of power electronics and use of large scale systems for utility applications have determined further, dramatic development of new battery technologies.

2.1 Lead acid

Lead acid represent more than 90% of the whole batteries market. Main constituents are the lead negative electrode, the lead dioxide positive one and the separator, with sulphuric acid as the electrolyte. The flooded type requires the filling up with distilled water, whereas the sealed, maintenance-free type has an absorbed electrolyte.
At present, typical lead batteries for transportation have a polypropilene external case, high performance separators based on glass fibre, polyethylene or polyvinyl chloride matrices, thin grids with extremely low Sb content or based on Pb-, Ca-, Sn- Al alloys. The effect of additives has been investigated and optimized, allowing to reduce or cancel the necessity of maintenance.
The continuous development of the technology has caused important improvements in its capabilities, resulting in low costs and high reliability. The main drawback remains, however, its low energy density due to the high density of lead.
The so-called "advanced lead" uses carbon additives in the negative electrode in order to increase its life-time; this type is still at a developmental stage.

The trend to produce lead batteries having superior performance, cyclability characteristics and life-time are dictated by the increasing demand of electrical energy for distributed generation, for hybrid vehicles, for auxiliaries in internal combustion engines. Future lead batteries will be characterized by the use of new materials with impact on the design, recovery, recycle, disposal and refining. The optimization of the whole cycle including production, use, recovery and recycle, driven by an appropriate life-cycle analysis, will be of fundamental importance.

2.2 Nickel Cadmium

Nickel-Cadmium (Ni-Cd) batteries represent a good compromise between specific energy, specific power, life cycle and reliability. They are constituted by a positive plate of nickel oxy-hydroxide (NiOOH), a negative one of cadmium (Cd) and an aqueous solution of potassium hydroxide (KOH) as the electrolyte. During discharge NiOOH is converted to $Ni(OH)_2$, which is restored during charge. Cd is converted to $Cd(OH)_2$ during discharge, and is restored during charge. The toxicity of Cd has led to the development of Ni-MH batteries, based on metal hydrides. These are similar to the Ni-Cd, the main difference being the composition of the negative plate made of hydrogen adsorbed on a metallic alloy. This can be of the type AB5 (mixture of nickel and rare earths) and AB2 (mixture of nickel and titanium, vanadium and zirconium). This type of battery is at present used in some hybrid vehicles.

Two designs have been developed for this technology, namely the pocket plate and the sintered plate. In the first, the active material is held between steel plates; the sintered plate configuration uses different materials as current collector, and the active substance is sintered onto these. The manufacturing process is more expensive than for lead acid, resulting in a higher cost for this technology. Furthermore, this suffers from "memory effect", resulting in a full charge only after a series of full discharges. This effect can be mitigated by an appropriate management procedure. A significant advantage over lead acid is represented the lower maintenance requirements.

2.3 Lithium ion

Historically, Li-ion were the first Li batteries developed for portable electronics (cellular phones, cordless, notebooks). They have the highest power density of commercial batteries. The anode is made of metallic Li, Li alloys or Li-C intercalation compounds. The electrolyte is based on non aqueous compounds, such as LiClO4 or LiPF6. The next generation batteries will use a polymer electrolyte containing Li dissolved in a polar polymer such as polyethylene oxide (PEO). The cathode is made of intercalation structures containing void channels or layers, able to accept Li ions, and a mixed valence in the host framework, able to receive or give electrons. TiS_2 has such a structure; it is able to receive Li ions in the void spaces separating the TiS_2 adjacent layers. To maintain the electroneutrality electrons must enter into the material. The intercalation reaction can be so written:

$$Ti_4 + S_2 + xLi = Li + xTi_4 + 1\text{-}x\ Ti_3 + xS_2.$$

Many others intercalation compounds can work as a cathode. The most commercially diffused is $LiCoO_2$, which has alternated layers of Li and Co ions in octahedral sites inside the ccp lattice of the oxide ions. The $LiCoO_2$ formula is referred to the completely intercalated form when the battery is discharged. During recharging Li ions are removed and the oxidation of Co^{3+} to Co^{4+} occurs. The Co^{3+}/Co^{4+} couple supplies a cell voltage of about 4.0V vs. metallic Li.

New configurations have been developed for stationary applications, with titanate at the anode and manganese at the cathode. Cost is at present very high; its reduction is linked to an eventual increase in production. Current energy density values reach 175 Wh/kg, with cycle lives as high as 20,000 cycles.
Recently, great interest has raised for sustainable mobility (HEVs, EvS, HEV&FC, scooters, motorcycles, electrical bicycles). In spite of a remarkable growth of the market, the scientific development of Li-ion batteries is often exposed to criticism because of its slow development in comparison with other technologies like semiconductors. The world market of Li-ion batteries is controlled by a limited number of great companies, most of which located in Asia (Japan, China, Korea, more than 90% of those commercialized in the world). The main users (Europe and USA) produce only 5-10% worldwide, USA being leader in R&D due to projects funded by DoE, NIST and DoD (Fig.2).

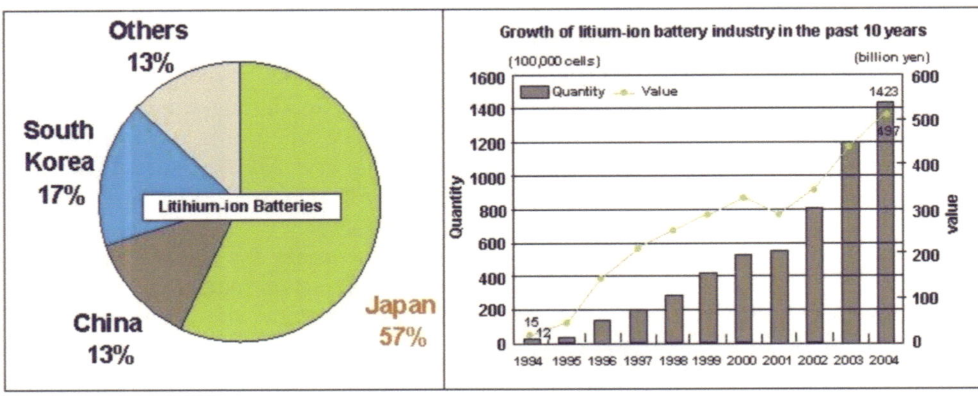

Fig. 2. World production of Lithium batteries

Fig. 3 shows the market trend of Li batteries from 2004 to 2012. The yearly growth is 15-20%, with even more promising scenarios in the case of introduction of HEV into the global market. Fig. 4 shows the market trend of conversion and storage of energy (source: Lux Research, 2008). Its value in 2011 is about 50 bln dollars (10 bln for Li batteries only, 20 bln for lead and 2 bln for NiMH).

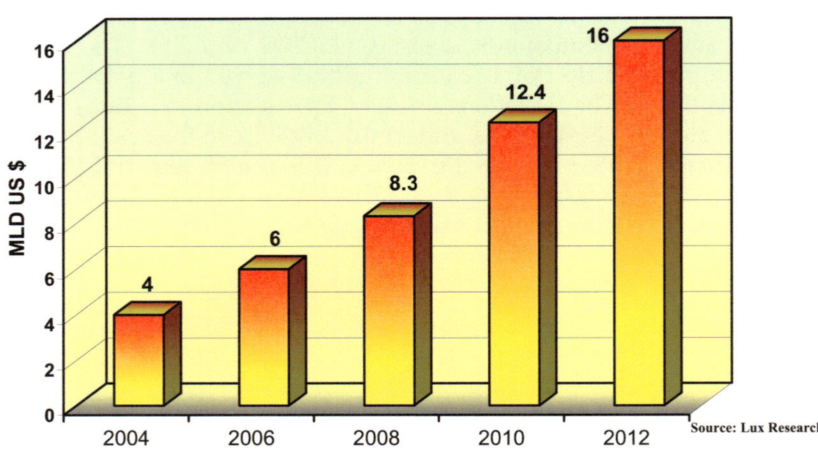

Fig. 3. World market of Lithium batteries

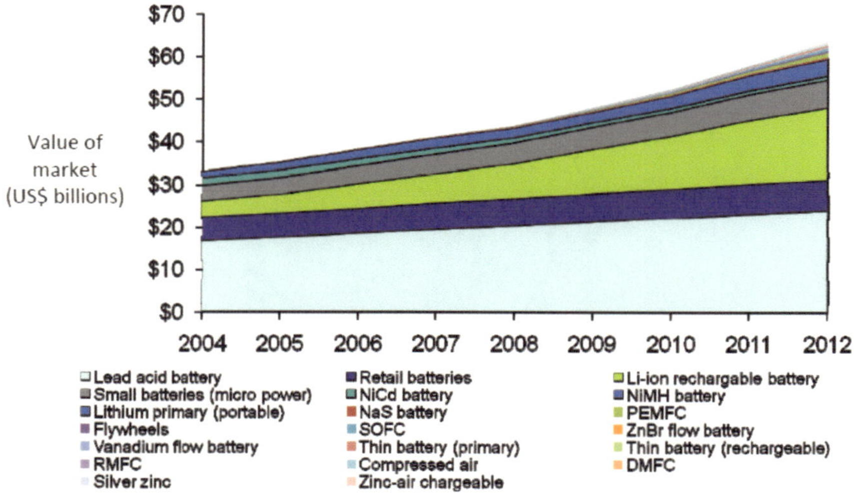

Fig. 4. Storage systems: world market values

2.4 High temperature batteries

2.4.1 Sodium sulphur

The battery consists in a positive electrode constituted by fused sulphur and a negative one of fused sodium separated by a ceramic electrolyte (sodium ions of beta alumina). The discharge process consists in the formation of sodium sulphide (sodium ions cross the electrolyte and the electrons go to the external circuit). The process is obviously reversible (charge phase). The discharge reaction is: 2 Na + xS = Na_2Sx, where x is dependent on the charge level of the cell. During the first steps of the discharge process, x is estimated to be equal to 5, approximately corresponding to the sodium sulphide formula more rich in sodium, Na_2S_5. The OCV depends on the charge level and on the temperature (max value: 2.08V).

The working temperature is about 300°C. This results in improved performance in comparison with room temperature systems, requiring yet efficient insulation to prevent heat losses, as well as a heat source using the stored energy of the battery itself. As no self-discharge occurs, efficiencies near 90% are usually reached. Yet, they have room for further development.

2.4.2 Sodium nickel chloride

Sodium-nickel chloride batteries ("ZEBRA", Zero Emission Battery Research Activity) possess a high energy density (120 Wh/kg) with high performance (180 W/kg). The operation temperature range is 270-350°C; they work in a thermal chamber. The nickel chloride allows high capacity (500 Ah). Furthermore, the solid electrolyte allows tolerance towards possible short circuits and self-discharge. The substitution of fused nickel chloride with fused iron chloride would make the system more promising but, to maintain $FeCl_2$ in the fused state, the cathode ($Fe/FeCl_2$) has to be impregnated with fused $NaAlCl_4$. The reactions occurring in the cell are:

$$\begin{aligned} 2Na + NiCl_2 &= 2NaCl + Ni \quad E = 2.58\ V \\ 2Na + FeCl_2 &= 2NaCl + Fe \quad E = 2.35\ V \end{aligned} \tag{1}$$

In comparison with sodium sulphur they have better safety characteristics and higher cell voltage. The disadvantages consist in the actual lower energy and power density.

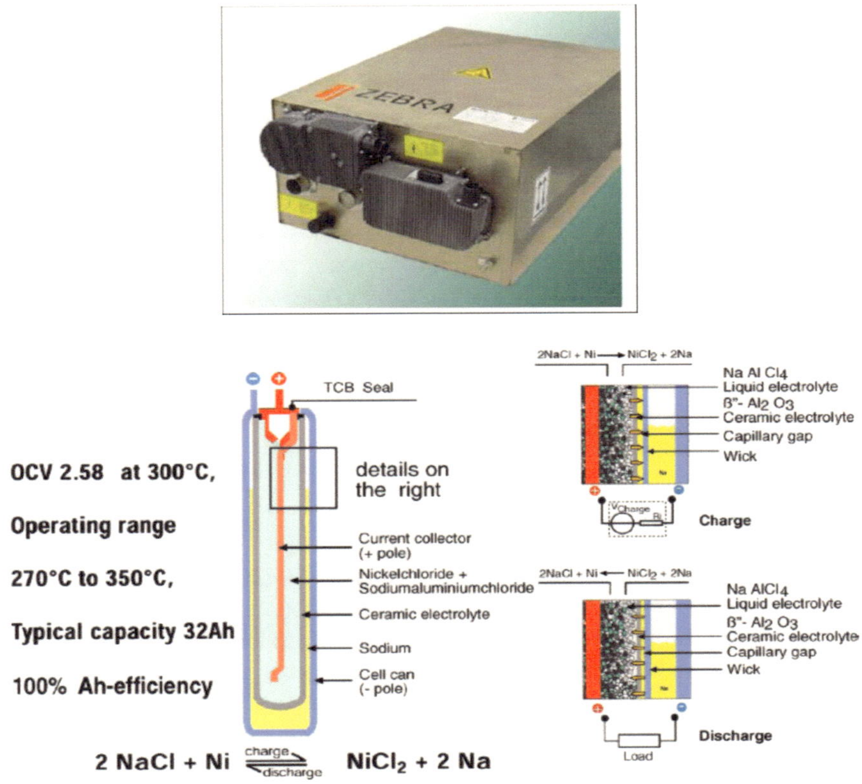

Fig. 5. Zebra battery

3. Electrochemical flow (redox) cell systems

Flow or redox batteries are electrochemical devices halfway between secondary batteries and fuel cells. The energy is stored in two electrolytes, separated by an ion exchange membrane. There is no electrochemical reaction between the electrolytes and the electrodes. They can supply power as long as they are supplied with electrolytes.

There is a conceptual difference between power and energy properties. The delivered power depends upon electrodes dimension and number of cells, whereas the storage capacity is determined by the volume of the reservoirs that contain the electrolytes flowing through the cells. For a given power absorption, the energy capacity can be increased by introducing more electrolyte (that is, using more capacious reservoirs). Therefore, flow batteries are particularly advantageous for high energy (long term) applications, when several hours of storage are required. The recharge capacity is equivalent to the rapidity of refilling, therefore this kind of battery appears particularly promising for mobile applications. Compared to other types of batteries, redox systems have an outstandingly long life (unlimited in theory) and no operative limitations (no memory effects or problems connected with discharge).

3.1 Vanadium Redox Battery (VRB)

The vanadium technology is based on the four oxidation states of V, exploiting the V^{2+}/V^{3+} transition one side of the membrane and the V^{5+}/V^{4+} transition of the other. The use of V sulphates or, more recently, V halides on both sides considerably simplifies the electrolyte management.
Reactions occurring during charging and discharging are:

$$V^{4+} \underset{\text{discharge}}{\overset{\text{charge}}{\rightleftarrows}} V^{5+} + e- \tag{2}$$

$$V^{3+} + e- \underset{\text{discharge}}{\overset{\text{charge}}{\rightleftarrows}} V^{2+} \tag{3}$$

VRB can be suitable for different applications, such as enhanced power quality, UPS, peak shaving, increased security of supply and integration with RES. Due to their low energy density, major development is addressed to stationary applications.

3.2 Zinc bromine

The negative zinc electrode and the positive bromine electrode are separated by a microporous separator. Circulation of zinc solution and bromine compounds occurs through the two cell compartments from the respective reservoirs. In the charging phase zinc is electroplated on the cathode while bromine evolves at the anode and stored at the bottom of the positive electrode reservoir. On discharge, zinc is oxidized to its ions, while bromine reduces to bromide ions.

Positive electrode:
$$2Br^- \underset{\text{discharge}}{\overset{\text{charge}}{\rightleftarrows}} Br_2 + 2e- \tag{4}$$

Negative electrode:
$$Zn^{2+} + 2e- \underset{\text{discharge}}{\overset{\text{charge}}{\rightleftarrows}} Zn \tag{5}$$

Discharge times range from few seconds up to several hours. Demonstration projects are primarily focused to on-grid utility applications for load levelling and RES optimization.

3.3 Polysulphide bromide (Regenesys ^{TM})

This system utilizes the reversible electrochemical reaction between two salt electrolyte solutions:

$$3\,NaBr + Na_2S_4\,(\text{charging}) \leftrightarrow 2Na_2S_2 + NaBr_3\,(\text{discharging}) \tag{6}$$

Previously developed over the past fifteen years, the system has been marketed as a grid-linked utility storage system for more than 5 MWe. Actually, no development or commercialization programs are foreseen.

4. Hydrogen storage systems

The storage of electricity is feasible by producing hydrogen through water electrolysis. The as produced hydrogen is stored in compressed gas or liquid form or through solid adsorption at low pressure. The system is constituted by an electrolyzer with a fuel cell, separated, or by a regenerative electrolyzer, which works as EL or FC by inverting the polarity. Such an approach allows a direct integration of distributed generation from RES with zero impact mobility.

Advantages of hydrogen-based storage systems include: i) the high energy density of hydrogen; ii) the possibility of realizing systems over a large size range (multi kW to MWs); modularity; environmental benignity. Main disadvantages are represented by the high cost and low round-trip efficiency.

4.1 Regenerative electrolyzers

Regenerative or reversible cells (RFC) are electrochemical systems able to both convert chemical to electrical energy (fuel cell mode) and carry out the reverse process, that is to dissociate the water molecule into hydrogen and oxygen (electrolyzer mode). They are composed of two distinct sections, with electrodes dedicated to only one function; in particular, the oxygen-side electrodes have different catalysts, generally platinum for oxygen reduction in the fuel cell section and a metal oxide for water ionization in the electrolyzer section.

Differently, Unitized Reversible Cells (URFC) combine the two functions in an unique device; in this case, the oxygen electrode contains both catalysts for water ionization and oxygen reduction. Also the flow-field, supported by conductive plates, must be able to operate both for gas evolution and electricity production.

Both RFC and URFC are used when the characteristics of rechargeable batteries (Pb, Li, NiMH) are needed; for example, in stationary plants for power generation connected to wind or photovoltaic systems, in the automotive, aeronautic or space sectors. In all cases, duty of the cells is to store the energy coming from an external source in the form of hydrogen and to return it when needed.

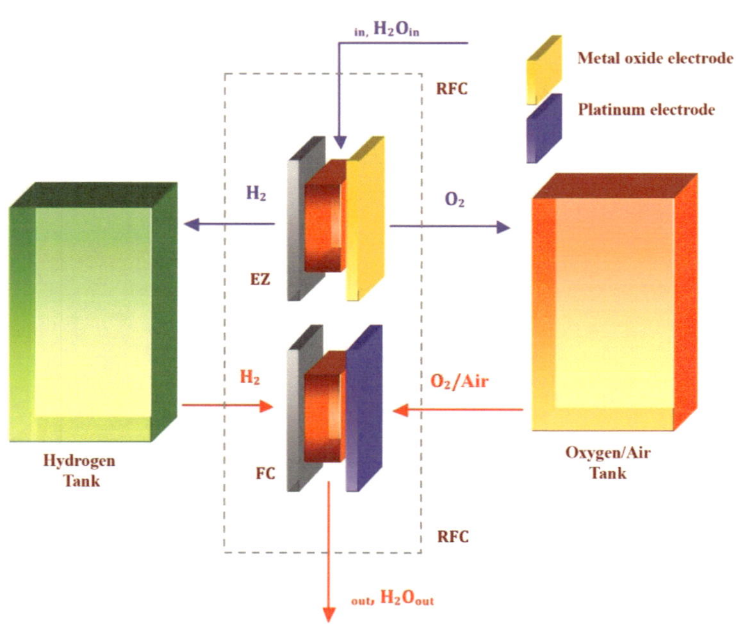

Fig. 6. Regenerative fuel cell (RFC) outline

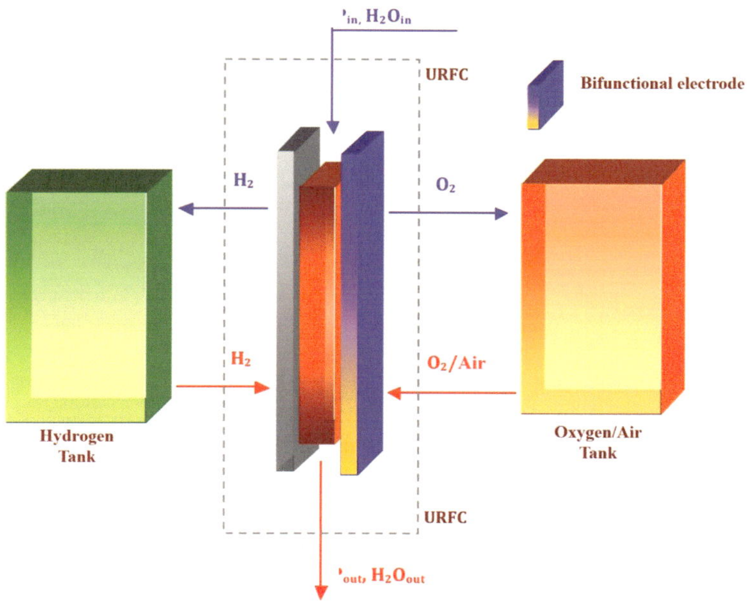

Fig. 7. Unitized regenerative fuel cell outline

5. Capacitors

The electrochemical capacitors, or supercapacitors, store energy through ions adsorption at the electrode-electrolyte interface (electrical double layer capacitors) or through redox reactions occurring on the electrode surface (pseudo-capacitors).They can supplement or replace batteries in applications where is necessary supplying or taking up high electrical power. Great performance enhancements have been reached through the recent comprehension of mechanisms of charge accumulation and the development of nanostructured materials. The use of nanomaterials with pseudo-capacitive characteristics, such as oxides, nitrides and polymers, as well as lithium nanostructured electrodes, has raised the energy density of electrochemical capacitors, approaching the values of batteries. The use of carbon nanotubes has allowed the fabrication of electrochemical micro-capacitors much more flexible.

Supercapacitors can be completely charged or discharged in a few seconds; as a consequence, their energy density (about 5 Wh kg^{-1}) is lower than that of batteries but the supply or absorption of power is much higher (10 kW kg^{-1}) even if for short times (few seconds). Mathematical modelling and simulation will be the key for the future design of devices with high energy and power. Electrochemical capacitors have an important role in supplementing or substituting batteries in some fields of energy storage, such as back-up devices for protection from current interrupting or load levelling. Several types of capacitors have been developed, being different in the mechanism of accumulation charge or active materials used.

Their energy density (about 5 Wh kg^{-1}) is lower than that of batteries but the supply or absorption of power is much higher (10 kW kg^{-1}) even if for short times (few seconds).

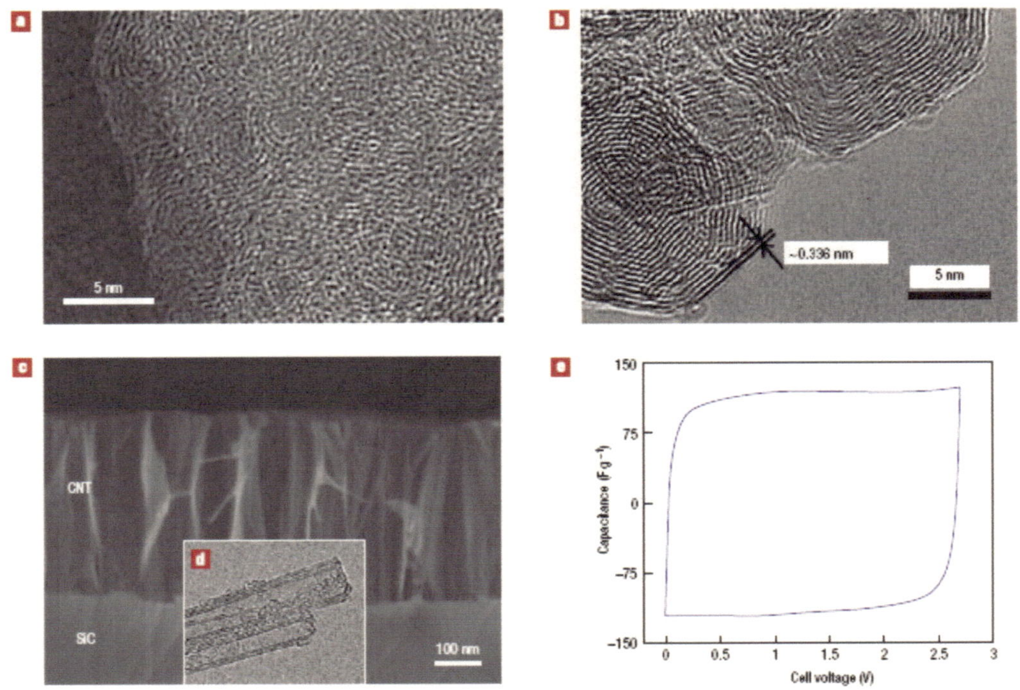

Fig. 8. Carbon structures for double-layer capacitors: a) TEM image of a microporous carbon (SiC-derived, three-hours chlorination at 1000 °C); b) TEM image of onion-like carbon (Elsevier, 2007); c) SEM image of CNT matrix (sinterization at 1700°C six hours) on SiC; inset (d): CNT TEM image; e) cyclic voltammetry of a two-electrodes EDLC cell in 1.5 M tetraethylammonium tetrafluoroborate (Net4+BF4) in acetonitrile with active carbon on Al-coated current collectors.

In pseudo-capacitors, or redox supercapacitors, fast redox reactions occur on the surface for charge accumulation. Metal transition oxides and electron conductive polymers are examples of materials with pseudo-capacitive properties. Hybrid capacitors, that combine a capacitive or pseudo-capacitive electrode with a battery-type electrode, take advantage of both capacitor and battery properties.
Electrochemical capacitors bridge the gap between batteries and conventional solid state electrolytic capacitors. They accumulate a charge two-three order of magnitude higher (tens of farad per gram) with respect to the electrolytic capacitor, due to a much more large specific surface (1000-2000 $m^2\ g^{-1}$). Yet, they have a lower energy density with respect to batteries, and this characteristic decrease to less than a minute the discharge time (many applications clearly need energy for a longer time).

5.1 Mechanism of double-layer capacity

EDLC are electrochemical capacitors which electrostatically memorize the charge with reversible absorption of electrolyte ions on electrochemically active materials with high specific surface area. During polarization, charge separation occurs at the electrode-electrolyte interface, producing the capacitive double layer (C):

$$C = \varepsilon_r \varepsilon_0 A / d \text{ or } C / A = \varepsilon_r \varepsilon_0 / d \quad (7)$$

where ε_r is the electrolyte dielectric constant, ε_0 is the vacuum dielectric constant, d is the actual thickness of double layer (charge separation distance), A is the electrode surface. The double layer capacity is comprised between 5 and 20 microfarad.cm^{-2} depending upon the electrolyte. The specific capacities obtained with alkaline or acid aqueous solutions are generally higher than those in organic solutions, but organic electrolytes are more widely used because they are able to sustain a higher working voltage (up to 2.7 V in symmetrical systems). As the stored energy is proportional to the square voltage:

$$E = \tfrac{1}{2} CV^2 \quad (8)$$

a three times increase of V results in an increase of about one order of magnitude of E for a capacitor having equal capacity.

The consequence of conservation of the electrostatic charge is that there is not a faradaic reaction (redox) at EDLC electrodes. The electrode of a supercapacitor can be considered as a blocking electrode from the electrochemical point of view. This difference with the batteries means that there is not any limitation by the kinetics of the electrochemical reaction. Furthermore, this accumulation mechanism on the surface allows the fast absorption and the supply of energy, therefore better power performance. The absence of faradaic reactions avoids also the swelling of the active material, as occurs in batteries during the charge/discharge cycles. EDLC are able to sustain billions of cycles, whereas batteries few thousands only. Furthermore, the electrolyte solvent is not involved in the mechanism of charge accumulation, differently from Li-ion batteries that contributes to the formation of the solid-electrolyte interface when graphite anodes or high cathode potentials are used. EDLC can work also at very low temperatures (up to -40°C) with some particular electrolytes and solvents. Yet, as a consequence of the mechanism of surface electrostatic charge, these devices have a low energy density. This explains because the research is mainly addressed to the improvement of their energy efficiency and to the enlargement of temperature range, where batteries cannot work.

5.2 Active high surface area (SSA) materials

The use of blocking, high SSA electrodes and good electronic conductors is the way to obtain high capacities. Graphitic carbon satisfies these requisites. Active carbons, template carbons, as well as carbon tissues, fibres and nanotubes have been largely tested for EDLC applications. Active carbons are today the mostly used because of their high specific surface area (SSA) and low cost.

Active carbons are obtained from carbon-rich organic precursors by thermal treatment in inert atmosphere and subsequent selective oxidation in CO_2, water vapour atmosphere or KOH in order to increase the pore volume and SSA. Natural materials, such as coconut shells, wood, resins, or synthetic materials, such as polymers, can be used as precursors. After activation a porous structure is formed inside the carbon particles, including micropores (< 2 nm), mesopores (2-50 nm) and macropores (> 50 nm). The average pore dimension increases as the activation time or the temperature are increased. The double-layer capacity of active carbons reaches 100-120 F g^{-1} with organic electrolytes and 150-300 F g^{-1} in aqueous electrolytes but at a lower cell voltage, because the voltage window of the electrolyte is limited by water decomposition. Carbon nanotubes or nanofibres show lower capacity (50-80 F g^{-1}), which can be little increased by functionalization with oxygen-rich surface groups, but their presence decreases the stability capacitor.

Fig. 9. a) Schematic representation of a double-layer spiral capacitor; b) assembled device (500 grams, 2600 F); Batscap, Groupe Bollorè; c) button cell (1.6 mm height, 5 F capacity); Y-Carbon.

It has been demonstrated that a pore size distribution between 2 and 5 nm is the more effective in increasing energy and power density, because this range is more accessible to solvated ions.

5.3 Redox electrochemical capacitors (pseudo-capacitors)

Pseudo-capacitors make use of fast and reversible reactions on the surface of the active materials, producing the so-called pseudo-capacitive behaviour. Metal oxides, such as RuO_2, Fe_3O_4 or MnO_2, as well as some electron conducting polymers, have been investigated in the last decades. The specific pseudo-capacitance is higher than that of carbons (based on the double-layer charge), and this justifies the interest for such systems. The drawback is that the redox reactions, similarly to what happens in batteries, often reduce the stability during operation.

Ruthenium oxide, RuO_2, has been largely investigated, as it is a good electronic conductor and has three distinct oxidation states below 1.2 V. The reversible process can be described like a fast electrons transfer together with an electro-absorption of protons on the particle surface of RuO_2, where the Ru oxidation states can vary from (II) to (IV):

$$RuO_2 + xH^+ + xe^- = RuO_{2-x}(OH)_x \quad (9)$$

Where x is comprised between 0 and 2. The continuous change of x during the entry or the exit of protons occurs in a window potential of about 1.2V, and leads to a capacitive behaviour with ion absorption of the Frumkin isotherm type. Specific capacities of more than 600 F g^{-1} have been observed. Unfortunately, based-Ru EC are not cheap, and the 1V voltage window limits their application to small electronic devices. To increase this value, non-protonic organic electrolytes (for example, Li^+ containing) could be used. Less expensive are the oxides of iron, vanadium, nickel and cobalt. In particular, Mn oxide has been largely investigated. Its charge accumulation mechanism is based on the surface absorption of the electrolyte cations (K^+, Na^+, H^+...) according to the reaction:

$$MnO_2 + xC^+ + yH^+ + (x+y)e^- = MnOOC_xH_y \quad (10)$$

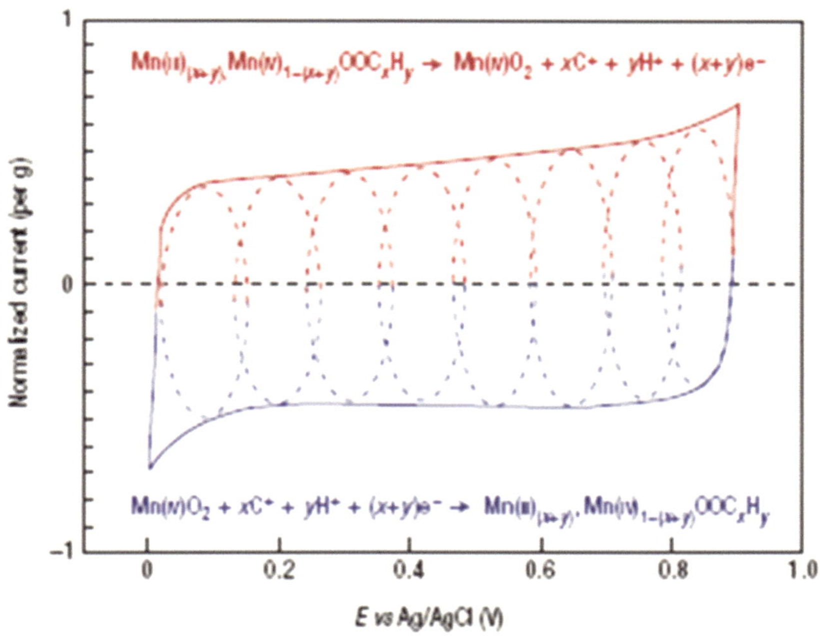

Fig. 10. Voltammogram of a MnO_2 electrode in neutral aqueous electrolyte

Fig.10 shows the voltammogram of a MnO_2 electrode in neutral aqueous electrolyte; the surface redox reactions (fast, reversible, with a regular sequence) define the voltammogram shape, which appears to be similar to that of ELDC. MnO_2 micropowders or micrometric films show capacities of about 150 F g^{-1} in neutral aqueous electrolytes in a voltge range of 1V, that are appropriate for hybrid systems (see below).

5.4 Nanostructured redox materials

Since in pseudo-capacitors the charge accumulates on the first nanometers of the surface, by reducing the particle dimension the active surface of the material increases. For example, MnO_2 and RuO_2 films deposited on different supports (metal collectors, carbon nanotubes, active carbon) have shown specific capacitance values higher than 1300 F g^{-1}. It has been demonstrated that the synthesis of thin films or the decoration of high SSA capacitive materials with pseudo-capacitive materials increases the energy density.

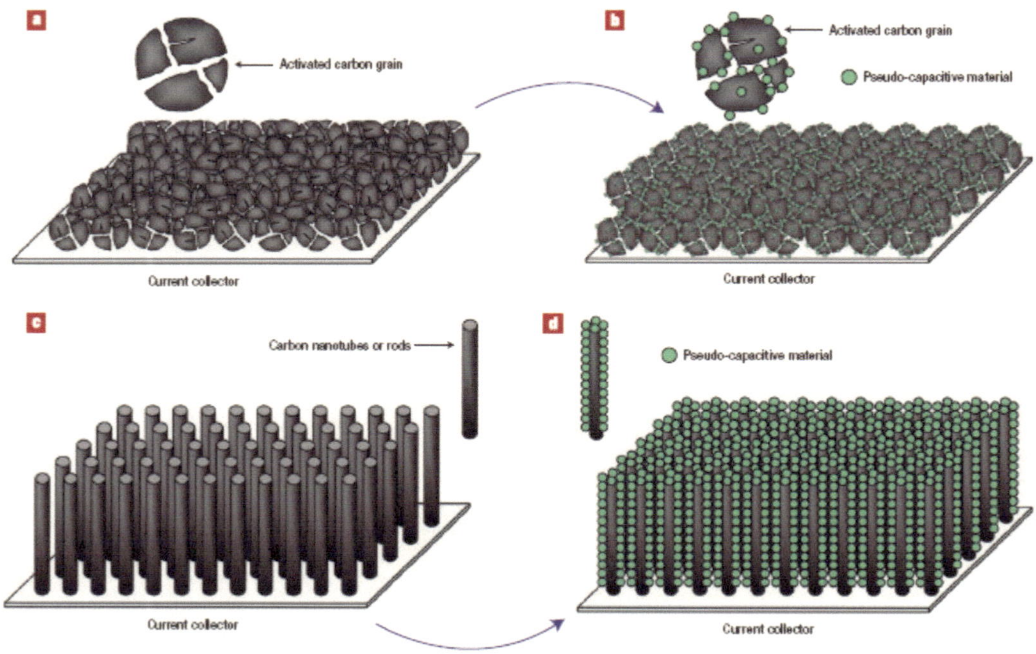

Fig. 11. Possible strategies to enhance energy and specific power of electrochemical capacitors: a, b) decoration of active carbon grains (a) with pseudo capacitive materials (b). c,d) attainment of an uniform deposit of pseudo-capacitive material (d) on high ordered, high SSA CNT layer (c).

5.5 Hybrid systems

Hybrid systems offer a valid alternative to traditional pseudo-capacitors or EDLC, allowing to combine in the same cell one battery-type electrode (energy source) with one capacitor-type electrode (power source). At present, two different approaches have been adopted: i) metal oxide electrode having pseudo-capacitive properties with a capacitive carbon electrode and ii) lithium-inserted electrode with capacitive carbon electrode.

Several combinations of positive and negative electrodes have been investigated in aqueous and organic electrolytes. In most cases, the faradic electrode produced an increase in energy density but a decrease in cyclability. This is certainly the main drawback of hybrid devices in comparison with EDLC.

MnO_2 represents a valid alternative to RuO_2 due to its low cost. Its pseudo-capacitance originates from the change in oxidation state, III/IV, at the MnO_2 particle surface. The association of a negative EDLC-type electrode with a positive MnO_2 one produces a 2V cell in aqueous electrolyte, thanks to the overvoltage of water decomposition on MnO_2 and to the high surface area of carbon. The low cost of the carbon-MnO_2 hybrid system, together with the high capacitance in neutral aqueous electrolyte and the high cell voltage, make this system a valid, environmental friendly alternative to EDLC, which use solvents like acetonitrile or fluoride salts. A challenge for these systems is the use of organic electrolytes, in order to reach higher cell voltages, thus improving the energy density.
The combination of a carbon electrode with a PbO_2 battery-type one produces a low cost EC for those applications in which weight has a minor role.
Furthermore, several studies have been carried out on different combinations of Li-inserted electrodes with a capacitive carbon electrode. Energy density of over 15 Wh kg^{-1} and 3.8V have been obtained.

6. Conclusions

Sustainable distributed power-supply systems are a realistic option to ensure a sustainable, affordable and secure energy supply. Nevertheless, because different renewable energy sources can complement each other, multi-source hybrid alternative energy systems have great potential to provide higher quality and more reliable power to customers than a system based on a single resource, but systemic, technical and institutional adaptations are required for their further development and incorporation in local electricity networks. Understanding the leading storage technologies and how they can affect grid operations is an important first step in this assessment and it is perceived as the pre-condition for extensive development of renewable energy systems. It is increasingly recognized that a large market for distributed energy systems will depend on significant technology advances in energy storage systems but as said above to reach commercialization, both distributed generation and energy storage must address a number of key issues such as lower installed costs, increased lifetime, and low maintenance costs.

7. References

Antonucci P.L., Internal Report 003/10/MecMat UniRc, Dept of Mechanics and Materials Mediterranea, University of Reggio Calabria, March 2010.
Antonucci V., Internal Report 006/11/ ITAE, CNR Institute of Advanced Energy Technologies, Messina, January 2011.
Makansi J. & Abboud J., "Energy Storage", Energy Storage Council White Paper, May 2002.
Strbac G. & Black M., DTI, "Review of Electrical Energy Storage Technologies and Systems of their Potential for the UK", London 2004.

Stuntz L., Electricity Advisory Committee Report,"Bottling Electricity: Storage as a Strategic Tool for Managing Variability and Capacity Concerns in the Modern Grid", 2008.

The Authors are indebted with the copyright holders of the figures included in the chapter for the use without previous permission due to difficulty in tracing their origin.

2

Supercapacitor-Based Electrical Energy Storage System

Masatoshi Uno
Japan Aerospace Exploration Agency, Japan

1. Introduction

Supercapacitors (SCs), also known as electric double-layer capacitors or ultracapacitors, are energy storage devices that store electrical energy without chemical reactions. Energy storage mechanisms that do not require chemical reactions provide several advantages over traditional secondary batteries such as lead-acid, Ni-Cd, Ni-MH and lithium-ion batteries (LIBs) in terms of cycle life performance, power capability, coulombic efficiency and low-temperature performance. In addition to these superior electrical properties, it is easier to estimate the state of charge (SoC) for SCs than that for secondary batteries because the terminal voltage of SCs is inherently proportional to the SoC.

In order to meet load variations, SCs are widely used as auxiliary power sources that complement main energy sources such as secondary batteries and fuel cells. In such applications, SCs act as electrical power buffers with large power capability. SCs are currently considered to be unsuitable as main energy storage sources because their specific energy values are lower than those of secondary batteries. However, with the emergence of new technologies and new chemistries that can lead to increased specific energies and reduced cost, they are considered to be attractive alternatives to main energy storage sources, especially because of their long life.

However, SCs have some major drawbacks originating from their inherent electrical properties. These are as follows:

1. The specific energy of SCs is lower than that of traditional secondary batteries.
2. Cell/module voltages of SCs in a series connection need to be eliminated since cell/module voltage imbalance may result in premature irreversible deteriorations and/or decrease in available energy.
3. Since the specific energy of SCs is low, energy stored by SCs should be delivered to loads as efficiently as possible in order to avoid energy wastage.
4. Terminal voltages of SCs vary widely with charging/discharging processes. Power converters having wide voltage ranges are required to power loads within a particular voltage range.

This chapter presents the SC-based electrical energy storage systems as alternatives to traditional battery-based systems. In the following sections, the above-mentioned issues are addressed in detail. In Section 2, the potential of SCs as alternative main energy storage sources is discussed on the basis of comparisons with specific energy and cycle life performance of a lithium-ion battery. In Section 3, cell/module voltage equalizers that are operable with a single switch or even without switches are introduced and compared with conventional topologies in terms of the number of components. Section 4 presents high-efficiency power converters suitable for SCs.

2. Supercapacitors as main energy storage sources

In general, the specific energy of SCs is lower than that of traditional secondary batteries. For example, specific energies of lead-acid and alkaline batteries (such as Ni-Cd and Ni-MH batteries) are 20–40 and 40–80 Wh/kg, respectively, and those of LIBs are at least 150 Wh/kg. On the other hand, the specific energy of conventional SCs does not exceed 10 Wh/kg. Lithium-ion capacitors (LICs), which are newly emerging SCs having a new chemistry, offer values less than 30 Wh/kg, which are comparable to those of lead-acid batteries but remain lower than other battery chemistries. LICs can match lead-acid batteries but their costs are not comparable. Meanwhile, there is still a large gap between LIBs and SCs (including LICs) in terms of specific energy, and therefore, SCs are usually considered unsuitable as main energy storage sources. However, SCs are considered to be potential alternative main energy storage sources considering their net specific energy, which is defined as

$$\text{Net Specific Energy} = \text{Specific Energy} \times \text{Depth of Discharge}, \qquad (1)$$

as well as their cycle life performance. For example, in low-Earth orbit satellite applications, where a minimum service life of three years is required for energy storage systems, three types of energy storage sources, (i) alkaline batteries, (ii) LIBs and (iii) SCs, are compared in terms of specific energy, depth of discharge (DoD) and net specific energy. The comparisons are shown in Table 1. Traditional secondary batteries for such satellites are operated with relatively shallow DoD of 20%–25%, allowing the life requirement to be fulfilled. Therefore, the net specific energies of alkaline batteries are 8–20 Wh/kg, and similarly, those of LIBs are 30–50 Wh/kg, although LIBs offer high specific energies of 150–200 Wh/kg. On the other hand, SCs can be cycled with deep DoD values even for such long-term applications because their cycle life performance is inherently excellent and is independent on DoD (as shown later). For LICs, the net specific energy reaches <24 Wh/kg for a DoD of 80% and the gap between secondary batteries and SCs (especially LICs) can, therefore, be bridged.

	Alkaline Battery (Ni-Cd, Ni-MH)	LIB	Supercapactor	
			Conventional	LIC
Specific Energy	40–80 Wh/kg	150–200 Wh/kg	< 10 Wh/kg	< 30 Wh/kg
Depth of Dischar	20–25%	20–25%	< 80%	< 80%
Net Specific Ener	8–20 Wh/kg	30–50 Wh/kg	< 8 Wh/kg	< 24 Wh/kg

Table 1. Specific energy, depth of discharge and net specific energy for traditional secondary batteries and SCs for low-Earth orbit satellite applications.

Fig. 1 shows an example of cycle life performance test results for a 3-Ah-class LIB and 2000-F-class LICs cycled with 20% and 80% (40%) DoD, respectively, at 25°C. A single cycle consists of a 65-min charge and 35-min discharge, and 10000 cycles are equivalent to approximately 1.9 years of service. The LIB deteriorated by 30% at the 10000th cycle while the LICs retained more than 96% of their initial capacitance, as shown in Fig. 1(a). The degradation of the LICs was almost independent on DoD, although that of LIBs, in general, significantly depends on DoD (Yoshida, et al., 2010). The deeper the DoD, the greater will be the deterioration experienced by the LIBs. Fig. 1(b) shows cycle life performance as a function of the square root of cycle number, using which the cycle life performance can be depicted linearly. The cycle life performances of the LIB and LICs can be predicted by extrapolating with straight lines (Mita, et al., 2010). The capacitance retention is expressed as a function of number of cycles and is expressed as

$$\text{Capacitance Retention} = 100 - K \times (\text{Number of Cycles})^{0.5} \tag{2}$$

where K is the degradation rate constant. From the results shown in Fig. 1(b), the values of K for the LIB and LICs were calculated to be 0.3 and 0.04, respectively. From Eq. (1), the cycle life of LIC is expected to be approximately 56 times longer than that of LIB under a given condition. For the LIB to achieve a cycle life that is as long as that of the LIC, the DoD must be shallower in order to alleviate degradations due to cycling. However, a lower DoD also results in a decrease in the net specific energy of the LIB, as determined by Eq. (1). Thus, from two aspects, the net specific energy and the cycle life performance, SCs (especially LICs) can be used as main energy storage sources and are suitable alternatives to traditional secondary batteries for shallow DoD applications.

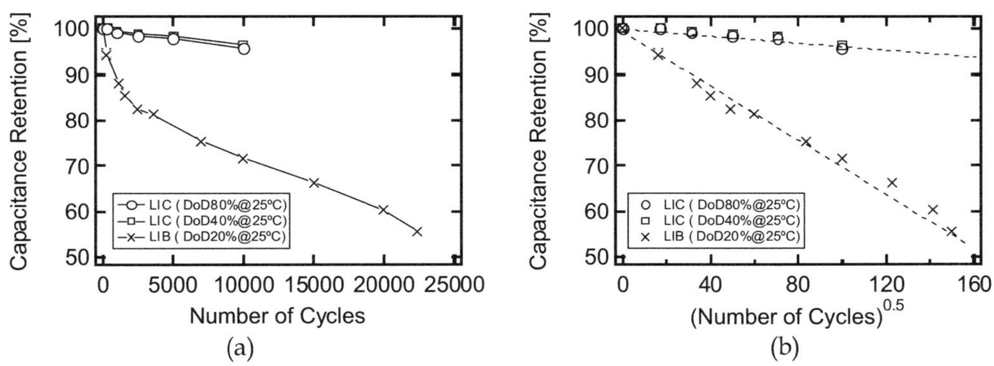

Fig. 1. Cycle life performances of a lithium-ion battery and lithium-ion capacitors as a function of (a) number of cycles and (b) square root of number of cycles.

The above comparison focuses on alternative applications for the batteries with shallow DoD for long-term cycle life. However, for deep DoD applications where the batteries are almost fully discharged, SCs cannot match the batteries from the perspective of net specific energy and cannot be an alternative energy storage source. Thus, SCs are practical and most suitable as main energy storage sources for applications where the batteries are used with shallow DoDs to achieve long cycle lives.

3. Cell/module voltage equalizer

3.1 Conventional cell/module voltage equalizer

Cell/module voltage equalizers are commonly used for SCs and LIBs. Voltage imbalances among cells/modules may result in not only reduced available energy but also premature deterioration caused by overcharging and over-discharging. In this section, representative conventional cell/module voltage equalizers are presented and technical concerns regarding their circuit complexity and reliability are addressed.

Various cell voltage equalizers, including dissipative and nondissipative approaches, have been proposed, demonstrated and reviewed (Cao, et al., 2008; Guo, et al., 2006). Fig. 2 shows the basic topologies of four examples of conventional dissipative and nondissipative equalizers. Various derivatives have also been proposed but are not shown here. As discussed in Section 2, the specific energy of SCs is lower than that of LIBs, so a larger number of cells/modules may be needed to constitute an SC-based energy storage system. The greater the number of cells/modules connected in series, the greater will be the number of voltage equalizers required. However, the system's complexity is prone to increase as the number of voltage equalizers increases, and hence, simple equalizers are desirable for SC-based energy storage systems.

The most prevalent topology is a shunting equalizer (Fig. 2(a)) (Isaacson, et al., 2001; Uno, 2009) that is a dissipative equalizer. Several battery management ICs containing dissipative equalizers are currently available. Dissipative equalizers typically consist of a series combination of a transistor and a current-limiting resistor. Excess stored energies of cells or charge currents are shunted to the transistor and resistor when the cell voltage exceeds a certain value. In other words, the excess energy or charge current is dissipated at the transistor and resistor, and this process generates heat, which is not desirable as it negatively impacts the energy efficiency and thermal management of the system.

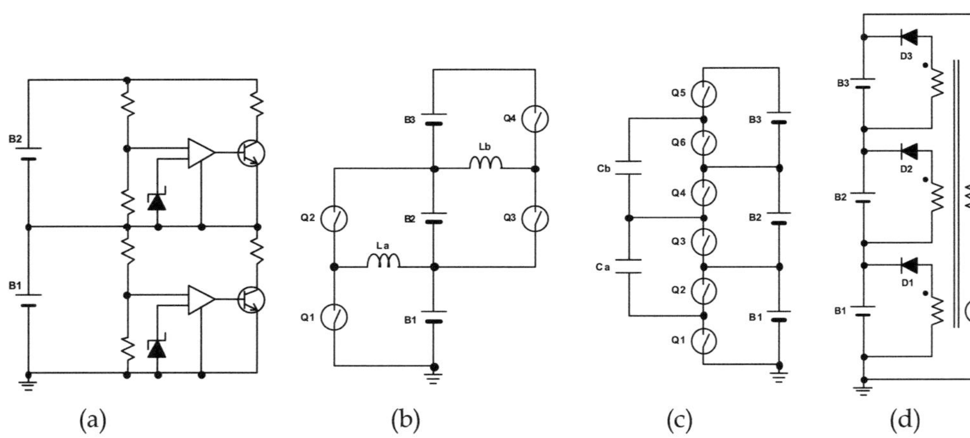

Fig. 2. Conventional cell/module voltage equalizers: (a) shunting equalizer, (b) buck-boost converter equalizer, (c) switched capacitor converter equalizer and (d) multi-winding flyback-based equalizer.

Conventional nondissipative equalizers are typically based on multiple individual dc–dc converters such as buck-boost converters (Nishijima, et al., 2000) and switched capacitor converters (Pascual & Krein, 1997), as shown in Figs. 2(b) and (c), respectively. In these topologies, the charges or energies of the series-connected cells can be exchanged between adjacent cells to eliminate cell voltage imbalance.

In the equalizers shown in Figs. 2(a), (b) and (c), the number of switches needed is proportional to the number of series connections of the cells. The number of switches is a good index for representing a circuit's complexity because switches require drivers and/or ancillary components. Hence, the circuit complexity and cost are prone to increase as the number of series connections increases, especially for applications where numerous series connections of cells are necessary.

In a transformer-based equalizer incorporating flyback- and forward-based topologies, the energies of series-connected cells can be redistributed via a multi-winding transformer (Kutkut, et al., 1995) to the cell having the lowest voltage. Fig. 2(d) depicts the flyback-based equalizer. The number of switches required are significantly less than those required with other topologies. However, this topology needs a multi-winding transformer that must be customized according to the number of series connections, and hence, the modularity is not good. In addition, the design and parameter matching for multiple windings are considered difficult (Cao et al., 2008).

As mentioned in Section 2, the specific energy of SCs is lower than that of traditional secondary batteries, so an SC-based energy storage system may require a larger number of cells to be connected in series and/or parallel than secondary batteries, although SCs have potentials to match or outperform the traditional batteries in terms of net specific energy for particular applications. In other words, the number of series connections of SCs is prone to be larger than that of secondary batteries. Hence, using multiple switches or transformer windings, which leads to increased cost and circuit complexity, is undesirable for an SC-based energy storage system. In addition, conventional topologies are undesirable because of their complexity, since electrical circuits should be as simple as possible in order to mitigate risks of failure, especially for applications that require long-term use, i.e., SC-based energy storage systems.

3.2 Voltage equalizer using single-switch multi-stacked SEPICs

3.2.1 Circuit configuration and major benefits

Fig. 3 shows a single-switch cell/module voltage equalizer for four series-connected SCs. This topology operates as a charger with an equalization function. V_{in} is the external power source, and the circuit consisting of V_{in}, C_{in}, L_{in}, Q, C_1, L_1, D_1 and SC_1 is identical to a conventional single ended primary inductor converter (SEPIC). The circuits consisting of C_i-D_i-L_i (i = 1…4) are identical and multi-stacked; inductor–diode pairs are stacked in series while all the capacitors are connected to Q and L_{in}. Hence, this equalizer may be regarded as a multi-stacked SEPIC.

This circuit contains a single active device (i.e., switch) and multiple passive components. This single-switch circuit configuration contributes to a significant reduction in circuit complexity when compared to the conventional topologies illustrated in Fig. 2. This equalizer is also advantageous with regards to its drive circuits. The conventional topologies shown in Figs. 2(b) and (c) require floating gate drivers in cases where N-channel MOSFETs are used for high-side switches (even-numbered switches in Figs. 2(b) and (c)). The equalizer shown in Fig. 3, on the other hand, does not require a floating gate drive circuit because the switch is connected to the ground. Moreover, since the basic topology of this equalizer is SEPIC, commercially available control ICs for SEPICs can be employed. Therefore, this equalizer reduces not only the number of switches but also the complexity of the gate drive circuit. Furthermore, this equalizer also offers good modularity because the number of series connections can be arbitrarily extended by stacking the circuit of C_i-D_i-L_i, without the need for additional active components such as switches or control ICs.

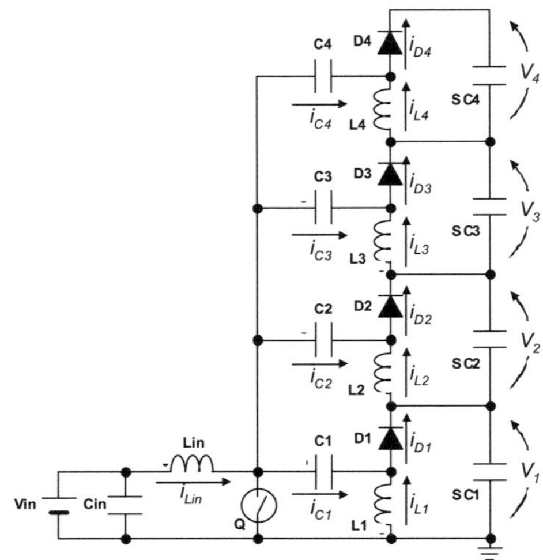

Fig. 3. Single-switch cell/module voltage equalizer using multi-stacked SEPICs.

3.2.2 Fundamental operation

The fundamental operation of this equalizer is similar to that of a conventional SEPIC. Fig. 4 shows the theoretical operating waveforms and current flow directions. When the switch is turned on (T_{on} period), all the inductor currents increase and the corresponding energies are stored in each inductor. When the switch is tuned off (T_{off} period), the diodes are turned on and all the inductor currents decrease. The current in L_{in} is distributed to each capacitor and SC depending on each cell voltage. As long as the cell voltages are uniform and cell impedances are negligible, the current in L_{in} is uniformly distributed to each capacitor.

The average voltage of inductors under a steady-state condition is zero. The voltages of the capacitors C_1–C_4, referred to as V_{C1}–V_{C4}, respectively, are

$$\begin{cases} V_{C1} = V_{in} \\ V_{C2} = V_{in} - V_1 \\ V_{C3} = V_{in} - (V_1 + V_2) \\ V_{C4} = V_{in} - (V_1 + V_2 + V_3) \end{cases} \tag{3}$$

where V_{in} is the input voltage and V_1-V_4 are voltages across SC_1-SC_4 denoted in Fig. 3, respectively. The voltage–time product of inductors in a single cycle under a steady-state condition is also zero. Therefore,

$$\begin{cases} DV_{C1} &= (1-D)(V_1 + V_{D1}) \\ D(V_{C2} + V_1) &= (1-D)(V_2 + V_{D2}) \\ D(V_{C3} + V_1 + V_2) &= (1-D)(V_3 + V_{D3}) \\ D(V_{C4} + V_1 + V_2 + V_3) &= (1-D)(V_4 + V_{D4}) \end{cases} \tag{4}$$

where D is the duty cycle and V_{D1}–V_{D4} are forward voltages of D_1-D_4, respectively. From Eqs. (3) and (4), we get

$$V_i = \frac{D}{1-D} V_{in} - V_{Di} \tag{5}$$

Eq. (5) indicates that this equalizer outputs uniform voltages to all SCs as long as the diodes' forward voltages are uniform. In the case where cell voltages are imbalanced, D can be controlled in order to regulate the output voltages higher and lower than the lowest and other SC voltages, respectively. This allows the cell having the lowest voltage to be charged preferentially.

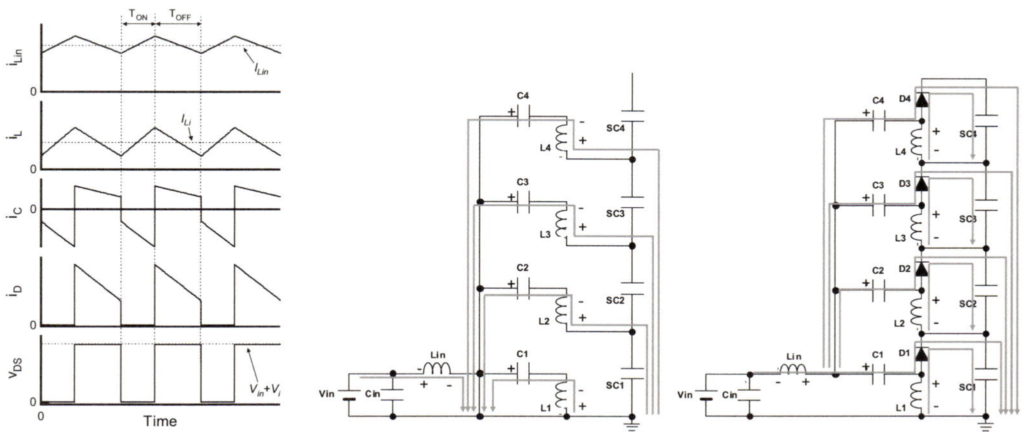

(a) (b) (c)

Fig. 4. (a) Theoretical operating waveforms and current directions during (b) T_{on} and (c) T_{off}.

3.2.3 Experimental equalization performance

Four SC modules with capacitance of 220 F each were connected in series and charged from an initially voltage-imbalanced condition by using a 40 W prototype shown in Fig. 5(a).The voltage input to the equalizer was 28 V, and by employing PWM control using a switching regulator IC (LTC1624) operating at 200 kHz, the input current and charge voltage were regulated to be 1.5 A and 14.5 V, respectively.

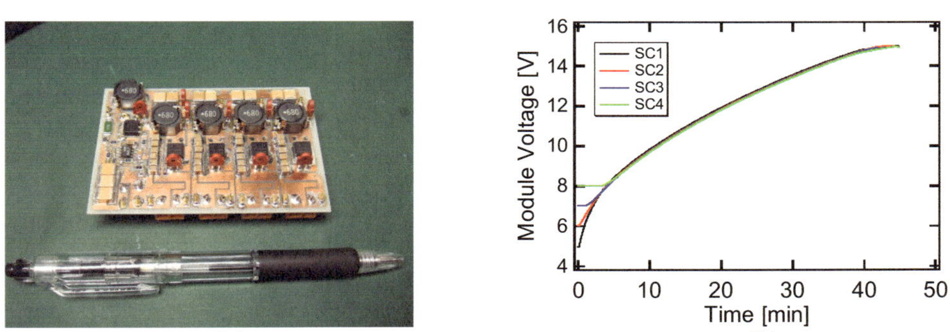

(a) (b)

Fig. 5. (a) Photograph of the 40 W prototype of the equalizer using multi-stacked SEPICs, and (b) experimental charge profiles of four series-connected SC modules charged by the prototype from an initially voltage-imbalanced condition.

The SC module(s) having the lowest voltage was(were) charged preferentially at each instant and the voltage imbalance was eliminated as the charging progressed, as shown in Fig. 5(b). After the voltage imbalance was eliminated, all the SC voltages increased uniformly. Eventually, at the end of the charge, all the SCs was charged to the uniform voltage of 14.5 V.

3.3 Switchless voltage equalizer

3.3.1 Circuit configuration and major benefits

A switchless voltage equalizer for three series-connected SCs is shown in Fig. 6. This topology also operates as a charger with an equalization function; the charge is provided by an ac power source. Two series-stacked diodes are connected to each SC and the junctions of stacked diodes are connected to the ac power source via energy transfer capacitors C_1–C_3.

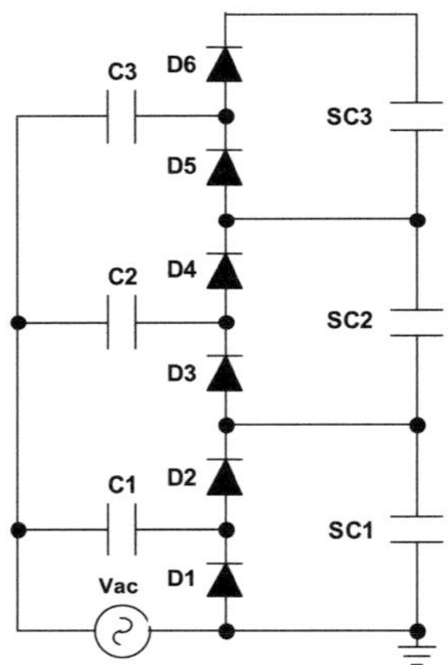

Fig. 6. Switchless cell/module voltage equalizer.

This equalizer consists of passive components only, resulting in reduced circuit complexity and improved equalizer reliability when compared with those of conventional ones. Similar to the single-switch equalizer presented in the previous section, this equalizer also exhibits good modularity. The number of series-connected SCs can be easily extended by adding a capacitor and stacked diodes.

3.3.2 Fundamental operation

The equalizer operates in two modes, and the current flow direction in each mode is shown in Fig. 7. Each SC can be charged to a uniform voltage level by the ac power source while alternating between the two modes.

In mode A, odd-numbered diodes are turned on and C_1–C_3 are charged by the ac power source and SC_1–SC_2. The voltages of C_1–C_3 in mode A are V_{C1A}–V_{C3A}, respectively, and are given by

$$\begin{cases} V_{C1A} = E_A - V_D \\ V_{C2A} = E_A + V_{SC1} - V_D \\ V_{C3A} = E_A + V_{SC1} + V_{SC2} - V_D \end{cases} \tag{6}$$

where E_A is the peak voltage of the ac power source in mode A and V_D is the forward voltage of the diodes.

In mode B, C_1–C_3 discharge to the SCs via even-numbered diodes and the voltages of C_1–C_3 in mode B are V_{C1B}–V_{C3B}, respectively, and are given by

$$\begin{cases} V_{C1B} = E_B + V_{SC1} + V_D \\ V_{C2B} = E_B + V_{SC1} + V_{SC2} + V_D \\ V_{C3B} = E_B + V_{SC1} + V_{SC2} + V_{SC3} + V_D \end{cases} \tag{7}$$

where E_B is the bottom voltage of the ac power source in mode B.
In general, the average current through capacitor, I_{Ci}, is given by

$$I_{Ci} = C_i \times f \times \Delta V_{Ci} \tag{8}$$

where C_i is the capacitance of C_i (i = 1…3), f is the frequency and ΔV_{Ci} is the voltage variation across C_i. ΔV_{Ci} is obtained by subtracting Eq. (7) from Eq. (6). Substituting the result into Eq. (8) gives

$$I_{Ci} = C_i f \left\{ (E_A - E_B) - 2V_D - V_{SCi} \right\} \tag{9}$$

$(E_A - E_B)$ is equivalent to the peak-to-peak voltage of the ac power source. This equation implies that all the SCs are charged to the uniform voltage level of $(E_A - E_B) - 2V_D$ at the end of the charge at which I_{Ci} becomes zero. The charge rate is determined by $C_i f$ whose dimension is the inverse of resistance (i.e., conductance). The greater the capacitance of C_i and f, the quicker the SCs will be charged. The inverse of $C_i f$, R_{Cf}, can be used as an index to represent the charging speed.

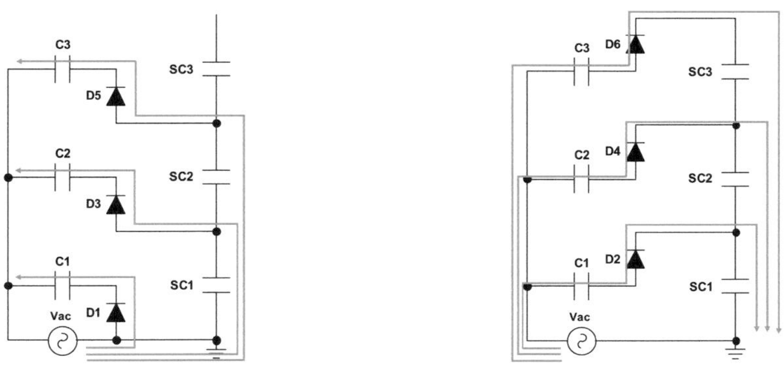

Fig. 7. Current flow directions in (a) mode A and (b) mode B.

3.3.3 Experimental equalization performance

Three SC modules with capacitance of 60 F each were connected in series and charged from an initially voltage-imbalanced condition using a prototype with R_{Cf} of 42 Ω (Fig. 8(a)). Al electrolytic capacitors having capacitance of 470 μF each were used for C_1–C_3. The ac voltage for the equalizer was 17 Vac (peak-to-peak voltage of 48 V) and was provided by a 50 Hz utility power source via a transformer.

The SC modules were charged at different charge rates, as indicated by Eq. (9), and are shown in Fig. 8(b). The voltage imbalance was gradually eliminated as the charging progressed. Eventually, all the SCs were charged to a uniform voltage of approximately 47

V, which is 1 V lower than the peak-to-peak voltage of the ac power source. This difference is attributed to diode voltage losses.

(a)

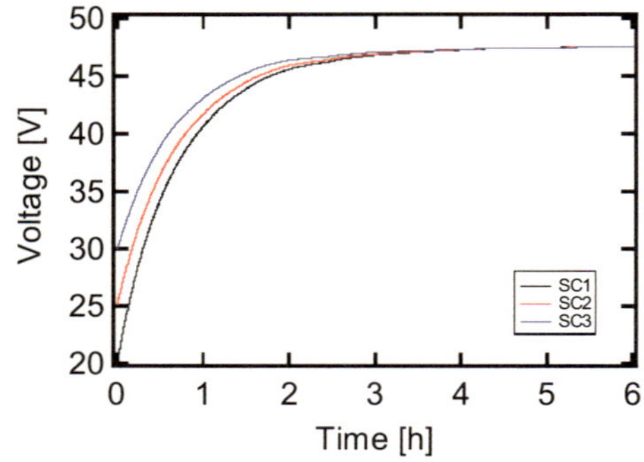

(b)

Fig. 8. (a) Photograph of the prototype of the switchless equalizer, and (b) experimental charge profiles of three series-connected SC modules charged by the prototype from an initially voltage-imbalanced condition.

3.4 Comparison for circuit complexity and number of components

In Table 2, two topologies presented in the previous sections are compared with the conventional equalizers. As mentioned in Section 3.1, the number of switches is a good index for representing a circuit's complexity. The number of switches required for conventional equalizers such as shunting, buck-boost and switched capacitor equalizers is proportional to the number of series-connected cells/modules, n. The transformer-based (flyback) equalizer can operate with a single switch, but the need for $(n + 1)$ windings results in poor modularity, design difficulties and cost penalty. However, the two topologies described in previous sections require neither multiple switches nor transformer windings, although they require multiple passive components. Therefore, these topologies have advantages over conventional ones in terms of circuit complexity, modularity and cost, especially for applications where numerous series connections of cells/modules are necessary. Moreover, employing fewer active components leads to reduced risks of failure and improved reliability.
These topologies can be applied for both SCs and LIBs.

Topology	Switch	R	L	C	D	Transformer
Multi-Stacked SEPICs	1	-	$n + 1$	n	n	-
Switchless Equalizer	-	-	-	n	$2n$	1 (1 core with 2 windings)
Shunting Equalizer	n	n	-	-	-	-
Buck-Boost	$2n - 1$	-	$n - 1$	-	-	-
Switched Capacitor	$2n$	-	-	$n - 1$	-	-
Transformer (Flyback)	1	-	-	-	n	1 (1 core with $(n + 1)$ windings)

(Smoothing capacitor is excluded)

Table 2. Comparison of the number of components required for each equalizer.

4. High-efficiency power converters with wide voltage range

4.1 Conventional power converters

As discussed in Section 1, power converters in SC-based energy storage systems are required to operate over a wide voltage range because SC voltages vary significantly due to charge-discharge processes. For the SC-based electrical energy storage systems as alternatives to traditional battery-based systems, the converters need to operate over a wide input voltage range and provide power to loads within a voltage range that is at least comparable to battery voltage variations. In addition, the power converters should operate as efficiently as possible. With traditional switching power converters, the stored energies of SCs can be provided to loads at a constant voltage, and the SCs can be discharged sufficiently deep. However, designing traditional converters to operate over a wide voltage range leads to increase and decrease in size and efficiency, respectively. An increase in the size of magnetic components such as inductors and transformers, which are relatively large components in converters, is significant. Since available energies of SCs are proportional to the converter efficiency, a decrease in the converter efficiency results in either a decrease in available energy or an increase in size, weight and cost of SCs.

Although emphasis on chargers is necessary, this section focuses on dischargers, which are especially important for SC-based energy storage systems, because the energy requirement as well as size and weight of SCs are directly proportional to the discharger efficiency.

4.2 Discharger using cascaded switched capacitor converters with selectable intermediate taps

4.2.1 Conventional switched capacitor converter and conceptual derivation of cascaded switched capacitor converters with selectable intermediate taps

Switched capacitor converters (SCCs) that do not require magnetic components have been proposed for non-isolated intermediate bus converters and automotive applications (Oraw & Ayyanar, 2007; Peng, et al., 2003; Xu, et al., 2006). SCCs achieve both high efficiency and high power density, but their input-output voltage ratio is usually uncontrollable and is a fixed value that is determined by the number of capacitors stacked in series.

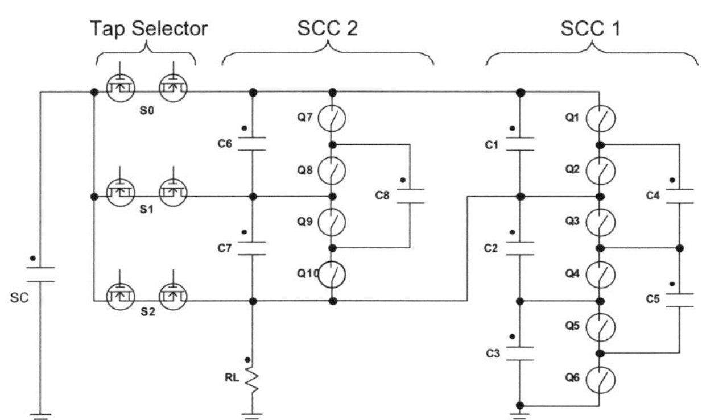

Fig. 9. SC Discharger using cascaded switched capacitor converters with selectable intermediate taps.

SCCs operate as nondissipative voltage dividers that produce M discrete voltage levels when M capacitors are stacked in series. In other words, voltage levels that can be provided by SCCs to a load that incorporates an SC having a voltage of V_{SC} are $V_{SC}(1/M)$, $V_{SC}(2/M)$ ...V_{SC}. As V_{SC} varies with charge-discharge processes, these voltage levels also vary. However, by selecting one of these voltage levels in accordance with the variation in V_{SC}, the load voltage can be maintained within a desired voltage range.

On the basis of the concept of selecting one of the multiple voltage levels, SCCs having selectable intermediate taps are derived as shown in Fig. 9. Two SCCs, referred to as SCCs 1 and 2, are cascaded to produce fine discrete voltage levels, and selectable intermediate taps are connected between SC and SCC 2. The SC voltage is divided via two stages using two SCCs, and the load voltage can be maintained by selecting one of the intermediate taps.

4.2.2 Operating principle

Each SCC consists of switches, stationary capacitors (C_1–C_3 and C_6–C_7) and energy transfer capacitors (C_4–C_5 and C_8). Odd- and even-numberd switches in each SCC alternate with a duty cycle of almost 50% to transfer charges among capacitors. Ideally, the capacitor voltages in each SCC become uniform. The load is directly connected to C_2–C_3 in SCC 1, and SCC 2 is cascaded to SCC 1 via C_1. The relationship between the load voltages (V_{Load}) and the capacitors in SCC 1 and SCC 2, (V_{C1} and V_{C2}), is given by

$$V_{Load} = 2V_{C1} = 4V_{C2} \tag{10}$$

One of the intermediate taps (MOSFET relays S_0–S_2) is selected accordingly, and V_{SC} is equal to the voltage level of the selected intermediate tap. For example, when S_0 is on, V_{SC} is the sum of the voltages across C_6–C_7 and V_{Load}. On the other hand, when S_1 is on, V_{SC} is the sum of the voltages across C_7 and V_{Load}. Therefore, the relationship between V_{SC} and V_{Load} can be generalized as

$$V_{Load} = \frac{4V_{SC}}{6-N} \tag{11}$$

where N is the subscript of S_0–S_2.

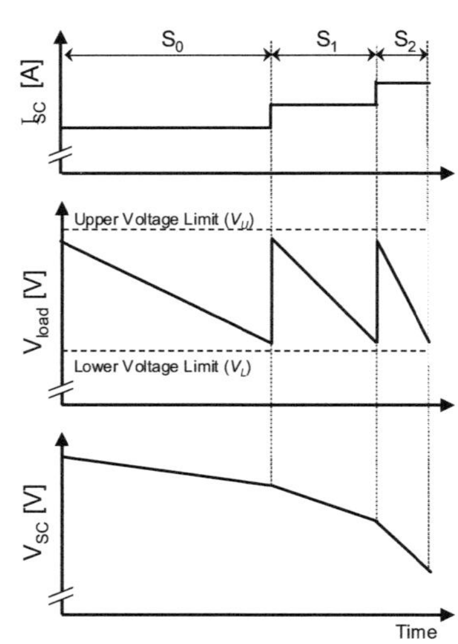

Fig. 10. Discharge characteristics of an SC with the discharger using cascaded switched capacitor converters with selectable intermediate taps.

The discharging characteristics of an SC when using the discharger shown in Fig. 9 are illustrated in Fig. 10. At the beginning of the discharging process of the SC, S_0 is turned on and the SC discharges via S_0 to the SCCs. V_{Load} is four-sixth of V_{SC}, as determined by Eq. (11). By assuming that the SCCs operate ideally without power conversion losses, the SC current, I_{SC}, is four-sixth of the load current, I_{Load}. As discharging progresses, V_{Load} and V_{SC} decrease. When V_{Load} decreases to the lower voltage limit level, V_L, S_0 and S_1 are turned off and on, respectively, in order to raise V_{Load}. After S_1 is turned on, V_{Load} increases to four-fifth of V_{SC}. Simultaneously, I_{SC} also increases to four-fifth of I_{Load}. The gradient of V_{SC} during S_1-on period becomes steeper than that during S_0-on period because of the larger I_{SC}. With further discharging, V_{Load} decreases further until it reaches V_L again. Then, S_2 is turned on in order to raise V_{Load} again. During S_2-on period, V_{Load} and I_{Load} are equal to V_{SC} and I_{SC}, respectively, because the load and SC are connected directly via S_2. Stepwise increases in I_{SC} result in inflection points in V_{SC}.

The above-mentioned sequence is repeated to maintain V_{Load} within a desired voltage range bounded by upper and lower limits, V_U and V_L, respectively. The greater the number of selectable intermediate taps and capacitors stacked in series, the finer will be the variation in load voltage that can be achieved. This section explains the sequence that takes place during discharging. A similar sequence can be applied for the charging process as long as the intermediate taps of S_0–S_2 are bidirectional switches.

4.2.3 Experimental discharge performance

With the goal of achieving an alternative application for a 42-V-battery with an operating voltage range of 30–42 V, a 200 W discharger prototype was designed for a V_{Load} of 30–40 V, a maximum load current of 5 A and an SC module at 60 V in a fully charged state. Capacitors in each SCC must be designed to be capable of RMS currents determined by the load current and switching frequency. A detailed design procedure for the capacitors has been reported elsewhere (Oraw & Ayyanar, 2007; Seeman & Sanders, 2008). The capacitances (number of capacitors) of each capacitor were determined on the basis of the reported design procedure, as shown in Table 3. Multiple ceramic capacitors were connected in parallel to satisfy the required capacitance.

Capacitor	Capacitance
C_1	22 μF
C_2	44 μF (22 μF × 2)
C_3	44 μF (22 μF × 2)
C_4	110 μF (22 μF × 5)
C_5	66 μF (22 μF × 3)
C_6	22 μF
C_7	44 μF (22 μF × 2)
C_8	66 μF (22 μF × 3)

Table 3. Capacitance of each capacitor.

A photograph of a 200 W prototype is shown in Fig. 11(a). N-channel MOSFETs with low on-resistance (9.2 mΩ for Q_1–Q_6, 1.8 mΩ for Q_7–Q_{10}) were used for switches in each SCC. The switches were operated at a fixed switching frequency of 300 kHz with 50% duty cycle.

Synchronous MOSFET gate driver ICs (ADP3120A) were used, and the ICs were powered by the stationary capacitors C_1–C_3 and C_6–C_7 via low-dropout linear regulator ICs. The intermediate taps, S_0–S_2, were composed of two MOSFETs with 9.2 mΩ on-resistance connected back-to-back.

Power conversion efficiencies when S_0, S_1 and S_2 were on and when V_{Load} = 40 V are shown in Fig. 11(b). In the low power region, the efficiencies were relatively low because the power consumption at the MOSFET gate driver ICs accounted for a relatively large part of the input power. However, in the region above 60 W, efficiencies higher than 96% were achieved. The highest efficiency of approximately 98% was observed at 200 W.

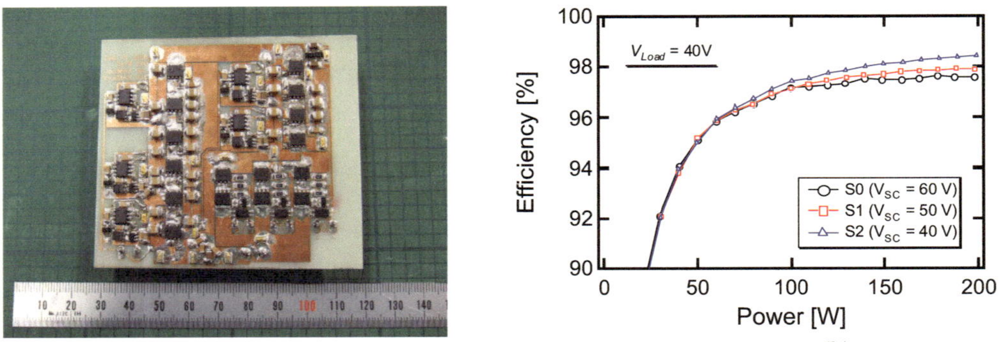

(a) (b)

Fig. 11. (a) Photograph of a 200 W prototype and (b) power conversion efficiencies when V_{Load} was 40 V.

With the prototype, an SC module with capacitance 55 F was discharged at a constant current of 4 A. The intermediate taps were shifted in the order of S_0, S_1 and S_2 to maintain V_{Load} within 30–40 V. Experimental discharge profiles are shown in Fig. 12. The SC was discharged from 60 to 30 V while V_{Load} was successfully maintained within the 30–40 V range. Discharging the SC to half of its initial voltage resulted in 75% energy utilization.

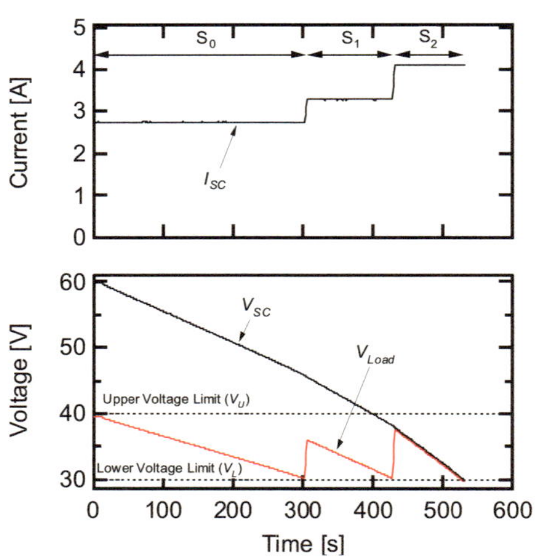

Fig. 12. Experimental discharge curves of the SC module with the prototype discharger using cascaded switched capacitor converters with selectable intermediate taps.

4.3 Series-parallel reconfiguration technique

4.3.1 Conventional reconfiguration techniques

In conventional switching converters including the SCCs shown in the previous section, switching losses caused by high-frequency switching operations are a major factor in reducing power conversion efficiencies. Series-parallel reconfiguration techniques that do not require high-frequency switching operations have been proposed (Sugimoto, et al., 2003). Because of the negligible switching loss, the conduction loss is the only major factor affecting the efficiency; hence, these techniques achieve higher efficiencies (99.5% efficiency has been reported) than switching power converters.

Fig. 13(a) shows a series-parallel changeover circuit. Two SCs are connected in parallel at the beginning of the discharging process via Q_1 and Q_2. After the SC voltages decrease to a predetermined lower voltage level, Q_1 and Q_2 turn off, and Q_3 turns on to connect the SCs in series. Hence, the output voltage (load voltage) increases and the SCs can be discharged deeply. However, the voltage variation caused by the reconfiguration is as high as 50%.

Another approach employs the shift-type changeover circuit shown in Fig. 13(b). SC_1-SC_2 and SC_3-SC_4 are initially connected in parallel via Q_1 and Q_2, and hence, the whole system is two series two parallel. SC voltages decrease with discharging, and Q_1 and Q_2 turn off while Q_3 and Q_5 turn on when the SC voltages decrease to a predetermined level. At that moment, only SC_1 and SC_4 are connected in parallel while SC_2 and SC_3 are connected in series. Therefore, at this instant, the number of series connections in the entire system is three. The SC voltages decrease further with discharging, and Q_3 and Q_5 turn off, and Q_4 turns on. All the SCs are connected in series via Q_4, so the system becomes four series one parallel. This system can increase the number of series connections one by one. As a result, the voltage variation caused by the reconfiguration can be mitigated compared with the series-parallel changeover circuit shown in Fig. 13(a). However, while Q_3 and Q_5 are on, a voltage imbalance inevitably occurs because the SCs are discharged unequally (Sugimoto, et al., 2003). The unequal discharging results in poor utilization of the stored SC energies, and in the worst case, overcharging and over-discharging.

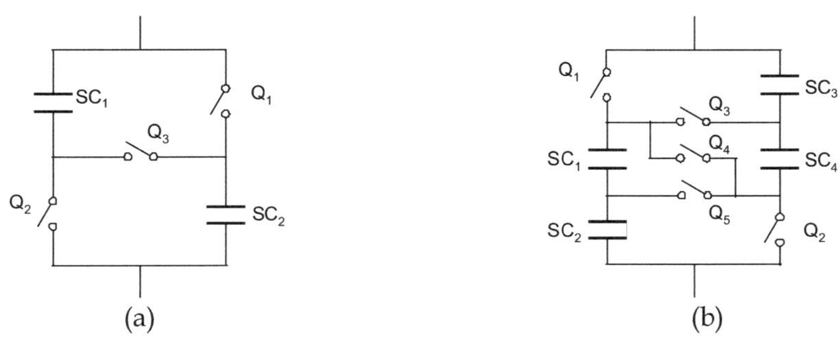

Fig. 13. Conventional reconfiguration techniques using (a) series-parallel and (b) shift-type changeover circuits.

4.3.2 Series-parallel reconfigurable SC unit

Fig. 14 shows a reconfigurable series-parallel SC unit that operates in two modes, modes A and B. This unit can be used in combination with the conventional circuits shown in Fig. 13. The following experimental section presents the discharge characteristics of the SC system based on a combination of the units in Fig. 14 and the changeover circuit shown in Fig. 13(a). This section focuses on the fundamental operation of the unit in Fig. 14.

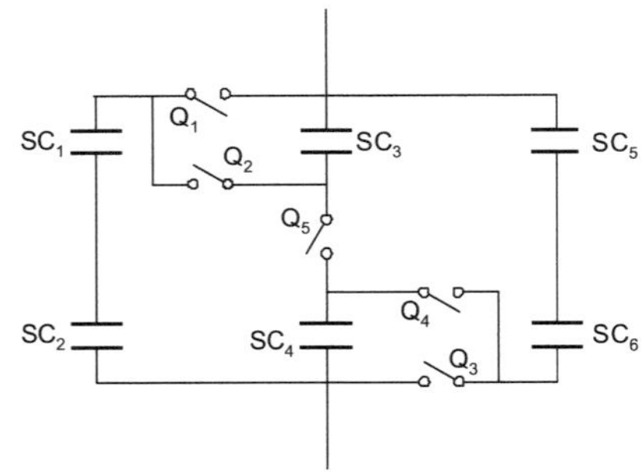

Fig. 14. Reconfigurable series-parallel SC energy storage unit.

The current flow directions and voltage curves during discharging are shown in Fig. 15. At the beginning of discharging, the unit operates as a two series three parallel system in mode A, in which three strings consisting of SC_1–SC_2, SC_3–SC_4 and SC_5–SC_6 are connected in parallel via odd-numbered switches, as shown in Fig. 15(a). As long as the capacitance of each SC is uniform, all the SCs discharge uniformly. The voltages across the SCs decrease as the discharging progresses. When the unit voltage falls below the predetermined level V_L, the series-parallel connections of the SC unit are reconfigured by turning off and on the odd- and even-numbered switches, respectively, as shown in Fig. 15(b).
In mode B, SC_1–SC_2 and SC_3, and SC_4 and SC_5–SC_6 are connected in series, respectively, through even-numbered switches and the unit is a three series two parallel configuration. Hence, the unit voltage at the beginning of mode B is 1.5 times higher than that at the end of mode A. In other words, the voltage variation caused by reconfiguration is 33%. This value is smaller than that for the conventional technique shown in Fig. 13(a). The unit and SC voltages in mode B decrease at a faster rate than that in mode A because of fewer parallel connections in mode B. The change in the current rate results in an inflection point in the SC voltage curve.

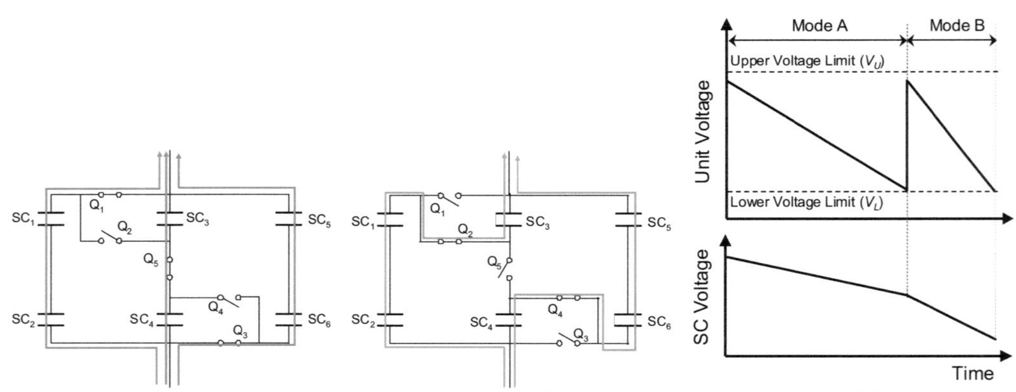

Fig. 15. Current flow directions in (a) mode A and (b) mode B, and (c) cell and unit voltage curves during discharging.

With the discharge sequence shown in Fig. 15, the SCs in the unit can be discharged deeply by reconfiguring the series-parallel configuration, while the unit voltage can be maintained within a particular voltage range. This reconfigurable unit can achieve very high efficiencies that are similar to those of the conventional techniques because the switching loss is negligible and efficiency is only affected by the conduction losses of switches.

This section explains the operating sequence for the discharging process. A similar reconfiguration sequence can be applied to a charging process. Since the reconfigurable SC unit consists of two or three strings in parallel, the unit is considered most suitable for relatively large-scale applications where parallel connections are usually required.

4.3.3 Experimental discharging characteristics

An SC-based energy storage system combining the reconfigurable units shown in Fig. 14 with the changeover circuit shown in Fig. 13 is illustrated in Fig. 16(a). SC_1 and SC_2 in Fig. 13 are replaced with the unit shown in Fig. 14. The system consisting of 500 F SCs was discharged at a constant current of 1.5 A, and the resultant discharge curves are shown in Fig. 16(b). Table 4 shows the operating status of the units and switches and the system configuration during the discharging experiment.

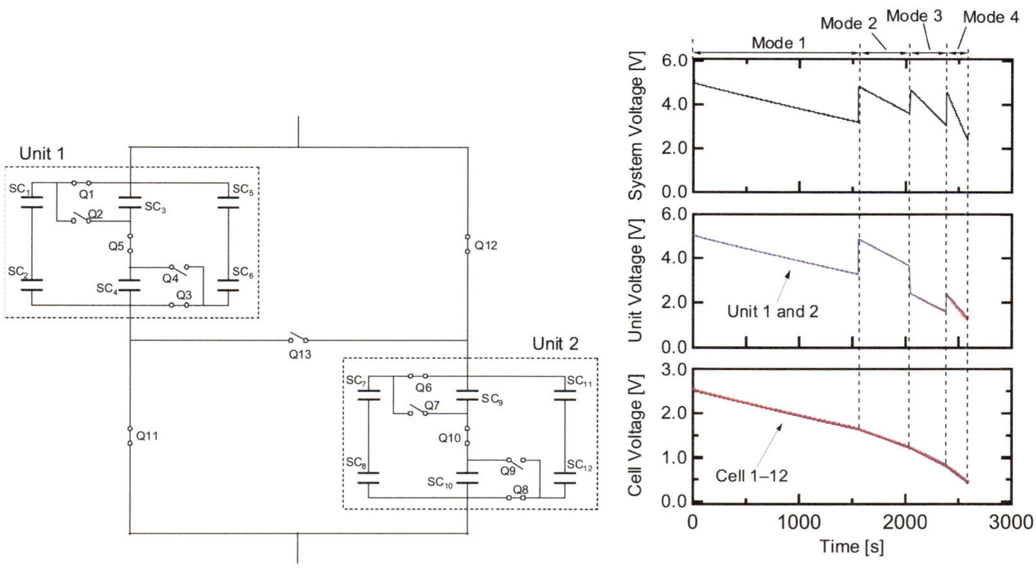

(a) (b)

Fig. 16. (a) Reconfigurable series-parallel SC energy storage system and (b) resultant discharging curves of system, unit and cell voltages.

The system configuration was modified from mode 1 to mode 4. When a system voltage lower than a predetermined lower voltage level (approximately 3.2 V in this case) was detected, the system configuration was modified by changing the operating statuses of the units and/or switches. As the discharging progressed, the number of series connections was increased consistently, whereas the number of parallel connections was decreased. The cell voltages decreased with discharging, but the system voltage was maintained within a desired voltage range. All the cells were uniformly and deeply discharged. The experimental results demonstrated that the system can achieve high energy utilization without causing voltage imbalance.

	Units 1 and 2	Q11, Q12	Q13	Configuration
Mode 1	Mode A	ON	OFF	two series six parallel
Mode 2	Mode B	ON	OFF	three series four parallel
Mode 3	Mode A	OFF	ON	four series three parallel
Mode 4	Mode B	OFF	ON	six series two parallel

Table 4. Operating statuses of units and switches during discharging, and system configuration in modes 1–4.

5. Conclusions

SCs offer numerous benefits over traditional secondary batteries. However, SCs are usually considered to be unsuitable as main energy storage sources because of their inherent low specific energy. In addition, several major issues need to be addressed before SCs can become main energy storage sources. These include the need for cell/module voltages to be balanced and for SCs to be discharged as efficiently and deeply as possible in order to maximise the use of the stored energies.

With regards to DoD and cycle life performance, SCs can match or outperform traditional secondary batteries in terms of net specific energy. In Section 2, the potential of SCs as alternatives to secondary batteries for energy storage applications was discussed from the perspective of net specific energies and cycle life performance. SCs can be used as alternative energy storage sources to traditional secondary batteries in applications where batteries have been cycled with shallow DoDs in order to achieve long cycle lives.

In Section 3, two types of cell/module voltage equalizers that can operate with either a single switch or without a switch were presented. SC-based energy storage systems may require numerous series connections of cells/modules. Therefore, single-switch and switchless equalizers are considered to be more suitable in terms of circuit complexity, modularity, cost and reliability.

Section 4 presented two types of high-efficiency voltage converters—cascaded switched capacitor converters with selectable intermediate taps and series-parallel reconfigurable SC systems. Neither type maintains its output voltage at a constant level but the output voltage can be maintained within a desired voltage range. For the former, we experimentally demonstrated power conversion efficiencies as high as 98% at 200 W. For the latter, even higher efficiencies can be achieved since the only losses that occur are conduction losses, while switching losses are negligible.

6. References

Cao, J., Schofield, N. & Emadi, A. (2008). Battery Balancing Methods: A Comprehensive Review, *Proceedings of IEEE Vehicle Power and Propulsion Conference*, ISBN 978-1-4244-1848-0, Harbin, China, September 3-5, 2008

Guo, K. Z., Bo, Z. C., Gui, L. R. & Kang, C. S. (2006). Comparison and Evaluation of Charge Equalization Technique for Series Connected Batteries, *Proceedings of IEEE Applied Power Electronics Conference and Exposition*, ISBN 0-7803-9716-9, Jeju, South Korea, June 18-22, 2006

Isaacson, M. J., Hollandsworth, R. P., Giampaoli, P. J., Linkowaky, F. A., Salim, A. & Teofilo, V. L. (2000). Advanced Lithium Ion Battery Charger, *Proceedings of Battery Conference on Applications and Advances*, ISBN 0-7803-5924-0, Long Beach, California, USA, January 11-14, 2000

Kutkut, N. H., Divan, D. M. & Novotny, D. W. (1995). Charge Equalization for Series Connected Battery Strings. *IEEE Transaction on Industry Applications,* Vol. 31, No. 3, (May & June 1995), pp. 562-568, ISSN 0093-9994

Nishijima, K., Sakamoto, H. & Harada, K. (2000). A PWM Controlled Simple and High Performance Battery Balancing System, *Proceedings of IEEE Power Electronics Specialist Conference*, ISBN 0-7803-5692-6, Galway, Ireland, June 18-23, 2009

Mita, Y., Seki, S., Terada, N., Kihira, N., Takei, K. & Miyashiro, H. (2010). Accelerated Test Methods for Life Estimation of High-Power Lithium-Ion Batteries. *Electrochemistry*, Vol. 78, No. 5, pp. 384-386

Oraw, B. & Ayyanar, R. (2007). Load Adaptive, High Efficiency, Switched Capacitor Intermediate Bus Converter, *Proceedings of IEEE International Telecommunications Energy Conference*, ISBN 978-1-4244-1627-1, Rome, Italy, October 30-November 4, 2007

Pascual, C. & Krein, P. T. (1997). Switched Capacitor System for Automatic Series Battery Equalization, *Proceedings of IEEE Applied Power Electronics Conference and Exposition*, ISBN 0-7803-3704-2, Atlanta, Georgia, USA, February 23-27, 1997

Peng, F. Z., Zhang, F. & Qian, Z. (2003). A Magnetic-Less DC-DC Converter for Dual-Voltage Automotive Systems. *IEEE Transaction on Industry Applications*, Vol. 39, No. 2, (May and April 2003), pp. 511-518, ISSN 0093-9994

Seeman, M. D. & Sanders, S. R. (2008). Analysis and Optimization of Switched-Capacitor DC-DC Converters. *IEEE Transaction on Power Electronics*, Vol. 23, No. 2, (May 2008), pp. 841-851, ISSN 0885-8993

Sugimoto, S., Ogawa, S., Katsukawa, H., Mizutani, H. & Okamura, M. (2003). A Study of Series-Parallel Changeover Circuit of a Capacitor Bank for an Energy Storage System Utilizing Electric Double Layer Capacitors. *Electrical Engineering in Japan*, Vol. 145, No. 3, (November 2003), pp. 33-42

Uno, M. (2009). Interactive Charging Performance of a Series Connected Battery with Shunting Equalizers, *Proceedings of IEEE International Telecommunications Energy Conference*, ISBN 978-1-4244-2490-0, Incheon, South Korea, October 18-22, 2009

Xu, M., Sun, J. & Lee, F. C. (2006). Voltage Divider and its Application in the Two-Stage Power Architecture, *Proceedings of IEEE Applied Power Electronics Conference and Exposition*, ISBN 0-7803-9547-6, Dallas, Texas, USA, March 19-23, 2006

Yoshida, H., Imamura, N., Inoue, T., Takeda, K. and Naito, H. (2010). Verification of Life Estimation Model for Space Lithium-Ion Cells. *Electrochemistry*, Vol. 78, No. 5, pp. 482-488

3

Rotor Design for High-Speed Flywheel Energy Storage Systems

Malte Krack[1], Marc Secanell[2] and Pierre Mertiny[2]
[1]*Institute of Dynamics and Vibration Research, Gottfried Wilhelm Leibniz Universität Hannover*
[2]*Department of Mechanical Engineering, University of Alberta*
[1]*Germany*
[2]*Canada*

1. Introduction

1.1 Kinetic energy storage using flywheels

Devices employing the concept of kinetic energy storage date back to ancient times. Pottery wheels and spinning wheels are early examples of systems employing kinetic energy storage in a rotating mass. With the advent of modern machinery, flywheels became commonplace as steam engines and internal combustion engines require smoothing of the fluctuating torque that is produced by the reciprocating motion of the pistons of such machines.

More recently, flywheel systems were developed as true energy storage devices, which are also known as mechanical or electromechanical batteries. A remarkable example of such a system was the sole power source of the 'Gyrobus' - a city bus that was developed by the Maschinenfabrik Oerlikon in Switzerland in the 1930's, see Motor Trend (1952). This vehicle contained a rotating flywheel that was connected to an electrical machine. At regular bus stops, power from electrified charging stations was used to accelerate the flywheel, thus converting electrical energy to mechanical energy stored in the flywheel. When traveling between bus stops, the electrical machine gradually decelerated the flywheel and thus converted mechanical energy back to electricity, which was used to power the electrical motor driving the bus. The disk-shaped flywheel rotor was made of steel, had a mass of about 1.5 metric tons and reached a maximum angular velocity of 314 rad/s or 3000 rounds per minute (rpm). In regular operation, deceleration of the flywheel was limited to about half of the maximum disk speed. The amount of energy thus made available allowed the Girobus to travel for a distance of up to 6 km in regular traffic.

Contemporary flywheel energy storage systems, or FES systems, are frequently found in high-technology applications. Such systems rely on advanced high-strength materials as flywheels usually operate at speeds exceeding 10,000 rpm. Vacuum enclosures and magnetic bearing systems are frequently employed to minimize energy losses due to friction. Only through the use of advanced technology have FES systems become commercially viable for a range of applications, causing FES research and development to be an active and rapidly evolving field.

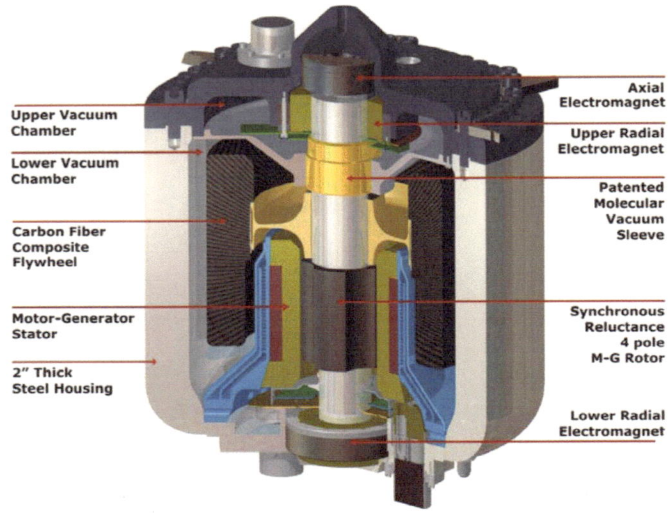

Fig. 1. Example of a commercial flywheel energy storage system (courtesy of POWERTHRU, Livonia, MI, USA - www.power-thru.com)

1.2 FES applications and industrial significance

FES systems and electrochemical batteries can be designed to have comparable energy storage capacities. But, FES systems offer superior energy discharge rates which are considerably higher than in comparable electrochemical battery systems. This characteristic makes FES systems attractive for certain applications. FES systems may provide a cost-effective means in cases where stand-by and rapidly engaging power supply is required. For example, FES units can be used as an uninterruptable power supply (UPS) to protect data centers from grid power failures, cf. Brown & Chvala (2005). For the same reasons FES technology is becoming increasingly popular as a means of ensuring reliable electricity supply to consumers (Bornemann & Sander (1997); Tarrant (1999)). An example of a commercial flywheel energy storage system is shown in **Figure 1**. The installation of clusters of FES units provides for power capacity in the megawatt-level, which enables electrical utilities to perform fast-response regulation of the grid frequency. FES technology lends itself to a range of similar applications, such as peak power support in off-grid industrial systems and energy supply management infrastructure involving renewable energy sources (wind and solar power).

Examples of FES systems given in the preceding section can be categorized as stationary applications. Interest in such systems is presently considerable, yet even greater attention has been given to the development mobile FES systems over the last decades. Even more so than rapid energy charge and discharge capabilities, the comparatively high specific power, i.e. power per unit mass makes FES systems highly attractive for applications in which the mass of the energy storage unit is of substantial importance. Space applications have traditionally been at the forefront of research and development activities in this context, see e. g. Christopher & Donet (1998). More recently, emerging segments in the automotive field, such as highly energy efficient and hybrid vehicles have become another area of applications. So-called kinetic energy recovery systems (KERS) are currently under development for use primarily in motorsports. A cutaway model of a KERS unit is shown in **Figure 2**, which reveals the flywheel rotor. In conjunction with an advanced mechanical transmission this

Fig. 2. Flywheel in a Kinetic Energy Recovery System (KERS) (courtesy of Flybrid Systems LLP, Silverstone, Northamptonshire, England

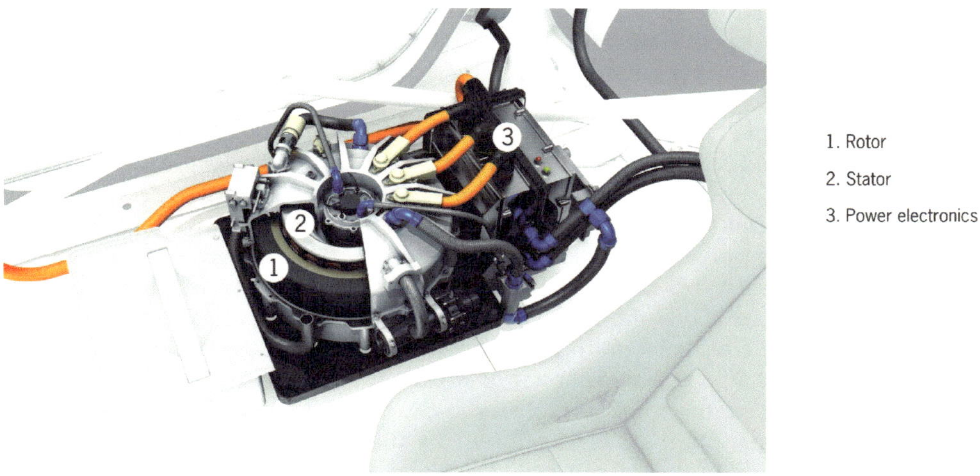

Fig. 3. FES system in a high-performance hybrid automobile (courtesy of Dr. Ing. h.c. F. Porsche AG, Stuttgart, Germany)

flywheel rotor is able to reach top speeds around 60,000 rpm. The energy storage and power capacity of the shown unit with mass of 25 kg is 400 kJ and 60 kW respectively. It is important to note that this and other KERS devices do not necessarily involve energy conversion from electrical to mechanical, and vice versa; instead, mechanical energy is transferred directly to the flywheel rotor using advanced transmission systems. Conversely, electrical-mechanical energy conversion is often required for hybrid vehicles. Shown in **Figure 3** is the electromechanical battery unit for a high-performance hybrid automobile.

1.3 Advantages and disadvantages of FES systems

Several advantages are associated with the use of FES systems compared to electrochemical batteries. Most commonly cited are the superior power and excellent energy capacity per system mass of FES units. The specific power of many FES systems ranges between 5 and 10 kW/kg whereas values for electrochemical batteries are typically smaller by one order of magnitude. The specific energy of advanced FES systems may exceed 200 Wh/kg (Arnold et al. (2002)), and values of 100 Wh/kg are commonly achieved. Specific energies of electrochemical systems are usually around 30 Wh/kg for lead-acid batteries and in excess of 100 Wh/kg for lithium-ion batteries. The situation is different when comparing FES technology to electrochemical batteries on a cost basis. Compared to lead-acid battery systems, an up to eight times higher purchase cost per amount of energy stored can be expected for FES systems (Hebner et al. (2002)). However, the considerably higher price of FES systems is offset by their significant longer life, which may exceed that of electrochemical batteries by the same factor. In this context it is often emphasized that FES units can sustain a practically unlimited number of charge/discharge cycles without reductions in energy storage capacity, whereas for electrochemical batteries the number of charge/discharge cycles is limited due to decreasing battery performance. Cost consideration also must include the storage system's energy recovery efficiency. Modern FES units are 90 to 95% efficient whereas corresponding values for electrochemical batteries are typically much lower, i.e. 60 to 70% for lead-acid batteries. Other advantages of FES technology are a lesser environmental impact due to the absence of harmful chemicals that are usually part of electrochemical batteries, and the ability of FES units to operate effectively over a wide temperature range; electrochemical batteries perform effectively only within a relatively narrow temperature band.

1.4 Basic working principle

The central part of every FES unit is the flywheel rotor. When set in rotation the rotor acquires angular momentum and stores mechanical energy. The rotor is accelerated or decelerated by an electrical machine, usually a combined motor/generator unit. Note that also mechanical systems are used for this purpose in some applications (e.g. the aforementioned KERS). As mentioned above, flywheel rotors usually rotate at high angular velocities. Magnetic bearings and vacuum enclosures are therefore often used to minimize frictional losses that occur in the bearings and with the air surrounding rotating components, respectively. The typical components of advanced FES cells are shown in **Figure 1**. The energy stored in the rotor is increased by accelerating the rotor to higher speeds, i.e. the FES is being charged. In electromechanical systems, the rotor is accelerated by the electrical machine operating in motor mode. When required the energy stored in the rotor can be released by operating the electrical machine in generator mode producing electricity. Conditioning of the electrical power to or from the motor/generator unit is achieved by power electronic systems. **Figure 4** illustrates the power flow affecting flywheel rotor rotation for the charge and discharge cycle.

1.4.1 Kinetic energy

The kinetic energy stored in a flywheel rotating with an angular velocity ω is given by the following equation:

$$E_{\text{kin}} = \frac{1}{2} I \omega^2 , \tag{1}$$

where I is the mass moment of inertia of the rotating components. Assuming a cylindrical rotor made from a single material with density ϱ, and having an inside and outside radius of

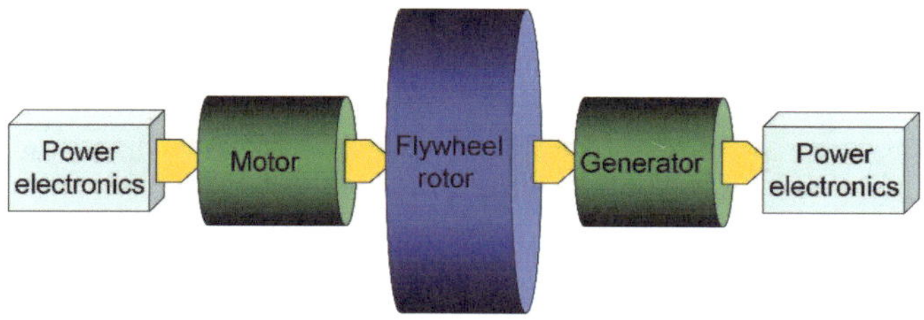

Fig. 4. Schematic showing power flow in FES system

r_i and r_o and a height of h, a further expression for the kinetic energy stored in the rotor can be determined as

$$E_{kin} = \frac{1}{4}\varrho\pi h(r_o^4 - r_i^4)\omega^2 . \quad (2)$$

From the above equation it can be deduced that the kinetic energy of the rotor increases quadratically with angular velocity, and hence high rotational speeds are desirable. The equation also shows that the rotor diameter has an even greater influence on kinetic energy. One must however consider the constraint that mechanical strength of the rotor material imposes on the rotor diameter and angular velocity. Considering a thin rim rotor as an approximation, it can be shown that the product of maximum rotor radius and angular velocity is dependent on the square root of the specific strength of the rotor material, i.e.,

$$r_{o,max}\omega = \sqrt{\frac{S_{\Theta\Theta}}{\varrho}} , \quad (3)$$

where $S_{\Theta\Theta}$ is the strength of the material in circumferential or hoop direction. Hence, rotors are preferably made from low density, high strength fiber-reinforced polymer composite (FRPC) materials that are filament-wound in circumferential direction.

1.4.2 Flywheel configurations

An important aspect of flywheel design, in addition to rotor material selection and dimensioning, is the structure connecting the rotor to the electrical machine. Such a structure also needs to support the rotor within the housing of the FES unit. It is common to connect the rotor via a hub to a rotating shaft that is supported by bearings. In these cases the electrical machine is either directly connected to the hub, or it drives the rotor using the common shaft. In other designs the electrical machine is integrated into the rotor hub structure.

During rotor rotation, centrifugal forces generate stresses in the circumferential as well as in the radial direction. As hoop tensile stresses are dominant, high material strength is required for this direction. In composite flywheel rotors this is accomplished by circumferentially aligning the fiber reinforcement phase. Hoop stresses are not uniform along the rotor radius. For rotors made from a single material the maximum tensile hoop stress is to be expected on the inside of the rotor, see e. g. Krack et al. (2010c). The rotation induced loading thus causes a mismatch in rotor growth creating tensile stresses in the radial direction. Since fiber composite rotors generally lack reinforcement in the radial direction, causing low strength transverse to the fiber direction, rotors are more likely to fail due to cracking of the polymer phase (delamination) rather than fiber breakage (Tzeng et al. (2005)). A central part of rotor design is therefore the reduction of radial tensile stresses. Researchers have proposed and implemented several solutions to this design problem based on material selection and manufacturing considerations, which will be discussed in subsequent sections.

An appropriately designed hub structure may also reduce radial stresses in flywheel rotors. This is accomplished by providing a compliant hub structure that due to the action of centrifugal forces superimposes compressive loading in the radial direction. Ha et al. (2006) proposed a hub structure that is split at multiple locations along its length. Hub segments are therefore able to impose the desired radial compressive load on the inside of the rotor.

1.5 Objectives in flywheel design

The design of FES units usually follows two different yet related objectives. For applications in which the FES unit is part of a mobile system, e.g. automobile or spacecraft, designers strive to maximize specific kinetic energy of the flywheel for given constraints and requirements. Such an approach implies incorporating rotor materials with maximum specific strength, which is usually associated with high cost. For stationary FES systems in e.g. UPS and utility applications, the energy storage capacity and not the mass of the flywheel are typically an important design factor. Such systems are usually of considerable size, requiring large volumes of material, and material cost is therefore an additional key design consideration. Flywheel design based on cost considerations is a rather new approach, which will be described further in subsequent parts of this article.

2. Flywheel rotor manufacturing

2.1 Filament winding

Filament winding is the most common manufacturing technique for flywheel rotors. In this method filamentous reinforcement is wrapped circumferentially onto suitable mandrels creating a rotor rim with high hoop strength. The reinforcement phase is embedded in a polymer matrix, which is applied along with the fibers either as a liquid phase (wet winding), semisolid or solid phase (i.e. winding of towpreg or prepreg material). In the latter cases the polymer phase must usually be liquefied upon deposition by heating to enable fiber consolidation. Thermoset polymer matrices such as epoxies are common for these processes, but alternative polymers such as elastomers have also been employed successfully (Gabrys & Bakis (1997)). Thermoset polymer phases further require curing, typically at elevated temperatures, which may occur after or even during the winding process. Elevated temperature curing may induce residual thermal stresses, which can affect the radial tensile strength of a composite rotor. Hence, thermal effects need to be considered during rotor design. Following the filament winding process the rotor rim may require removal of the winding mandrel and machining to specific tolerances.

2.2 Rotor composition and assembly

Flywheel rotors can be filament-wound as a single material, single rim rotor. Rotor and hub may be assembled by interference fit to mitigate aforementioned radial tensile stresses during flywheel operation. An interference fit is frequently produced by appropriately heating and cooling the rotor and hub to exploit thermal expansion effects for the assembly. By varying fiber tensioning during filament winding some additional compressive radial pre-stressing of the rotor may be achieved.

Although the manufacture of thick single rim rotors is feasible, such designs generally result in suboptimal energy storage capacity. Rotors with a large ratio of outside to inside

radius were found to provide only limited radial tensile strength (Arvin & Bakis (2006)). Better performance can be achieved by assembling a rotor from several individual rims of the same material by mechanical press-fit, thermal shrink-fit and pressurized adhesion (Ratner et al. (2003)). In this manner a compressive radial pre-stressing of the rotor can be tailored that enables the flywheel to operate at higher rotational speeds without failure; greater energy storage capacity is thus achieved.
A hybrid rotor structure consisting of multiple rings made from different fiber and/or resin materials may further reduce radial tensile stresses, see Genta (1985); Portnov (1989). Rotor performance was shown to improve by placing rims with higher material density and/or lower YOUNG modulus on the inside radius of the rotor (promoting increased expansion of these rims in radial direction during flywheel rotation). Conversely, rims with progressively lower density and higher modulus should be located toward the outside of the rotor (constraining radial displacement) (Ha, Kim & Choi (1999)). Similarly, a tailoring method called 'ballasting' was proposed for reducing radial tensile stresses and attaining a more uniform stress field in the rotor. In this method, the ratio of hoop YOUNG modulus to mass density, $\frac{E_{\Theta\Theta}}{\varrho}$, is increased along the outward radial direction of the flywheel rotor by tailoring fiber winding angles and/or the type of fiber material with respect to the radial position. 'Ballasting' may even include low modulus regions that contain resin with little or no reinforcement. To further mitigate the risk of failure, a polymer phase with high strain to failure and reduced YOUNG modulus may be considered. However, highly radial compliant rotors may become susceptible to adverse dynamic behavior. Incorporating compliant elastic or elastomeric interlayers between rims was also shown to improve rotor performance. This approach aims at inhibiting the transmission of radial stress between rims, thus relieving radial stresses.
To overcome the problem of circumferential crack propagation in filament-wound flywheel rotors, Gowayed et al. (2002) discussed flywheel configurations containing reinforcement fibers in the circumferential as well as the radial direction. In the same manner, rims may be made of so-called polar woven composites (Huang (1999)). In this approach, rims are composed of a circular or spiral weave incorporating fibers in the radial and circumferential directions to achieve a balance between radial and hoop strength. Limitations for this type of rotor were found to arise from failures in resin-rich zones and matrix cracking induced by fiber kinking and ensuing high local stresses.

3. Modeling

During service, the flywheel rotor undergoes large rotational speeds, typically in the range of 10,000 to 100,000 rmp. Moreover, the rotor is subjected to fast charge and discharge operations. Thus, the load spectrum of a flywheel rotor is characterized by large static centrifugal loading as well as transient accelerating/braking load. In order to perform an accurate stress analysis, the interaction with the surrounding components, in particular the flywheel hub, is also of importance.
The stress distribution has a significant impact on the failure criteria that represent nonlinear constraints of the design optimization problem formulated in Section 4. Simplified analytical and finite element (FE) approaches exist in order to calculate the stress distribution within the rotor and they are presented in Subsections 3.1 and 3.2, respectively. Remarks concerning the benefits of either approach are given in Subsection 3.3. Since multi-rim rotors of different fiber reinforced polymer composites are state-of-the-art technology, this chapter focuses on this type of rotor design.

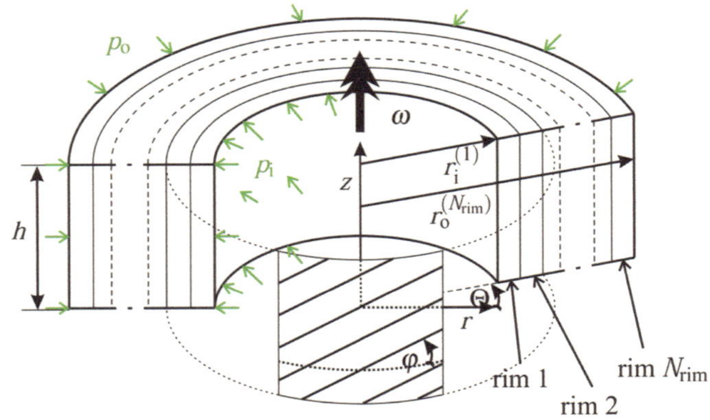

Fig. 5. Multi-rim setup of the flywheel rotor

3.1 Analytical approaches

In **Figure 5**, a multi-rim flywheel rotor is illustrated. Its geometry is typically modeled as axially symmetric. This assumption appears sound since the balancing in terms of achieving axisymmetry is an important objective in the manufacturing of a flywheel rotor. Danfelt et al. (1977) was one of the first to publish an analytical method of analysis for a hybrid composite multi-rim flywheel rotor with rim-by-rim variation of transversely isotropic material properties. The method presented in this subsection generalizes DANFELT's approach in terms of its various extensions. Thorough validation of the method by means of FE analysis and experiments is given in references Ha et al. (2003); Ha & Jeong (2005); Ha et al. (2006).

To the authors' knowledge all publications regarding analytical solutions to the described problem assume a constant rotational velocity. Hence, the transient behavior of charging and discharging operations which might indirectly limit the allowable maximum rotational speed, cannot be accounted for. The local equation of equilibrium in the radial direction of the cylindrical coordinate system for purely centrifugal loading due to the rotational velocity ω reads as

$$\frac{\partial \sigma_{rr}}{\partial r} + \frac{1}{r}\left(\sigma_{rr} - \sigma_{\Theta\Theta}\right) + \varrho\omega^2 r = 0\,. \tag{4}$$

For typical strains in flywheel applications, the nonlinearity of the FRPC material behavior can be neglected. Thus, a linear relationship between stress $\boldsymbol{\sigma}$, strain $\boldsymbol{\varepsilon}$ and temperature ΔT can be stated,

$$\boldsymbol{\sigma} = \boldsymbol{Q}\left(\boldsymbol{\varepsilon} - \boldsymbol{\alpha}\Delta T\right)\,. \tag{5}$$

Herein, $\boldsymbol{\alpha}$ is the vector of thermal expansion coefficients and $\boldsymbol{Q}$ is the global stiffness matrix. The stresses and strains are written as vectors of generally six elements of the symmetric stress tensor in cylindrical coordinates. The stress vector therefore comprises the three normal stresses $\sigma_{rr}, \sigma_{\Theta\Theta}, \sigma_{zz}$ and the the shear stresses $\sigma_{\Theta z}, \sigma_{zr}, \sigma_{r\Theta}$. Using the temperature difference ΔT, the effect of residual stresses from the curing process can be studied, see Ha et al. (2001). Viscoelasticity can also be considered by means of the analytical modeling. This effect may have a significant influence on the long-term stress state within the flywheel rotor. Tzeng et al. (2005); Tzeng (2003) investigated this effect by transforming the thermoviscoelastic problem into its corresponding thermoelastic problem in the LAPLACE space. The resulting thermoelastic relationship is similar to Eq. (5) and can thus be solved in an analog manner, cf. reference Tzeng (2003) for details. It was shown, however, by Tzeng et al. (2005)

that stress relaxation occurs when time progresses. Thus, the constraining state which has to be considered in the optimization procedure is the initial state so that effects of thermoviscoelasticity are not considered in the following.
Only unidirectional laminates shall be studied. Thus, transversely isotropic material behavior is assumed. Ha et al. (1998) were one of the first authors to investigate effects of varying fiber orientation angles on optimum rotor design. For this type of lay-up, the fiber direction does not coincide with the circumferential direction so that the local and the global coordinate systems are not identical. The global stiffness matrix $\boldsymbol{Q}$ then has to be computed from the local stiffness matrix $\bar{\boldsymbol{Q}}$ by means of a coordinate transformation,

$$\boldsymbol{Q} = \boldsymbol{T}^{\mathrm{T}}(\psi)\bar{\boldsymbol{Q}}\boldsymbol{T}(\psi)\,. \tag{6}$$

The local stiffness matrix $\bar{\boldsymbol{Q}}$ only depends on the material properties and can be assembled e. g. using the well-known five engineering constants for unidirectional laminates (Tsai (1988)),

$$\bar{\boldsymbol{Q}} = \bar{\boldsymbol{Q}}(E_1, E_2, G_{12}, \nu_{12}, \nu_{23})\,. \tag{7}$$

Typically, the rotor geometry qualifies for a reduction of the independent unknowns in terms of a plain stress or a plain strain assumption. It is thus possible to obtain a closed-form solution of the structural problem (Ha et al. (1998); Krack et al. (2010c); Fabien (2007)). The assumption of plain stress is valid only for thin rotors ($h \ll r_\mathrm{i}$), whereas thick rotors ($h \gg r_\mathrm{i}$) can be treated with a plain strain analysis.
Assuming small deformations, the quadratic terms of the deformation measures can be neglected, resulting in a linear kinematic. The relationship between the radial displacement distribution u_r and the circumferential and radial strains holds,

$$\varepsilon_{\Theta\Theta} = \frac{u_r}{r}\,, \quad \varepsilon_{rr} = \frac{\partial u_r}{\partial r}\,. \tag{8}$$

Substitution of Eqs. (5)-(8) into Eq. (4) yields the governing equation for u_r, which represents a second-order linear inhomogeneous ordinary differential equation with non-constant coefficients. A closed-form solution is derived in detail in reference Ha et al. (2001). Since the governing equation depends on the material properties, the solution is only valid for a specific rim.
The unknown constants of the homogeneous part of the solution for each rim are determined by the boundary and compatibility conditions, i. e. the stress and the displacement state at the inner and outer radii of each rim j, ${r_\mathrm{i}}^{(j)}$ and ${r_\mathrm{o}}^{(j)}$ respectively. Regarding compatibility, it has to be ensured that the radial stresses are continuous along the rim interfaces of the N_rim rims, whereas the radial displacement may deviate by an optional interference $\delta^{(j)}$,

$$\sigma_{r_\mathrm{i}}^{(j+1)} = \sigma_{r_\mathrm{o}}^{(j)}\,, \qquad \text{for } j = 1(1)N_\mathrm{rim} - 1 \quad \text{and} \tag{9}$$

$$u_{r_\mathrm{i}}^{(j+1)} = u_{r_\mathrm{o}}^{(j)} + \delta^{(j)}\,, \quad \text{for } j = 1(1)N_\mathrm{rim} - 1\,. \tag{10}$$

The effect of interference fits $\delta^{(j)}$ was studied in reference Ha et al. (1998).
It has to be noted that the continuity of radial stresses implies that the rims are bonded to each other. This is generally not the case for an interference fit since mating rims are usually fabricated and cured individually. Hence, no tensile radial stresses can be transferred at the

interface. A computed positive radial stress would mean detachment failure in this case. Therefore, the general analytical model does not take care of implausible results so that the results have always to be regarded carefully.
The required last two equations are obtained from the radial stress boundary conditions at the innermost and outermost radius of the rotor

$$\sigma_{r_\mathrm{i}}^{(1)} = p_\mathrm{in}\,, \quad \sigma_{r_\mathrm{o}}^{(N_\mathrm{rim})} = -p_\mathrm{out}\,. \tag{11}$$

The pressure at the outermost rim p_out is typically set to zero, the inner pressure p_in can be used to consider the interaction with the flywheel hub. The conventional ring-type hub can simply be accounted for as an additional inner rim. It should be noted that the typically isotropic material behavior of a metallic hub can easily be modeled as a special case of transversal isotropy. A split-type hub was studied in reference Ha et al. (2006). Therefore, the inner pressure was specified as the normal radial pressure caused by free expansion of the hub,

$$p_\mathrm{in} = \omega^2 \varrho_\mathrm{hub} \frac{\left(r_\mathrm{i}^{(1)}\right)^3 - \left(r_\mathrm{i}^{(\mathrm{hub})}\right)^3}{3\left(r_\mathrm{i}^{(1)}\right)}\,. \tag{12}$$

In the generalized modified plain strain assumption used in reference Ha et al. (2001), a linear ansatz for the axial strain was chosen. Thus, two additional constraints were introduced: The resulting force and moment caused by the axial stress for the entire rotor was set to zero. According to Ha et al. (2001), the linear ansatz for the axial strain yielded better results than its plain stress or plain strain counterparts in comparison to the FE analysis results.
In conjunction with the solution, the compatibility and boundary conditions can be compiled into a real linear system of equations for the $N_\mathrm{rim} + 1$ unknown constants of the solution. It can be shown that the system matrix is symmetric for a suitable preconditioning described in reference Ha et al. (1998). Once solved, the displacement and stress distribution can be evaluated at any point within the rotor.

3.2 Numerical approaches

In comparison to the analytical approaches, finite element (FE) approaches offer several benefits in terms of modeling accuracy. For a general three dimensional or two dimensional axisymmetric FE analysis, a plain stress or strain assumption is not necessary. Furthermore nonlinearities can be accounted for, including the contacting interaction of rotor and hub, the nonlinear material behavior and the nonlinear kinetmatics in case of large deflections. Also, more complicated composite lay-ups other than the unidirectional laminate could be modeled. Another advantage is the capability of examining the effect of transient accelerating or braking operations on the load configuration of the rotor.
In order to provide insight into the higher accuracy of the numerical model, the radial and circumferential stresses for a two-rim rotor similar to the one presented by Krack et al. (2010b) is illustrated in **Figures 6(a)-6(b)**. The rotor consists of an inner glass/epoxy and an outer carbon/epoxy rim and is subjected to a split-type hub (not shown in the figure).
It should be noted that apart from the non-axisymmetric character of the stress distributions, the stress minima and maxima are no longer located at the same height. This indicates that optimization results that are only based on plain stress or strain assumptions and axial symmetry should at least be validated numerically. It has to be remarked that the normal stress in the axial direction and the shear stresses, which are not depicted, are generally

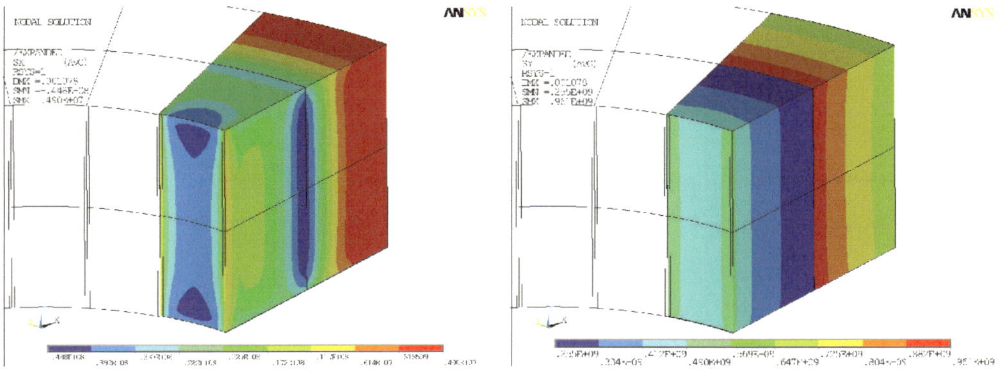

(a) Radial stress σ_{rr} in $\mathrm{N/m^2}$ (b) Circumferential stress $\sigma_{\Theta\Theta}$ in $\mathrm{N/m^2}$

Fig. 6. Stress distributions in the finite element sector model for a rotational speed of $n = 30000\ \mathrm{min}^{-1}$

non-zero which cannot be accurately predicted by the analytical model.
Despite the higher accuracy of the numerical model, comparatively few publications can be found in the literature concerning the design of hybrid composite flywheels using numerical simulations. Ha, Kim & Choi (1999) developed an axisymmetric finite element and employed it to find the optimum design of a flywheel rotor with a permanent magnet rotor. Takahashi et al. (2002) examined the influence of a press-fit between a composite rim and a metallic hub employing a contact simulation technique in an FE code. Gowayed et al. (2002) studied composite flywheel rotor design with multi-direction laminates using FE analysis. In Krack et al. (2010b), both an analytical and an FE model were employed in order to predict the stress distribution within a hybrid composite flywheel rotor with a nonlinear contact interaction to a split-type hub.

3.3 Remarks on the choice of the modeling approach

The main benefit of the analytical model is that it is much less computationally expensive. Since there are typically several orders of magnitude between the computational times of analytical and numerical approaches, this advantage becomes a significant aspect for the optimization procedure (Krack et al. (2010b)). Some optimization strategies, in particular global algorithms require many function evaluations and would lead to an enormous computational effort in case of using an FE model. The choice of the model thus not only affects the optimum design but also facilitates optimization. On the other hand, the FE approach facilitates a greater modeling depth and flexibility, since there is no need for the simplifying assumptions that are necessary to obtain a closed-form solution in the analytical model.
Owing to the capability of greater modeling depth, numerical methods gain importance for the design optimization of flywheel rotors. If effects such as geometric, material and contact nonlinearity or complex three-dimensional loading need to be accounted for in order to achieve a sufficient accuracy of the model, the FE analysis approach renders indispensable. Furthermore, increasing computer performance diminishes the significant disadvantage of more computational costs in comparison to analytical methods. Methods that combine the benefits of both approaches are discussed in Subsection 4.4.

4. Optimization

Various formulations for the design optimization problem of the flywheel rotor have been published. A generalized formulation reads as

$$\begin{array}{lll} \text{Maximize} & f(\boldsymbol{x}) = f(E_{\text{kin}}(\boldsymbol{x}), M(\boldsymbol{x}), D(\boldsymbol{x}), \cdots) & \\ \multicolumn{3}{l}{\text{with respect to } \boldsymbol{x} = \{\text{set of geometric variables, rotational speed, material properties}\}} \\ \text{subject to} & \text{structural constraints} & \text{and} \\ & \boldsymbol{x}_{\min} \leq \boldsymbol{x} \leq \boldsymbol{x}_{\max} & . \end{array} \tag{13}$$

Thus, the objective of the design problem is to maximize a function generally depending on the kinetic energy stored E_{kin}, the mass M and the cost D. The design variables can be any subset of all geometric variables, rotational speed and material properties. The optimum design is always constrained by the strength of the structure. In addition, bounds for the design variables might have to be imposed. The concrete formulation of the design problem strongly depends on the application, manufacturing opportunities and other design restrictions. Different suitable objective function(s) are discussed in Subsection 4.1, common design variables are addressed in Subsection 4.2 and constraints are the topic of Subsection 4.3. Depending on the actual formulation of the design problem, an appropriate optimization strategy has to be employed, see Subsection 4.4.

4.1 Objectives

Regardless of the application, all objectives for FES rotors are energy-related. The total kinetic energy stored in the rotor can be expressed as

$$E_{\text{kin}} = \frac{1}{2} I_{zz} \omega^2 , \tag{14}$$

where I_{zz} is the rotational mass moment of inertia. It was assumed that the rotation of the flywheel is purely about the z-axis with a rotational velocity ω.
For small deflections, I_{zz} can approximately be calculated considering the undeformed structure only,

$$I_{zz} = \frac{1}{2} \sum_{j=1}^{N_{\text{rim}}} m_j \left[\left(r_{\text{o}}^{(j)}\right)^2 + \left(r_{\text{i}}^{(j)}\right)^2 \right] = \frac{\pi}{2} h \sum_{j=1}^{N_{\text{rim}}} \varrho_j \left[\left(r_{\text{o}}^{(j)}\right)^4 - \left(r_{\text{i}}^{(j)}\right)^4 \right] , \tag{15}$$

with the masses m_j, the rotor height h and the constant density ϱ_j of each rim. It becomes evident from Eq. (14) that the kinetic energy increases quadratically with the rotational speed ω and only linearly with the inertia I_{zz}. The inertia of the outer rims has more influence on the kinetic energy than the one in the inner rims. It should be noted that in typical FES applications the total energy is not the most relevant parameter, instead the difference between the maximum energy stored and the minimum energy stored, i. e. the energy that can be obtained by discharging the FES cell from its bound rotational velocities $\omega_{\max}$ and $\omega_{\min}$ is relevant.
Another important aspect is the minimization of the rotor weight. This is particularly significant for mobile applications. The total mass M of the rotor reads as

$$M = \sum_{j=1}^{N_{\text{rim}}} m_j = \pi h \sum_{j=1}^{N_{\text{rim}}} \varrho_j \left[\left(r_{\text{o}}^{(j)}\right)^2 - \left(r_{\text{i}}^{(j)}\right)^2 \right] . \tag{16}$$

In case of stationary applications, it might be even more critical to minimize the rotor cost. Therefore, the total cost D (Dollar) has to be calculated,

$$D = \pi h \sum_{j=1}^{N_{\text{rim}}} d_j \varrho_j \left[\left(r_{\text{o}}^{(j)} \right)^2 - \left(r_{\text{i}}^{(j)} \right)^2 \right] . \tag{17}$$

Herein, the weighting factors d_j are the price per mass values of each material. Thus, it is assumed that the total cost can be split up into partitions that can directly be associated with the mass of each material. It should be noted that these prices are often hardly available in practice and are subject to various influences such as the manufacturing expenditure and the required quantities. The former aspect is usually strongly influenced by the complexity of the rim setup, i. e. the number of rims and optional features such as interference fits. Conclusions directly drawn from an optimization for an arbitrarily chosen set of prices should therefore be regarded as questionable. In Krack et al. (2010c) and Krack et al. (2010a), the optimization is therefore performed with the price as a varying parameter.

Naturally, trade-offs between the main objectives have to be made. A large absolute energy value can only be achieved by a heavy and expensive rotor. Minimizing the cost or the weight for a given geometry would result in selecting the cheapest or lightest material only. However, the benefits of hybrid composite rotors, i. e. rim setups using different materials in each rim have been widely reported.

In order to obtain a design that exhibits both requirements, i. e. a large storable energy and a low mass or cost, it is intuitive to formulate the optimization problem as a dual-objective problem with the objectives energy and mass or energy and cost. As an alternative, the ratio between both objectives can be optimized in order to achieve the largest energy for the smallest mass/cost, resulting in a single-objective problem. The ratio between energy and mass is also known as the specific energy density SED,

$$SED = \frac{E_{\text{kin}}}{M} . \tag{18}$$

The energy-per-cost ratio reads as follows:

$$ECR = \frac{E_{\text{kin}}}{D} . \tag{19}$$

The following discussion regarding single- and multi-objective design problem formulations addresses the trade-off between storable energy and cost. However, the statements generally also hold for the goal of minimizing the mass instead of the cost.

Solving optimization problems with multiple objectives is common practice for various applications with conflicting objectives, (e. g. Secanell et al. (2008)). The solution of a multi-objective problem is typically not a single design but an assembly of so called PARETO-optimal designs. In brief, PARETO-optimality is defined by their attribute that it is not possible to increase one objective without decreasing another objective. The dual-objective approach thus covers a whole range of energy and cost values associated to the optimal designs. This is the main benefit compared to a single-objective optimization with the energy-per-cost ratio as the only objective, which only has a single optimal design. It is generally conceivable that this design with the largest possible energy-per-cost value might exceed the maximum cost, or its associated kinetic energy could be too low for a practical application.

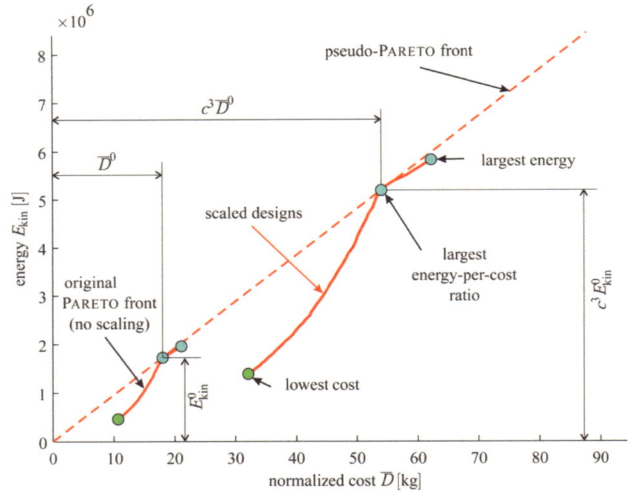

Fig. 7. Reduction of the multi-objective to a single-objective design problem using the scaling technique

For the particular mechanical problem of a rotor with a purely centrifugal loading and linear materials, however, Ha et al. (2008) showed that any flywheel design can be linearly scaled in order to achieve a specified energy or cost/mass value. Due to the linearity of Eqs. (4)-(8), the stress distribution remains the same if all geometric variables are scaled proportionally and the rotational velocity inversely proportional to an arbitrary factor c. After scaling, the energy, ${E_{\text{kin}}}^0$, and cost, D^0, of the original optimal design would increase by the factor c^3 so that the energy-per-cost value ${E_{\text{kin}}}^0/D^0 = c^3{E_{\text{kin}}}^0/(c^3D^0)$ is also constant. This design scaling is illustrated in **Figure 7**. If scaling is possible, i. e., the total radius of the rotor is not constrained, then, scaling can be used in order to achieve a rotor that always has the maximum energy-per-cost ratio. Therefore, if scaling is possible, all other points in the PARETO fronts in **Figure 7** would be suboptimal compared to scaling the design in order to achieve the maximum energy-per-cost ratio. A new PARETO front for the dual-objective design problem in conjunction with the scaling technique would therefore be a line through the origin with the optimal energy-per-cost value as the slope. This pseudo-PARETO front is also depicted in **Figure 7** (dashed line). If size is constrained, other points in the PARETO set will have to be considered for the given geometry. It should be noted that it is assumed that scaling opportunity still holds approximately also for nonlinear materials and large deformations within practical limits. It is also important to remark that there are more established and computationally efficient numerical methods for the solution of single-objective design problems than for multi-objective problems. Therefore, the single-objective problem formulation should be preferred if the mechanical problem and the constraints of the problem Eq. (13) allow this. In the following, it shall be assumed that this requirement holds. Hence, the specific energy density or the energy-per-cost ratio can be applied in a single-objective design problem formulation. For problems where mass and cost are of inferior significance, it is also common to optimize the total energy stored as the only objective, $\boldsymbol{f} = E_{\text{kin}}$.

It should be noted that there is generally no set of design variables that maximizes all of the objectives but there are different solutions for each purpose (Danfelt et al. (1977)). The total energy stored was considered as objective in Ha, Yang & Kim (1999); Ha, Kim & Choi (1999); Ha et al. (2001); Gowayed et al. (2002). The trade-off between energy and mass, i. e. maximization of the specific energy density SED was addressed in the following publications: Ha et al. (1998); Arvin & Bakis (2006); Fabien (2007); Ha et al. (2008). Particularly for stationary energy storage applications, the aspect of cost-effectiveness might be more relevant. Krack et al. (2010c); Krack et al. (2010b); Krack et al. (2010a) addressed this economical aspect by maximizing the energy-per-cost ratio ECR.

4.2 Design variables

Various design variables have been investigated for the optimization of the composite rotor and hub design for FES. A list of the most relevant design variables is given below:

- Rotational speed
- Material properties (E_{ij}, ν_{ij}, ϱ)
- Interferences
- Fiber direction angle
- Rim thicknesses
- Rotor height
- Hub design

Many design variables directly influence the rim setup. It was shown in Ha, Yang & Kim (1999) that a lay-up with radially increasing hoop stiffness to density ratio $\frac{E_{\Theta\Theta}}{\varrho}$ is most beneficial in terms of energy capacity. An increasing value $\frac{E_{\Theta\Theta}}{\varrho}$ ensures that the outer part of the rotor prevents the inner part from expanding. Thus, the radial stresses tend to be compressive during operation, and the more critical tensile stresses across the fiber are reduced.

Apparently this type of rim setup can be achieved by designing the material properties in a suitable manner. Discrete combinations of rims with piecewise constant material properties, i. e. hybrid composite rotors are state-of-the art. By using different materials in the same rotor, the hoop stiffness as well as the density can be varied. A continuously varying fiber volume fraction is also conceivable but more complex in terms of design and manufacturing. Due to anisotropy, the hoop stiffness can also be decreased by winding the fibers not circumferentially but with a non-zero fiber angle (fiber angle variation).

The overall radial stress level can also be decreased by introducing interferences between adjacent rims. It should be noted that interferences are also necessary in order to accomplish compressive interface stresses for the torque transmission within the rotor. By adapting the hub design, e. g. by employing a split-type hub, the strength of the rotor can also be increased, as it will be shown later in this subsection.

Naturally the rotational speed is also a common variable that influences not only the kinetic energy stored but also increases the centrifugal loading. Thus, there exists a critical rotational speed for any type of rotor. However, the rotational speed is different from the design variables discussed above in that it varies with service conditions. Consequently, the rotational speed can be treated as a design variable or a constant parameter that determines the size of the flywheel design in terms of the scaling technique as in Ha et al. (2008), see Subsection 4.1. In fact, for the case of a single-material rotor with constant inner and outer radii, the rotational speed could also be treated as an objective in order to optimize the kinetic

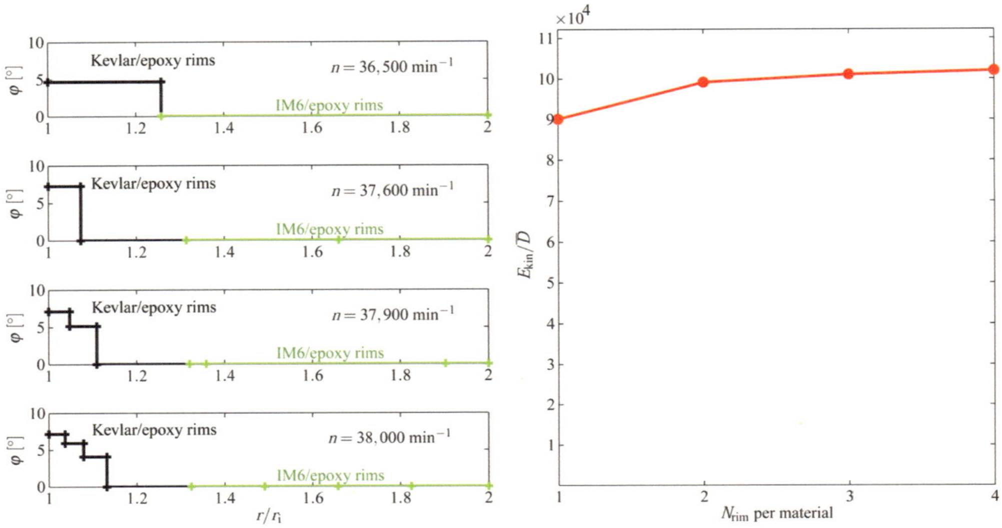

(a) Optimal designs for different numbers of rims (b) Optimal energy-per-cost ratio depending on the number of rims

Fig. 8. Influence of the number of rims per material

energy, cf. Ha et al. (1998).
In Danfelt et al. (1977), the POISSON ratio, the YOUNG modulus and the density were considered as design variables for a flywheel rotor with rubber in between the composite rims. Ha et al. (1998) optimized the design of a single-material multi-rim flywheel rotor with interferences and different fiber angle in each rim. They were able to increase the energy storage capacity by a factor of 2.4 compared to a rotor without interferences and purely circumferentially wound fibers. They also concluded that interferences had more influence on the increase of the overall strength than fiber angle variation. In a following publication, Ha, Yang & Kim (1999) studied the design of a hybrid composite rotor with up to four different materials and optimized the thickness of each rim for different material combinations. Fiber angle variation was also addressed in Fabien (2007). The authors considered the optimization of a continuously varying angle between the radial and the tangential direction for a stacked-ply rotor.
It should be noted that it is also conceivable to optimize the rotor profile, i. e. to vary the height along the radius, see Huang & Fadel (2000a). However, the winding process impedes this type of design optimization in case of an FRPC rotor. Consequently, the height optimization is uncommon to FES using composite materials and instead the ring-type architecture is widely accepted.
In what follows, two design optimization case studies will be presented: (1) The optimization of the discrete fiber angles for a multi-rim hybrid composite rotor and (2) the investigation of the influence of the hub design on the optimum design of a hybrid composite rotor.

4.2.1 Optimum fiber angles for a multi-rim hybrid composite rotor

The effect of fiber angle variation on the optimum energy-per-cost value for a multi-rim hybrid composite rotor with inner Kevlar/epoxy and outer IM6/epoxy rims has been studied. The optimization was carried out for different numbers of rims per material. Due to increased complexity in manufacturing and assembly the potential for increased expenditure exists with increasing number of different rims. However, such cost-increasing effects were not considered in the modeling. Thus, it is interesting to study the influence of the number of rims on the optimal energy-per-cost value. In **Figures 8(a)** and **8(b)** the results are depicted with (a) their corresponding optimal designs and (b) optimal objective function values. There are only rims with nonzero fiber angles for the Kevlar/epoxy material. The fiber angle is decreasing for increasing radius. The optimal fiber angle for the IM6/epoxy rims is zero. The reason for this is probably that the critical tensile radial stress level in the Kevlar/epoxy rims would be increased by more compliant outer rims. Hence, a non-zero value for the IM6/epoxy fiber angle might lead to delamination failure in this case. Theoretically, it is thus not necessary to increase the number of rims for the IM6/epoxy material to obtain the optimal energy-per-cost ratio. In order to show that the fiber angle still vanishes for additional rims, however, the redundant rims have not been removed in **Figure 8(a)**.

It can be postulated that there is an optimal continuous function for the fiber angle with respect to the radius. In that case, the optimization method would try to fit the discontinuous fiber angle to this continuous function by adjusting the thicknesses and fiber angles of the discrete rims. This assumption is supported by the results of Fabien (2007) which include the computation of an optimal continuous fiber angle distribution. In that reference, however, the fibers are aligned in the radial direction so that the optimization results cannot be compared to the ones in this paper.

As expected, the objective function value increases monotonically with additional design variables. The energy-per-cost value for the configuration with four rims per material exceeds the corresponding value for the single rim configuration by 13%. Since the total thickness of each material remains approximately constant, the normalized cost does not decrease significantly. Thus, the increase in the energy-per-cost ratio is mainly due to the increase of the energy storage capacity. However, it can be seen well from **Figure 8(b)** that the optimal objective converges with increasing numbers of rims per material. Hence, additional manufacturing complexity is not necessarily worthwhile considering the comparatively slow decrease of the energy-per-cost ratio with respect to the number of rims.

4.2.2 Optimization of the hub geometry

The optimization of the hub geometry connected to a two-rim glass/epoxy, carbon/epoxy rotor with $r_i = 120$ mm and $r_o = 240$ mm was examined for two common hub types: The conventional ring-type hub and the split-type hub as proposed in Ha et al. (2006). The basic idea of the split-type hub is to interrupt the circumferential stress transmission by splitting up the hub into several segments, facilitating the radial expansion during rotation of the split-ring. This expansion causes compressive hub/rim interface stresses, which makes interference fits or adhesives unnecessary in terms of torque transmission. Furthermore, the compressive hub/rim interface stresses reduce the magnitude of radial tensile stress within the composite rims. Since the radial tensile stress is often the speed-limiting constraint for rotating filament wound composite rings, the energy storage capability can thus be increased. On the other hand, the pressure loading causes increased hoop stresses within the composite rims, which also have the potential of limiting the energy storage capability. Thus, there exists an optimum thickness of the ring part of the hub, as shown in Ha et al. (2006); Krack et al. (2010b). Both hub configurations were considered in the optimization of a hybrid two-rim rotor with prescribed inner and outer rotor diameter. The design variables were the rotational speed n, the inner rim thickness t_1 and the hub thickness t_{hub}.

	ring-type hub	split-type hub
$\frac{t_1^{\text{opt}}}{t_{\text{all}}}$ [%]	58.28	66.91
n^{opt} [min^{-1}]	46846	44872
$t_{\text{hub}}^{\text{opt}}$ [mm]	0.00	3.80
$\frac{f^{\text{opt}}}{f_{\text{no hub}}^{\text{opt}}}$ [%]	100	103.7

Table 1. Optimization results for different hub architectures with an optimized hub thickness

The optimal hub thickness became zero in the case of the ring-type hub. This means that a ring-type hub generally weakens the strength of the rotor for the given material properties. However, a minimum thickness for the hub ring would be necessary in order to avoid failure and to transmit torque between rotor and shaft. Hence, the results for the optimized ring-type hub with vanishing hub thickness have to be regarded as only theoretical extremal values. For this extreme case, the optimum energy-per-cost value is identical to the one for the case without any hub, i. e., the relative value equals 100%.

On the other hand, an optimal hub thickness of $t_{\text{hub}}^{\text{opt}} = 3.80$ mm was ascertained for the split-type hub. With this optimal design, the energy-per-cost value for the split-type hub is 3.7% higher than for the model with an optimized ring-type hub in this example. Therefore, it is proven that a split-type hub with an optimized thickness enhances the strength of the hybrid composite rotor and thus increases the optimal energy-per-cost value.

4.3 Constraints

The design problem stated in Eq. (13) is constrained by the strength limits of the structure, geometrical bounds and dynamical considerations. Geometrical bounds arise from the design of the surrounding components. A given shaft, hub or casing geometry can restrict the dimensions of the rotor, i. e. the inner and outer radii as well as the axial height. The aspect ratio and the absolute size in conjunction with the bearing properties can also necessitate size constraints in terms of dynamic stability for large rotational speeds, cf. Ha et al. (2008).

The most critical constraints are, however, the structural ones. Various failure criteria have been studied for the design of flywheel rotors. The most common criteria are the Maximum Stress Criterion, the Maximum Strain Criterion and the TSAI-WU Criterion. As the constraints represent the boundary of the feasible region and the optimal designs can typically be found at this boundary, cf. Danfelt et al. (1977), the choice of the failure criterion is essential to the solution of the design problem. The influence of the failure criterion on the optimum design was investigated by Fabien (2007) and Krack et al. (2010c). The stress state in a typical flywheel rotor is dominated by the normal stresses. Thus, the deviations between these failure criteria are often not crucial.

In **Figure 9**, the feasible region for the two design variables, rotational speed n and inner rim thickness $\frac{t_1}{t_{\text{all}}}$ for a two-rim glass/epoxy and carbon/epoxy rotor is illustrated. The feasible region is composed of the nonlinear structural constraints in terms of the Maximum Stress Criterion and the bounds of the thickness. The structural constraints are labeled by their strength ratio R between actual and allowable stress for each composite (glass/epoxy or carbon/epoxy). The first index of the strength ratio corresponds to the coordinate direction ('1' for across the fiber, '2' for in the fiber direction), the second index denotes the sign of the stress ('t' for tensile, 'c' for compressive).

In case of concavely shaped constraint functions, it was shown in Krack et al. (2010c) that

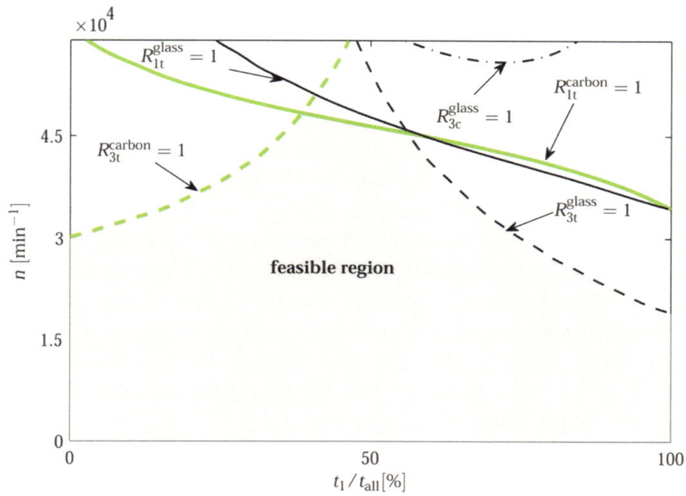

Fig. 9. Composition of the nonlinear constraint for the Maximum Stress Criterion

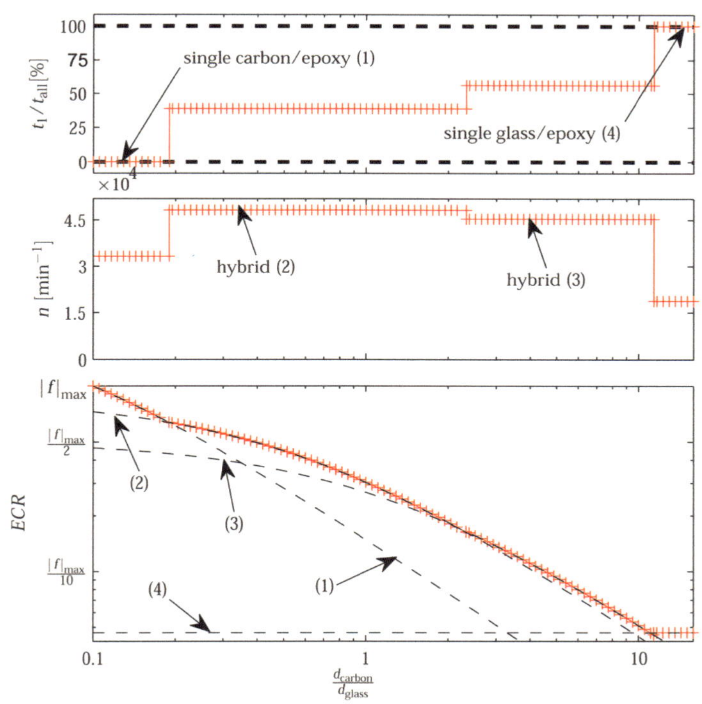

Fig. 10. Optimal designs and objective function values dependent on the cost ratio

in particular the intersecting points of different strength limits that bound the feasible region are candidates for optimal designs. **Figure 10** shows the value of the design variables and objective function at different cost ratios for the hybrid composite flywheel rotor described above. The rotor design was optimized in terms of the energy-per-cost ratio objective ECR, cf. Eq. (19). It is remarkable that the optimum design variables turn out to be discontinuous over

the cost ratio. At specific cost ratios, the optimum thicknesses $\frac{t_1}{t_{all}}$ and the rotational speed n jump between two different values. Between these jumps, i. e. for wide ranges of the cost ratio, the optimum design variables remain constant in this case.
Four different optimal design sets have to be distinguished according to **Figure 10** depending on the cost ratio interval. At very high or very low cost ratio values, i. e. relatively expensive carbon or glass based composite materials respectively, a single rim rotor with the correspondingly cheaper material is preferable. Hence, a value of $\frac{t_1}{t_{all}} = 0\ \%$ or $\frac{t_1}{t_{all}} = 100\ \%$ corresponding to a full carbon/epoxy or a full glass/epoxy material rotor respectively, is obtained. In between these trivial solutions, two additional optimal designs exist.
While the total energy stored and the specific energy density have discrete values for a varying cost ratio, the actual objective, i. e. the optimal energy-per-cost value ECR changes continuously with the cost ratio as illustrated in **Figure 10**. In this figure, the objective function for each of the four design sets is depicted dependent on the cost ratio. Note that the discontinuities of the optimal design variables coincide with intersections of the design-dependent objective function graphs.
It can be concluded from this section that the constraints are essential to the design problem but the decision which design is optimal also depends significantly on the shape of the objective function with respect to the design variables.

4.4 Optimization strategies

Based on the previous discussion, the flywheel design problem in Eq. (13) is a multi-objective, multi-variable nonlinear constrained optimization problem. This section of the chapter discusses possible optimization algorithms that can be used in order to solve such optimization problems. Subsection 4.2 outlined the design variables for the problem which include lay-up materials, fibre angles and thickness, hub geometry and rotational speed. Most of these variables are real variables; therefore this section will focus on optimization strategies for optimization problems with real design variables.
The solution of multi-objective, multi-variable nonlinear constrained optimization problems is a challenging endeavor. First, in a nonlinear optimization problem, there are usually many designs that satisfy the Karush-Kuhn-Tucker (KKT) optimality conditions, see A. Antoniou & W.-S. Lu (2007). All these designs, known as local optima, meet the necessary requirements for optimality, but usually one of these designs will provide better performance than the others. Therefore, the optimization algorithm needs to search not only for an optimal design, but for the optimal design among optimal designs. In addition to the nonlinear nature of the optimization problem, since there are multiple criteria to be optimized, the most optimal design will depend on the relative importance of each one of the design objectives. Therefore, a methodology needs to be used to identify the different trade-offs between design objectives. Finally, optimization problems usually involve a large number of complex numerical simulations, e. g., a detailed multi-dimensional FE simulation of the flywheel. Therefore, it is necessary to select optimization strategies that can minimize the computer resources necessary to solve the design problem.
Subsection 4.4.1 will discuss the advantages and disadvantages of the optimization algorithms that can be used to solve nonlinear constraint optimization problems. Subsection 4.4.2 provides an overview of multi-objective optimization and presents two alternative methods that can be used to solve such problems. Finally, Subsection 4.4.3 will present several methodologies that have recently been used in order to reduce computational resources.

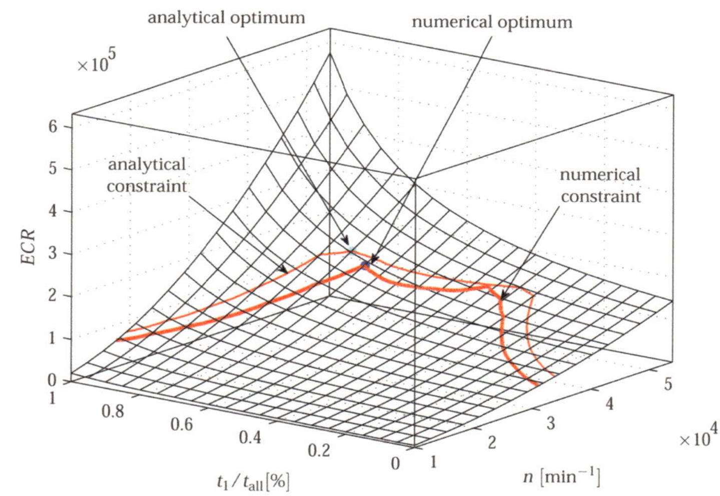

Fig. 11. Objective function and analytical and numerical nonlinear constraints depending on the relative inner rim thickness t_1/t_{all} and the rotational speed n

4.4.1 Constraint optimization algorithms

As discussed, for many nonlinear optimization design problems, multiple local optima may exist which makes solving the optimization problem more difficult. **Figure 11** shows the design space for the flywheel optimization problem solved by Krack et al. (2010c). It can be observed in **Figure 11** that there are two points that can be considered optimal solutions, i.e. $(n, t_1/t_{all}) = (4.25 \times 10^4, 0.4)$ and $(n, t_1/t_{all}) = (4.0 \times 10^4, 0.7)$. Therefore, even for monotonic objective functions and a small number of design variables multiple local optima occur due to the introduction of strongly nonlinear constraints. Hence, it is important to verify that the optimum detected by a specific method is a global optimum and not only a local one.

Nonlinear constraint optimization algorithms can be classified as local methods and global methods. Local methods aim to obtain a local minimum, and they cannot guarantee that the minimum obtained is the absolute one. These methods are usually first-order methods, i.e. they require information about the gradient of the objective function and the constraints. The most commonly used local methods include the method of feasible directions (MFD) and the modified method of feasible directions (MMFD) (see Arora (1989); Vanderplaats (1984)); sequential linear programming (SLP) (see Arora (1989); Lamberti & Pappalettere (2000); Vanderplaats (1984)); sequential quadratic programming (SQP) (see A. Antoniou & W.-S. Lu (2007)); nonlinear interior point methods (see A. Antoniou & W.-S. Lu (2007); El-Barky et al. (1996)), and; response surface approximation methods (RSM) (see Rodríguez et al. (2000); Wang (2001)). Local methods are prone to finding an optimum in the nearby region of the initial starting guess; however, these methods work very efficiently in the vicinity of the optimum.

Global methods aim at obtaining the global minimum. These methods do not require any information about the gradient, and they employ primarily either a stochastic-based or an heuristic-based algorithm. Therefore, the use of global methods can reduce the likelihood of missing the global optimum. (Albeit there is no guarantee of finding the global optimum.) Global methods, however, have the disadvantage of requiring far more function evaluations. Particularly in the case of computationally expensive function evaluations, e. g. nonlinear FE analyses with a large number of elements, global methods are often not applicable in practice. Global methods either solve the constraint nonlinear problem directly, or they transform the problem into an unconstrained problem using a penalty method (see Vanderplaats (1984) for a description of common penalty methods). Common optimization algorithms that solve the constrained problem directly include covering methods and pure random searches. If the constrained optimization problem is transformed into an unconstrained one, common unconstrained global optimization problems include genetic algorithms (see Goldberg (1989)), evolutionary algorithms (see Michalewicz & Schoenauer (1996)) and simulated annealing (see Aarts & Korst (1990)).

Although local methods do not aim at obtaining a global optimum, several approaches can be used to continue searching once a local minimum has been obtained, thereby enabling the identification of all local minima. Once all local minima have been obtained, it is easy to identify the global minimum. Some of these methods are: random multi-start methods (e.g., He & Polak (1993); Schoen (1991)), ant colony searches (e.g., Dorigo et al. (1996)) and local-minimum penalty method (e.g., Ge & Qin (1987)).

Another approach to obtaining a global solution when the computational resources are limited is to combine a global and a local optimization algorithm. Global optimization algorithms are usually relatively quick at obtaining a solution that is near the global optimum; however, they are usually slow at converging to an optimal solution that meets the optimality conditions. In order to reduce computational resources during the later stages of finding an optimal solution, a global optimization algorithm can be used during the initial stages of the solution of the design problem. Then, the sub-optimal solution obtained by the global optimization algorithm can be used as the initial guess to the local method. Since local optimization algorithms usually converge very quickly to the optimal solution, a reduction in computational resources can usually be achieved. Further, since the initial solution was already in the vicinity of the global optimum, it is likely that the local optimization algorithm will converge to the global optimum. This approach was recently used to design a hybrid composite flywheel by Krack et al. (2010c). In order to show the benefits of the proposed multi-strategy scheme, the optimization problem was solved with a global method, i.e. an evolutionary algorithm (EA), a local method, i.e. a nonlinear interior-point method (NIPM), and the multi-strategy scheme, i.e. start with EA algorithm and switch to the interior-point method after a relatively flexible convergence criteria was achieved. Krack et al. (2010c) showed that the multi-strategy scheme was 35% faster than the global method.

4.4.2 Multi-objective optimization algorithms

The optimization formulation in Eq. (13) contains multiple objectives that need to be optimized simultaneously such as kinetic energy stored, mass and cost. In the late-nineteenth-century, Edgeworth and Pareto showed that, in most multi-objective problems, an *utopian* solution that minimizes all objectives simultaneously cannot be obtained because some objectives are conflicting. Therefore, the scalar concept of optimality does not apply directly to design problems with multiple objectives that need to be optimized simultaneously.

A useful notion in multi-objective problems is the concept of *Pareto optimality*. A design, $\vec{x}$, is a *Pareto optimal solution* for problem (13), if and only if the solution $\vec{x}^*$ cannot be changed to improve one of the objectives without adversely affecting at least one other objective (Ngatchou et al. (2005)). Based on this definition, Pareto optimality solutions, $\vec{x}^*$, are non-unique. The *Pareto optimal set* is defined as the set that contains all Pareto optimal solutions. Furthermore, the *Pareto front* is the set that contains the objectives of all optimal solutions.

Since all Pareto optimal solutions are good solutions, the most appropriate solution will depend only upon the trade-offs between objectives; therefore, it is the responsibility of the designer to choose the most appropriate solution. It is sometimes desirable to obtain the complete set of Pareto optimal solutions, from which the designer may then choose the most appropriate design.

There is a large number of algorithms for solving multi-objective problems, see e.g. Das & Dennis (1998); Kim & de Weck (2005; 2006); Lin (1976); Messac & Mattson (2004); Ngatchou et al. (2005). These methods can be classified between: a) classical approaches; and, b) meta-heuristic approaches as proposed by Ngatchou et al. (2005). Classical approaches are based on either transforming the multiple objectives into a single aggregated objective or optimizing one objective at a time, while the other objectives are treated as constraints. Examples of classical methods are the weighted sum method and the ε-constraint method (see Ngatchou et al. (2005)). In the weighted sum method (e.g., Kim & de Weck (2006)), the multiple objectives are transformed into a single objective function by multiplying each objective by a weighting factor and summing up all contributions such that the final objective is:

$$F_{weighted\,sum} = w_1 f_1 + w_2 f_2 + \cdots + w_n f_n \tag{20}$$

where f_i are the objective functions, w_i are the weighting factors and $\sum_i w_i = 1$. Each single set of weights determines one Pareto optimal solution. A Pareto front is obtained by solving the single objective optimization problem with different combinations of weights. The weighted sum method is easy to implement; however it has two drawbacks: 1) a uniform spread of weight parameters rarely produces a uniform spread of points on the Pareto set; 2) non-convex parts of the Pareto set cannot be obtained, (see Das & Dennis (1997)).

Meta-heuristic methods are population-based methods using genetic or evolutionary algorithms. Meta-heuristic methods aim at generating the Pareto front directly by evaluating, for a given population, all design objectives simultaneously. For each population, all designs are ranked in order to retain all Pareto optimal solutions. The main advantage of these methods is that many potential solutions that belong to the Pareto set can be obtained in one single run. Examples of multi-objective meta-heuristic methods include the multi-objective genetic algorithm (MOGA), the non-dominated sorting genetic algorithm (NSGA) and the strength Pareto evolutionary algorithm (SPEA). A detailed description of these methods can be found in Ngatchou et al. (2005) and Veldhuizen & Lamont (2000).

Multi-objective optimization of flywheels has recently been attempted by Huang & Fadel (2000b) and Krack et al. (2010b). In both cases, the weighted sum method was used in order to solve the optimization problem. Huang and Fadel aimed at maximizing kinetic energy storage while minimizing the difference between maximum and minimum Von Mises stresses for an alloy flywheel with different cross-sectional areas. The flywheel was divided into several rims and the design variables were the height of each rim in the flywheel. Krack et al. (2010b) aimed at maximizing kinetic energy storage while minimizing cost. Stress within the flywheel was included as a constraint in the optimization problem. In their case, the flywheel was a composite flywheel with several rims and the design variables were the thickness of each rim and the flywheel rotational speed.

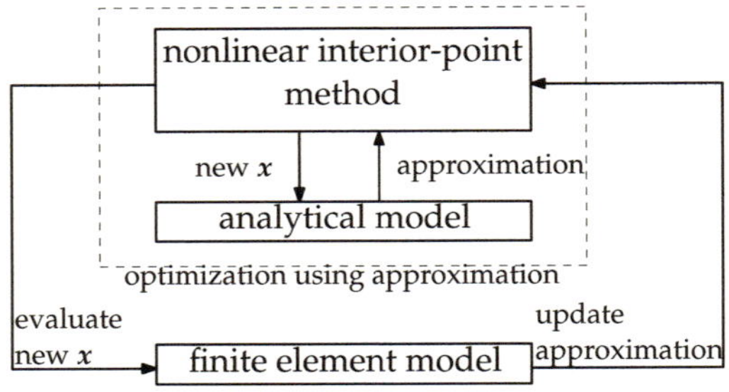

Fig. 12. Schematic of a multi-fidelity simulation. The high-fidelity finite element simulation is called to correct the lower-fidelity analytical model

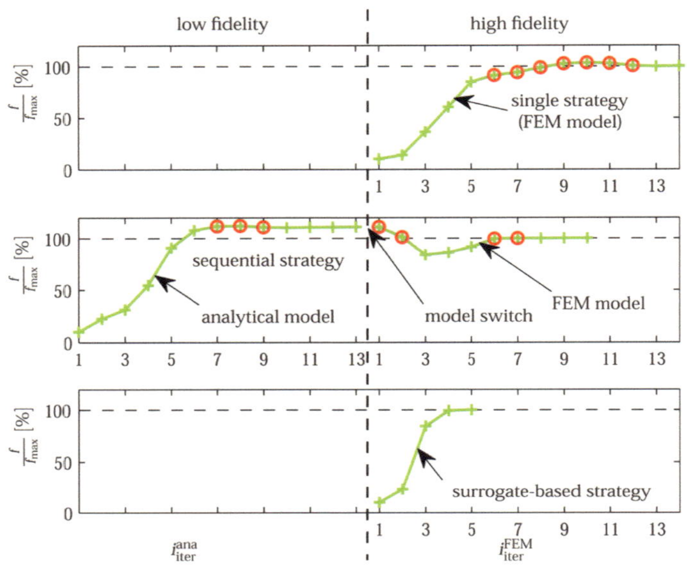

Fig. 13. Convergence histories of the cost optimization of a hybrid composite flywheel rotor with a split-type hub for different optimization strategies

4.4.3 Multi-fidelity and surrogate-based optimization

Accurate predictions of stress and strain in variable geometry flywheels and hubs require solving a set of complex multi-dimensional partial differential equations (PDEs). The system of PDEs is usually solved using the finite element method (FEM). Multi-dimensional FEM simulations of complex geometries require a substantial amount of computational resources. Further, since in order to solve a flywheel optimization problem many flywheel designs will need to be evaluated, the computational expense associated with flywheel design and optimization is a major challenge for solving such problems.

In order to reduce the computational resources associated with solving optimization problems, optimization strategies based on combining analysis tools of different accuracy have emerged in the literature (see Alexandrov et al. (2000); Forrester & Keane (2009); Simpson et al. (2001)). In multi-fidelity and surrogate-based optimization strategies, the optimization method only iterates on an approximate model. The multi-dimensional flywheel model is then used sporadically in order to apply a correction to the approximation. In multi-fidelity models, the approximation is usually a simplified version, i.e. a lower fidelity model, of the original problem such as a one-dimensional simplification of the multi-dimensional flywheel problem. In surrogate-based optimization, the approximation or meta-model, called a surrogate, is simply a fit to numerical or experimental data and, therefore, it is not based on the physics of the problem. Various approaches exist to construct a surrogate model, including the commonly used polynomial response surface models (RSM) and neural networks. Many of them are described in great detail in references Forrester & Keane (2009); Simpson et al. (2001).

Krack et al. (2010b) used a multi-fidelity approach to minimize the computational time required to solve a flywheel optimization problem. A variant of the approximation model management framework (AMMF) proposed by Queipo et al. (2005) was used in order to solve the problem. In this case, the optimization is performed using the low fidelity model and the FEM model is used to correct the low fidelity model for accuracy. The correction, a first order polynomial that is added to the solution of the low fidelity model, is obtained using the FEM model. The correction guarantees that the low fidelity model matches the FEM predictions for the design objective and constraints and its gradients at a specified design point. A schematic of the interaction between the low and high fidelity model is shown in **Figure 12**. The optimization algorithm uses information from the low fidelity model to obtain the optimal solution. After the optimal solution using the low fidelity model has been obtained, a correction polynomial is obtained using FEM and a new optimization problem is solved in the corrected low fidelity model. This process is repeated until both FEM and low fidelity model result in the same optimal design. In reference Krack et al. (2010b), using the multi-fidelity approach the computational resources were reduced three fold from 3,025 sec. to 1,087 sec. **Figure 13** compares the convergence history of three different strategies to solving the problem: a) using only a high-fidelity model; b) using the low- and high-fidelity models sequentially, i.e. solve the optimization problem using the low-fidelity model and then, use the solution as the initial design for a new optimization problem with the high-fidelity model; and, c) the multi-fidelity approach. Red circles indicate infeasible designs. Using the multi-fidelity model involves the least number of evaluations of the high-fidelity model.

5. Conclusion

An overview of rotor design for state-of-the-art FES systems was given. Practical design aspects in terms of manufacturing have been discussed. Typical analytical and FE modeling approaches have been presented and their suitability for the design optimization process regarding accuracy and computational efficiency has been investigated. The design of a hybrid composite flywheel rotor was formulated as a multi-objective, multi-variable nonlinear constrained optimization problem. Well-proven approaches to the solution of the design problem were presented and thoroughly discussed. The capabilities of the suggested methodology were demonstrated for various numerical examples.

6. References

A. Antoniou & W.-S. Lu (2007). *Practical Optimization: Algorithms and Engineering Applications*, Springer.

Aarts, E. & Korst, J. (1990). *Simulated annealing and Boltzmann machines*, Vol. 7, Wiley Chichester.

Alexandrov, N., Lewis, R., Gumbert, C., Green, L. & Newman, P. (2000). Optimization with variable-fidelity models applied to wing design, *AIAA paper* 841(2000): 254.

Arnold, S. M., Saleeb, A. F. & Al-Zoubi, N. R. (2002). Deformation and life analysis of composite flywheel disk systems, *Composites Part B: Engineering* 33(6): 433–459.

Arora, J. (1989). *Introduction to optimum design*, McGraw-Hill.

Arvin, A. & Bakis, C. (2006). Optimal design of press-fitted filament wound composite flywheel rotors, *Composite Structures* 72(1): 47–57.

Bornemann, H. J. & Sander, M. (1997). Conceptual system design of a 5 mwh/100 m w superconducting flywheel energy storage plant for power utility applications, *IEEE Transactions on Applied Superconductivity* 7(2 PART 1): 398–401.

Brown, D. R. & Chvala, W. D. (2005). Flywheel energy storage: An alternative to batteries for UPS systems, *Energy Engineering: Journal of the Association of Energy Engineering* 102(5): 7–26.

Christopher, D. A. & Donet, C. (1998). Flywheel technology and potential benefits for aerospace applications, *IEEE Aerospace Applications Conference Proceedings*, Vol. 1, pp. 159–166.

Danfelt, E., Hewes, S. & Chou, T. (1977). Optimization of composite flywheel design, *International Journal of Mechanical Sciences* 19: 69–78.

Das, I. & Dennis, J. (1997). A closer look at drawbacks of minimizing weighted sums of objectives for Pareto set generation in multicriteria optimization problems, *Structural Optimization* 14: 63–69.

Das, I. & Dennis, J. (1998). Normal-boundary intersection: a new method for generation pareto optimal points in multicriteria optimization problems, *SIAM Journal of Optimization* 8: 631–657.

Dorigo, M., Maniezzo, V. & Colorni, A. (1996). Ant system: optimization by a colony of cooperating agents, *Systems, Man, and Cybernetics, Part B: Cybernetics, IEEE Transactions on* 26(1): 29–41.

El-Barky, A., Tapia, R., Tshchiya, T. & Zhang, Y. (1996). On the formulation and theory of the newton interior point method for nonlinear programming, *Journal of Optmization Theory and Applications* 89(3): 507–541.

Fabien, B. (2007). The influence of failure criteria on the design optimization of stacked-ply composite flywheels, *Structural and Multidisciplinary Optimization* 33(6): 507–517.

Forrester, A. I. & Keane, A. J. (2009). Recent advances in surrogate-based optimization, *Progress in Aerospace Sciences* 45(1-3): 50–79.

Gabrys, C. W. & Bakis, C. E. (1997). Design and manufacturing of filament wound elastomeric matrix composite flywheels, *Journal of Reinforced Plastics and Composites* 16(6): 488–502.

Ge, R. & Qin, Y. (1987). A class of filled functions for finding global minimizers of a function of several variables, *Journal of Optmization Theory and Applications* 54(2): 241–252.

Genta, G. (1985). *Kinetic energy storage: theory and practice of advanced flywheel systems*, Butterworth and Co. Ltd, London.

Goldberg, D. (1989). *Genetic algorithms in search, optimization, and machine learning*, Addison-Wesley.

Gowayed, Y., Abel-Hady, F., Flowers, G. & Trudell, J. (2002). Optimal design of multi-direction composite flywheel rotors, *Polymer composites* 23(3): 433–441.

Ha, S., Jeong, H. & Cho, Y. (1998). Optimum design of thick-walled composite rings for an energy storage system, *Journal of composite materials* 32(9): 851–873.

Ha, S. & Jeong, J. (2005). Effects of winding angles on through-thickness properties and residual strains of thick filament wound composite rings, *Composites Science and technology* 65(1): 27–35.

Ha, S., Kim, D. & Choi, S. (1999). Optimal design of a hybrid composite flywheel rotor using finite element methods, *Evolving and revolutionary technologies for the new millennium* 44: 2119–2132.

Ha, S., Kim, D. & Sung, T. (2001). Optimum design of multi-ring composite flywheel rotor using a modified generalized plane strain assumption, *International journal of mechanical sciences* 43(4): 993–1007.

Ha, S., Kim, H. & Sung, T. (2003). Measurement and prediction of process-induced residual strains in thick wound composite rings, *Journal of Composite Materials* 37(14): 1223–1237.

Ha, S., Kim, J. & Han, Y. (2008). Design of a Hybrid Composite Flywheel Multi-rim Rotor System using Geometric Scaling Factors, *Journal of Composite Materials-Lancaster* 42(8): 771–786.

Ha, S., Kim, M., Han, S. & Sung, T. (2006). Design and Spin Test of a Hybrid Composite Flywheel Rotor with a Split Type Hub, *Journal of Composite Materials* 40(23): 2113–2130.

Ha, S., Yang, H. & Kim, D. (1999). Optimal design of a hybrid composite flywheel with a permanent magnet rotor, *Journal of Composite Materials* 33(16): 1544–1575.

He, L. & Polak, E. (1993). Multistart method with estimation scheme for global satisfying problems, *Journal of Global Optimization* 3: 139–156.

Hebner, R., Beno, J. & Walls, A. (2002). Flywheel batteries come around again, *IEEE spectrum* 39(4): 46–51.

Huang, B. C. (1999). Polar woven flywheel rim design, *International SAMPE Symposium and Exhibition (Proceedings)*, Vol. 44, p. II/.

Huang, J. & Fadel, G. (2000a). Heterogeneous flywheel modeling and optimization, *Materials and Design* .

Huang, J. & Fadel, G. (2000b). Heterogeneous flywheel modeling and optimization, *Materials and Design* 21(2): 111–125.

Kim, I. & de Weck, O. (2005). Adaptive weighted sum method for bi-objective optimization: pareto front generation, *Structural and Multidisciplinary Optimization* 29: 149–158.

Kim, I. & de Weck, O. (2006). Adaptive weighted sum method for multiobjective optimization: a new method for pareto front generation, *Structural and Multidisciplinary Optimization* 31: 105–116.

Krack, M., Secanell, M. & Mertiny, P. (2010a). Advanced optimization strategies for cost-sensitive design of energy storage flywheel rotors, *International SAMPE Symposium and Exhibition (Proceedings)* .

Krack, M., Secanell, M. & Mertiny, P. (2010b). Cost optimization of a hybrid composite flywheel rotor with a split-type hub using combined analytical/numerical models, *Structural and Multidisciplinary Optimization* pp. 1–17.

Krack, M., Secanell, M. & Mertiny, P. (2010c). Cost optimization of hybrid composite flywheel rotors for energy storage, *Structural and Multidisciplinary Optimization* 41(5): 779–796.

Lamberti, L. & Pappalettere, C. (2000). Comparison of the numerical efficiency of different sequential linear programming based algorithms for structural optimization problems, *Computers and Structures* 76(6): 713–728.

Lin, J. G. (1976). Multiple-objective problems: pareto-optimal solutions by method of proper equality constraints, *IEEE Transactions on Automatic Control* 21(5): 641–650.

Messac, A. & Mattson, C. A. (2004). Normal constraint method with guarantee of even representation of complete pareto frontier, *AIAA Journal* 42(10): 2101–2111.

Michalewicz, Z. & Schoenauer, M. (1996). Evolutionary algorithms for constrained parameter optimization problems, *Evolutionary computation* 4(1): 1–32.

Motor Trend (1952). the GYROBUS: Something New Under the Sun?, *Motor Trend* p. 37.

Ngatchou, P., Anahita Zarei & El-Sharkawi, M. (2005). Pareto Multi Objective Optimization, *Proceedings of the 13th International Conference on Intelligent Systems Application to Power Systems* art. no. 1599245: 84-91.

Portnov, G. G. (n.d.). *Composite flywheels, In: Handbook of Composites, Structures and Design Vol. 2*, Elsevier Science Publishers, Amsterdam.

Queipo, N. V., Haftka, R. T., Shyy, W., Goel, T., Vaidyanathan, R. & Tucker, P. K. (2005). Surrogate-based analysis and optimization, *Progress in Aerospace Sciences* 41(1): 1–28.

Ratner, J. K. H., Chang, J. B. & Christopher, D. A. (2003). Composite flywheel rotor technology - a review, *ASTM Special Technical Publication*, pp. 3–28.

Rodríguez, J., Renaud, J., Wujek, B. & Tappeta, R. (2000). Trust region management is multidisciplinary design optimization, *Journal of Computational and Applied Mathematics* 124: 139–154.

Schoen, F. (1991). Stochastic techniques for global optimization: A survey of recent advances, *Journal of Global Optimization* 1(3): 207–228.

Secanell, M., Songprakorp, R., Suleman, A. & Djilali, N. (2008). Multi-objective optimization of a polymer electrolyte fuel cell membrane electrode assembly, *Energy and Environmental Science* 1(3): 378–388.

Simpson, T., Peplinski, J., Koch, P. & Allen, J. (2001). Metamodels for computer-based engineering design: Survey and recommendations, *Engineering with Computers* 17(2): 129–150.

Takahashi, K., Kitade, S. & Morita, H. (2002). Development of high speed composite flywheel rotors for energy storage systems, *Advanced Composite Materials* 11(1): 40–49.

Tarrant, C. (1999). Revolutionary flywheel energy storage system for quality power, *Power Engineering Journal* 13(3): 159–163.

Tsai, S. (1988). Composite design, *Think Composites* 1: 1–19.

Tzeng, J. (2003). Viscoelastic analysis of composite rotor for pulsed power applications, *IEEE Transactions on Magnetics* 39(1): 384–388.

Tzeng, J., Emerson, R., Moy, P. & MD, A. R. L. A. P. G. (2005). *Composite Flywheel Development for Energy Storage*, Defense Technical Information Center.

Vanderplaats, G. (1984). *Numerical optimization techniques for engineering design with applications*, McGraw-Hill.

Veldhuizen, D. & Lamont, G. (2000). Multiobjective evolutionary algorithms: analyzing the state-of-the-art., *Evolutionary computation* 8(2): 125–147.

Wang, G. (2001). Improvement on the Adaptive Response Surface Method for High-Dimensional Computation-Intensive Design Problems, *ASME Design Engineering Technical Conferences–Design Automation Conference.*

4

An Application of Genetic Fuzzy Systems to the Operation Planning of Hydrothermal Systems

Ricardo de A. L. Rabêlo[1], Fábbio A. S. Borges[1], Ricardo A. S. Fernandes[1], Adriano A. F. M. Carneiro[1] and Rosana T. V. Braga[2]

[1]*Engineering School of São Carlos / University of São Paulo (USP)*
[2]*Institute of Mathematical and Computer Sciences / University of São Paulo (USP)*
Brazil

1. Introduction

The operation planning of hydrothermal systems aims to specify how the set of power plants should be operated so that the resources available for power generation are used efficiently. In hydrothermal systems with great participation of hydroelectric generation, as is the case of the Brazilian system, the operation planning intends to establish Reservoir Operation Rules (RORs) to replace, whenever possible, the thermoelectric generation by the hydroelectric generation (Christoforidis et al., 1996). Due to their peculiar characteristics, the operation planning of the Brazilian hydrothermal system can be classified as a problem coupled in time (dynamic) and space (not separable), nonlinear, nonconvex, stochastic and large scale (Leite et al., 2002; Oliveira & Soares, 1995; Silva & Finardi, 2001).
It is worth mentioning that the RORs are present in some stages of the operation planning of hydrothermal systems, such as:

- Obtaining the equivalent reservoir of energy(Arvanitidis & Rosing, 1970a;b);
- Breakdown of the goals of hydraulic generation of the equivalent reservoir (Soares & Carneiro, 1993) and;
- Performance evaluation of the operation of hydroelectric system (Silva & Finardi, 2003).

An operation rule widely adopted in practice, including computational models of the Brazilian electric power system, known as the rule of parallel operation (RORP) (Marques et al., 2005), determines that all the reservoirs of the hydroelectric system should keep the same percentage of their useful volume. The greatest advantage of this rule is its simplicity, however, it does not conform to the principles of optimal operation of reservoirs for the electric power generation (Lyra & Tavares, 1988; Read, 1982; Sacchi, Nazareno, Castro, Silva Filho & Carneiro, 2004; Sjelvgren et al., 1983; Soares & Carneiro, 1991; Yu et al., 1998).
In order to have RORs inspired by the optimized behavior of the reservoirs, an optimization algorithm, inspired in (Carneiro et al., 1990; Carvalho & Soares, 1987), is initially applied for the operation of the hydroelectric system. As a result of the optimization, a set of operating points is obtained, which relate the energy stored in the hydroelectric system to the storage status of each reservoir. In order to make the set of points able to be used as an indication for

obtaining a ROR for a given hydro plant, it is necessary to set mathematical functions. These sets give a function that represents the rule of operation of each hydropower plant.
It is worth mentioning that several papers, from related literature, refer to obtaining RORs, differing only in the technique used for setting the points and implementing the obtained RORs. In (Soares & Carneiro, 1993), the authors use third-degree polynomial functions to set the points. The obtained RORs were applied and compared to the RORP in simulations of the operation of hydroelectric systems. The authors in (Cruz Jr & Soares, 1996; 1999; 1995) use the method of least squares to set the polynomial, exponential and linear functions. However, the obtained RORs were applied in a computational model that adopts the representation of the equivalent reservoir and compares them with the RORP. In (Carneiro & Kadowaki, 1996), the authors do the settings of the points through an algorithm that used the method of least squares, obtaining polynomial and exponential functions to express the RORs. The obtained RORs were used to simulate the operation of hydroelectric systems and compared with the ROR-P. In (Sacchi, Carneiro & Araújo, 2004a;b), Artificial Neural Networks (ANN) are used, more specifically SONARX networks. The obtained RORs are integrated into an algorithm of operation simulation and compared with the RORP. In (Rabelo et al., 2009b) the authors present a methodology based on Takagi-Sugeno fuzzy inference systems (Takagi & Sugeno, 1985) to obtain RORs, and the application of these rules in the simulation of the operation of hydroelectric systems and compares them with the RORP. In the latter case, the representative points of the optimal operation of reservoirs are used to set the parameters of consequents of the fuzzy production rules.
Therefore, this paper intends to use some principles that govern the optimized behavior of the reservoirs in order to assist the implementation of RORs for hydroelectric systems. The proposed methodology for specifying RORs combines Mamdani fuzzy inference systems (Mamdani, 1974) and Genetic Algorithms (GAs) (Goldberg, 1989). Mamdani fuzzy inference systems are used to determine the operation rule of each reservoir, i.e., estimate the operating volume of hydroelectric power plants, using the value of the energy stored in the system as input parameter. Thus, our goal is to generate RORs through the heuristic knowledge of the relationship between the global storage status of the hydroelectric system (energy stored in the system) and the operating volume of each hydroelectric power plant. Genetic Algorithms are used to find the optimal setting of the membership functions associated with each primary term of the consequent of the fuzzy production rules. Importantly, the GAs are global optimization algorithms, based on mechanisms of natural selection and genetics, which have proven effective in a variety of problems, because they overcome many of the limitations found in the traditional methods of search/optimization (Haupt & Haupt, 1998). The systems obtained from the integration between models of fuzzy inference and Genetic Algorithms are called Fuzzy-Genetic Systems (FGSs) (Cordón et al., 2004; Cordon, Herrera, Hoffman & Magdalena, 2001; Herrera, 2005; 2008).
Another fuzzy model broadly used is the Takagi-Sugeno fuzzy inference system. This model was proposed as an effort to develop a systematic approach to generate fuzzy production rules from a set of input and output data (Mendel, 2001). The fuzzy rules, in a Takagi-Sugeno fuzzy inference system, have linguistic variables only in their antecedents, and the definition of its consequents, usually based on the method of least squares, requires numeric data. On the other hand, the production rules in a Mamdani fuzzy inference model have linguistic variables in both their antecedent and in their consequent. Therefore, the basis of rules in the Mamdani fuzzy model can be defined solely in linguistic form, without the need for numeric input/output data. However, the need to adjust the membership functions of the linguistic

variables of the consequent requires an additional effort by the designer in developing the system.

2. Operation planning of hydrothermal systems

2.1 Mathematical formulation

The operation planning of hydrothermal systems, with individualized representation of the hydroelectric plants and deterministic inflows can be formulated as the following optimization problem:

$$\min \quad \sum_{t=1}^{T} CVP_t \cdot 0,5 \cdot \Phi(D_t - H_t)^2 + V(x_T) \tag{1}$$

$$s.a. \quad D_t = E_t + H_t, \tag{2}$$

$$H_t = \sum_{i=1}^{N} k_i \cdot hl(x_{i,t}^{avg}, u_{i,t}) \cdot min[u_{i,t}, q_{i,t}^{max}], \tag{3}$$

$$x_{i,t} = x_{i,t-1} - x_{i,t}^{evap}$$

$$+(y_{i,t}^{inc} + \sum_{k \in \Omega_i} u_{k,t} - u_{i,t}) \cdot [\tfrac{\Delta t_t}{10^6}], \tag{4}$$

$$u_{i,t} = q_{i,t} + v_{i,t}, \tag{5}$$

$$x_{i,t}^{min} \leqslant x_{i,t} \leqslant x_{i,t}^{max}, \tag{6}$$

$$u_{i,t}^{min} \leqslant u_{i,t} \leqslant u_{i,t}^{max}, \tag{7}$$

$$q_{i,t}^{min} \leqslant q_{i,t} \leqslant q_{i,t}^{max}, \tag{8}$$

$$x_{i,0} \text{ given}, \tag{9}$$

where:

- T: number of intervals of the planning horizon;
- N: number of hydroelectric plants;
- CVP_t: coefficient of present value associated with the interval t;
- E_t: complementary generation (thermal generation, imports of energy and load shortage) [MW];
- H_t: total hydroelectric generation [MW];
- D_t: demand (electricity market) [MW];
- $x_{i,t}$: volume stored in the reservoir i at the end of the interval t [hm^3];
- $x_{i,t}^{avg}$: average volume stored in the reservoir i at the interval t [hm^3];
- $x_{i,t}^{evap}$: volume evaporated in the reservoir i during the interval t[hm^3];
- $hl_{i,t}$: height of the net fall of the plant i in the interval t [m];
- $y_{i,t}^{inc}$: incremental inflow to the reservoir of the plant i in the interval t [m^3/s];
- $q_{i,t}$: water discharge (through turbines) of the plant i in the interval t [m^3/s];
- $u_{i,t}$: flow released of the plant i in the interval t [m^3/s];
- $v_{i,t}$: flow spilled from the plant i in the interval t [m^3/s];

- $x_{i,t}^{max}, x_{i,t}^{min}$: maximum and minimum of volume stored for the reservoir of the plant i at the end of the interval t [hm^3];
- $u_{i,t}^{max}, u_{i,t}^{min}$: maximum and minimum flow released of the plant i in the interval t [m^3/s];
- $q_{i,t}^{max}, q_{i,t}^{min}$: maximum and minimum water discharge through turbines of the plant i in the interval t [m^3/s];
- Δt_t: number of seconds in the interval t [s];
- Ω_i: set of indexes of the plants immediately upstream of the plant i.

The objective function consists of two parts: the operational cost during the planning horizon (Φ) and the future costs associated with the state of final storage of the hydroelectric reservoir (V). For our purposes, the complementary generation system is being represented by an equivalent thermoelectric plant that replaces all the non-hydraulic system (Huang, 2001). Thus, the operating cost is given by the cost of fuel used in the operation of the thermoelectric power plants, the cost of importing energy from other systems, and the cost of "deficit" (penalty associated with the lack of power supply). Therefore, the operating cost is represented by a single cost function of the non-hydraulic sources (Carvalho & Soares, 1987). The future cost is a terminal condition, used in the optimization models of the operation, to balance the water use during the planning horizon and its future use (Martinez & Soares, 2002).

Equality (2) represents the restriction of meeting the demand of electricity in the time range t. The total generation from the hydroelectric system is represented by equation (3), given by the sum of the functions of hydraulic production of each hydroelectric power plant. Equation (4) represents the equation of water balance in the reservoirs. The flow released is equal to the sum of the turbine discharge flow with the spilled flow, and is shown in equation (5). The restrictions (6) and (7) represent the limits of storage and flow released of the hydroelectric plants, respectively. These limits vary over time, as they reflect the operational constraints of the power plants and other constraints associated with the multiple uses of water such as irrigation, flood control, navigation, etc. The restriction (8) represents the minimum and maximum turbine discharge, which may be associated with physical restrictions of the plant itself or electrical constraints. The values of the initial volumes of the reservoirs are given (9).

2.2 Hydroelectric system used

Figure 1 illustrates the set of hydroelectric plants used in this chapter. The hydroelectric system is comprised of seven hydroelectric plants of the Brazilian system (Emborcação, Itumbiara, São Simão, Furnas, Marimbondo, Água Vermelha e Ilha Solteira) represented individually. The set of plants chosen forms a complex system because it contains large plants, connected in parallel and in cascade. It is worth mentioning that the representation of the hydroelectric plants is done individually. The simplification of the equivalent representation of the hydroelectric system causes the generating capacity not to be utilized as efficiently as possible, since the energy equivalent reservoir cannot represent the operating characteristics of individual plants and hence their respective hydraulic coupling (Oliveira et al., 2009; Silva & Finardi, 2001). These factors, therefore, lead to an inefficient use by the hydraulic generator park (Cruz Jr & Soares, 1995). Thus, the individualized representation of each hydroelectric power plant is important since the main objective of the operation planning is to determine the generation targets for each plant, at each interval.

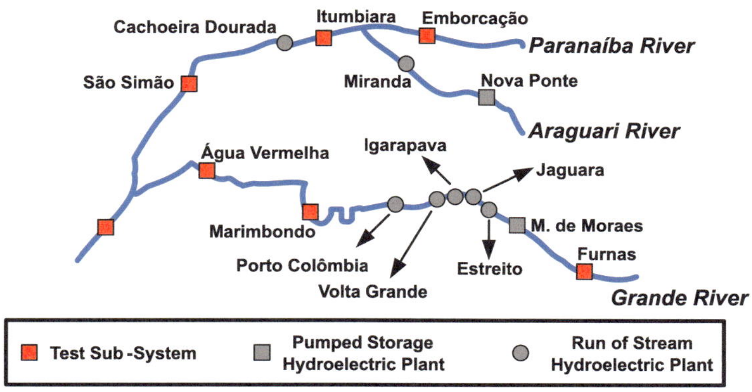

Fig. 1. Hydroelectric System Used.

2.3 Planning horizon

The complexity of the operation planning cannot be accommodated by a single mathematical model, therefore the use of models chains with different planning horizons and degrees of detail in the representation of the hydrothermal generation system is necessary (Pereira, 1985). In this chapter, the operation planning with the five-year horizons, discretized on a monthly basis, was adopted, which implies a horizon composed of 60 intervals.

2.4 Computational models of optimization and simulation

In this study, we used a computational model for optimization and simulation of the operation of hydroelectric systems. The optimization model is used to determine the optimal operation of reservoirs and is inspired by optimization algorithms specifically designed for the operation planning of hydrothermal systems (Carneiro et al., 1990; Carvalho & Soares, 1987). The simulation model includes a simulation algorithm which enables the evaluation of the performance of reservoir operation rules. The simulation algorithms aim to replicate the operating behavior of the power plants of the hydroelectric system under certain operating conditions. It is noteworthy that the computational models used are part of a computational tool that has been developed by the authors to conduct studies related to the operation planning of hydrothermal systems (Rabelo et al., 2009a). It should be stressed that the authors applied a process of development (UML Components) (Cheesman & Daniels, 2001) based on software components (Szyperski, 2002) for building the computer models mentioned above, in order to guide the development of the tool, with the possibility to add or change requirements in an orderly manner, even when the application is running.

2.5 Optimized operation of the reservoirs for the power generation

Figures 2, 3 and 4 show representative points of the optimized operation of some plants of the hydroelectric system.

Despite the dispersion in the points, we can see a different behavior of the hydroelectric plants in the optimal operation. It appears that the volume of the reservoirs upstream is reduced when the energy stored in the system decreases. The plants further downstream aim to keep their reservoirs full, and only reduce the volume when the energy stored in the system is critical. On the other hand, intermediate plants have variations not as severe as in reservoirs upstream, or as soft as in downstream reservoirs. Thus, the relative location

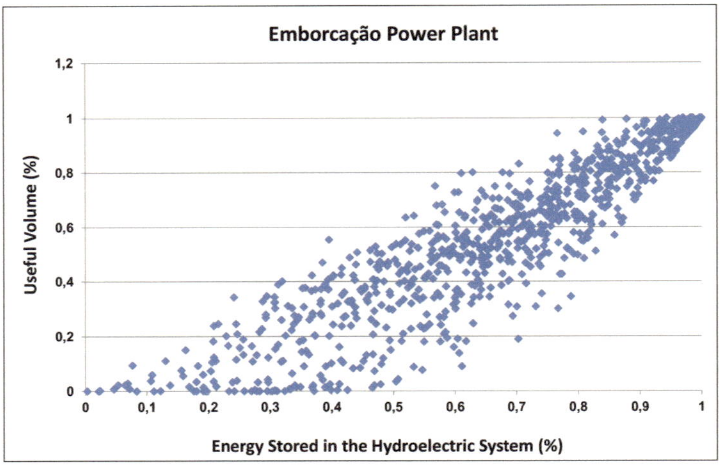

Fig. 2. Representative Points of the Optimized Operation of Emborcação Power Plant.

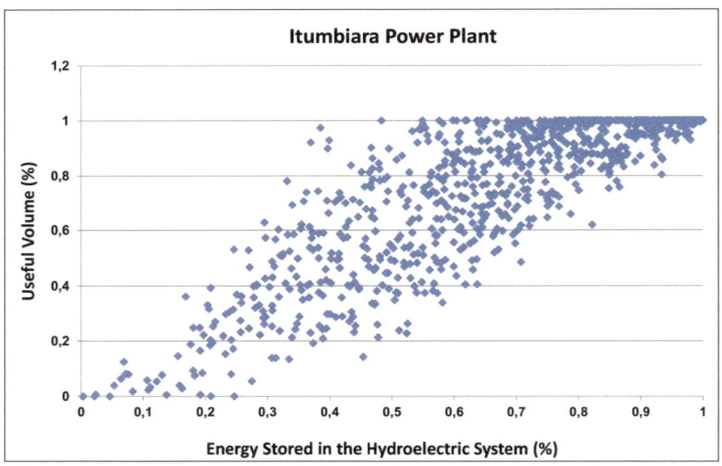

Fig. 3. Representative Points of the Optimized Operation of Itumbiara Power Plant.

of the hydroelectric plants influences the optimized operating behavior of the reservoirs. (Soares & Carneiro, 1991).

2.6 Reservoirs operation rules

The operation rules are functions that determine the operating volume of each reservoir to establish a coupled behavior between the hydroelectric power plants. To implement the coupling on the operation of the hydroelectric plants set, a global parameter called coupling factor of the operation of the hydroelectric system is defined, denoted by λ_t. The coupling factor represents the storage percentage of the system in a given interval t, and is calculated (Equation 10) as the ratio between the energy stored in the system (ESS_t) and maximum energy that can be stored in the system (ESS^{max}), resulting in values in the interval $0 \leq \lambda_t \leq 1$.

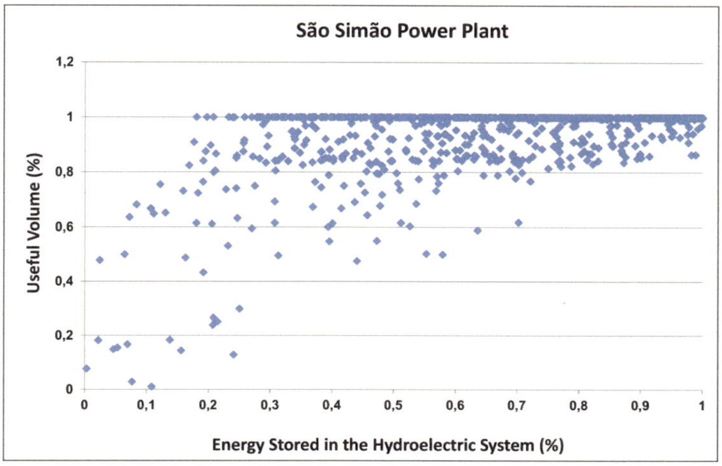

Fig. 4. Representative Points of the Optimized Operation of São Simão Power Plant.

$$\lambda_t = \frac{ESS_t}{ESS^{max}} \tag{10}$$

With the value of λ, the volume of each plant reservoir can be defined by the following equation:

$$x_{i,t}(\lambda_t) = x_{i,t}^{min} + f_i(\lambda_t) \cdot (x_{i,t}^{max} - x_{i,t}^{min}) \tag{11}$$

where:

- $f_i(\lambda_t)$: reservoir operation rule of the plant i in relation to the parameter λ_t. It expresses the operating behavior (emptying/filling) of the reservoir together with the other reservoirs of the hydroelectric system.

3. Fuzzy-genetic systems

3.1 Fuzzy inference systems

The fuzzy inference systems are based on linguistic production rules like "if ... then" in which the fuzzy set theory (Zadeh, 1965) and the fuzzy logic (Zadeh, 1996) provide the mathematical foundations needed for dealing with very complex processes, based on inaccurate, uncertain and qualitative information. These rule-based systems are more suitable for complex problems where it is very difficult to describe the problem (behavior of the process) quantitatively. Additionally, fuzzy rule-based systems are able to yield good results with reasonably simple mathematical operations (Ross, 2004).

The fuzzy inference systems are based on three steps: fuzzification, inference procedures and defuzzification. The fuzzification is a mapping from the domain of the input variable to the fuzzy domain, representing the allocation of primary terms (linguistic or qualitative values), defined by membership functions, to the input variables. The fuzzy inference procedure is responsible for assessing the primary terms of the input variables by applying the production rules in order to obtain the fuzzy output value of the inference system. The defuzzification is used to associate a numeric value to the fuzzy output set, which is obtained from the fuzzy inference procedure.

3.2 Genetic algorithms

Genetic Algorithms (GAs) are search/optimization algorithms based on the mechanisms of genetics and natural selection. Its operation follows the biological inspiration, which implies that in a given population, individuals with "good" genetic characteristics are more likely to survive and to generate individuals increasingly stronger (able), while the less fit individuals tend to disappear during the evolutionary process. When using GAs, each individual in the population, called chromosome, represents a potential solution to the problem to be solved.
The basic operation of GA is to generate an initial population formed by a set of individuals. During the evolutionary process, an evaluation function is applied for each individual, to assign it an ability (fitness) index that characterizes the quality of the individual as solution of the problem. Based on the ability index, a part of the individuals is selected randomly, while others are discarded. Individuals chosen by the selection process are subject to form descendants for the next generation through changes in their genetic characteristics by applying the genetic operators of mutation and crossover (recombination). This iterative process continues until a satisfactory solution to the problem is found. Each of the iterations of the process is denominated a generation of GA. To prevent the most capable individuals to disappear from the population by applying genetic operators, an elitist strategy can be applied (Goldberg, 1989), which is to automatically put the best individuals in the next generation.
Seemingly simple, due, in part, to its bio inspired nature, GAs are capable of solving complex problems in a very elegant manner. Moreover, they are not affected by assumptions about differentiability or continuity of the objective function of the problem. This implies that the GAs can be very appropriate for dealing with problems with non-differentiable and discontinuous functions. Additionally, GAs operate on a population of individuals in order to explore different points of the search space in parallel.

3.3 Aspects of the implementation of genetic fuzzy system

The implemented fuzzy inference systems have a linguistic input variable, the energy stored in the system (ESS_L), defined on the set of linguistic terms (Very Low, Low, Medium, High and Very High) (Figure 5) and a output linguistic variable, the useful volume, also defined in the set of linguistic terms (Very Low, Low, Medium, High and Very High).
In this paper, a Mamdani fuzzy inference system, consisting of a rule base with 5 disjunctive rules of inference was specialized for each hydroelectric plant (Figure 1). The syntax of the rule base of the implemented fuzzy systems is represented by the following linguistic conditional statements:

- Rule 1: If (ESS_L *is Very Low*) Then (*Useful Volume is Very Low*), or
- Rule 2: If (ESS_L *is Low*) Then (*Useful Volume is Low*), or
- Rule 3: If (ESS_L *is Medium*) Then (*Useful Volume is Medium*), or
- Rule 4: If (ESS_L *is High*) Then (*Useful Volume is High*), or
- Rule 5: If (ESS_L *is Very High*) Then (*Useful Volume is Very High*)

GAs were used to set (adjust) the membership functions associated with the linguistic variable Useful Volume in each of the seven fuzzy systems. The differentiated setting in the linguistic output variable is made to represent the different behavior of each reservoir in optimal operation of the system. After setting all fuzzy systems, they can make inferences from numerical values of the input variable, to obtain the value of the output variable, the operating volume of the reservoirs in the interval t. For this, the rules are inferred in parallel. The inference of each rule consists in evaluating the antecedent, then the application of the

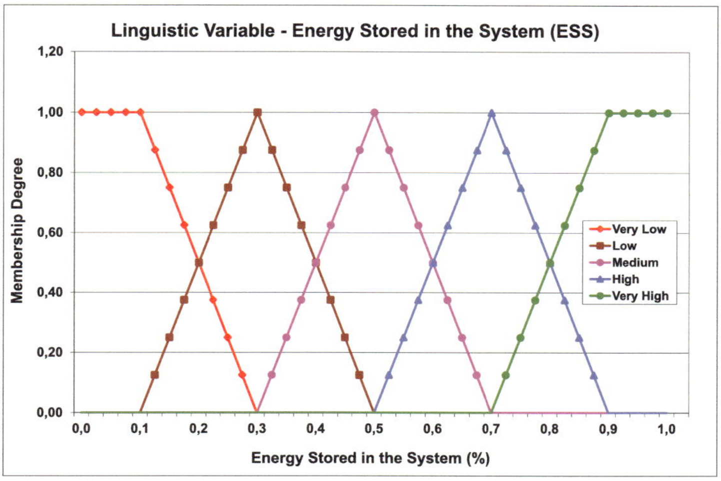

Fig. 5. Linguistic Variable representing the Energy Stored in the System.

implication operator to determine the implication of the fuzzy set of the consequent. The aggregation of the consequent consists of aggregating, or combining, the consequent obtained by the inference of each rule. The defuzzification procedure gets the numeric value of the useful operating volume of each of the plants of the hydroelectric system. The implication operator used was the minimum operator of Mamdani, the aggregation of the consequent was performed by the maximum operator, and the defuzzification method used was the method of center of area (Ross, 2004; Zimmermann, 2001).

The first aspect to be considered in the use of GAs for solving a problem is the chromosome representation, that is, map the information of the parameters of the problem in a way viable to be handled by the GA. When designing a fuzzy system using a GA, the fuzzy system within a chromosome must be mapped. A fuzzy system is specified only when the rule base and the definition of membership functions of each linguistic variable are specified (Shi et al., 1999). In this paper, as the rule base has been previously defined, based on the authors' experience regarding the optimal behavior of the reservoirs, the chromosomes contain only information about the distribution of the membership functions of the linguistic output variable (useful volume) against to universe of discourse, to reflect the specialized behavior of each reservoir in the coupled operation of the system. It is noteworthy that some studies have shown that the performance of the fuzzy inference system is much more sensitive to the choice of fuzzy database (fuzzy sets, membership functions) than to the composition of the fuzzy production rule base (Bonissone et al., 1996; Cordón et al., 2000; Cordón, Herrera & Villar, 2001; Zheng, 1992). Since the problem has continuous parameters that need high precision, the chromosome representation was done by using real numbers (Srikanth & kamala, 2008) instead of binary representation (Holland, 1975) traditionally used. The membership functions used for the linguistic variables are triangular and trapezoidal. The output linguistic variables of each of the seven fuzzy systems have five membership functions, two trapezoidal representing the very low and very high fuzzy sets, and three triangular representing low, medium and high sets. Due to the distribution adopted by the fuzzy sets, each trapezoidal membership function requires two parameters to represent it, and each triangular membership function needs 3, so 13 genes are used for each fuzzy inference

system, which implies a chromosome with 91 genes, due to the fact that each chromosome stores information from all seven plants of the hydroelectric system. The values of the genes are real numbers ranging between 0 and 1 and the population is composed of 80 individuals.
After defining the chromosome representation, the design of GA focuses on the specification of an evaluation function. The evaluation function assigns a numerical value (fitness, ability index) that reflects how well the parameters represented in the chromosome adapt and thus it is the way used to determine the quality of an individual as a solution to the problem. As the availability of water in a given interval depends on the degree of its former use, this study used as evaluation function the difference between the maximum stored energy that can be achieved in the system (ESS^{MAX}) and the energy stored in the system regarding the last interval of the planning horizon (ESS_{60}). Since the decisions taken at interval of the planning depends on the decisions taken in the past and determine the future development of the hydroelectric system, the use of stored energy in the last interval of the horizon is feasible because it takes the link between operational decisions in time into account, commonly known as temporal coupling (problem coupled in time). Numerically, the evaluation function is represented by (12), where 60 indicates the index of the last interval of the planning horizon:

$$\text{Evaluation Function} = ESS^{MAX} - ESS_{60} \tag{12}$$

Therefore, there is a minimization problem, whose goal is to find a value ESS_{60}, so as to minimize the difference from ESS^{MAX}.
After calculating the evaluation function for every individual of the chromosomes population, the selection process chooses a subset of individuals of the current population, to compose an intermediate population in order to apply the genetic operators. The selection method adopted in this study was the method of the tournament (Eiben et al., 1999). It is worth mentioning that the tournament size adopted was equal to 2. In combination with the selection module, an elitist strategy was used, keeping the best individual from one generation to another.
Genetic operators are applied to make the population go through an evolution. The genetic, crossover and mutation operators are used to transform the population through successive generations in order to extend the search/optimization to a satisfactory result. The crossover is the operator responsible by the genetic recombination of the parents, in order to enable the next generation to inherit these characteristics. In this study we used the discrete crossover (Herrera et al., 2003; 2005). This operator includes the main crossover operators for the binary representation, which are directly applicable to the real representation. The mutation genetic operator (Hinterding et al., 1995) is necessary to introduce and maintain genetic diversity of the population through random change of genes within the chromosomes, which provides a means to incorporate new genetic characteristics in the population. Therefore, the mutation ensures the possibility of reaching any point in the search space, and helps overcome the problem of local minima. However, the mutation is applied less frequently than the crossover, in order to preserve the relationship exploration-exploitation (Herrera et al., 1998). In this study, the random mutation was used (Michalewicz, 2011).
Table 1 sumarizes the values of the parameters used in the implementation of the Fuzzy System. The Table 2 sumarizes the values of the parameters used in the implementation of the Genetic Algorithm responsible by the adjustment of the Fuzzy System.
Several criteria can be applied to finalize the implementation of a GA. In this paper, a maximum limit of 100 generations was set. The stop criterion was set for this value

Parameters of Fuzzy System	
Membership Functions	Trapezoidal and Triangular
Implication Operator	Minimum of Mamdani
Agregation Operator	Maximum
Defuzzification	Center of Area

Table 1. Main Parameters used in the Fuzzy System.

Parameters of Genetic Algorithm	
Representation	Real
Selection	Tournament
Crossover	Discrete
Probability of Crossover	100%
Mutation	Random
Probability of Mutation	10%

Table 2. Main Parameters used in the Genetic Algorithm.

of generations, so there is a balance between computational effort and the result of the optimization.
As a result of the GAs operation in setting the fuzzy systems, Figures 6, 7 and 8 show the membership functions associated with the linguistic variable useful volume of plants Furnas, Água Vermelha and Ilha Solteira. One can observe a different distribution of fuzzy sets (Very Low, Low, Medium, High and Very High) for each reservoir, where the positioning of the membership functions is done according to the Genetic Algorithm.

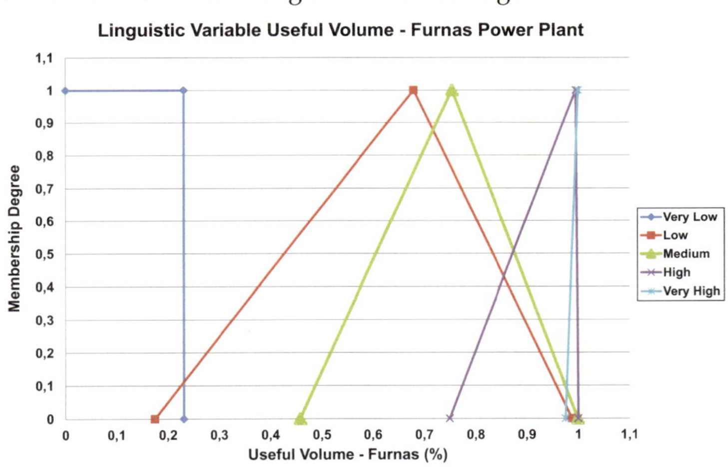

Fig. 6. Linguistic Variable Representing the Useful Volume of Furnas Power Plant.

4. Results and discussions

The simulation of the operation aims to verify the operating behavior of a hydroelectric system subject to certain operating conditions (electric power market, operating rules, water inflow, operational constraints, initial volume, etc.). So to make the comparison between the proposed Reservoir Operation Rules, based on Genetic Fuzzy Systems (RORGFS), the

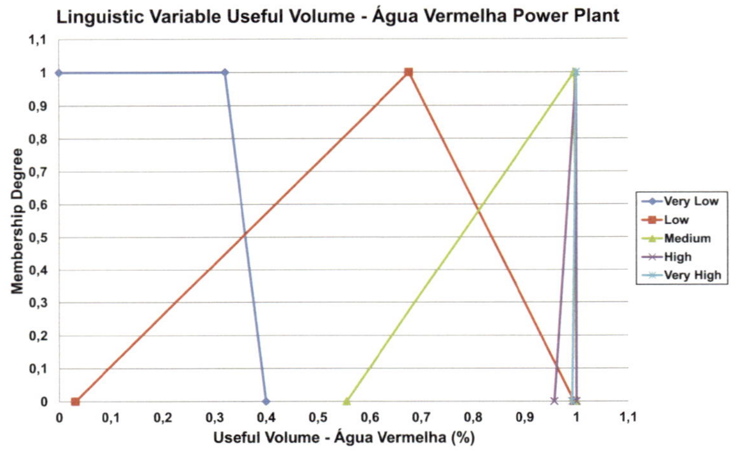

Fig. 7. Linguistic Variable Representing the Useful Volume of Água Vermelha Power Plant.

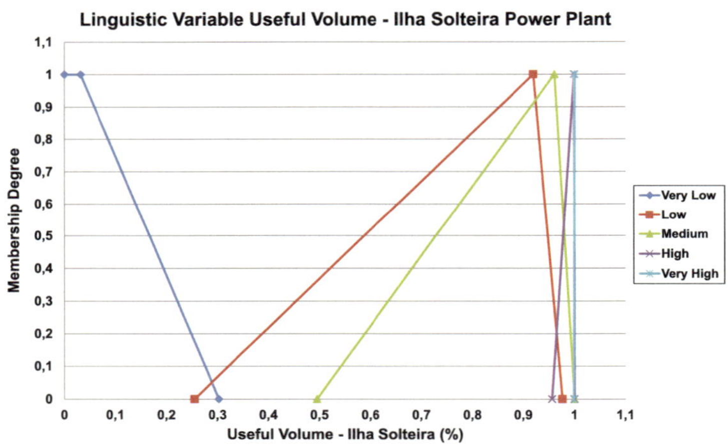

Fig. 8. Linguistic Variable Representing the Useful Volume of Ilha Solteira Power Plant.

operating rules based on mathematical polynomial and exponential functions (RORMF) (Carneiro & Kadowaki, 1996; Soares & Carneiro, 1993), the rule of parallel operation (RORP) (Marques et al., 2005) and the operation rule based on Takagi-Sugeno fuzzy systems (RORTS) (Rabelo et al., 2009b); the operation simulations are performed considering the same remaining operating conditions. Therefore, differences in behavior in the operation of the hydroelectric system will result only from the operational rules used. In this study, the computer model of operation simulation of hydroelectric systems was used, to evaluate the performance of RORs (Rabelo et al., 2009a).

Computer models of optimization and simulation, as well as the various rules of operation of reservoirs were implemented using the programming language C++ (Stroustrup, 2000). The developed software was run on an Intel Core 2 Duo 1.83 GHz, 3.00 GB of RAM on a Microsoft Windows Vista operating system with 32 bits.

4.1 Operating conditions

Five case studies were carried out, considering the water inflow of plants for the periods from 1936 to 1941, from 1951 to 1956, from 1971 to 1976, from 2000 to 2005 and with data from LTA (Long Term Average), in order to make a comparison between the RORs implemented in the simulation model under various hydrological conditions. To determine the target of hydraulic generation (demand or electric power market), the optimization of the energy operation of the hydroelectric system was performed with the actual water inflows occurred during the periods in order to obtain the solution with the perfect knowledge of water inflows for the entire planning horizon. The natural water inflows used in the operational simulations correspond to the flow rates recorded for the same periods of history. The month of May was adopted (dry season for the river basin of the system) as the starting month for all case studies. In all case studies, the initial volume stored in the reservoirs was considered as being equal to the maximum operating volume.

4.2 Results

The results illustrated by Figures 9 and 10 show fluctuations in the volume of the reservoirs depending on the location of the plant in the cascade through the application of RORGFS. With the predominant influence of the head effect (Read, 1982), the plant of Furnas, located upstream of Grande River, presented the highest levels of fluctuations in the reservoir, causing the reservoir to be operated at lower levels when compared to other plants in the cascade, such as Água Vermelha and Ilha Solteira. Ilha Solteira plant is operated with its reservoir full during most of the planning horizon. As the energy stored in a system is valued by the productivity of the plants further downstream, the power plant Ilha Solteira behaves like a run-of-river plant and appreciates all the water of the hydroelectric system, to be operated with maximum productivity. Água Vermelha plant, with an intermediate location in the cascade, has milder fluctuations in the reservoir storage than the Furnas plant, however exhibits more severe oscillations when compared to Ilha Solteira plant. Thus the application of RORGFS emphasized the filling of the reservoirs downstream to upstream, and the emptying of reservoirs upstream to downstream.

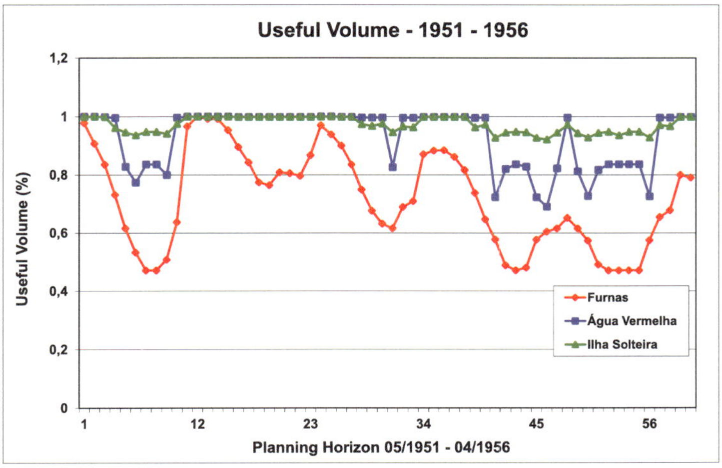

Fig. 9. Trajectories of Volume of some Reservoirs (1951 - 1956).

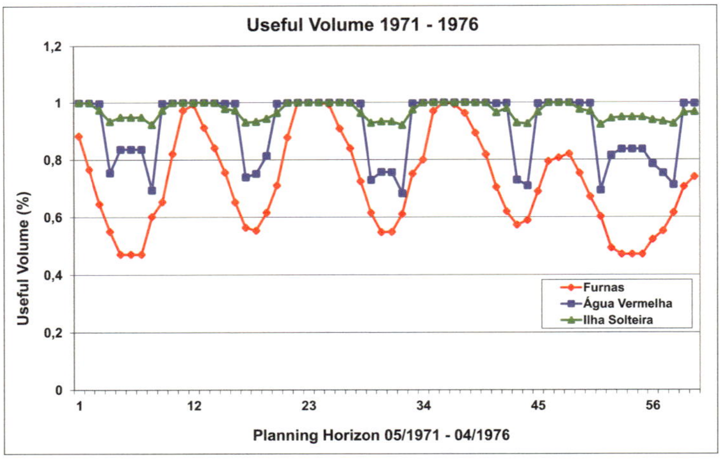

Fig. 10. Trajectories of Volume of some Reservoirs (1971 - 1976).

The operation rules based on the implementation of fuzzy genetic systems have established a specialized profile for all reservoirs set so as to maximize the stored energy in the hydroelectric system. This different behavior is obtained by different settings in the linguistic output variable in each of the seven fuzzy inference systems. The results presented by Figures 11, 12 and 13 illustrate the most efficient use of the generation hydroelectric resources by the operation rule based on genetic fuzzy systems. A more severe depletion of all the reservoirs can be verified when using RORP, RORMF and RORTS, which implies a more efficient use of water from reservoirs by RORFGS. It can also be pointed out that, throughout the planning horizon, the RORGFS always showed higher values of energy stored in the system, confirming that the operation rule for the reservoirs need to use less water to meet the same electricity market. Additionally, at the end of the planning horizon, one can see that RORP, RORMF and RORTS do not reach the storage levels achieved by RORGFS, making the reliability and the cost of operation extremely committed to the continued operation of the system. Therefore, RORGFS allows that the operation simulation of the hydroelectric system is consistent with the continuity of operation of the system, since it does not cease to be operated at the end of the planning horizon.

Thus, one can verify that RORGFS can ensure a more reliable and economic supply of electricity. It is economical because it requires less generation hydraulic resources (water) than the RORP, RORMF and RORTS. And it is reliable because it allows the operation of the hydroelectric system with higher levels of storage in the reservoirs, reducing the possibility of hydraulic deficits of the hydrothermal generation system. Therefore, the potential of RORGFS on optimizing the use of water resources, aimed at generating electricity can be verified. Moreover, RORGFS is quite consistent with the objectives of the planning of the energetic operation of hydrothermal systems as the optimization of water resources seeks to minimize additional generation. Thus, the higher the performance of the operation rules of the reservoirs for the use of hydroelectric generation resources, the lower necessary complementation to supply the electric power market.

Table 3 shows the average of energy stored in the system, during the planning horizon, to allow a numerical verification of the efficiency of each rule in the simulation of the operation of the plants in the hydroelectric system.

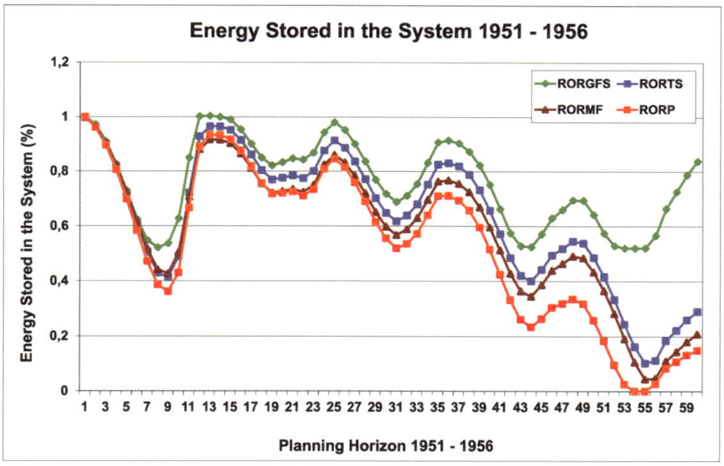

Fig. 11. Trajectories of Energy Stored in the System (1951 - 1956).

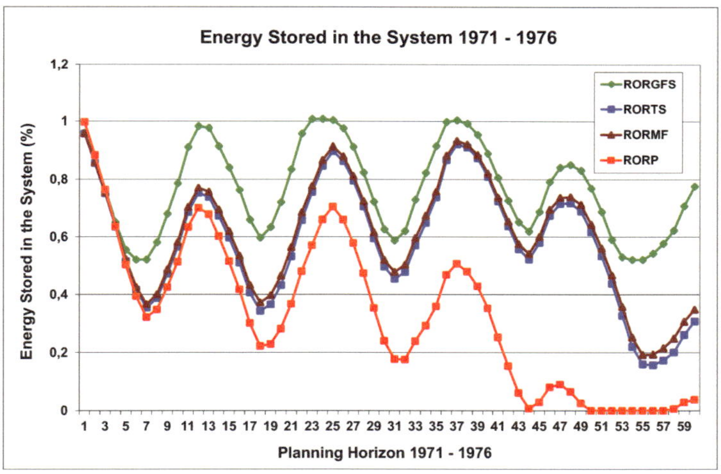

Fig. 12. Trajectories of Energy Stored in the System (1971 - 1976).

Planning Horizon	RORP	RORMF	RORTS	RORGFS
1936-1941	27865.82	29773.74	32299.55	35858.53
1951-1956	24232.82	26817.27	28851.86	34674.15
1971-1976	14329.13	27517.44	26544.69	34791.76
2000-2005	18151.44	21761.86	25847.11	36068.96
MLT	17171.52	25950.09	27437.61	36881.12

Table 3. Average of Energy Stored in the System [MW].

The reservoir operation rules based on the implementation of Genetic Fuzzy Systems have established a specialized profile for all reservoirs so as to maximize the stored energy in the hydroelectric system. This different behavior is obtained by different settings in the linguistic

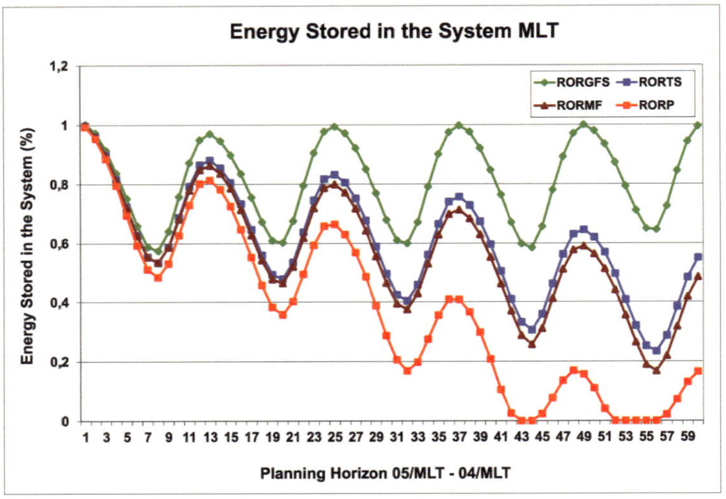

Fig. 13. Trajectories of Energy Stored in the System (MLT).

output variable in each of the seven fuzzy inference systems. With the predominant influence of the head effect, the plants where the volume of the reservoir have no great influence on the productivity of the system have drawdown priority. On the other hand, the plants whose operating volume of the reservoir has great influence on the productivity of the system have filling priority. As the energy stored in the system is valued by the productivity of the plants further downstream, the operating rules emphasize the filling of the reservoir downstream to upstream, and the drawdown of the reservoir from upstream to downstream. Thus, the reservoirs upstream, with the additional function of regulating the seasonal nature of water inflows, are those who present higher fluctuations in their level of storage. As for the reservoirs downstream, with the function of maintaining maximum productivity, they do not usually show high fluctuations being operated as run of river plants.

5. Conclusions

This chapter emphasized the specification of reservoir operation rules by means of Genetic Fuzzy Systems. Mamdani fuzzy inference systems were used to estimate the operating volume of each hydroelectric plant based on the value of the energy stored in the hydroelectric system. For this, a fuzzy system for each hydroelectric plant was specialized, to represent the different behavior of each reservoir in the optimal operation of the system. Genetic Algorithms were applied to tune the membership functions of the linguistic variable of the consequent of the production rules of the $N = 7$ fuzzy systems.

The reservoir operation rule proposed was implemented and compared, through some case studies, with the rule of parallel operation, and with the operation rule based on mathematical functions, and with the operation rule based on Takagi-Sugeno fuzzy system. The results showed the efficiency of the proposed rule when used in the simulation of energy operation of hydroelectric systems. With respect to the energy stored in the system, the tests illustrated that the proposed operation rule requires less water resources under the same operating conditions than the other implemented rules. With the operation rule based on Genetic Fuzzy Systems, power plants downstream, where possible, remain full in order to keep high productivity and enhance the volume of water flowing through them. Thus, the membership functions of the consequent of the fuzzy inference systems prioritize increasingly higher levels of storage in reservoirs upstream to downstream in the cascade of power plants. With the specialization of a fuzzy inference system for each reservoir plant, the operation of each plant reflects the role that it plays in the hydroelectric system, according to its location in the cascade. Therefore, the hydroelectric system is able to maintain higher levels of stored energy. It can be stated that the simulation of the operation using RORGFS maximizes the hydroelectric benefits of the hydrothermal generation system, because it serves the same electricity market, using less hydroelectric resources. It is noteworthy that at the end of the planning horizon, RORP, RORMF and RORTS were not able to keep the storage levels of reservoirs of the system close to the storage levels established by RORGFS, implying that the reliability and the cost of generation of the hydrothermal system will be severely compromised in the future operation of the system.

When a Mamdani fuzzy inference system is chosen to determine the operation rules of the plants of the hydroelectric system, an action/control strategy is obtained which can be monitored and interpreted by the linguistic point of view. Because the fuzzy inference systems are potentially able to express and manipulate qualitative information, another advantage in the application of Mamdani fuzzy systems is due to the fact that domain experts are able to map their experience and decision-making process, both qualitatively. Thus, the strategy of action/control of the Mamdani fuzzy inference system can be regarded as justified and as consistent as the strategy of domain experts.

6. References

Arvanitidis, N. V. & Rosing, J. (1970a). Composite Representation of a Multireservoir Hydroelectric Power System, *IEEE Transactions on Power Apparatus and Systems* PAS-89(2): 319–326.

Arvanitidis, N. V. & Rosing, J. (1970b). Optimal Operation of Multireservoir Systems Using a Composite Representation, *IEEE Transactions on Power Apparatus and Systems* PAS-89(2): 327–335.

Bonissone, P., Khedkar, P. & Chen, Y. (1996). Genetic Algorithms for Automated Tuning of Fuzzy Controllers: A Transportation Application, *Proceedings of the 1996 5th IEEE International Conference on Fuzzy Systems*, Vol. 1, Citeseer, pp. 674–680.

Carneiro, A. A. F. M. & Kadowaki, M. (1996). Operation rules for great hydroelectric systems in cascade, *11o. Automatic Brazilian Conference*.

Carneiro, A. A. F. M., Soares, S. & Bond, P. S. (1990). A Large Scale of an Optimal Deterministic Hydrothermal Scheduling Algorithm, *IEEE Transactions on Power Systems* 5(1): 204–211.

Carvalho, M. F. & Soares, S. (1987). An Efficient Hydrothermal Scheduling Algorithm, *IEEE Transactions on Power Systems* 2(3): 537–542.

Cheesman, J. & Daniels, J. (2001). *UML components*, Addison-Wesley Boston.

Christoforidis, M., Aganagic, M., Awobamise, B., Tong, S. & Rahimi, A. (1996). Long-term/mid-term Resource Optimization of a Hydrodominant Power System using Interior Point Method, *IEEE Transactions on Power Systems* 11(1): 287–294.

Cordón, O., Gomide, F., Herrera, F., Hoffman, F. & Magdalena, L. (2004). Ten years of genetic fuzzy systems: Current framework and new trends, *Fuzzy Sets and Systems* 41: 5 – 31.

Cordon, O., Herrera, F., Hoffman, F. & Magdalena, L. (2001). *Genetic Fuzzy Systems - Evolutionary Tuning and Learning of Fuzzy Knowledge Bases*, World Scientific.

Cordón, O., Herrera, F. & Villar, P. (2000). Analysis and Guidelines to Obtain a Good Uniform Fuzzy Partition Granularity for Fuzzy Rule-Based Systems Using Simulated Annealing*, *International Journal of Approximate Reasoning* 25(3): 187–215.

Cordón, O., Herrera, F. & Villar, P. (2001). Generating the Knowledge Base of a Fuzzy Rule-Based System by the Genetic Learning of the Data Base, *IEEE Transactions on Fuzzy Systems* 9(4): 667–674.

Cruz Jr, G. C. & Soares, S. (1996). Non-Uniform Composite Representation Hydroelectric Systems for Long-Term Hydrothermal Scheduling, *IEEE Transactions on Power Systems* 11(2): 702–707.

Cruz Jr, G. C. & Soares, S. (1999). General Composite Representation of Hydroelectric Systems, *Power Industry Computer Applications, 1999. PICA'99. Proceedings of the 21st 1999 IEEE International Conference*, pp. 177–182.

Cruz Jr, G. & Soares, S. (1995). Non-parallel composite representation of hydroelectric systems for long-term hydrothermal scheduling, *IEEE Power Industry Computer Applications Conference*, pp. 566–571.

Eiben, A. E., Hinterding, R. & Michalewicz, Z. (1999). Parameter Control in Evolutionary Algorithms, *IEEE Transactions on Evolutionary Computation* 3(2): 124–141.

Goldberg, D. E. (1989). *Genetic Algorithms in Search, Optimization and Machine Learning*, Addison-Wesley Longman Publishing Co., Inc. Boston, MA, USA.

Haupt, R. L. & Haupt, S. E. (1998). *Practical Genetic Algorithms*, Wiley New York.

Herrera, F. (2005). Genetic fuzzy systems: Status, critical considerations and future directions, *International Journal of Computational Intelligence Research* 1(1): 59 – 67.

Herrera, F. (2008). Genetic Fuzzy Systems: Taxonomy, Current Research Trends and Prospects, *Evolutionary Intelligence* 1(1): 27–46.

Herrera, F., Lozano, M. & Sánchez, A. M. (2003). A taxonomy for the crossover operator for real-coded genetic algorithms: An experimental study, *International Journal of Intelligent Systems* 18(3): 309–338.

Herrera, F., Lozano, M. & Sánchez, A. M. (2005). Hybrid crossover operators for real-coded genetic algorithms: an experimental study, *Soft Computing-A Fusion of Foundations, Methodologies and Applications* 9(4): 280 – 298.

Herrera, F., Lozano, M. & Verdegay, J. (1998). Tackling Real-Coded Genetic Algorithms: Operators and Tools for Behavioural Analysis, *Artificial Intelligence Review* 12(4): 265–319.

Hinterding, R., Gielewski, H. & Peachey, T. (1995). The Nature of Mutation in Genetic Algorithms, *Proceedings of the Sixth International Conference on Genetic Algorithms*, Citeseer, pp. 65–72.

Holland, J. H. (1975). *Adaptation in Natural and Artificial Systems*, Ann Arbor: The University of Michigan Press.

Huang, S. J. (2001). Enhancement of hydroelectric generation scheduling using ant colony system based optimization approachs, *IEEE Transactions on Energy Conversion* 16: 296–301.

Leite, P. T., Carneiro, A. A. F. M. & Carvalho, A. C. P. L. F. (2002). Energetic Operation Planning Using Genetic Algorithms, *IEEE Transactions on Power Systems* 17(1): 173–179.

Lyra, C. & Tavares, H. (1988). A Contribution to the Midterm Scheduling of Large Scale Hydrothermal Power Systems, *IEEE Transactions on Power Systems* 3(3): 852–857.

Mamdani, E. H. (1974). Application of Fuzzy Algorithms for Control of Simple Dynamic Plant, *Proceedings of IEE Control and Science* 121(12): 1585–1588.

Marques, T. C., Cicogna, M. A. & Soares, S. (2005). Assessment of energy availability of hydro system under different operation policies, *Proceedings of IEEE Power Tech* .

Martinez, L. & Soares, S. (2002). Comparison between closed-loop and partial open-loop feedbackcontrol policies in long term hydrothermal scheduling, *IEEE Transactions on Power Systems* 17(2): 330–336.

Mendel, J. M. (2001). *Uncertain Rule-Based Fuzzy Logic Systems: Introduction and New Directions*, Prentice Hall.

Michalewicz, Z. (2011). *Genetic Algorithms + Data Structures = Evolution Programs*, Springer.

Oliveira, E. J., Marcato, A. L. M., Rocha, R. S., Oliveira, L. W., Silva Jr., I. C. S. & Pereira, J. L. R. (2009). Influência da variação da produtividade das usinas hidrelétricas no cálculo da energia firme, *SBA: Controle & Automação*. 20(2): 247–255.

Oliveira, G. G. & Soares, S. (1995). A Second Order Network Flow Algorithm For Hydrothermal Scheduling, *IEEE Transactions on Power Systems* 10(3): 1635–1641.

Pereira, M. V. F. (1985). Optimal scheduling of hydrothermal systems - an overview, *IFAC Symposium on Planning and Operation of Electric Energy Systems*.

Rabelo, R. A. L., Carneiro, A. A. F. M. & Braga, R. T. V. (2009a). Component-based development applied to energetic operation planning of hydrothermal power systems, *Proceedings of IEEE Bucharest PowerTech*.

Rabelo, R. A. L., Carneiro, A. A. F. M. & Braga, R. T. V. (2009b). An energetic operation policy using fuzzy controllers for maximization of benefits in the brazilian hydrothermal power system, *Proceedings of IEEE Bucharest PowerTech*.

Read, E. G. (1982). Economic Principles of Reservoir Operation I: Perfect Foresight, *CBER Working Paper, College of Business Administration, University of Tennessee* .

Ross, T. J. (2004). *Fuzzy Logic with Engineering Applications*, Wiley.

Sacchi, R., Carneiro, A. A. F. M. & Araújo, A. F. R. (2004a). Operation policies for hydropower systems: Using the unsupervised sonarx neural network, *IEEE - Power Systems Conference & Exposition* .

Sacchi, R., Carneiro, A. A. F. M. & Araújo, A. F. R. (2004b). A rbf network trained by the sonarx model and applied to obtain the operation policies of the hydropower systems, *8th Brazilian Symposium on Artificial Neural Networks*.

Sacchi, R., Nazareno, J. S., Castro, M. A. A., Silva Filho, D. & Carneiro, A. A. F. M. (2004). Economics principles of the hidrothermal power systems operation, *IX Symposium of Specialists in Electric Operational and Expansion Planning*.

Shi, Y., Eberhart, R. & Chen, Y. (1999). Implementation of Evolutionary Fuzzy Systems, *IEEE Transactions on Fuzzy Systems* 7(2): 109–119.

Silva, E. & Finardi, E. (2003). Parallel Processing applied to the Planning of Hydrothermal Systems, *IEEE Transactions on Parallel and Distributed Systems* pp. 721–729.

Silva, E. L. & Finardi, E. C. (2001). Planning of Hydrothermal Systems Using a Power Plant Individualistic Representation, *2001 IEEE Porto Power Tech Proceedings*, Vol. 3.

Sjelvgren, S., Anderson, S. & Dillon, T. S. (1983). Optimal operations planning in a large hydro-thermal power system, *IEEE Transactions on Power Apparatus and Systems* 102(11): 3644 – 3651.

Soares, S. & Carneiro, A. A. F. M. (1991). Optimal Operation of Reservoirs for Electric Generation, *IEEE Transactions on Power Delivery* 6(3): 1101–1107.

Soares, S. & Carneiro, A. A. F. M. (1993). Reservoir Operation Rules for Hydroelectric Power System Optimization, *Athens Power Tech, 1993. APT 93. Proceedings. Joint International Power Conference*, Vol. 2.

Srikanth, T. & kamala, V. (2008). A real coded genetic algorithm for optimization of cutting parameters in turning, *IJCSNS International Journal of Computer Science and Network Security* 8(6): 189 – 193.

Stroustrup, B. (2000). *A Linguagem de Programação C++*, Bookman.

Szyperski, C. (2002). *Component Software - Beyond Object Oriented Programming*, Addison-Wesley.

Takagi, T. & Sugeno, M. (1985). Fuzzy identification of systems and its applications to modeling and control, *IEEE Trans. Syst., Man, Cybern.* 15: 116–132.

Yu, Z., Sparrow, F. T. & Bowen, B. H. (1998). A new long-term hydro production scheduling method for maximizing the profit of hydroelectric systems, *IEEE Transactions on Power Systems* 13(1): 66 – 71.

Zadeh, L. (1965). Fuzzy Sets*, *Information and control* 8(3): 338–353.

Zadeh, L. A. (1996). Fuzzy logic = computing with words, *IEEE Transactions on Fuzzy Systems* 4(2).

Zheng, L. (1992). A Practical Guide to Tune of Proportional and Integral (PI) Like Fuzzy Controllers, *IEEE International Conference on Fuzzy Systems*, IEEE, pp. 633–640.

Zimmermann, H. J. (2001). *Fuzzy Set Theory – and its Applications*, Kluwer Academic Publishers.

5

Lightning Energy: A Lab Scale System

Mohd Farriz Basar, Musa Yusop Lada and Norhaslinda Hasim
Universiti Teknikal Malaysia Melaka (UTeM), Malaysia

1. Introduction

This chapter which has six subchapters explains the energy storage system in harvesting a lightning return stroke for a lab scale system. Nowadays, the world is facing the energy crisis and consequently a renewable energy is required as an energy contributor to solve the crisis. Hence, it is believed that lightning return stroke has a good future to be a free electricity sources. The main difficulty in harnessing the lightning stroke is to attract and simultaneously to store the energy, which limited in a microsecond. Due to that, the computer simulation works using PSpice is done as the preliminary effort intended for the hardware development as well as to understand and verify the proposed system. A lab scale system is set up based on natural characteristics of lightning to determine the performance of the sample capacitor as energy storage accurately. Hence, the single stroke impulse voltage is used as a mock of lightning. Regarding the energy storage device, the capacitor is employed due to the reliability, cost-effective and it is the most common. In addition, the direct tapping system and the high speed switching is most wanted in order to make the whole system more realistic. The capacitors are subjected to 1.2/50μs, 4,200V single-stroke impulse voltages generated by a single stage impulse generator. In this chapter, the energy of impulse voltage that successfully transferred and stored in the storage capacitors is discussed. Basically, the efficiency of the energy transfer is depends on the capacitance values and the switching times. As a final point, the lab scale system explained in this chapter demonstrates the capability to capture the energy from lightning return strokes that can be a clean energy sources.
On the other side, lightning which have extremely high current and high voltage is a gratis electricity energy sources that can be replenished. The lab scales systems for harvesting the energy from lightning return stroke, which discussed in this chaper able to give a new contribution to solve the energy crisis and it will be very challenging. Up until now, the mature technology in harvesting the lightning stroke for the large-scale system is still not yet ready and the relevant scientific literature is not easily found. It noted that the final system proposed in this chapter would provide an understanding of the system principle and additionally provide a noteworthy contribution for further research.

2. Clouds and lightning

Lightning return stroke is a complex phenomenon. The large peak currents or the electromagnetic shock wave are capable to kill people and destroy the buildings, trees, animals as well as electrical appliances. As a result, the damage can be outstanding in term of cost.

2.1 Clouds

Most researches on the electrical structure of clouds have focused on the cumulonimbus, the familiar thundercloud or thunderstorm, because this cloud type produces most of the lightning. There have been limited studies of the electrical properties of other types of clouds such as stratus, stratocumulus, cumulus, nimbostratus and cirrus clouds [2].
Briefly, clouds carry positive and negative charges. Through the dynamic of nature, clouds distribute these charges and collect negative charges at its bottom as well as positive charges at the top. After going through all the processes, charge at the bottom of the cloud draws and equal in magnitude but opposite polarity charge at the ground level. This creates a look like capacitor system between the cloud and the ground where the dielectric is air [3]. During stormy weather, the dynamic mechanism of the cloud will increase the charge density at clouds until threshold is reached and air loses its dielectric. Subsequently, lightning discharge occurs where air becomes conductor and simultaneously the charge travel from cloud to ground.

Fig. 1. Thundercloud charge distribution of lightning between cloud and ground

Thundercloud charge distribution of lightning between cloud and earth have been identified which is shown in Figure 1. There are different types of lightning like example the upward-initiated flashes; are relatively rare and usually occur from mountain peaks and tall man-made structures. Cloud-to-ground lightning has been studied more comprehensively than other form of lightning because of it is happen regularly surrounding us. It is known that lightning strike involves very large and very fast impulse voltage and current. It is flow to the ground, which in turn produces the corresponding electromagnetic fields.
Previous studies on lightning as an electrical energy and the possibilities of harnessing the lightning energy have been since 1752 starting with Benjamin Franklin observation on characteristics of lightning behavior. The estimation the lightning strike to the surface of earth is 100 time every one second. The challenge with lightning is to suggest a storage device to distribute the lightning power that it can be extracted later and the critical aspect on safety capture need to be alert. Data from NASA's lightning imaging sensor shows that the lightning occurs frequently over the land compare to the water. About 90% of lightning phenomenon happens in the land in spite of 75% of earth cover by the water.

2.2 Mock lightning

In this lab scale system, it proposed to use single impulse voltage 1.2/50μs as a mock lightning. It can be obtained by using the single stage impulse voltage generator. An impulse voltage is a unidirectional voltage which characterized by two time intervals expressed in microseconds, µs which is wave front time, t_f and wave tail time, t_t. Figure 2 shows the impulse voltage waveform that rises rapidly to a maximum value and then decays slowly to zero.

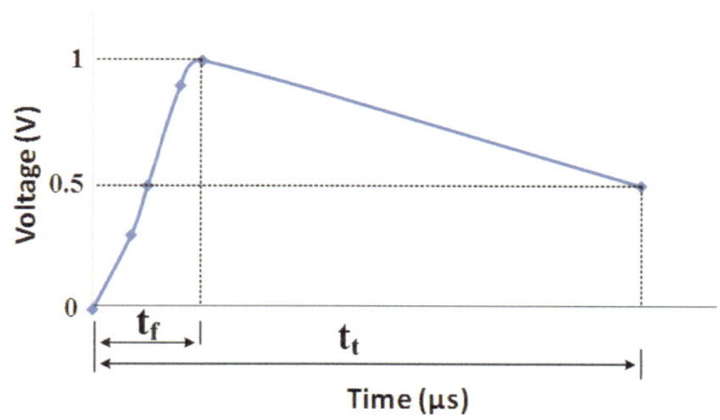

Fig. 2. Standardized impulse voltage wave shape

According to the standard wave shapes, the time to peak value or front time, t_f is set to be 1.2µs with the tolerances is ± 30%. Thus, the system proposed must capable to attract and stored the voltage at this peak time. Besides that, the tail time, t_t is set to be 50µs with the tolerances is ± 20%. The time to half value of the wave tail of an impulse voltage is the total time occupied by the impulse voltage in rising to peak value and declining there form to half the peak value of the impulse.

3. Energy storage

Energy storage technologies do not represent energy sources, but they provide valuable added benefits to improve power quality, stability and reliability of supply. In this modern power application, practicable storage technologies also known as viable storage technologies like batteries, flywheels, ultra capacitors and superconducting energy storage system was rapidly used. Figure 3 shows a specific energy ranges versus specific power.

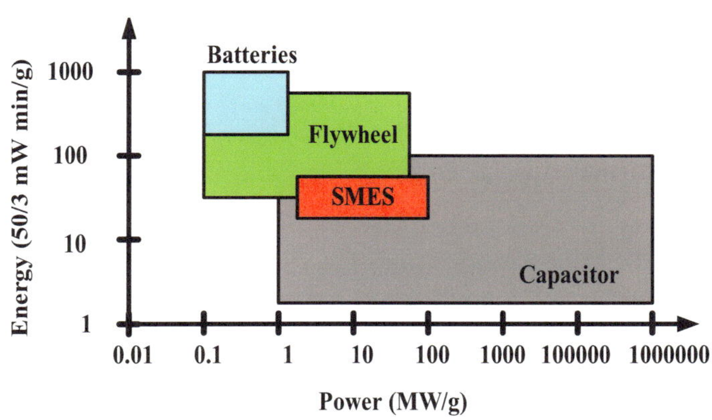

Fig. 3. Specific power versus specific energy ranges

Optimal Energy System (OES) that consist of flywheel based energy storage system currently be manufacture and design to provide pulse of energy for charging high voltage capacitor. This system has been used for electromagnetic air-craft launch system (EMALS) applications. Flywheel technology has been considered an attractive energy storage choice due to its potential for reduced weight and volume, high duty-cycle tolerance, and low maintenance requirements. Flywheel technology overcomes some of the shortcomings of today's energy storage systems by having an extremely high cyclic-life, limited temperature sensitivity, no chemical hazards, charge rate equal to discharge, and reduced weight and space.

They are a few benefits of adding energy storage to power electronic compensators for utility application such as improved system reliability, dynamic stability, enhanced power quality, transmission capacity enhancement and area protection. It shows that energy storage devices can be integrated to power electronics converters to provide power system stability, enhanced transmission capability, and improved power quality. Adding energy storage to power electronics compensators not only enhances the performances of the device, but can also provide the possibility of reducing the MVA ratings requirements of the front-end power electronics conversion system. This is an important benefit consideration when considering adding energy storage systems.

Supercapacitor was an advanced technology as compared to battery or electrostatic capacitor. The advantage of supercapacitor is high fast step response in term of charge and discharging. Effects of supercapacitor on power system application are absorbing high frequency power surges, reduce the degree of discharge and reduce the power losses.

In order to look into the capability of the lab scale system proposed for harvesting the lightning energy, two different types of sample capacitor are used. The first type is KNU 1910 Metallized Polypropylene Film Capacitors. The second type is CBB20 Axial-type Metallized Popypropylene Film Capacitor. These two types of sample capacitor used same dielectric, which is polypropylene film. The polypropylene capacitor was selected to be used in the testing because it is cheap, high temperature stability, readily available and it is widely used in high frequency, DC and pulse circuit's applications.

Many polypropylene film capacitors have a tolerance about 5% to 10%, which is adequate for many applications. In addition, there is very little change in capacitance when these capacitors are used in applications within frequency range of 100,000Hz. Moreover, the electrodes are vacuum evaporated metal on dielectric. So, the possibility for bad contact during the operation of capacitors is excluded. It also has a long life due to self-healing effect and suitable for high current.

4. High speed switching

As discussed in subchapter 2.2, the maximum value of the impulse voltage is occurred at 1.2μs. Consequently, the system proposed must able to draw out and store the energy of the impulse voltage (mock lightning) particularly at that extremely short time. Hence, the high speed switching is imperative to isolate the sample capacitor from any connection once the energy enters the sample capacitor. As a result, the potential voltage that can be retained or sustained in the sample capacitor can be investigate.

Figure 4 shows the block diagram of the high speed switching circuit that is used in the experimental work. The high speed switching controller consists of a microcontroller, a gate drive and a switching device and the orientation of components is shown in Figure 5.

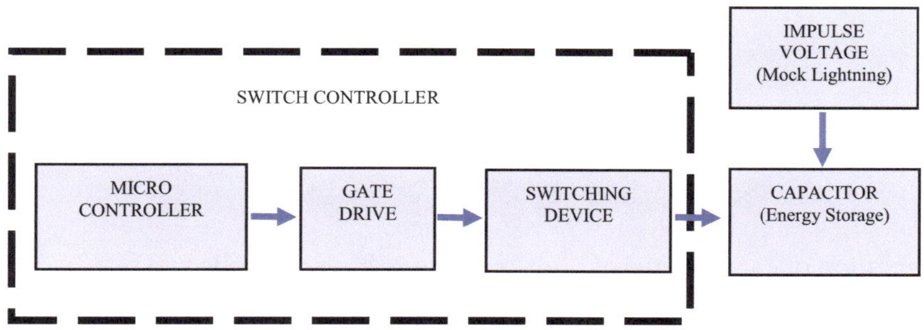

Fig. 4. Block diagram of the high speed switching circuit

4.1 High speed switching components

The advantages of this switching circuit are the circuit is simple and the cost is low. The main components that are used in the circuit are indicated in the Table 1. The high voltage IGBT is preferred to use as a switching device because it has a tremendous performance and widely used. The type of IGBT that be used are NPT (Non Punch Through) type, which is low efficient emitter and high carrier lifetime compare to others types of IGBT.

	Components	Description
Micro Controller	PIC 16F84A	• It is a peripheral interface controller • 16 series PIC, 14 bit (instructions) microcontroller • Maximum operating frequency is 20MHz
Gate Drive	NPN Transistor	• Part Number : BC548B • Manufacturer : Fairchild Semiconductor
Switching Device	High Voltage IGBT	• Part Number : IXGR 16N170AH1 • Manufacturer : IXYS

Table 1. Main components of switching controller

The micro controller circuit and gate drive requires 5Vdc of supply whilst the high voltage IGBT (switching device) requires 15Vdc to turn ON and 0Vdc to turn OFF as illustrated in Figure 6. Meanwhile, in Figure 7, to activate the switching circuit, two different level of voltage supply is required. The micro controller circuit and the gate drive requires 5Vdc of supply whilst the high voltage IGBT (switching device) requires 12Vdc.

It can see that there has an 4,200V impulse voltage at the right side of the circuit in Figure 7. At the beginning, the IGBT switch is in close position. Ideally, when the charges from the impulse voltage go into the capacitor, the IGBT switch will be open in order to isolate the capacitor to discharge. With the aid of IGBT switch, now the capacitor is like a battery.

Fig. 5. Orientation of components of high speed switching controller

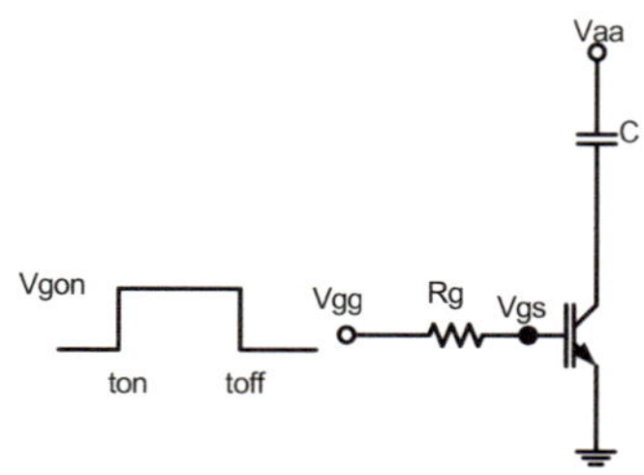

Fig. 6. Circuit diagram of IGBT with signal turn ON and OFF

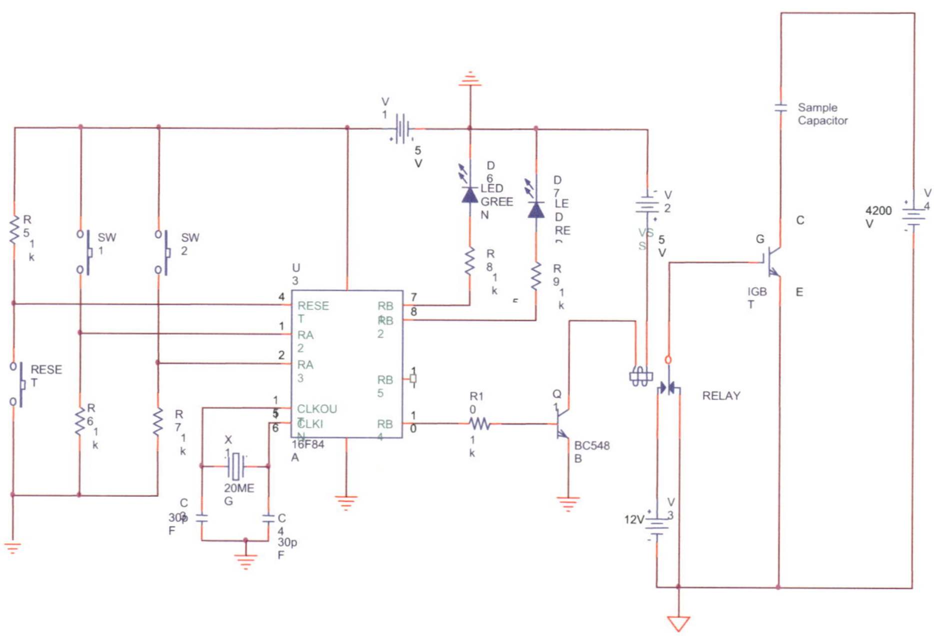

Fig. 7. Schematic diagram of the high speed switching circuit

4.2 Peripheral interface controller (PIC16F84A)

PIC is an abbreviation for Peripheral Interface Controller. It is not PLC or Programmable Logic Controller as most misunderstood, because it is an integrated chip (IC) based controller. When IC based controller is concerned, the logic levels, 0 and 1 are 0 volt and 5 volt respectively.

PIC is one of the micro technology generation and even more popular in industrial and hobbyists causing by the advantage on using it. Some of the advantage of PIC is low cost, widely available, large user base, extensive application, and serial programming capability which is the programming data of HEX (PIC data) can be write and re-write this is because PIC is made from flash memory.

A microcontroller is normally used for simple applications such as washing machines, rice cooker, air conditioner, keyboards, mouse, Liquid Crystal Displays (LCD) and much more. This means that the microcontroller is used for small applications and sometimes for standalone systems.

PIC is a brand of microcontroller, from its manufacturer, Microchip. The decision to use the PIC in this project is due to the vast employ of this device in many applications. It has Reduced Instruction Set Computer (RISC) architecture and moreover the assembly program is much simpler rather than other brands, such as Freescale (Motorola), Intel and many others. However, for the lab scale system proposed in this chapter used PIC16F84A as shown in Figure 8. Table 2 shows the function for every pin of PIC16F84A.

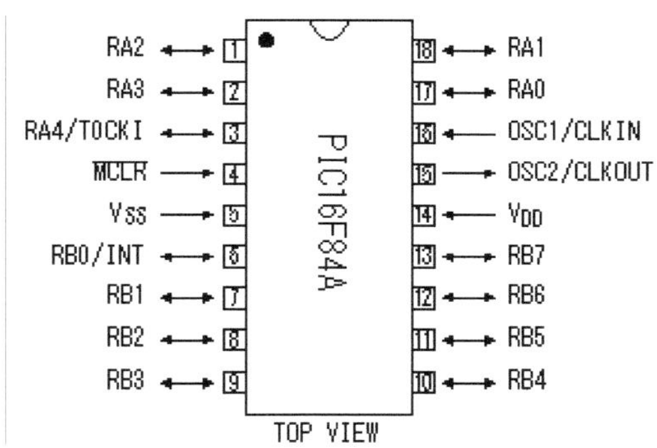

Fig. 8. Pin diagram for PIC16F84A

For programming, C language is used to write the program. Then compiler software is required to convert the C language to machine language (in zeros and ones). The compilers that have been used is mikroC. Next when the machine languages are generated, a downloader is required. The functions of this software is to transfer the machine codes of the program along with the settings to the PIC 16F84A. In simple words, the software installs the PIC with the machine codes of the program. In this project, WINPIC800 software is used because one of its advantages is that it can detect the type of PIC inserted automatically.

5. Computer simulation works

This section discuss about the computer simulation works using Pspice software. The purpose of the simulation is to obtain the testing circuit configuration that needs to be used in the experimental work. Moreover, the results of the computer simulation will verify the effort to harvest the lightning impulse voltage.

5.1 Single stage impulse voltage

The front time and the tail time of the impulse wave shape are dependently controlled by varying the value of R_D and R_E separately. The circuit arrangement for single stage impulse voltage generator involves the arrangement of high voltage dc supply from the rectifier, couple of sphere, one unit of wave shaping resistance which are wave front resistor, R_D and wave tail resistor, R_E, one unit charging capacitor C_S and one unit capacitor C_B as load capacitance. In order to obtain the single voltage impulse of 1.2/50µs, the value of resistance R_D and R_E can be verified by the equation 1 and equation 2.

$$t_f = 3R_D \frac{C_S C_B}{C_S + C_B} \tag{1}$$

$$t_t = 0.7(R_D + R_E)(C_S + C_B) \tag{2}$$

A capacitor C_S, previously charged to a particular dc voltage via a HV diode. It is suddenly discharged into the wave shaping network (through combination of R_D and R_E) after the breakdown or ignition of sphere gap. The sphere gap is acting as closing switch, where it is triggered by the voltage injected from the capacitor C_S.

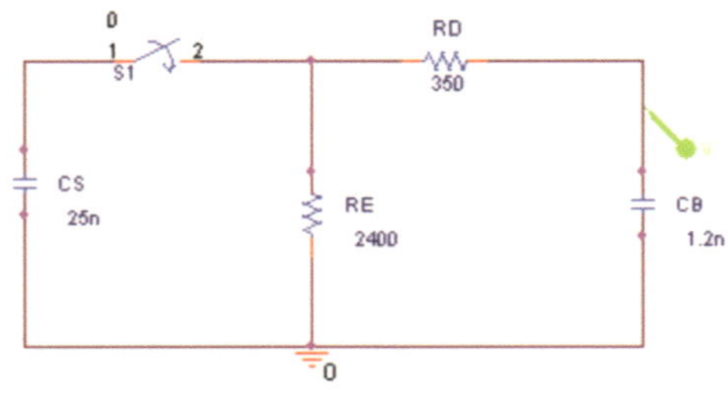

Fig. 9. Single stage impulse voltage generator simulation circuit

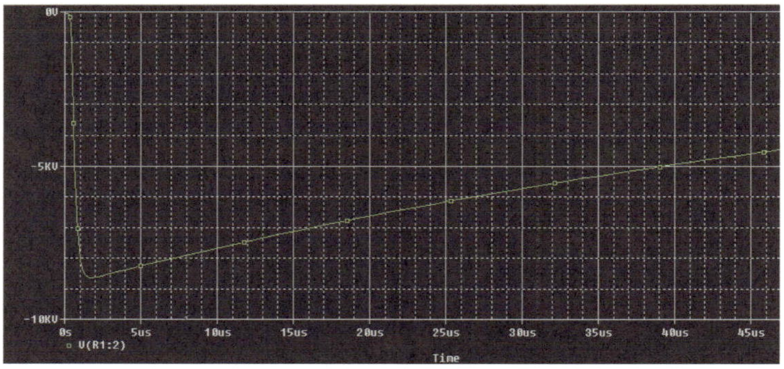

Fig. 10. Impulse voltage waveform 1.2/50µs

Afterward, the discharge voltage gives the desired voltage impulse wave shape. The front time and tail time can be controlled by changing the value of R_D and R_E. Switch S_1 is actually

a sphere gap. The rapid rise and slow decay impulse voltage wave shape can be generated by the activities of charging and discharging circuits with two energy storage elements, which are C_S and C_B. The C_B will be the capacitance of insulation to be tested or known as a load. As shown in Figure 9, it can see that C_S>>C_B and R_D>>R_E. Theoritically, the impulse voltage wave shape is composed by the superposition of two exponential functions.
As mentioned before, the charging capacitor C_S is charged via a high voltage dc supply through the rectifier. In the simulation, the initial condition for C_S is set to 8,600V and then it is discharged by closing switch S_1 (sparks in the sphere gap). Then, the simulation result is taken from the voltage across the load capacitor C_B. The simulation result in Figure 10 shows that, the peak impulse voltage is 8,600V with the front time t_f of 1.5µs and the tail time t_t of 47µs.

5.2 Additional switches

Figure 11 shows the circuit configuration that is almost similar with the previous circuit in Figure 9, except that there are two additional normally closed (NC) switches S_2 and S_3. Later, in the hardware development, both switches are replaced by the IGBT switch.

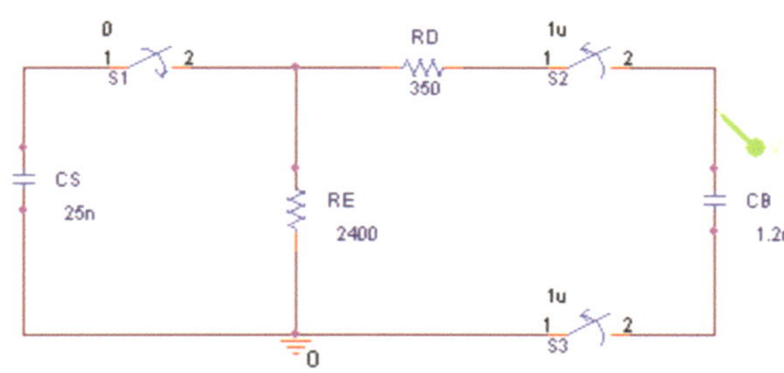

Fig. 11. Single impulse voltage generator circuit with additional two normally closed (NC) switches

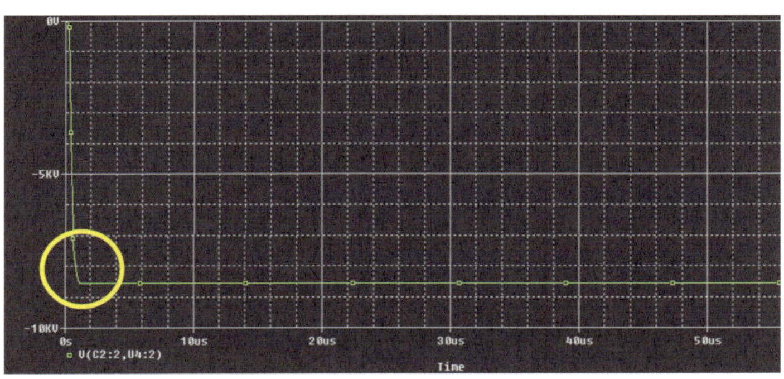

Fig. 12. Peak voltage 8.6kV is hold at t ≥ 1.5µs

The purpose of using these switches is to isolate C_B from any connection after the peak voltage has been attained in the capacitor. According to the Figure 2, the peak value of impulse voltage occurs at 1.5µs. Both NC (S_2 and S_3) switch are setting to be open at 1µs where the setting time is less than the time where the peak impulse voltage occurred. It is because the switches that have been used in the simulation have a delay about 0.5μs. For that reason, even though the switches are set at 1μs, the simulation result had showed that the both switches are operated at approximately 1.5μs.

Referring to the Figure 12, the voltage of the load capacitor C_B is not decaying after reach the peak value at 1.5μs (yellow circle). It proves that, when the both switches applied in the circuit, the voltage is slightly maintained. It shows that, the capacitor C_B is not able to discharge because it was isolated from any connection.

6. Laboratory experimental testing

In order to facilitate easy understanding, this subchapter will be divided into three stages. In Stage 1, the intention is to build up a 4.2kV single stroke impulse voltage. Therefore, the right combination of resistance value and capacitance value has to be clarified in order to get the standard impulse wave shape. The wave shape must follow the standard parameter, which is 1.2/50μs and when it is complete, now a mock lightning (source of lightning) in a small scale is ready and can be applied for the testing in Stage 2 and stage 3.

In Stage 2, the generated impulse voltage produced in Stage 1 is applied to a number of sample capacitors with different capacitance value. The important electrical parameter that need to gather is the voltage wave shape for each capacitor that has been tested. The wave shape will give an information about the sample capacitor characteristics such as the voltage capture, voltage store, voltage dissipated, the energy efficiency as well as charging and discharging time of the sample capacitor.

In Stage 3, high speed switching device is applied in the testing circuit. In the beginning, the sample capacitor will be charged by the impulse voltage. Then, it is isolated from any connection with the aid of high speed switching device. After approximately 15 minutes, the voltage waveform in the capacitor will be observed and this waveform is the data that required to be obtained.

6.1 Stage 1: impulse generation

As previously mentioned in section 5.1, the standard impulse lightning wave shapes is generated by using the single stage impulse voltage generator. The standard wave shapes has 1.2μs for the front time and 50μs for the tail time.

Fig. 13. Circuit arrangement for the testing in stage 1

The equipments that need to be used to setup the single impulse generator as shown in Figure 13 are consisted of Single Phase AC Voltage Test Transformer, HV rectifier, charging and load capacitor, wave shaping resistance, sphere gap, grounding rod, voltage divider, HV probe and oscilloscope.

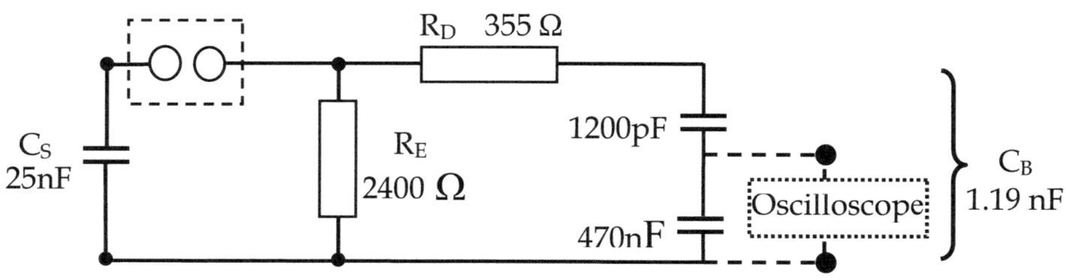

Fig. 14. Circuit diagram for stage 1 testing

By using the single stage impulse voltage generator circuit arrangement as shown in Figure 14, the desired impulse voltage wave shape has been achieved by choosing the correct resistance values for R_D and R_E as well as the capacitance values for C_S and C_B. In order to verify the impulse wave shape parameters, a digital oscilloscope is used to capture the experimental impulse voltage wave shapes with conjunction with the voltage divider. Besides that, the value of C_S, C_B, R_D and R_E are fixed in order to get a sustained 1.2/50μs wave shape. This is because the testing in the stage 2 and stage 3 will use the same magnitude of impulse voltage.

6.1.1 Front time, tail time and peak voltage of the experimental wave shape

Table 2 shows the right combination of the capacitance and resistance value for C_S, C_B, R_D and R_E that connected as shown in Figure 13 and Fiure 14. In order to obtain the resistance and capacitance value, equation (1) and equation (2) in section 5.1 is used. As a result, the front time t_f and tail time t_t for the generated impulse voltage are 1.3μs and 45μs respectively. This value is slightly different from the standard impulse wave shape. However, it is accepted for the reason that the experimental value is still under the tolerance of impulse wave shape parameter, which is +8.3% for the t_f and −10% for the t_t.

Standard Impulse Voltage Waveshape 1.2/50μs	
Value of charging capacitance, C_S	25 nF
Value of load capacitance, C_B	1.19 nF
Value of resistance, R_D	355 Ω
Value of resistance, R_E	2400 Ω
Front Time, $t_{f\ simulation}$	1.51 μs
Tail Time, $t_{t\ simulation}$	47.0 μs
Front Time, $t_{f\ calculation}$	1.21 μs
Tail Time, $t_{t\ calculation}$	44.0 μs
Front Time, $t_{f\ experimental}$	1.30 μs
Tail Time, $t_{t\ experimental}$	45.0 μs

Table 2. Value of C_S, C_B, R_D and R_E that was used to obtain the standard impulse

Figure 15 and Figure 16 are the experimental impulse wave shapes that were recorded by using a digital oscilloscope. This is the output produced from the single stage impulse voltage generator. According to the experimental wave shape, the peak of the impulse voltage is 4,200V.

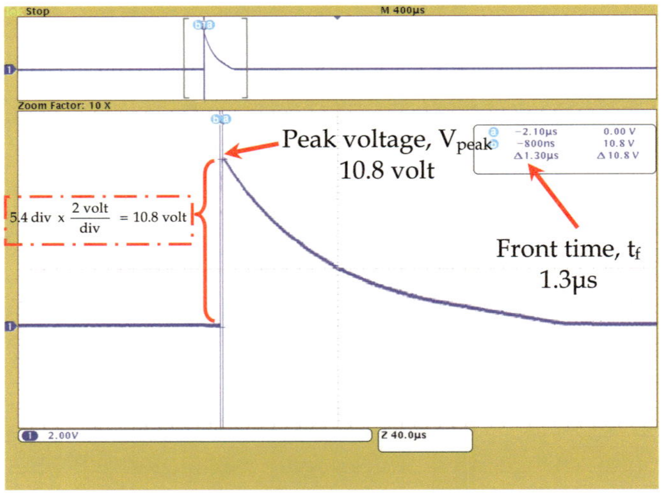

Fig. 15. Front time t_f of the experimental impulse voltage wave shape

In Figure 15, each division represents 2 volts and the peak voltage magnitude is 5.4 divisions. However, this is not the actual value of the peak voltage. To obtain the actual peak voltage, the reading is multiplied with the capacitive voltage divider ratio and finally recorded as an actual peak voltage. The capacitive voltage divider ratio is 392. As a result, the actual peak voltage is 10.8 volt x 392; equal to 4,200V.

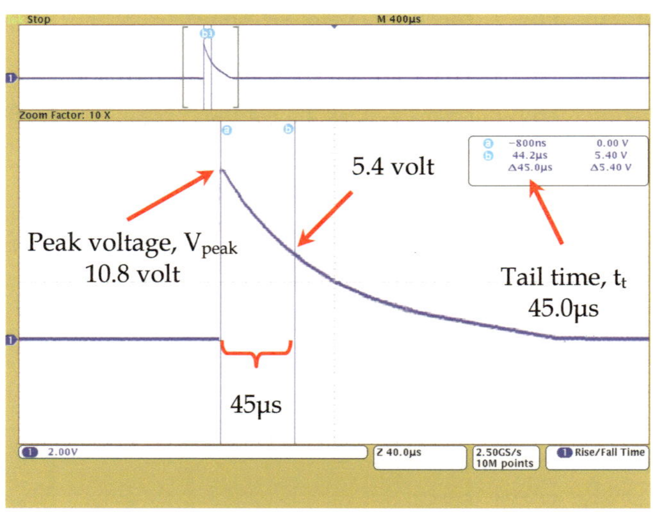

Fig. 16. Tail time t_t of the experimental impulse voltage wave shape

Meanwhile, Figure 16 illustrates the tail time, t_t of the experimental impulse voltage wave shape. It shows that the tail time, t_t is 45.0μs with the voltage at that time is half from the peak voltage of 10.8V, which is 5.4V.Finally, the lightning impulse voltage 4.2kV was generated. It is ready to be used in stage 2 and stage 3 for the laboratory experiment. This generated impulse voltage represents as a mock lightning in order to setup a small scale system for harvesting the lightning stroke.

6.2 Stage 2: energy collected In the capacitor

In this stage, it is intended to discover the electrical characteristics and the time response of the capacitor as energy storage element. In addition, the investigation is performed with varying the capacitance value and also increasing the number of sample capacitors. The characteristics of the sample capacitors have been described previously in section 3. If more than one unit sample capacitor is used in the experiment, the capacitors will be connected in parallel to acquire more capacitance.

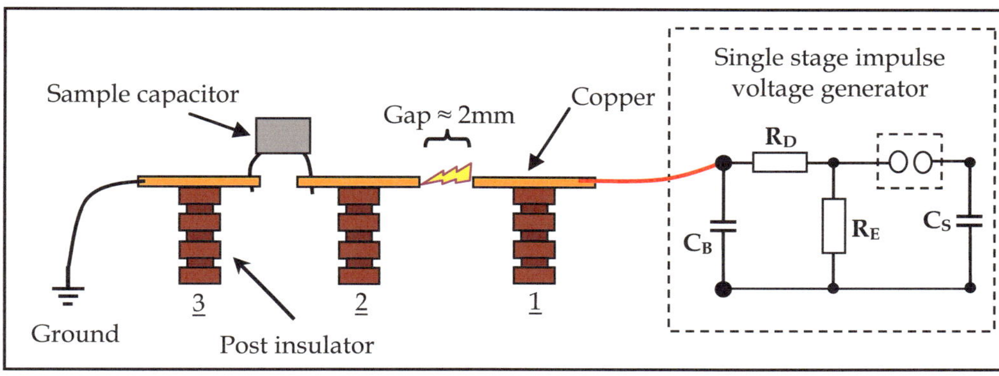

Fig. 17. Connection of sample capacitor with the impulse voltage generator.

In this stage, the process and the equipment involved in Stage 1 remain the same. Except, it has an additional circuit that is connected parallel with the load capacitor C_B. The additional circuit is consisted of three units of post insulator and sample capacitor.

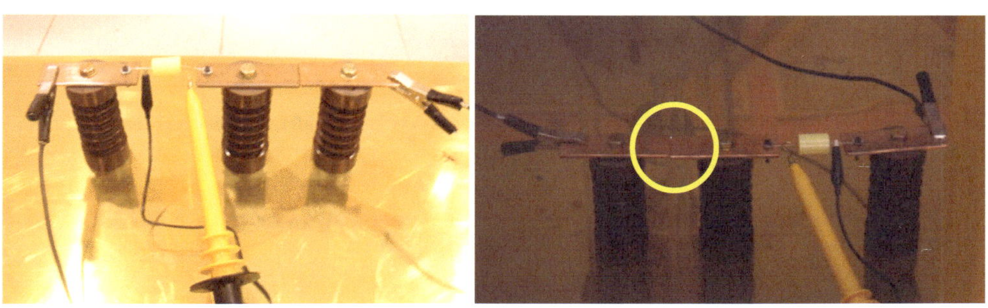

Fig. 18. Flashover occurred between the gap (spark in the yellow circle)

As shown in Figure 17, post insulator no 1 is connected to the load capacitor C_B. Meanwhile, one unit of sample capacitor is connected between post insulator no 2 and no 3. The copper bar at the post insulator no 1 and no 2 is not connected and it has a gap about 2mm. At the time when flashover occurs as shown in Figure 18, the gap will act as a closing switch and simultaneously completed the circuit and charging the sample capacitor. Throughout all these processes, the electrical characteristics and the time response of the sample capacitor can be obtained and measured via HV probe.

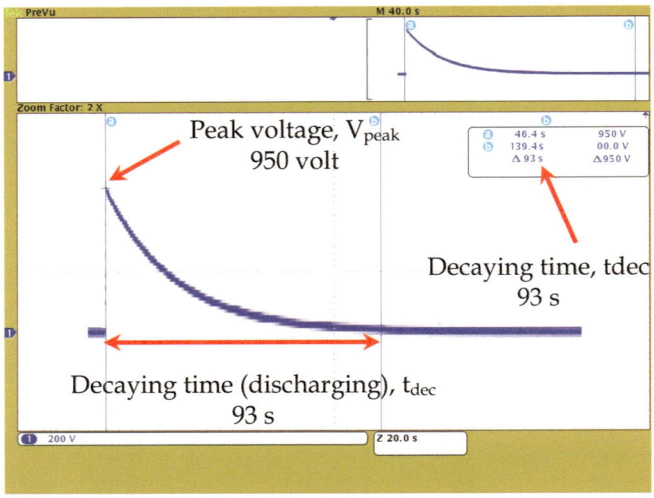

Fig. 19. Decaying time for one unit sample capacitor 0.22 µF

The detailed information regarding the decaying period for one unit capacitor 0.22uF (KNU1910) is illustrated in Figure 19. The decaying time from the peak voltage to the zero voltage is 93 seconds. After the data collection for the peak voltage, rising time and the decaying of one unit capacitor is completed, the process will be continued for two units as well as three units of capacitors connected in parallel.

In order to achieve a clearer picture regarding the voltage efficiency and dissipated voltage for all the sample capacitors used in the laboratory experiment, Figure 20, Figure 21 and Figure 22 can be used as references. In these figures, the pie charts illustrate the relationship between voltage and voltage efficiency for one unit, two units and three units of sample capacitor. It proves that the voltage efficiency has an inverse relation with the dissipated voltage, that is when the voltage efficiency increased, the voltage dissipated will be reduced and vice versa.

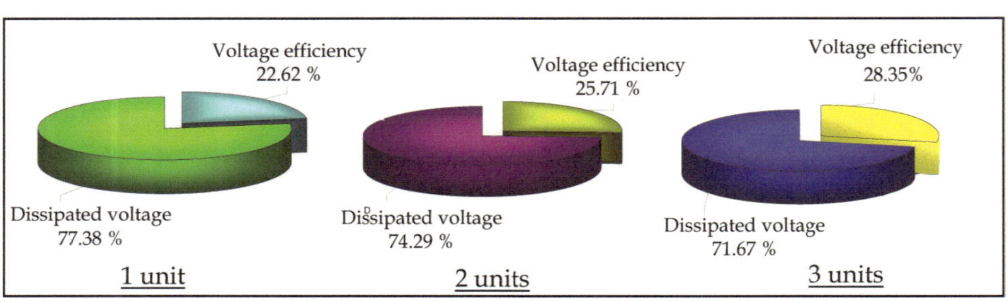

Fig. 20. Pie chart for sample capacitor 0.22µF (KNU 1910)

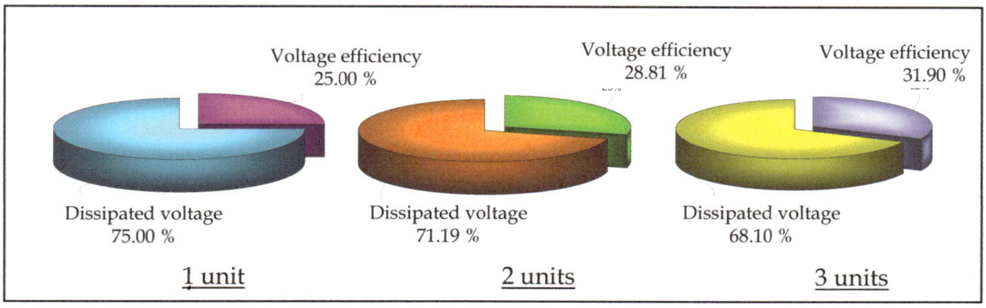

Fig. 21. Pie chart for sample capacitor 0.47μF (CBB20)

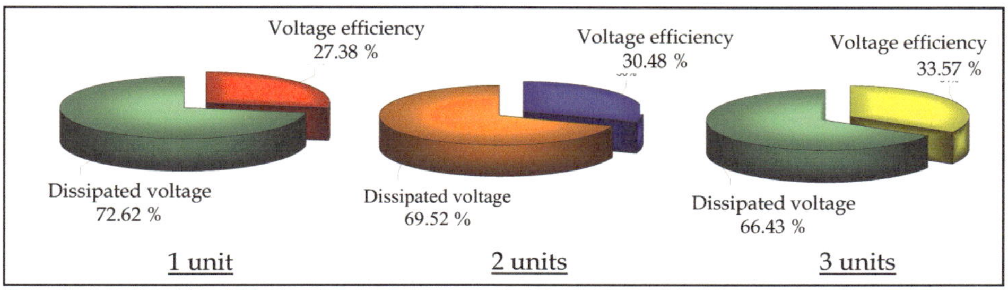

Fig. 22. Pie chart for sample capacitor 0.68μF (CBB20)

6.2.1 Experimental data analysis

In this section, the analysis is focused on the voltage stored V_{stored}, the charging and discharging time of the capacitor as well as the effect of capacitance value to the experimental result. Table 3 show the rough picture for the whole experimental result for stage 2.

Capacitor	0.22μF	0.47 μF	0.68 μF
Charging Voltage, V_s			Fix
Peak Voltage, V_{peak}			Increase
Voltage Efficiency			Increase
Dissipated Voltage			Reduced
Front Time, t_f			Increase
Decaying Time, t_{dec}			Increase

Table 3. Overall experimental data for stage 2

6.2.1.1 Analysis of experimental data for voltage stored V_{stored}

By referring to Figure 20, Figure 21 and Figure 22, the capacitors have the voltage efficiency approximately 22% to 33% and it has a dissipated voltage during the charging process. During the laboratory testing, the spark gap (gap between copper bar at post insulator no 1 and no 2) represents as a mock cloud. It is believed that the dissipated voltage was occurred at the spark gap and this situation is aligned with the theory of lightning energy. Most of the energy, dissipated in a sudden expansion that causes heating and ionization. The other fraction of energy will flow to the ground. According to that statement, that is the reason that the peak voltage V_{peak} in the sample capacitor is lower than the charging voltage V_s.

Besides that, all the capacitors used in the experiment have the capability to capture the incoming impulse voltage. Seemingly, if more sample capacitors are used in the testing, more incoming energy can be attained. However, from the theory and the experimental result, it is confirmed that the biggest energy that could be harvested cannot be equal to the source of lightning energy. And the fact warns that, these efforts still need a reliability energy storage with a very high standard safety system as well as a very strong structure even though the energy can be tapped was not 100% from the source.

6.2.1.2 Analysis of experimental data for time response and capacitance value

In term of time response, the increment of capacitance value has given an effect to the decaying time or discharging time of the sample capacitor. The decaying time is increased in proportion with the increment of the capacitance value. The condition where the capacitor is slowly decaying can be manipulated in order to give a good prospect for future works such as to charge batteries or other electrical appliances.

In the testing, if more that one unit sample capacitor is used, the sample capacitors will be connected in parallel. The purpose is to increase the capacitance value in order to store more charge and voltage. This is the basic approach in determining the raise size of the plate area. The greater the plate area, the greater value of capacitance will be obtained. It is because, the larger plates area will result in more field flux (charge collected on the plates) and at the same time more field force (voltage across the plates) appears. According to the testing result, it was proved that, when greater capacitance value is obtained, the peak voltage V_{peak}, the stored voltage and energy efficiency also will be greater and increased.

6.3 Stage 3: high speed switching and energy stored

Basically, the experiment conducted in Stage 3 is almost similar with the Stage 2. The different is only the additional of high speed switching circuit using IGBT device. The purpose of the IGBT switch is to isolate the capacitor from any connection. In the beginning, the IGBT switch is set to be in a close position. Then, the impulse voltage is applied to charge the sample capacitor. Then, the switch will be opened and simultaneously disconnect the capacitor from any connection in order to fetter the capacitor from discharging.

Before the switching circuit is on, the condition of the IGBT switch is normally open. When the circuit is on, the IGBT switch from normally open will be in close position. According to the program instructions, the PIC controller will start to count until a few microsecond. Then, the IGBT switch from close position will be opened and at the same time, the capacitor is disconnected from any connection. The capacitor is left for about 900 seconds (isolating time) before reconnect to the ground connection. Finally, the voltage across the capacitor as energy storage element is measured once again in order to investigate the remaining voltage. Figure 23 illustrate the location of IGBT switch in the testing circuit.

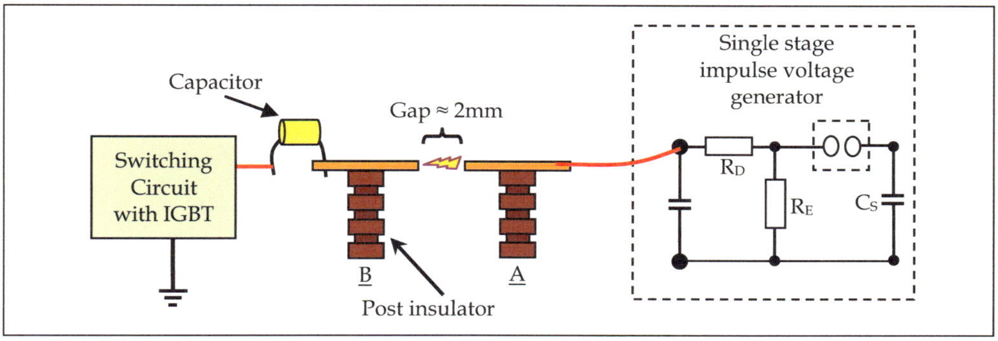

Fig. 23. IGBT switch is placed in series with the sample capacitors

6.3.1 Testing result: characteristics of the sample capacitor

Figure 24 and Figure 25 show the voltage waveform against the time across one unit of sample capacitor 0.22uF (KNU1910). These figures show the step by step explanation concerning the voltage behavior in the sample capacitor. Later, it can be seen that, different types of sample capacitors will give a dissimilar result in terms of the stored voltage, voltage losses and also the time response.

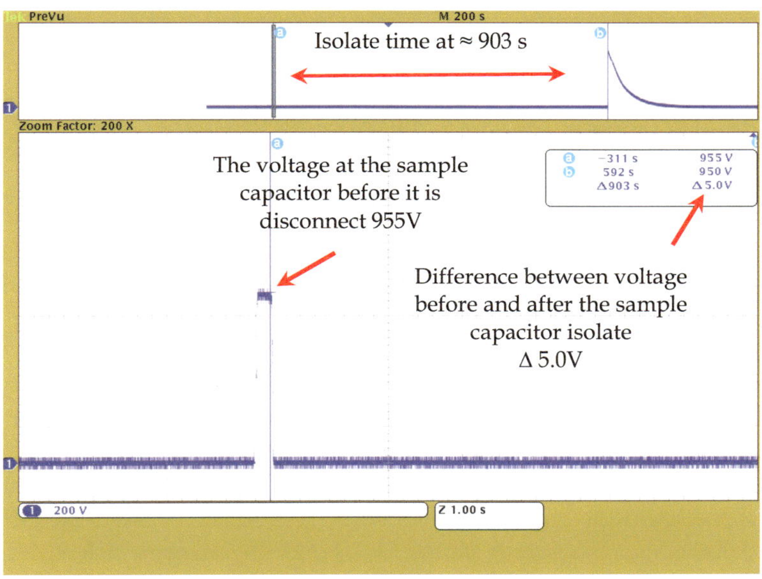

Fig. 24. The capacitor 0.22μF is isolated for 903 second after the switch opens

In Figure 24, the peak voltage in the sample capacitor is 955V. After 300 milliseconds, the switch opens and isolates the capacitor from any connections for approximately 900 seconds. Subsequently, after the isolating period (903 s ≈ 15 min), the voltage in the sample capacitor 0.22μF (KNU1910) is 950V. The difference between the voltages before and after the capacitor was isolated is very small, which is only 5V.

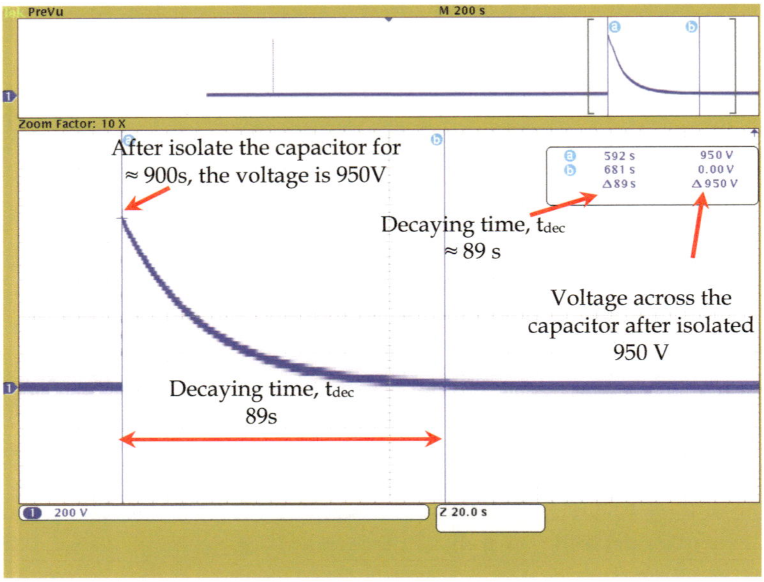

Fig. 25. Voltage waveform of the sample capacitor after isolating period.

Moreover, Figure 25 shows the voltage waveform after the isolating period where the sample capacitor 0.22uF (KNU1910) was reconnected with the ground connection. This figure illustrates that the decaying time is 87 second. The difference in terms of decaying time between stage 2 and stage 3 does not have much difference, where the discharging time for the stage 2 is 91 second and for stage 3 is 87 second.

However, the performance of the capacitor 0.46uF (CBB20) and 0.68uF (CBB20) was not as good as the capacitor 0.22uF (KNU1910). Both capacitors have no capability to retain and sustain the voltage during the isolating period. Besides that, there was a huge difference between the decaying time for stage 2 and stage 3.

6.3.2 Experimental data analysis

Table 4, Table 5 and Table 6 shows the step by step calculation in determining the energy stored and the energy efficiency of the sample capacitor. Thus, determining the decaying voltage function *v(t)* for every sample capacitor will be the basic step for the energy calculations. Once, the decaying voltage function *v(t)* is obtained, it will be used to calculate the average voltage V_{ave}, the energy stored E_{stored} and the energy efficiency $E_{efficiency}$ of the sample capacitor.

Throughout the testing in stage 3, the source voltage V_{source} is fixed at 4.2kV. It means that, the energy supplied by the impulse voltage generator to the sample capacitor is maintained. Herewith is the calculation for the energy supplied from the impulse voltage generator.

$$\text{Energy source, } E_S = 0.5\, C_s V_s^{\,2} = 0.5\,(2500\text{ pF})\,(4200\text{V})^2 = 0.2205\text{ Joule}$$

	0.22 µF (KNU1910)		
No of capacitor use	**1 unit**	**2 unit**	**3 unit**
Time constant , τ	31 second	59 second	92 second
Voltage Decay, v(t)	$950\, e^{\frac{-t}{31}}$	$1070\ e^{\frac{-t}{59}}$	$1180\, e^{\frac{-t}{92}}$
Average Voltage, V_{ave}	$\frac{1}{93}\int_0^{93} 950\ e^{\frac{-t}{31}}\, dt$ = 300.92 V	$\frac{1}{184}\int_0^{184} 1070\ e^{\frac{-t}{59}}\, dt$ = 327.93 V	$\frac{1}{271}\int_0^{271} 1180\, e^{\frac{-t}{92}}\, dt$ = 379.53 V
Energy Stored, E_{stored}	$\frac{1}{2}(0.22\,\mu F)(300.92V)^2$ = 9.96 x 10^{-3} J	$\frac{1}{2}(0.44\,\mu F)(327.93V)^2$ = 23.66 x 10^{-3} J	$\frac{1}{2}(0.66\,\mu F)(379.53\,V)^2$ = 47.53 x 10^{-3} J
Energy Efficiency $= \frac{E_{stored}}{E_{source}}$ x 100%	$= \frac{9.96\ x\,10^{-3}\ J}{220.50\ x\,10^{-3}\,J}\ x\,100\%$ = 4.52 %	$= \frac{23.66\ x\,10^{-3}\ J}{220.50\ x\,10^{-3}\,J}\ x\,100\%$ = 10.73 %	$= \frac{47.53\ x\ 10^{-3}\ J}{220.50\ x\ 10^{-3}\,J}\ x\ 100\ \%$ = 21.55%

Table 4. Energy efficiency and the energy stored of the capacitor 0.22µF

	0.47 µF (CBB20)		
No of capacitor use	**1 unit**	**2 unit**	**3 unit**
Time constant , τ	47 second	90 second	126 second
Voltage Decay, v(t)	$1050\ e^{\frac{-t}{90}}$	$1210\, e^{\frac{-t}{90}}$	$1340\ e^{\frac{-t}{126}}$
Average Voltage, V_{ave}	$\frac{1}{140}\int_0^{140} 1050\ e^{\frac{-t}{47}}\, dt$ = 334.57 V	$\frac{1}{268}\int_0^{268} 1210\ e^{\frac{-t}{90}}\, dt$ = 371.07 V	$\frac{1}{396}\int_0^{396} 1340\, e^{\frac{-t}{126}}\, dt$ = 407.96 V
Energy Stored, E_{stored}	$\frac{1}{2}(0.47\,\mu F)(334.57V)^2$ = 26.31 x 10^{-3} J	$\frac{1}{2}(0.94\,\mu F)(371.07V)^2$ = 64.72 x 10^{-3} J	$\frac{1}{2}(1.41\,\mu F)(407.96V)^2$ = 117.33 x 10^{-3} J
Energy Efficiency $= \frac{E_{stored}}{E_{source}}$ x 100%	$= \frac{26.31\,x\,10^{-3}\ J}{220.50\,x\,10^{-3}\,J}\,x\,100\%$ = 11.93 %	$= \frac{64.72\,x\,10^{-3}\ J}{220.50\,x\,10^{-3}\,J}\,x\,100\%$ = 29.53 %	$= \frac{117.33\,x\,10^{-3}\ J}{220.50\,x\,10^{-3}\,J}\,x\,100\%$ = 53.21 %

Table 5. Energy efficiency and the energy stored of the capacitor 0.47µF

	0.68 μF (CBB20)		
No of capacitor use	**1 unit**	**2 unit**	**3 unit**
Time constant, τ	68 second	121 second	171 second
Voltage Decay, v(t)	$1150\ e^{\frac{-t}{68}}$	$1280\ e^{\frac{-t}{121}}$	$1410\ e^{\frac{-t}{171}}$
Average Voltage, V_{ave}	$\frac{1}{205}\int_0^{205} 1150\ e^{\frac{-t}{68}}\ dt$ = 362.75 V	$\frac{1}{385}\int_0^{385} 1280\ e^{\frac{-t}{121}}\ dt$ = 385.59 V	$\frac{1}{556}\int_0^{556} 1410\ e^{\frac{-t}{171}}\ dt$ = 410.45 V
Energy Stored, E_{stored}	$\frac{1}{2}(0.68\mu F)(362.75V)^2$ $= 44.74 \times 10^{-3}$ J	$\frac{1}{2}(1.36\mu F)(385.59V)^2$ $= 101.10 \times 10^{-3}$ J	$\frac{1}{2}(2.04\mu F)(410.45V)^2$ $= 171.83 \times 10^{-3}$ J
Energy Efficiency $= \frac{E_{stored}}{E_{source}} \times 100\%$	$= \frac{44.74 \times 10^{-3} J}{220.50 \times 10^{-3} J} \times 100\%$ = 20.29 %	$= \frac{101.10 \times 10^{-3} J}{220.50 \times 10^{-3} J} \times 100\%$ = 45.85 %	$= \frac{171.83 \times 10^{-3} J}{220.50 \times 10^{-3} J} \times 100\%$ = 77.93%

Table 6. Energy efficiency and the energy stored of the capacitor 0.68μF

6.3.3 Analysis of experimental data for voltage stored and energy efficiency

As mentioned before, the energy supplied by the lightning impulse voltage generator is maintained at approximately 0.2205 Joule. According to Table 4, Table 5 and Table 6, the measurement data shows that the capacitors used in the testing have the energy efficiency in the range of 4.52% to 77.93%. The data also proved that the energy efficiency increases together with the increment of capacitance value.

Instead of the capacitance value, there are a few factors that will give a significance effect to the voltage stored as well as the energy efficiency. One of the factors is the type of capacitor that was used in the experiments. It knows that, the energy efficiency is proportional with the stored voltage. From the measurement result, only the capacitor 0.22μF (KNU1910) is capable of sustaining the voltage compared to other sample capacitors. In other words, even though the energy efficiency for this capacitor is lower from other sample capacitor, only a small voltage losses occurred. The opposite of that, the sample capacitors 0.47μF (CBB20) and 0.68μF (CBB20) suffer a huge voltage losses during the isolating period. It means that, there will be sometime when the voltage stored in both these capacitor reduces to the zero voltage and causes a zero energy efficiency.

Another factor that is responsible in obtaining the higher energy efficiency is the tapping method. It is because, the tapping system will be the primary role to attract and capture the lighting energy. It means that, this part have to captured as much as the energy supplied and also transfer the energy to the storage devices.

The last factor is the effectiveness of the high speed switching device. The main task of this switching device is to facilitate the capacitors to sustain the charge and the voltage during the separating period. By referring to the measurement data, the high speed switching applied in the testing circuit gave a good performance.

7. Discussion and conclusion

The present work has been directed towards a better understanding of harvesting the lightning energy in a small scale system. In order to achieve the above , the work was focused on the development of a small scale laboratory experiment. It was done by injecting the capacitor, which represents an energy storage element, with lightning impulse voltage as a mock lightning. Prior to this, a computer simulation using Pspice was done in order to develop a reliable testing circuit in the laboratory.

The simulation work provides an understanding on how to generate the single stroke impulse voltage. Besides that, it also verifies that the circuit that uses fast switching is capable of retaining the incoming lightning impulse voltage. In addition, the simulation circuit was considered as a reference for the circuit arrangement in the lab.

In the experimental work, it should be noted that the direct taping approach is suitable for the small scale system. This approach is capable of attracting about a quarter of the incoming lightning impulse voltage and simultaneously charging the sample capacitor. Besides that, the employment of the high speed switching in the testing circuit for the isolation purpose also gives a meaningful outcome. The switching circuit is able to separate the storage device from the circuit connection. However, it has voltage losses that occurred in a few of the sample capacitors during the isolating period. This happened not because of the switch, but due to the losses that came from the sample capacitor itself.

The effects of single stroke impulse voltage 1.2/50μs, 4.2kV on the electrical characteristics of the sample capacitor have been determined from the experimental work. There were significant effects on the electrical characteristics such as peak voltage, stored voltage, charging time, discharging time as well as the energy efficiency of the sample capacitor for a single stroke impulse voltage.

Clearly, the stored voltage and the energy efficiency of the sample capacitor is extremely increased in proportion with the increment of the capacitance value. By referring to the result and the analysis that have been achieved, the capacitor 0.22μF(KNU1910) has the capability to store energy compared to the other samples which are the 0.47μF(CBB20) capacitor, and the 0.68μF(CBB20) capacitor.

Moreover, the experimental results also confirmed that the small scale system constructed which consists of the tapping system and the high speed switching are successfully implemented. It gives significant effects on the electrical characteristics of the sample capacitor subjected to a single stroke impulse voltage of 1.2/50μs, 4.2kV. In particular, the electrical characteristics such as the voltage stored, the response time and the energy efficiency of the sample capacitor is proportional to the value of capacitance and the type of capacitor that have been used.

It proves that, the outcome can be improved as long as the energy storage devices, the fast switching circuit and the tapping system are correctly chosen. Generally, the way of testing that was conducted in this project may not be accurate enough to represent the model system for harvesting the lightning energy. However, from the results obtained, it is found to be acceptable.

The race to develop the technology in harvesting the lightning energy among the scientist and engineers is ongoing. Since, there are no mature studies of this technology and the scientific literature is not easily found, therefore this project is carried out to be as a sensible part of the research studies. In addition, the experimental result gives a meaningful indicator regarding the effort to create a small scale system for harvesting the lightning impulse energy. It should be noted that the final system of this project would provide an understanding of the system principle and furthermore give a significant contribution for further research Finally, it is hoped that this project will be able to contribute towards making the process of harvesting lightning energy a reality.

8. References

Abidin, H.Z. & Ibrahim, R. (2003). Thunderstorm Day and Ground Flash Density In Malaysia. *Proceedings of National Power and Energy Conference, (PECon 2003),* pp. 217-219, ISBN 0-7803-8208-0, Bangi, Kuala Lumpur, Malaysia, December 15-16, 2003

Karthick S. & Jason G., (2006). Lightning as Atmospheric Electricity. *Canadian Conference on Electrical and Computer Engineering, (CCECE '06)*, pp. 2258-2261, ISBN 1-4244-0038-4, Ottawa, Canada, May 7-10, 2006

Rebeiro, P.F.; Johnson, B.K. ; Crow, M.L. ; Arsov, A. & Liu, Y. ; (2001). Energy Storage System for Advanced Power Application, *Proceedings of the IEEE*, Vol.89, Issue.12, (December 2001), pp. 1744-1756, ISSN 0018-9219

Uman, M.A. (1994). Natural Lightning. *IEEE Transactions on Industry Applications,* Vol.30, Issue.3, (June 1994), pp. 785-790, ISSN 0093-9994

Yao, Y.Y.; Zhang, D.L. & Xu, D.G. (2006). A Study of Supercapacitor Parameters and Characteristics. *International Conference on Power System Technology, (PowerCon 2006),* pp. 1-4, ISBN 1-4244-0110-0, Chongqing, China, October 22-26, 2006

Kumar, B. (1992). Magnetic Energy Storage Devices for Small Scale Applications. *IEEE Aerospace and Electronic Systems Magazine,* Vol.7, Issue.11, (November 1992), pp. 12-17, ISSN 0885-8985

Chowdhuri, P. (2001). Parameter of Lightning Stroke and Their Effect on Power System. *IEEE/PES Transmission and Distribution Conference and Exposition,* pp. 1047-1051, ISBN 0-7803-7285-9, Atlanta, USA, October 28-November 02, 2001

Hefner, A.R. Jr. & Diebolt, D.M. (1994). An Experimentally Verified IGBT Model Implemented in the Saber Circuit Simulator. *IEEE Transactions on Power Electronics,* Vol.9, Issue.5, (September 1994), pp. 532-542, ISSN 0085-8993

Microchip Technology Inc. (2001). PIC16F84A Data Sheet, Available from http://ww1.microchip.com/downloads/en/devicedoc/35007b.pdf

Ibrahim, D. (2008). Advanced PIC Microcontroller Projects in C : From USB to RTOS with the PIC18F Series, In: C Programming Language, 119-165, ISBN 978-0-7506-811-2, Oxford UK, Newnes

Yao, Y. Y. ; Zhang, D. L. & Xu, D. G. (2006) A Study of Supercapacitor Parameters and Characteristics. *Power System Technology, 2006. PowerCon 2006. International Conference on.* pp. 1-4, ISBN 1424401100, Chongqing, China, October 22-26, 2006.

Wei, L. & JOOS, G. (2008) A Power Electronic Interface for a Battery Supercapacitor Hybrid Energy Storage System for Wind Applications. *Power Electronics Specialists Conference, 2008. PESC 2008. IEEE* , pp. 1762-1768, ISBN 978-1-4244-1667-7, Rhodes, Greece, June 15-19, 2008.

Pagano, M. & Piegari, L. (2004) Supercapacitor Flywheel for High Power Electrochemical Sources. *Power Electronics Specialists Conference, 2004. PESC 04. 2004 IEEE 35th Annual.* pp. 718-723 (Vol.1), ISBN 0-7803-8399-0, June 20-25, 2004..

Swett, D. W. & Blanche, J. G. IV. (2005) Flywheel Charging Module for Energy Storage Used in Electromagnetic Aircraft Launch System. *12th Symposium on Electromagnetic Launch technology, IEEE Transactions on,* Vol.41, pp. 525-528, ISBN 0-7803-8290-0, Snowbird, Utah, May 25-28, 2005.

6

Fabrication and Characterization of MicroPCMs

Jun-Feng Su
Institute of Materials Science and Chemical Engineering,
Tianjin University of Commerce,
P. R. China

1. Introduction

The increasing gap between the demand and supply of energy is an essential problem affecting the globe climate and economy. Energy efficiency improvement is one approach to reduce the mismatch between supply and demand. Energy conservation can be achieved through increased efficient energy use, in conjunction with decreased energy consumption and/or reduced consumption from conventional energy sources. Thermal energy conservation is easy to realize by storing thermal energy as latent heat in which energy is stored when a substance changes from one phase to another by either melting or freezing. The temperature of the substance remains constant during phase change. Phase change materials (PCMs) include organic (e. g. paraffin), inorganic (e. g. salt hydrates) and salt eutectics (e. g. $CaCl_2 \cdot MgCl_2 \cdot H_2O$) have been widely studied and applied in energy saving field. A good design for the PCMs requires that their phase-change processes, especially the melting and solidification processes perform in a container. It is also proposed that the heat transfer rate between PCM and source/sink can be increased by using microencapsulated PCMs (microPCMs). The reason is that the small microPCMs provide larger heat transfer area per unit volume and thus higher heat transfer rate. Moreover, the microPCMs bring in more advantages like less reaction of PCM with matrix material, ability to withstand volume change during phase change, etc.

To date, microPCMs have two application approaches: in emulsion and in solid matrix. Emulsion containing microPCMs can be used as thermal exchange medium to enhance the energy efficiency of thermal exchanger. MicroPCMs are widely applied in solid matrix as smart thermal-regulation composites, such as fibers, construction materials, blood temperature-controlling materials and anti-icing coats.

This chapter summarized our published works on fabrication and characterization of microPCMs, including shell compactability enhancement, thermal and chemical stability improvement, heat conductivity accelerating and phase change behavior controlling. Styrene-maleic anhydride (SMA) copolymer solid was used as a nonionic dispersant. These microPCMs have been applied in energy saving fields.

In addition, the strength of the bond at the microcapsule/matrix interface controls the fatigue life of the composites significantly. So by controlling the stress-strain response and ductility of the interface region, it is possible to control overall behavior of the composites. We applied a methanol-modified process to enhance the banding of microPCMs and the polymer matrix. The interface morphologies were investigated systemically to understand the effects of the average diameters, contents and core/shell ratios of microPCMs on the interface stability behaviors.

2. Fabrication and characterization MF-shell microPCMs

Melamine-formaldehyde (MF) resin [1, 2], urea-formaldehyde (UF) resin [3, 4] and polyurethane (PU) [5-7] were usually selected as microcapsule shell materials for the PCMs protection. In practical usages of microPCMs, the volume of PCM in shells will change obviously during absorbing and releasing thermal energy. The volume alternant changes original bring the liquefied PCMs leaking from the microcapsules. And the breakage of the shell will happen based on the mismatch of thermal expansion of the core and shell materials at high temperatures [5]. Thus, it is necessary to keep the shell stability and compact for a long-life with less cracks and lower permeability.

MF resin has been successfully applied as shell material of microcapsules in fields of carbonless copying paper, functional textiles, liquid crystals, adhesives, insecticides and cosmetics [8]. Furthermore, literatures have showed that MF has been applied for microPCMs encapsulating various solid-liquid PCMs with the size range of about 50 nm to 2 mm. Generally, MF resin is adsorbed and cured on surfaces of core particles though an in-suit polymerization with the help of a polymeric surfactant.

2.1 Materials and fabrication method

Styrene maleic anhydride copolymer solid (SMA, Scripset®520, Hercules, USA) was used as a dispersant. A small percentage of the anhydride groups have been established with a low molecular weight alcohol and it is fine, off-white, free flowing power with a faint, aromatic odor. Nonionic surfactant, NP-10 [poly (ethylene glycol) nonylphenyl] getting from Sigma Chemical was used as an emulsifier. The pre-polymer of melamine-formaldehyde (MF) resin which solid content was 50±2wt. %, was purchased from Shanghai Hongqi. Chem. Company (Shanghai, China). The *n*-octadecane purchased from Tianjin Fine Chemical Company (Tianjin, China) was encapsulated as the core material. All other chemical reagents were analytical purity and supplied by Tianjin Kermel Chemical Reagent Development Center (Tianjin, China).

Microcapsulation by coacervation proceeds along three main steps:

1. Phase separation of the coating polymer solution. SMA (10. 0 g) and NP-10 (0. 2 g) were added to 100 ml water at 50 °C and allowed stir for 20 min. And then a solution of NaOH (10%) was added dropwise adjusting its pH value to 4-5. The above surfactant solution and *n*-octadecane (32 g) were emulsified mechanically under a vigorous stirring rare of 3000 r·min^{-1} for 10 min using a QSL high-speed disperse-machine (Shanghai Hongtai Ltd, Shanghai, China.).
2. Adsorption of the coacervation around the core particles. The encapsulation was carried out in a 500 ml three-neck round-bottomed flask equipped with a condensator and a tetrafluoroethylene mechanical stirrer. The above emulsition was transferred in the bottle, which was dipped in steady temperature flume. Half of MF pre-polymer (16 g) was added dropwise with a stirring speed of 1500 r·min^{-1}. After 1h, the temperature was increased to 60 °C with a rate of 2 °C·min^{-1}. Then another half of pre-polymer (16 g) was dropped in bottle at the same dropping rate.

3. Solidification of the microparticles. Then the temperature was increased to 75 °C. After polymerization for 1h, the temperature was decreased slowly at a rate of 2 $^{\circ}C \cdot min^{-1}$ to atmospheric temperature.

At last, the resultant microcapsules were filtered and washed with pure water and dried in a vacuum oven. In addition, we could control the average diameter of microcapsules by stirring speed. Also, the OSC microPCMs was fabricated in this work according to the above process by adding the same amount (32 g) prepolymer shell material in one step.

2.2 Mechanism of TSC

Hydrolyzed SMA is a kind of polymeric surfactant that is soluble in water but sufficiently amphiphilic to be absorbed by surfaces and interfaces, particularly by dispersed solid or liquid phases [9]. In addition, hydrolyzed SMA plays two important roles during the formation of microcapsules: dispersant and anionic polyelectrolyte [10, 11].

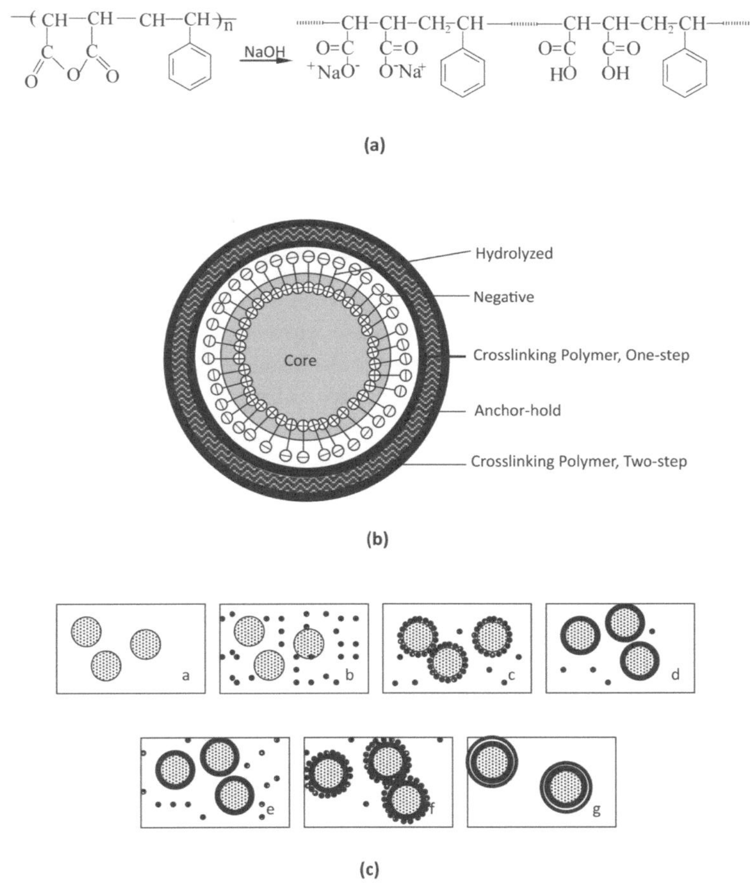

Fig. 1. Sketch mechanism of the fabrication process to TSC microcapsules: (a) Chemical structures of styrene maleic anhydride (SMA) alternating copolymer and hydrolysis polymer, (b) the structure of a TSC microcapsule, (c) the process of fabrication microcapsules by TSC.

Fig. 1 illustrates the complex TSC process for forming the microcapsules. Fig. 1(a) depicts the chemical structure of styrene-maleic anhydride (SMA) and hydrolysis polymer. As a kind of polymer dispersant, SMA molecules will be hydrolysis by NaOH and then the -COO group insert and directional arrange on oil droplet surface. In Fig. 1(b), hydrophilic groups of carboxyl arrange alternatively along the backbone chains of SMA molecules. When hydrolyzed SMA molecules are adsorbed at the interfaces of oil droplets, it is easy for the molecules to have such directional arrangement with hydrophobic groups oriented into oil droplet and hydrophilic groups out of oil droplet. This kind of molecular arrangement brings results in a relatively strong electron negative field on the surface of oil particles. Anionic polyelectrolyte hydrolyzed SMA has anionic carboxyl groups that can interact with positively charged MF-prepolymer below the ζ potential. The MF-prepolymer was affinities on these particulates by static. Then, the reaction of microencapsulation took place under acid and heat effect on the surface of oil particles of emulsion, which formed membrane of capsule in such a way. Fig. 1(c) shows the formation process of the TSC by another prepolymer (MF) addition at a slowly rate. Also, under the effect of heat the second coacervation will cross-linked to form another part of shells. That can be concluded that the twice-addition prepolymer and twice increasing-decreasing temperature courses lead to compact shells.

2.3 MicroPCMs in emulsion

In order to bring the coacervation process to a clear understanding, optical microphotographs of microcapsules were taken to illuminate the details. Fig. 2 (a-b) show optical microphotographs of core material dispersed by hydrolyzed SMA after 5 min and 10 min at room temperature. At the beginning 5 min, the hye size distribution of drolyzed SMA dispersed the core material into particles. However, these particles had not been separated each other directly due to the molecule linkage of the hydrolyzed SMA. Being emulsified for 10 min, the core particles were separated through the regulation of hydrolyzed SMA molecules.

In previous study [12], we have drawn a conclusion that the average diameter of 1μm-5μm, fabricated under stirring speed of 3000 $r{\cdot}min^{-1}$, is the perfect range of size insuring both of narrow size distribution and enough rigidity. Based on the experiment, MF prepolymer (32 g) was dropped into the above emulsion with a stirring speed of 3000 $r{\cdot}min^{-1}$. The prepolymer cured on core particles in 60 min by increasing the temperature to 60 °C slowly at a rate of 2 $°C{\cdot}min^{-1}$. Fig. 2(c) shows the optical microphotographs of microcapsules with fleecy or pinpoint morphologies. Imaginably, these incompact structures may lead cracking or releasing of core material such as Fig. 2 (d) showing.

Compared with OSC microcapsules, Fig. 2(e) shows the optical microphotograph of TSC ones. The prepolymer covered on particles without cracks and thparticles is uniform with global and distinct shape. Moreover, there is nearly no conglutination between each microcapsule in very stability solution system.

2.4 SEM morphologies of shells

Fig. 3(a) shows SEM morphology of dried OSC microcapsules with the size of 1-5 μm. These microcapsules have a rough morphology and a little polymer occupies the interspaces of microcapsules. It can be contributed to the unencapsulated core material and the uncovered shell material. Especially, the surfaces have many protrusions, which may be occurred by

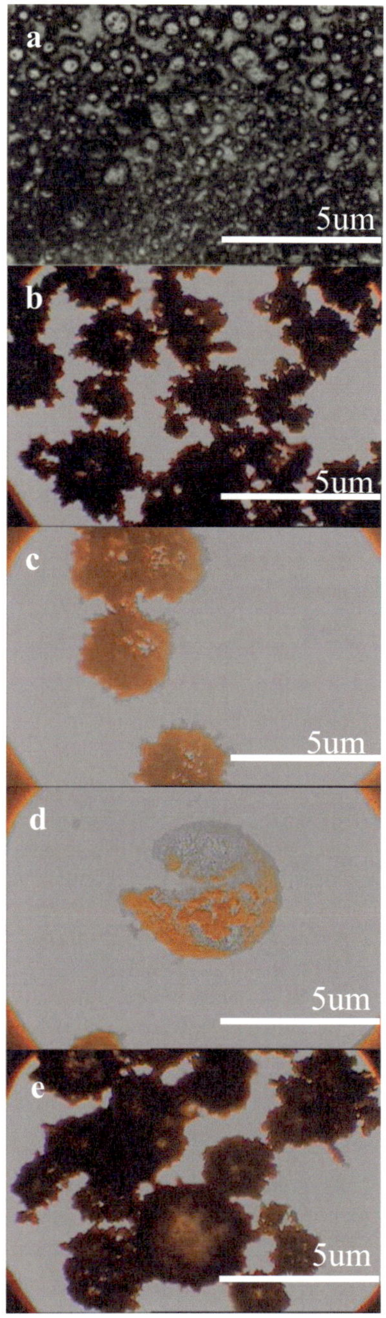

Fig. 2. Optical microphotographs of microcapsules: (a) core material dispersed by hydrolyzed SMA in water for 5 min; (b) core material dispersed by hydrolyzed SMA in water for 10 min; (c) microcapsules by OSC; (d) a crack OSC microcapsule (e) microcapsules by TSC.

not completely cross-linking or high-speed chemical reaction. In images Fig. 3(b-c) (×10000), it is clearly that the surfaces of microcapsules seem to be coarse and porous. Interestingly, there is a depressed center on a microcapsule reflecting the lower rigidity of shell in Fig. 3(d). We may draw a conclusion from these defects that OSC could not achieve a perfect coacervation on cores slowly and tightly in enough time under condition of mass shell material. Fig. 3(e) reflects the surface morphologies of piled microcapsules fabricated by TSC. It appears that nearly all these smooth microcapsules have a diameter about 2 μm with regularly globe shape. Moreover, not only there is nearly no concavo-convex and wrinkle in bedded in shell surfaces, but also little polymer is pilling between piled microcapsules. From these results, it could be imaged that the method of twice-dropping prepolymer has decreased the roughage through molecules regulation of the second-dropped polymer. At the same time, the flaws may be decreased by padding the second cross-linking on the previous coacervation.

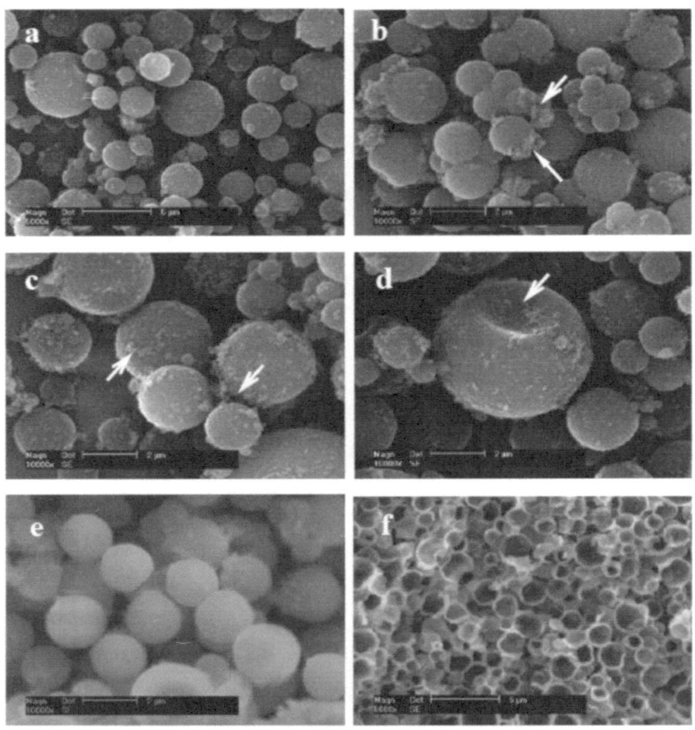

Fig. 3. SEM photographs of microcapsules dried in a vacuum oven at 40 °C for 24 h, (a) (b) surface morphology of piled OSC microcapsules, (c) (d) the rough surface morphology of OSC microcapsules, (e) surface morphology of piled TSC microcapsules, (f) cross section of TSC microcapsules.

2.5 Density and thickness of shells

Density and thickness of shells are useful parameters reflecting the compactness of shells. Originally, the thickness data can be measured from cross-section of SEM images as shown in Fig. 3(f). In this study, a series of microcapsules were fabricated with various weight ratios of

core (32 g) and shell materials from 1. 0 to 2. 0 (core: shell) by two kinds of coacervation methods to evaluate encapsulation effect. All the microcapsules had the same preparation materials and environmental conditions. At least of fifty shells for each sample were measured and the average data was recorded by a MiVnt Image analyze system automatically.
From the data in Fig. 4(a), it shows that the thickness of OSC and TSC shells are both decreased with the increasing of value of weight ratios. And at the same weight ratios, all the thickness of TSC shell is less than that of OSC. This may be attributed to two aspects. Firstly, the TSC may decrease the structure defects, such as holes and caves. Secondly, this method of TSC allows the prepolymer to regulate their molecules on core material with enough reaction time for higher cross-linking density.
The data of density affected by various weight ratios are shown in Fig. 4(b). At the same weight ratio of 1. 0, the densities of OSC and TSC are 1. 75 g/cm^3 and 1. 67g/cm^3, respectively. With the increasing of weight ratio, both densities of OSC and TSC shells are decreased. And at the weight ratio point of 2. 0, both kinks of microcapsules have nearly shell density of 1. 58 g/cm^3.

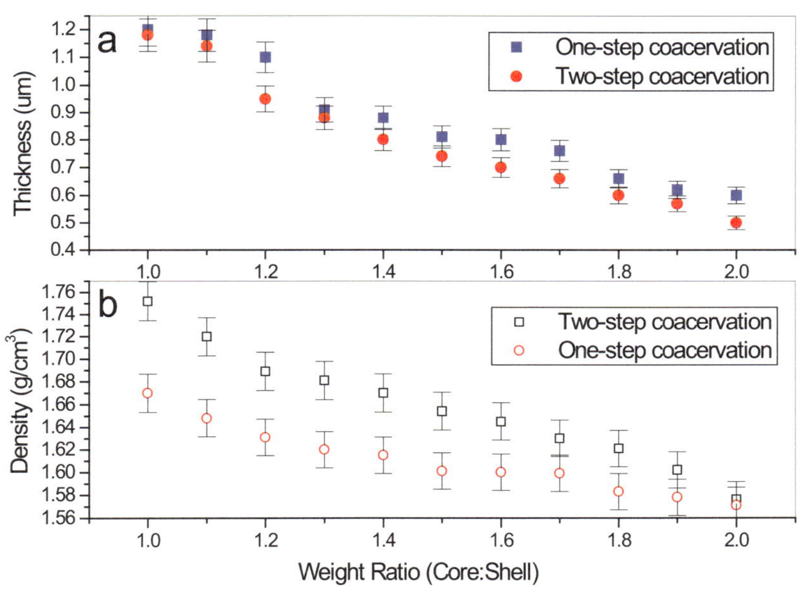

Fig. 4. The Data of shell thickness and density fabricated by OSC and TSC with carious core and shell weight ratio.

2.6 Shell stability in water

Usually, we may simply evaluate the compactability and stability of shell by observing the morphologic changes of shell in water during different times. This method will be helpful to understand the structures of shells. In this study, dried microcapsules with diameter of 1μm were applied for convenience to indicate the endurance of shells in water by means of TEM. Fig. 5 (a-b) show the dried OSC microcapsules after immersed in water for 60 min and 120 min respectively. It is found clearly that the microcapsules are not spherical in shape because of absorbing water. And after 150 min, the polymer shell peeled off from the core particles as shown in Fig. 5(c). The peeled shells are in spreading state and the core material has been separated without covering. We show particular interest to Fig. 5(d) depicting the compact TSC microcapsules after immersed in water for 150 min. Not only the capsules still keep the original global sphere and size nearly without peeling and expansion, but also the core material is safely protected avoiding releasing.

By referring back to Fig. 2 and Fig. 3, these above results are understandable in view of molecular structure of shells. When hydrolyzed SMA molecules were absorbed at the interfaces of the oil particles, the molecules had directional arrangement with hydrophobic groups oriented into oil droplet. It was easy for water to permeate in the shells through cracks and capillary. The force of interface adhesion between core and shell would reduce due to the static electricity force decreased by the effect of water molecular. Then, OSC microcapsules were swelled and destroyed with the time prolongation. Comparatively, shell of TSC microPCMs had less cracks and capillary, which also decreased the effect of water molecular.

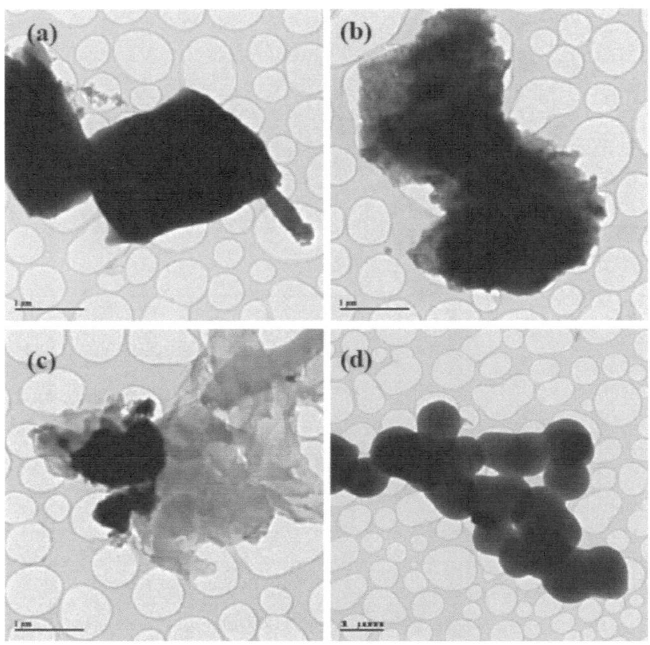

Fig. 5. TEM photographs of microcapsules in water, (a) OSC microcapsules in water for 30 min, (b) OSC microcapsules in water for 60 min, (c) OSC microcapsules in water for 90 min, (d) TSC microcapsules in water for 90 min.

2.7 Thermal stability of microcapsules

Fig. 6 shows thermogravimetric (TGA/DTG) curves of microcapsules prepared during various coacervation times. The blue and red lines are curves of TGA and DTG curves. Both axis of left and right are residual weight (%) of TGA curves and lose weight ratio of DTG curves. The microcapsules were decomposited with increasing temperature according to presenting residual weight (%). The curves may reflect thermal stability and structure of polymeric shell. Fig. 6 (a) shows that pure n-octadecane lost its weight at the beginning temperature of 137 °C and lost weight completely at 207 °C. In order to know the compactness of encapsulation effect, we compare TGA curves of the OSC microcapsules fabricated by prepolymer dropping rates of 0. 5 ml·min^{-1} (Fig. 6b) and 1. 0 ml·min^{-1} (Fig. 6c). Contrastively, both kinds of OSC microcapsules containing *n*-octadecane lost weight rapidly at the temperature of 100 °C. The lost weight in the beginning may be some water and other molecule ingredients. And then the weight decreased sharply from 160 to 350 °C because of the cracking of shells. The weight-loss speed of microcapsules was distinctly less than that of pure n-octadecane. Though it indicates that the OSC method can encapsulate the core material, we can draw a conclusion that the lower dropping speed of shell material has little effect on improving the shell compactness.

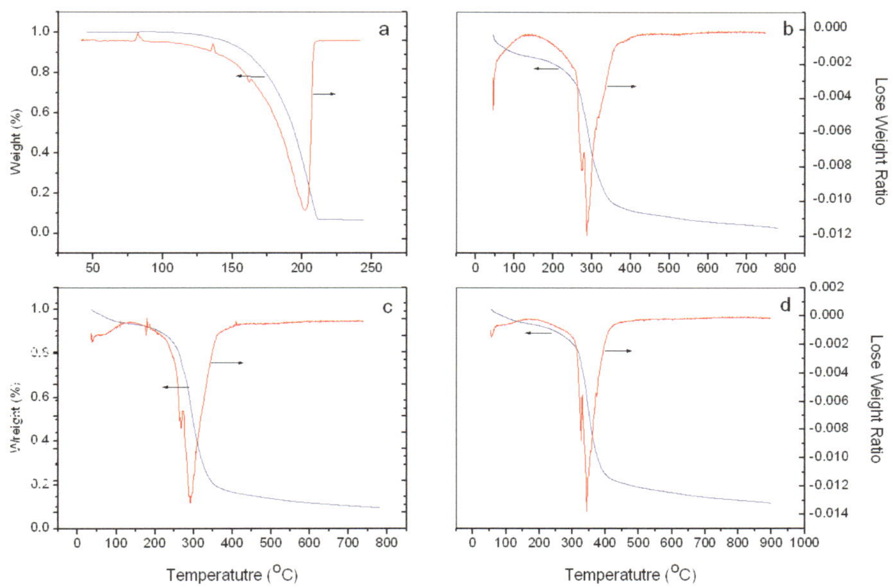

Fig. 6. TGA and DTG curves for (a) pure core material, (b) OSC microcapsules, (c) TSC microcapsules, made by 1. 0 ml· min^{-1} dropping rate of the second adding prepolymer, (d) TSC microcapsules, made by 0. 5 ml· min^{-1} dropping rate of the second adding prepolymer.

Fig. 6(d) shows TGA curve of the expected TSC microcapsules fabricated with dropping shell material speed of 0. 5 ml·min^{-1}. It losses weight between 200 °C to 400 °C. We also find that the beginning temperature of of TSC is higher than that of OSC, which proves that the method of TSC has a better effect on protecting of core material.

2.8 Permeability of microPCMs

Release rate depends largely on the polymer structure of shells, which in turn is influenced by the conditions employed in preparation. A typical SEM morphology of microcapsules after releasing is shown in Fig. 7. The arrows sign a broken shell-structure formed during releasing process. Moreover, the weight ratio of core and shell will greatly affect the permeability. For example, we have discussed that penetrability of microPCMs with average diameter 5 μm is lower than that of 1. 5 μm. And their penetrability with mass ratio of 1:1 (core:shell) is lower than that of 3:1 and 5:1 under the same core material emulsion speed

[12]. Considering the above results, only one kind of microPCMs fabricated with mass ratio of 1:1 and diameter of 5 μm were selected in this study to simplify the relationship between the fabrication process and the permeability. In addition, there different shell-structures were fabricated by controlling of pre-polymer dropping speeds of 0. 5, 1. 0 and 2. 0 ml·min^{-1}, respectively.

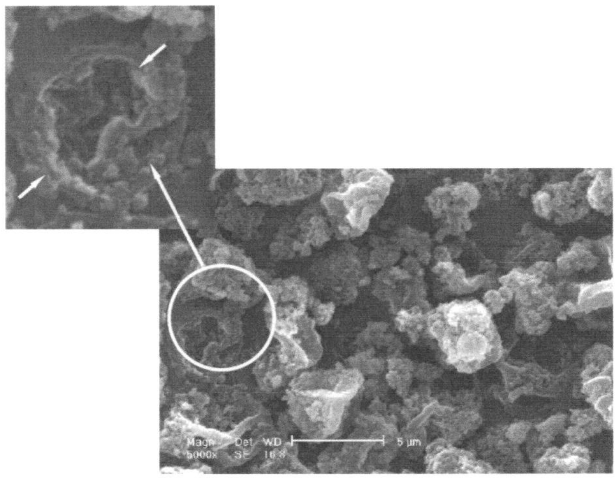

Fig. 7. A typical SEM morphology of broken microcapsules during releasing.

Fig. 8 shows curves of relationship between the percentage residual weight (wt. %) of core material in microcapsules and the time course of the transmittance. Five systems, coded as A-F, were measured in extraction solvent. The systems correspond to the following conditions of coacervation method and prepolymer dropping speed: A(■) OSC, 2. 0 ml·min^{-1}; B(●) OSC, 1. 0 ml·min^{-1}; C(▲) OSC, 0. 5 ml·min^{-1}; D(□) TSC, 2. 0 ml·min^{-1}; E(○) TSC, 1. 0 ml·min^{-1} and F(Δ) TSC, 0. 5 ml·min^{-1}, respectively. Although the resultant microcapsules had been filtered and washed with water, there was a little un-encapsulated *n*-octadecane and other fabrication materials attaching on shells. Therefore, the initial transmittances in media are 98%, 98%, 98%, 97% and 99%, which were nearly equality values. The rate of permeation of OSC microcapsules shell decreases in the order of system A, B and C. It can be concluded that the shell pre-polymer dropping rates affect the penetrability directly. The total PCM permeated time from shells is just in 45 min of system A, comparing to 90 min of system B and 125 min of system C. Especially, the release profile of system A is directly just in one step, but systems of B and C have multi-steps. Comparatively, the rates of permeation of TSC microcapsules decrease in the order of system D, E and F with multi-steps. At the same time, the rate of permeation of TSC is all less than that of OSC even at same dropping rate. Moreover, the data in systems of D, E and F nearly do not change in the beginning 50 min, and system of F begins to change rapidly at the time of 90 min.

The reason of above-mentioned phenomena may be attributed to two aspects. One is that the pre-polymer concentration in solution determined by the dropping rate, will affects the shells formation speed. Under rapider dropping rate, the shell will be formed faster with enough shell material, which brings disfigurements, such as micro-crack, micro-cavity and capillary on shells. These disfigurements will lead the core material to penetrate with low resistance. Contrarily, shells will form slowly under the situation of low pre-polymer concentration in solution. The pre-polymer molecules will adhere on core particles compactly. On the other hand, the channels and disfigurements of penetration in shells were decreased by the twice-dropping fabrication method. The core material penetrates the TSC shells need longer distance and more time. Thus, system of F presents the best resistance of core material.

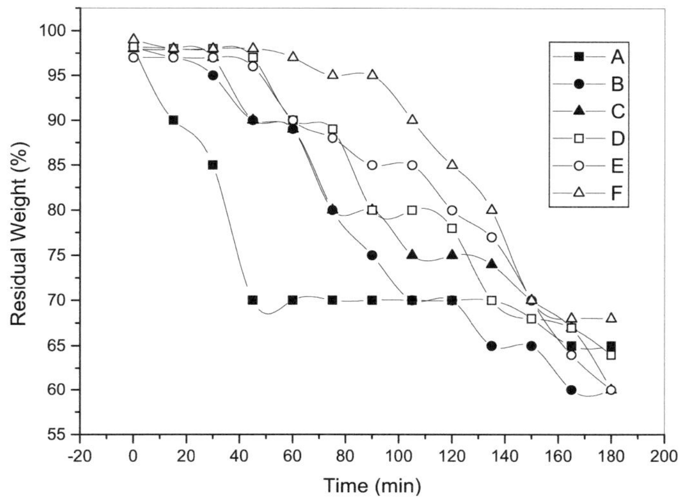

Fig. 8. Curves of the percentage residual weight (%) of core material in microcapsules in extraction solvent. The systems correspond to the conditions of coacervation methods and prepolymer dropping speed: A (■) OSC, 2. 0 ml·min-1; B(●) OSC, 1. 0 ml·min-1; C(▲) OSC, 0. 5 ml·min-1; D(□) TSC, 2. 0 ml·min-1; E(○) TSC, 1. 0 ml·min-1 and F(△) TSC, 0. 5 ml·min-1.

2.9 Permeability coefficient of the shell (k)

1ml of pure water suspension with the percentage weight of dried microPCMs being 10% was spread homogeneously with a wire bar on a polyethylene terephthalate (PET) film. Poly (vinyl alcohol) (PVA) played a role of a binder between the PET film and the microcapsules. The film was cut into squares of 1cm×1cm. The squire films were soaked in to glass vessels containing 30 ml of ethyl alcohol with a density of 0. 97 g ml^{-1}. The glass vessels were sealed avoiding volatility and with light stirring at a room temperature. The penetration property of different microcapsules was evaluated by an UV/visible spectrophotometry in ethyl alcohol. From changes of transmittance of light, we got the core material penetrating time and the residual weight (%) of core material. In this process, the optical density (OD) of the dispersing medium was measured and converted into the concentration of *n*-octadecane using a calibration curve,

$$\text{Residual weight (\%)} = \frac{OD_0 - OD_t}{OD_0} \times 100\% \quad (1)$$

where OD_0 is the optical density of all encapsulated core material in ethyl alcohol, OD_t is the optical density of encapsulated core material in ethyl alcohol at a permeation time (t).

Generally, the kinetic theory of penetrability is determined by the structure and characteristic of shell. Fundamentally, by comparing the permeability coefficient of the shells (k), the minimum can be chosen fabricated by different preparation processes. Most of release properties observed in experiments have been analyzed by kinetic theories based on Fick's Law with an assumption that the release rate is proportional to the concentration gradient of solutes [13, 14]. As each microcapsules system has different structure shell and core material, it is considered to be complex system [15] and required to employ different strict treatment to understand the release mechanism and to characterize such complex system clearly. In this study, we expect to get the compatible kinetic theory applied to the transfer of n-octadecane out of the TSC shell though the release curves of microcapsules in 20 wt. % ethanol.

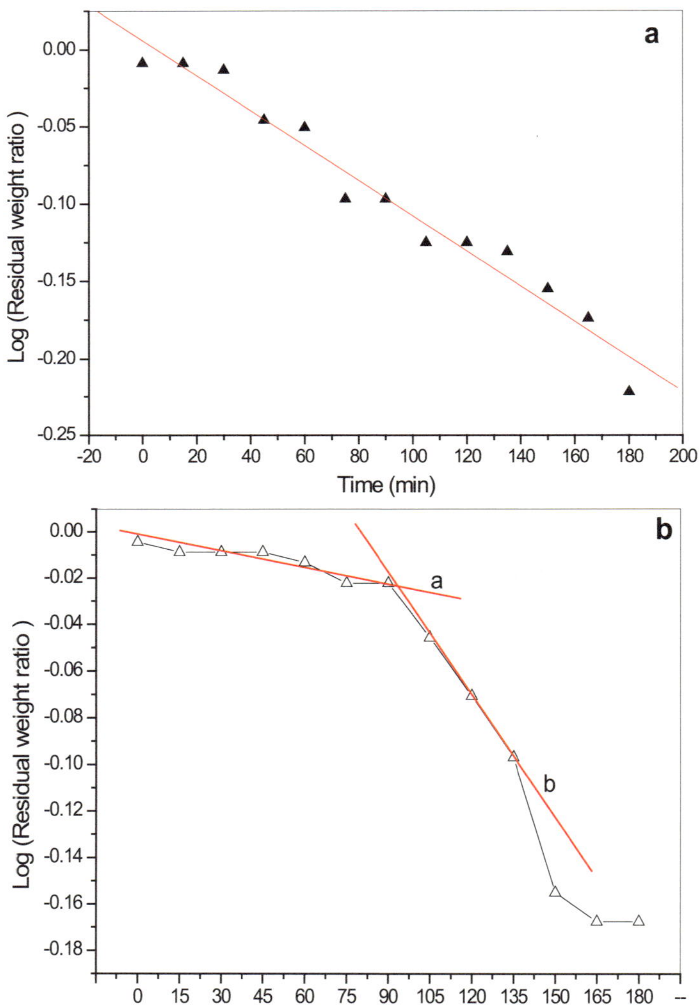

Fig. 9. The release curves of the OSC (a) and TSC (b) shells prepared by pre-polymer material dropping rates of 0. 5 ml·min^{-1}.

Fig. 9 shows the release curves of systems of C and F. It is linear relationship between time and logarithmic residual weight of core of system C in Fig. 9(a). Linear regression fit the first-order kinetic release theory,

$$2.303\log q_{mc} / q_{\infty} = kt \tag{2}$$

where q_{mc} is the residual weight of core material at time t; q_{∞} is the total weight of core material. The calculated value of k_1 is,

$$k_1 = -2.625 \times 10^{-3} \tag{3}$$

Fig. 9(b) shows the release curve of system F with a two decrease-linear step (line a and b) after curve regression. Lines of a and b separate at the release time of 90 min and the release rate of first step is lower that the of the second step. The TSC shell release curve of *n*-octadecane has a special release behavior, which reflects a complex shell structure. According calculation of equation (3), the value of k_2 and k_3 are,

$$k_2 = -3.333 \times 10^{-4} \tag{4}$$

$$k_3 = -5.8333 \times 10^{-4} \tag{5}$$

By comparing the values of k_1, k_2 and k_3, it shows

$$k_1 > k_2 > k_3 \tag{6}$$

3. Fabrication and characterization MMF-shell microPCMs

To date, microPCMs/polymer composites have been paid more attentions applying their thermo-regulation or thermo-saving abilities. A survey of literature shows that these composites are smart functional materials, such as thermosmart fibers, heat preservation building materials, solar heating materials and anti-icing coating, et al [16-19]. Although lots of investigations have been carried out, these composites have not yet been explored to a significant extent. Nearly all researches focused on the themes of selecting of PCMs, encapsulation methods, microPCMs characterization and thermal properties. However, we still have little knowledge about the interfacial morphology changes between microPCMs and matrix polymers, which will greatly affect the stability of microPCMs. Fig. 10 illustrates the mechanism of the interfacial morphology changes between microPCMs and matrix polymer. During a repeated phase change process with heat transmittance in application for microPCMs/polymer composites, expansibility will appear coming from both of microcapsules and polymeric resin with difference inflation coefficients. The volume of microPCMs can be affected by the encapsulated PCM when environmental temperature changes. Moreover, the above phenomena may occur micro-cracks or fractures in matrix resin during heat absorbing and resealing; then these structures may spoil the thin shells of microcapsule (broken or ruptured), the encapsulated PCM will lose the shells protection [20]. Hence, the mechanical properties of composites will decrease following with the internal cracks or microcapsules rupture [21].

These entire interface changes certainly influence and shorten the service-life of these microPCMs/polymer composites [1].

It can be imaged from Fig. 10 that the encapsulated PCMs characteristics, microPCMs shell properties (thickness, intension and average diameter), matrix and the interfacial adhesion structures (microPCMs and matrix) are four main factors affecting the interfacial stability of microPCMs/polymer composites. To simplify this complex problem in this study, the shell property could be regarded as the only one characteristic being considered because of the microPCMs/polymer composites with the same shell and matrix materials. Normally, the microcapsule shell properties are determined by the weight ratios of core and shell materials and emulsification rates [22].

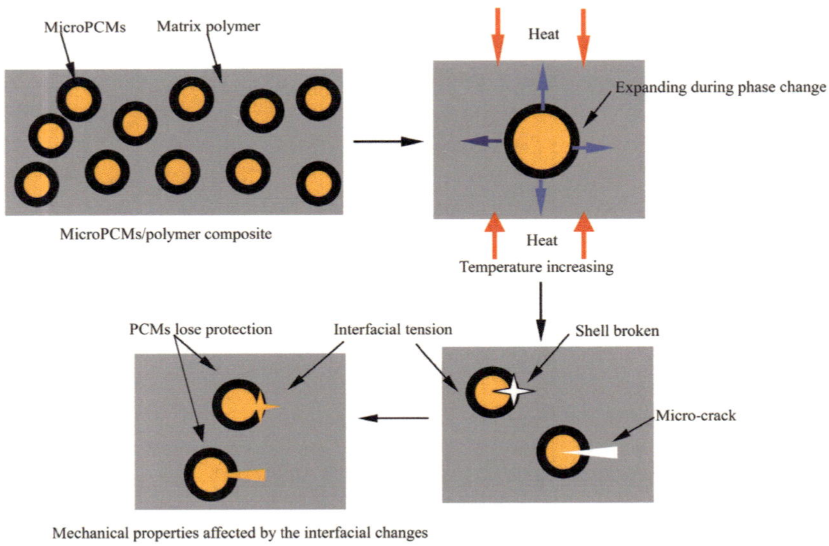

Fig. 10. Illustration of the mechanism of the interfacial morphologies changes between microPCMs and matrix polymers after repeated heat transmittance.

Based on these considerations, the purpose of this work was to fabricate novel microPCMs containing dodecanol by an in-situ polymerization using methanol-modified melamine-formaldehyde (MMF) as shell material and investigate the interfacial morphologies changes of microPCMs/epoxy composites treated with a simulant thermal process with a 10-times rapid temperature variation. A series of microPCMs with different core material (dodecanol) contents and diameters were prepared and embedded in epoxy resin to investigate the interfacial phenomena generated by the expansion or shrinking of the microPCMs during temperature change. Under this simulant and controlled thermal treatment conditions, the affects of average diameters and PCMs ratios of microPCMs on the interface morphologies of microPCMs/epoxy composite were analyzed applied the scanning electron microscopy (SEM). We believe these results will be guides for the fabrication and application of these functional composites.

3.1 Fabrication method

Dodecanol (Tianjin Kemel Chemical Reagent Development Center, Tianjin, China) was used as the PCM (core material). Its solid-liquid phase change temperature was about 21 °C. The shell material was prepolymer of melamine-formaldehyde modified by methanol (Solid content was 78. 0%, Aonisite Chemical Trade Co. , Ltd. , Tianjin). Styrene maleic anhydride (SMA) copolymer (Scripset® 520, USA) was applied as dispersant. Organic diluent (Butyl glycidyl ether), bisphenol-A epoxy resin (E-51) and curing agent (amine) were supplied by Tianjin Synthetic Material Research Institute (Tianjin, China).

The encapsulation was carried out in a 500 ml three-neck round-bottomed flask. 3. 0 g SMA and 0. 8 g NaOH were dissolved in 100 ml water (50 °C). The pH value was adjusted to 4-5 by acetic acid solution. 10. 0 g dodecanol was added to the aqueous SMA solution, and the mixture was emulsified mechanically under a vigorous stirring rate of 3000 r·min^{-1} for 10 min using QSL high-speed disperse-machine (Shanghai Hongtai Ltd. , Shanghai, China). Then dropped the emulsion in the bottle dipped in steady temperature flume and stirred at a speed of 1500 r·min^{-1}, and dropped 25 g mixture of prepolymer (12. 8 g) and deionized water at a rate of 0. 25 g·min^{-1}. The shell formed after 2. 5 h by heating slowly to temperature of 60 °C. After 30 min, the temperature was elevated to 75°C directly. After polymerization for 1. 5 h, temperature was dropped slowly at 2 °C·min^{-1} to room temperature. The resultant microcapsules were filtered and washed with water and dried in a vacuum oven.

The 5 g of dodecanol microcapsules was mixed with 1. 5 g organic diluent (Butyl glycidyl ether) by ultrasonic vibration for 5 min. The power of ultrasonic devices was 40 W. The 10 g bisphenol-A epoxy resin was dropped in the mixture prepared, and then the mixture was mixed by the same ultrasonic vibration for 5 min. 2. 5 g of amine curing agent was added and ultrasonic vibrated for 5 min. At last, the mixture prepared was casted in the PTFE mold. After cured for 24 h at room temperature, the samples of were demoulded out.

3.2 FT-IR analyses of microPCMs synthetic structure

Fig. 11(a,b) illustrates the chemical structural formula of the MMF formation and the in-situ polymerization process of microcapsule shells. Melamine-formaldehyde (MF) microPCMs have been widely fabricated by in-situ polymerization because of their high mechanical properties and compactability of shells [22]. However, it is found that the MF shells have relatively high brittleness. This disadvantage may bring brittle broken of microPCMs under

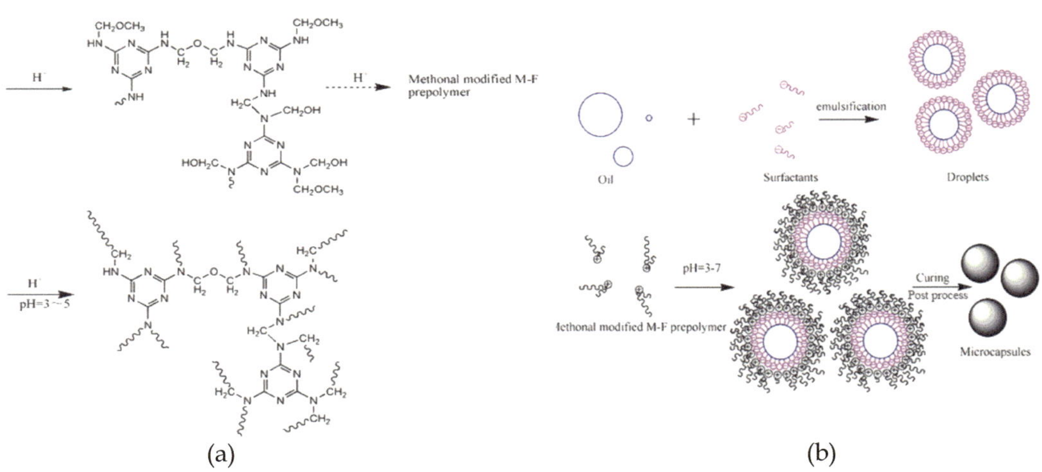

(a) (b)

Fig. 11. Illustration of the (a) chemical structural formula of the MMF formation and (b) the in-situ polymerization process of microcapsule shells.

an extreme temperature change process or an impact force. The methanol groups were grafted on melamine with long branched structure to enhance the flexibility of MF-shells in this study (Fig. 11a). As Fig. 11b shown, the MMF shell formation is an in-situ polymerization process with the steps of W/O emulsification, MMF prepolymer adsorbed on oil particles and prepolymer polymerization. This method has been described in our previous report [20].

FT-IR could be applied to compare the chemical structures of MF and MMF shells, and to confirm the encapsulation of core material. Fig. 12 (a-e) shows the FT-IR spectra of (a) dodecanol, (b) SMA, (c) MMF, (d) MF and (e) microPCMs, respectively. The strong and wide absorption peaks at approximately 3369 cm^{-1} in Fig. 12 (a) of core material is assigned to O-H stretching vibrations of dodecanol. The multiple strong peaks at 2925 cm^{-1} and 2854 cm^{-1} are associated with aliphatic C-H stretching vibrations of methyl and methylene groups. The moderate strong peak at 1057 cm^{-1} is related to C-OH stretching vibration of primary alcohol. In Fig. 12(b), the peaks at approximately 1494 and 1603 cm^{-1} are assigned to the C=C stretching vibrations of benzene ring and the strong peaks at approximately 1858 and 1777 cm^{-1} are the C=O stretching vibrations of anhydride. In Fig. 12(c,d), the strong and wide absorption peaks at approximately 3342 and 3350 cm^{-1} are attributed to the superposition of O-H and N-H stretching vibrations. According to the work of Salaün [23], the peaks at 1556 and 815 cm^{-1} in Fig. 12. (c, d) are assigned to the vibrations of triazine ring; and the corresponding peaks of cured MMF lie at 1559 and 815 cm^{-1}. The characteristic peaks of aliphatic primary alcohol dodecanol at approximately 2925, 2853, and 1057 cm^{-1} can be observed in Fig. 12 (e) indicating that dodecanol has been microcapsulated with MMF resin. In addition, the characteristic peaks of MF resin at approximately 1550 and 814 cm^{-1} in Fig. 12 (e) indicates that dodecanol has been encapsulated as core material successfully with MF resin as shell material.

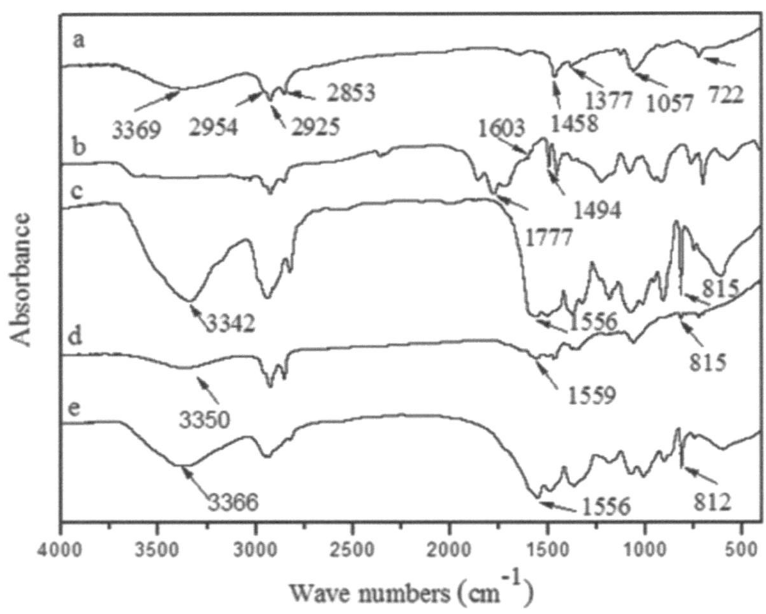

Fig. 12. FT-IR spectra of (a) dodecanol, (b) SMA, (c) MMF, (d) MF resin, (e) microPCMs.

3.3 Morphologies and average diameters of microPCMs

The contents of PCM in microcapsules can be estimated according to the Eq. (7) as the theoretical value (C_t, %),

$$C_t(\%) = \frac{m_{core}}{m_{core} + m_{shell}} \times 100\% \tag{7}$$

where m_{core} is the amount of core material (g) used and m_{shell} is the amount of shell material used (g) in synthesizing the microPCMs. The content of PCM in microcapsules (C_a, %) also can be estimated according to the measured melting heat according to Eq. (8),

$$C_a(\%) = \frac{|\Delta H_m|}{|\Delta H_{m_0}|} \times 100\% \tag{8}$$

$$E_e(\%) = \frac{C_a}{C_t} \times 100\% \tag{9}$$

where ΔH_m is the melting heat of microcapsules(J/g) and ΔH_{m_0} is the melting heat of PCM(J/g). The encapsulation efficiency (E_e, %) of microcapsules can be calculated as the ratio of the measured PCM content in microPCMs to the theoretical value depending on the amount of PCM and MMF prepolymer (Eq. 9) added in the system of fabrication.

The properties of microPCMs containing dodecanol synthesized by various conditions are listed in Tab. 1. The pure dodecanol has the melting enthalpy value of 206. 9 J/g [24]. As the SMA amounts and stirring rates greatly affecting the microPCMs morphologies and properties [20], we firstly investigated the average diameter, C_a and E_e values of microPCMs fabricated by different C_t with the same string rate and the amount of SAM. The results show that the average diameter, C_a and E_e values have increased with the increasing of C_t. The average diameter is in the range of 1. 55±0. 84 to 1. 99±1. 10 μm, which is not greatly affected by C_t. Comparatively, the average diameter of microPCMs, fabricated with the same SAM amount (2. 0 g) and C_t (50%) by different stirring rates (1000-4000 r·min^{-1}), is in the range of 1. 21±1. 08 to 16. 20±7. 82 μm. With the increasing of stirring rates, the average diameters are sharply decreased. Interestingly, C_a and E_e values of microPCMs both increased with the increasing of stirring rates.

Samples (g)	SMhA amount	H_2O (g)	C_t (%)	Stirring rate (r·min^{-1})	Average diameter (μm)	Melting enthalpy (J/g)	C_a (%)	E_e (%)
Dodecanol	-	-	-	-	-	206.9	-	-
1	3.0	100	40.0	3,000	1.55±0.84	45.5	22.0	55.0
2	3.0	100	50.0	3,000	1.63±0.91	76.3	36.9	73.8
3	3.0	100	60.0	3,000	1.83±1.40	124.5	60.2	90.3
4	3.0	100	70.0	3,000	1.99±1.10	141.5	68.4	91.2
5	2.0	100	50.0	4,000	1.21±1.08	100.7	48.7	97.4
6	2.0	100	50.0	3,000	6.30±3.58	93.5	45.2	90.4
7	2.0	100	50.0	2,000	10.67±6.82	61.6	29.8	59.6
8	2.0	100	50.0	1,000	16.20±7.82	49.6	24.0	48.0

Table 1. Core contents and encapsulation efficiencies of microPCMs

Fig. 13 shows the SEM morphologies (a-d) and diameter number fractions (a'-d') of microPCMs fabricated with different C_t values (40, 50, 60 and 70%) under the same stirring rate of 3000 r·min^{-1}. It can be seen that all the microcapsules have the irregular spherical shape. Some microPCMs are shrunken because of pressure or phase change. Their average diameter is about 1-2 µm, and the different amounts of MMF have little affect on diameter of microPCMs. Fig. 14 shows the SEM morphologies (a-d) and diameter number fractions (a'-d') of microPCMs fabricated with different stirring rates (1000, 2000, 3000 and 4000 r·min^{-1}; C_t=50%). The average diameter was about 1. 2 µm when the emulsification rate was 4000 r·min^{-1}. With the increasing of stirring rates from 1000 to 4000 r·min^{-1}, the average diameters decreased sharply. Also, some depressions on the surface of microcapsules were observed due to the liquid-solid phase change induced by the temperature decreasing in the process of synthesis of microcapsules.

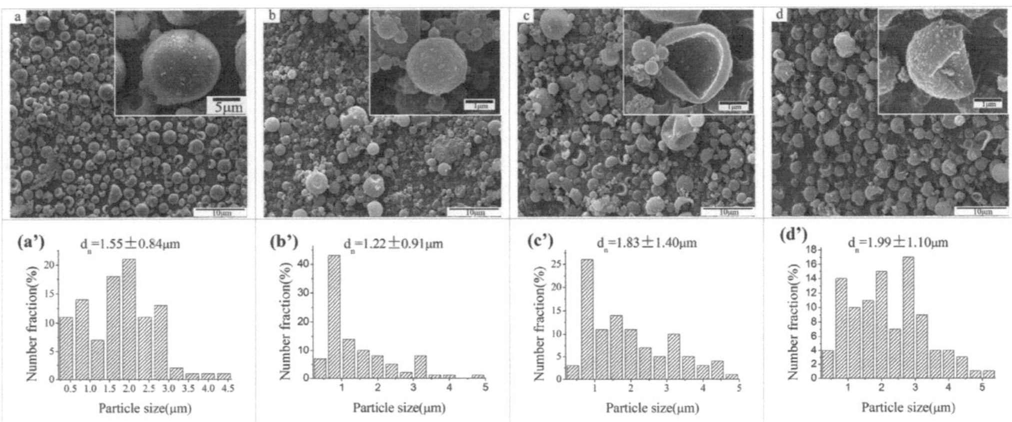

Fig. 13. SEM morphologies (a-d) and diameter number fractions (a'-d') of microPCMs fabricated with different Ct values (40, 50, 60 and 70%) under the same stirring rate of 3000 r·min^{-1}.

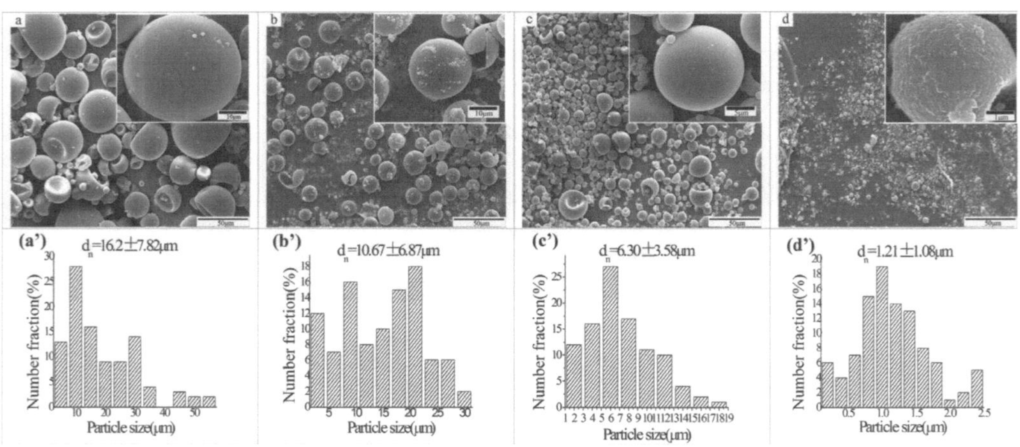

Fig. 14. SEM morphologies (a-d) and diameter number fractions (a'-d') of microPCMs fabricated with different stirring rates (1000, 2000, 3000 and 4000 r·min^{-1}; C_t=50%).

3.4 DSC analyses of microPCMs

Fig. 15 (a) shows the DSC curves (a-e) of pure dodecanol and the microPCMs synthesized with different C_t values of 40, 50, 60 and 70%. For pure PCM of dodecanol, the strongest endothermic peak at 24. 3 °C is its phase change temperature. For each microPCMs sample, it has one obvious endothermic peak (curve b-e). With the C_t values increasing from 40 to 70% from microPCMs, their phase change temperatures are 20. 9, 22. 6, 23. 5 and 24. 3 °C, respectively. The shells of microPCMs do not greatly affect the phase change temperature of pure dodecanol.

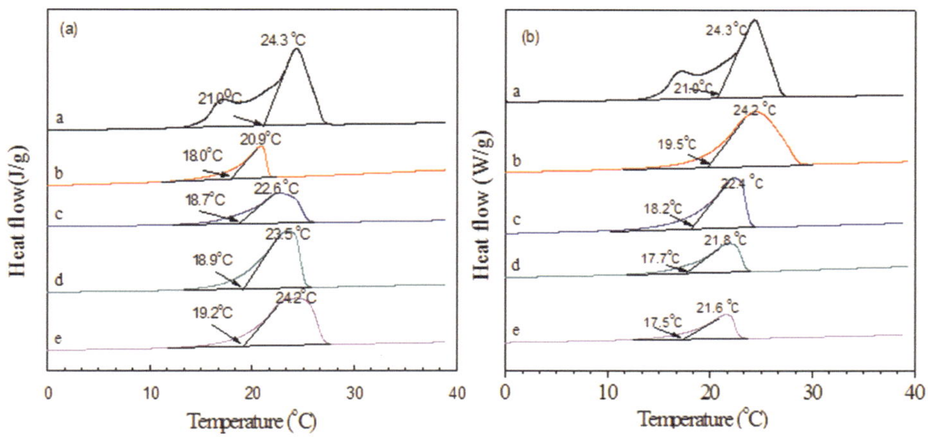

Fig. 15. DSC curves of (a) pure dodecanol and the microPCMs synthesized with different Ct values (40, 50, 60 and 70%); DSC curves of (b) pure dodecanol and the microcapsules synthesized with different stirring rates (1000-4000 r·min^{-1}; SMA=2. 0 g, core/shell (wt/wt)=1/1)).

In Fig. 15 (b), the phase change temperature and endothermic peak increased with increasing of emulsification rates for high core encapsulation because of better distribution of emulsifier in the emulsion system. The melting enthalpy and encapsulation efficiency (E_e, %) of dodecanol microcapsules synthesized at different conditions can be seen in Table 1. And the melting enthalpy and E_e were increasing with the increasing of C_t values and emulsification rates, which accorded with the DSC measurement results.

3.5 Interface morphologies of microPCMs/epoxy composites

A simulant temperature change process was designed in this study to lead the microPCMs in composites phase change. The temperature change was in the range of 15-50 °C, which could suffice the phase change of dodecanol. As the epoxy had the heat-resistance ability, this maximal temperature of this simulant process was higher than the phase change temperature of dodecanol. At the top temperature of 50 °C, the composites were retaining 10 min to ensure the absolute phase change of microPCMs. In order to evoke the interface variation distinctly, this temperature change process was fast (2 °C·min^{-1}) and frequent (repeated 10 times). As C_t and the average diameter were two main parameters affecting the stability of microPCMs, we paid more attention to their effects on the interface morphologies of composites before and after the thermal treatments.

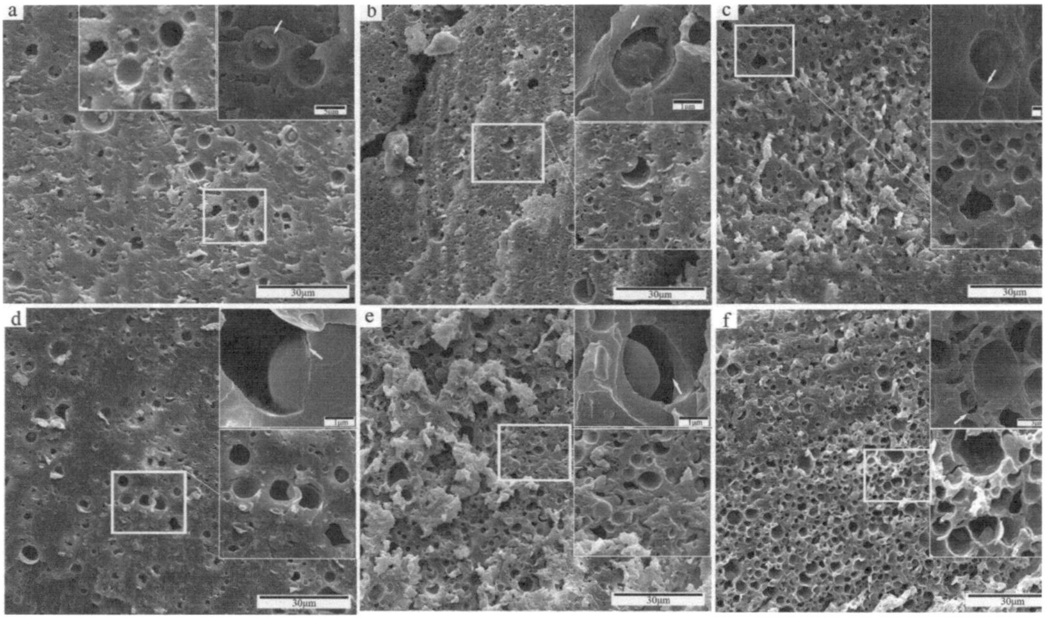

Fig. 16. SEM morphologies of microPCMs/epoxy composites (microPCMs/E-51, wt/wt)=1/1) before (a, b, c) and after (d, e, f) a thermal treatment process with different C_t (%) values: (a) 40%, (b) 50%, (c) 60%.

Fig. 16 shows the SEM interface morphologies of microPCMs/epoxy composites (microPCMs/E-51, wt/wt=1/1) before (a, b, c) and after (d, e, f) the simulant thermal treatment process with different C_t (%) values: (a, d) 40%, (b, e) 50% and (c, f) 60%. All these morphologies indicate that the microPCMs could be dispersed homogeneously in epoxy resin through an ultrasonic vibration method. It would ensure the isotropic thermal transmission in composites. In Fig. 16 (a-c), more imbedded microPCMs will make the composites more loose. However, the interfaces of microPCMs/epoxy are compact without gaps. The shells of microPCMs and matrix adhere tightly keeping the microPCMs steadily. Comparatively, micro-cracks and gaps occurred after a thermal treatment in the interface of microPCMs and epoxy matrix obviously, as shown in Fig. 16 (d-f). With the increasing of C_t, the absorbing-releasing of latent heat will more intensively in composites. This will give great actions to interfaces between microPCMs and matrix polymers because of their different expand coefficients and thermal transmission abilities. At the same time, the inherent properties of volume expanse for PCMs in tiny microcapsules induced the internal stress in the composites leading to separation of microPCMs and matrix. And the interface morphologies changes will be easily produced under the repeated expand-shrink affects and the internal stress actions.

Fig. 17 shows the SEM morphologies of microPCMs/epoxy composites (MicroPCMs/E-51, wt/wt=1/1; C_t=50%) before (a, b, c) and after (d, e, f) a thermal treatment. These microPCMs had different average diameters fabricated by controlling the emulsion rates of 1000, 2000 and 3000 r·min^{-1}, respectively. It can be seen from Fig. 17 (a-c) that higher stirring rate will make microPCMs with smaller average diameter, and the number of microPCMs will also increase with the accelerating of stirring. The interfaces of microPCMs/epoxy are also compact without gaps before the thermal treatment similar to the phenomena in Fig. 16 (a-c). In Fig. 17 (d), some microPCMs are deformed under the repeated expand-shrink affects and the internal stress actions. The interface separation occurred between microPCMs and matrix as shown in Fig. 17 (e, f). Moreover, more micro-cracks appeared in the microPCMs/epoxy composites containing microPCMs fabricated with smaller average diameter.

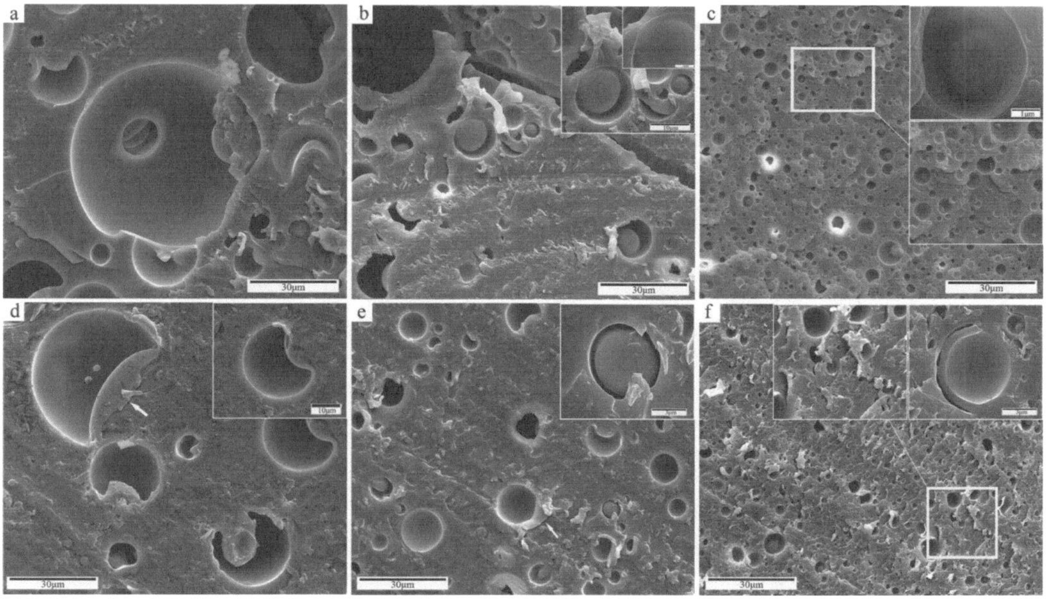

Fig. 17. SEM morphologies of microPCMs/epoxy composites (MicroPCMs/E-51, wt/wt=1/1; C_t=50%) before (a, b, c) and after (d, e, f) thermal treatment containing microPCMs prepared by different emulsion rates: (a, d) 1000 r·min^{-1}, dn=16. 20±7. 82 μm; (b, e) 2000 r·min^{-1}, d_n=10. 67±6. 82 μm; (c, f) 3000 r·min^{-1}, d_n=6. 30±3. 58 μm.

Normally, both of C_t and the average diameter parameters will greatly affecting the shell thickness of microcapsules. Although the interface gaps or deformation of microPCMs had been detected in the composites treated with a thermal process (Fig. 16 and 17), there was no shell broken phenomenon for microPCMs embedded in matrix. The microPCMs with MMF shells have enough mechanical properties to resist the internal stress. It can be concluded that the internal stress generated by the expansion or shrinking of the microPCMs is the main factor leading to the interface morphology changes and damaged of composites. And these interface changes will affect the mechanical properties of these microPCMs/polymer composite in application. Therefore, these above results mean that we should balance C_t, the average diameter and mechanical properties of microPCMs/polymer composites systemically to satisfy the long service time of these composites.

4. Fabrication and characterization PU-shell microPCMs

A survey of literature indicates that melamine-formaldehyde (MF) resin, urea-formaldehyde (UF) resin and polyurethane (PU) are usually selected as microcapsule shell materials for the PCMs protection. However, there may exist ineluctable remnant formaldehyde after forming the shell through polymerization, such as using MF and UF resins, which causes environmental and health problems. Generally speaking, it is hard to find an effective means to eliminate formaldehyde that is dissociative and can be continuously releasing from the products.

In this regard, PU is the promising polymer material in the field of microPCMs. Usually, a convenient technology is applied by polycondensating two complementary monomers to obtain PU-shell microcapsules. This method has been widely used in making flame-retarded, functional oil and artificial organ PU shell microcapsules [25–29]. It starts with making an emulsion of two immiscible phases (i. e. , oil and water, O/W). Each phase contains a dissolved specific monomer that is able to react with the other monomer present in the other phase. In earlier studies of PU-shell microcapsules (TDI) [26], hexamethylene diisocianate (HMDI) [30] and isophorone diisocyanate (IPDI) [31] are employed as the oil soluble monomers. In order to make the emulsion stable, certain surfactants or polymers are usually added to the continuous and/or the dispersed phase. Subsequently, these additives form an adsorption layer at the liquid–liquid interface, which may affect the polycondensation polymerization more or less [30]. After forming the nascent shell, the following polycondensation reaction is a diffusioncontrolled process since it must occur at the O/W interfaces. As a result, capsule-like structured particles can formed with the particle size depending on the initial emulsion conditions and formation process parameters, such as emulsion rate, reaction condition and molar ratio of monomers.

However, there was litter information available focusing on the PU-shell microPCMs, especially relationship between properties and the dispersant, which is a main factor for stability and application of microPCMs. The aim of this research was to prepare a series of novel PU-shell microPCMs containing *n*-octadecane, which were fabricated by interfacial polycondensation between toluene-2, 4-diisocyanate (TDI) and diethylene triamine (DETA). *n*-Octadecane is usually a PCM desirable for usage in thermal storage and releasing for its availability in a reasonable phase change temperature range and its large amount latent heat [5]. The latent heat of *n*-octadecane is 241. 2 J/g and its melting point (T_m) is 28 °C. Styrene-maleic anhydride copolymer solid (SMA) was used as a nonionic dispersant in this preparation. To investigate the effect of dispersant (SMA) on the properties of microPCMs is another focus point of this paper. Especially, in order to understand the thermal prosperities of microPCMs, the factors influencing thermal stability were examined with regard to the amount of SMA.

4.1 Fabrication method

The microcapsule shellwas synthesized from TDI and DETA monomers, which were both purchased from Nankai University Special Reagent Co. for forming the microcapsule shell by using an interfacial polymerization process. *n*-octadecane (purity 98% minimum)was the core material as PCM purchased from Tianjin Chem. Reagent Co. Nonionic surfactant, SMA (Scripset®520, Hercules, USA) was used as a dispersant. All the chemicals were of reagent grade and used without further purification.

The concept of the microcapsules preparation method is based on the polycondensation reaction of hydrophilic and lipohilic monomers at the interface of an O/W submicronic emulsion. Initially, *n*-octadecane and lipophilic monomers TDI were dissolved in 20 ml of acetone as an oil phase. Different weight of SMA solid was dissolved in 40 ml of distilled water to adjusted pH value to 10 by using 0. 1 M NaOH solution to form a serious microPCMs. The oil phase was then slowly dipped into the aqueous phase under magnetic stirring speed of 4000 r·min^{-1} for 2 min. Then a measured DETA was diluted in 20 ml of distilled water, and then slowly added into the emulsion system to ignite the interfacial polycondensation. The microcapsules were formed by progressive interfacial polycondensation reaction between DETA and TDI at the interface of hydrolysis SMA in water submicronic emulsion. After 2 h, the colloidal solution was then concentrated by evaporation under reduced pressure at 40 °C to about 20 ml.

4.2 Morphologies of PU-shell microPCMs

Fig. 18 shows the SEM morphologies of treated and dried microcapsules. In Fig. 18(a), the final microcapsules are found to be piled with a thickness of about 100 μm formed by deposition mechanism. It is very interestingly noted that the PU-shell microcapsules are individually distributed without excessive conglutination among each other. In Fig. 18(b), the surface of microcapsules is mostly smooth and the shape is regularly spherical with a diameter of about 5–10 μm. Some polymer filling between microcapsules can be observed, possibly due to that the core material could not have been encapsulated completely or the shell material could not absolutely cover the cores.

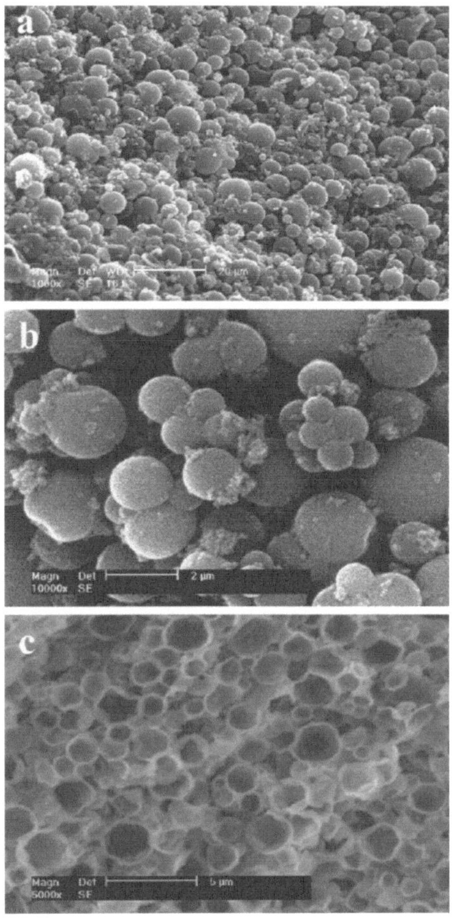

Fig. 18. SEM morphologies of treated and dried microcapsules: (a) piled microPCMs, (b) smooth and the regular global shape with a diameter of about 5–10 μm and (c) a typical cross-sectional morphology of shells.

In order to achieve the pure PU-shell material avoiding the influence of core material, the piled PU-shell microcapsules were blended in paraffin. After the blend was dried in room temperature, it was carefully cut to obtain the cross-section by an ultramicrotome (RMCMT-7000, USA). The half-shelled microcapsules were put into alcohol for removing PCM and paraffin. Then the shells were washed again with distilled water. In addition, it is very interesting that we can achieve the morphologies of the inside of the shell and the shell thickness after the pure shell material has been dried at 40 °C for 2 h. Fig. 18(c) shows the typical cross-sectional view of microcapsules. Obviously, it is seen that inside of the shell is rather smooth and the shell thickness is about 0. 3–0. 5 µm.

4.3 FTIR of shell material

FTIR analysis of shell material enables to determine the chemical structure of the shell polymer containing chemical bonds. Also, it will help us to understand the reality reaction degree of TDI. This information can be achieved from the corematerial in shells and the shell material. The liquid core material in shells at the temperature of 40 °C was separated from a half of the cross-section as abovementioned method by a centrifugal machine (Biofuge®Pico Barkey, Germany). The FTIR spectrum is shown in Fig. 19(a). It has absorption bands at 3367 and 1573–1542 cm^{-1}, which are assigned to the C–H stretching vibrations, C–H, CH_2 vibrations. The absorption bands at 2262 cm^{-1} is the unreacted–NCO group coming from TDI. Fig. 19(b) shows an absorption band at 3471–3381 cm^{-1} for the N–H stretching vibration, which may result in the formation of the strong hydrogen bonds on the shell side. C–H stretching vibrations of aliphatic diamine are shown at 2868, 1660 and 1726 cm^{-1} for the C O stretching of urethane. The sharp C–H stretching vibration for case (b) indicates that the SMA chain associated on core material may have reacted with TDI and formed a part of shell material. The chemical schematic is showed in Fig. 20. These FTIR results have confirmed that the shell of microcapsules is polyurethane.

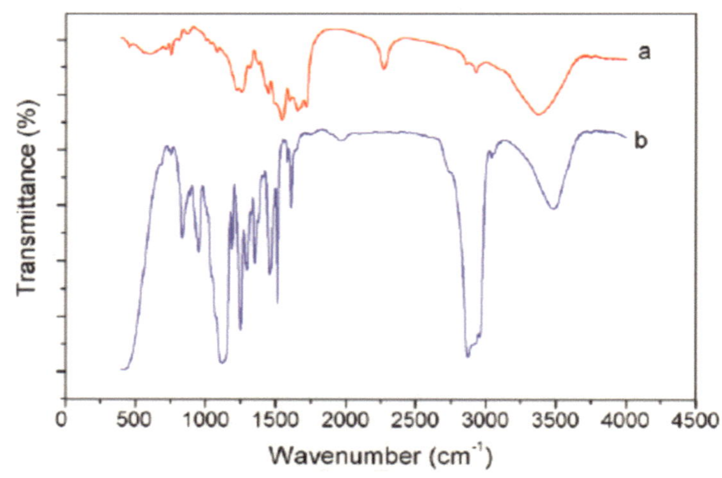

Fig. 19. FTIR curves of (a) liquid core material in shells and (b) shell material.

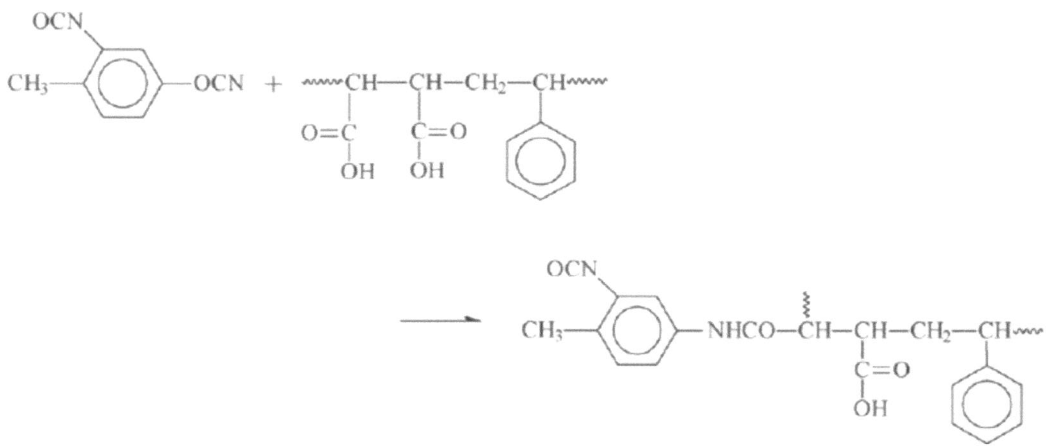

Fig. 20. The reaction scheme of SMA molecular and TDI.

5. Conclusions

1. In order to increase the compact properties of MF shell microPCMs containing *n*-octadecane, a method of TSC encapsulation through an in-situ polymerization was applied by dropping the shell material twice. All the evidences demonstrate that twice coacervations reduced the cracks and on shells and increase the compactness of shells. Especially, we confirm that the shell resistance permeability is increased by TSC shell technology from the values of permeability coefficient k.
2. MMF prepolymer can be applied as shell material to fabricate the microPCMs. The microPCMs containing dodecanol have the irregular spherical shapes and their average diameter can be controlled by regulating the stirring rates. The melting temperatures of these microPCMs were not greatly larger than the phase change temperature of pure dodecanol. With the increasing of stirring rates, the average diameters of microPCMs are sharply decreased. C_a and E_e values of microPCMs both increased with the increasing of stirring rates. As two main parameters affecting the stability of microPCMs, both C_t and the average diameter affected the interface morphologies of microPCMs/epoxy composites after the thermal treatments. Micro-cracks and gaps occurred after a thermal treatment in the interface of microPCMs and epoxy matrix obviously. The MMF-shells were flexible without broken. And the internal stress generated by the expansion or shrinking of the microPCMs is the main factor leading to the interface morphology changes and damaged of composites.
3. A series of PU-shell microPCMs containing *n*-octadecane is successfully prepared with particle size distribution properly controlled by the emulsion-stirring rate at

around 4000 r·min^{-1}. FTIR results have confirmed that the PU-shell microcapsules are successfully fabricated by interfacial polymerization between TDI and DETA with SMA as disperse.

6. Acknowledgements

The author is grateful to the financial support of the National Natural Science Foundation of China (No. 50803045).

7. References

[1] J. F. Su, L. X. Wang, L. Ren, Preparation and characterization of double-MF shell microPCMs used in building materials, J. Appl. Polym. Sci. 97(2005) 1755 - 1762.
[2] M. N. A, Uddin M. S, Zhu H. J, Preparation and evaluation of a novel solar storage. material: Microencapsulated paraffin, Int. J. Solar Energy 20 (2000) 227-238.
[3] L. Y. Wang, P. S. Tsai, Y. M. Yang, Preparation of silica microspheres encapsulating phase-change material by sol-gel method in O/W emulsion, J. Microencap. 23 (2006) 3 -14.
[4] X. X. Hu, Y. P. Zhang, Novel insight and numerical analysis of convective heat transfer enhancement with microencapsulated phase change material slurries: laminar flow in a circular tube with constant heat flux, International J. Heat Mass Transfer 45 (2002) 3163 - 3172.
[5] X. X. Zhang, X. M. Tao, K. L. Yick, X. C. Wang, Structure and thermal stability of microencapsulated phase-change materials, Colloid Polym. Sci. 282 (2004) 330 - 336.
[6] J. W. Kim, J. Y. Ko, J. B. Jun, I. S. Chang, H. H. Kang, K. D. Suh, Multihollow polymer microcapsules by water-in-oil-in-water emulsion polymerization: morphological study and entrapment characteristics, Colloid Polym. Sci. 281 (2003) 157-163.
[7] K. Hong, S. Park, Preparation of polyurethane microcapsules with different soft segments and their characteristics, React. Funct. Polym. 42 (1999) 193-200.
[8] X. X. Zhang, X. M. Tao, K. L. Yick, Y. F. Fan Expansion space and thermal stability of microencapsulated n-octadecane, J. Appl. Polym. Sci. 97 (2005) 390-396.
[9] J. S. Hwang, J. N. Kim, Y. J. Wee, J. S. Yun, H. G. Jang, S. H. Kim, H. W. Ryu, Preparation and characterization of melamine-formaidehyde resin microcapsules containing fragrant oil, Biotech. Bioprocess Engine. 11 (2006) 332-336.
[10] G. Sun, Z. Zhang, The mechanical properties of melamine-formaldehyde (M-F) microcapsules were studied using a micromanipulation technique, J. Microencap. 18 (2001)593-602.
[11] Y. Rong, H. Z. Chen, D. C. Wei, J. Z. Sun, M. Wang, Microcapsules with compact membrane structure from gelatin and styrene–maleic anhydride copolymer by complex coacervation, Colloids and Surfaces A: Physicochem. 242 (2004) 17 - 20.
[12] L. X. Wang, J. F. Su, L. Ren, Preparation of thermal energy storge microcapsules by phase change, Polym. Mater. Sci. Engine. 21 (2005) 276-279.
[13] X. X. Zhang, X. C. Wang, X. M. Tao, K. L. Yick, Energy storage polymer/MicroPCMs blended chips and thermo-regulated fibers, J. Mater. Sci. 40 (2005) 3729-3734.

[14] N. Nihant, C. Grandfils, R. Jerome, P. Teyssie, Microencapsulation by coacervation of poly(lactide-co-glycolide) IV. Effect of the processing parameters on coacervation and encapsulation, J. Controlled Release 35 (1995) 117-125.

[15] K. Makino, M. Umetsu, Y. Goto, A. Kikuchi, Y. Sakurai, T. Okano, H. Ohshima, Two-Layer structure of microcapsule membrane as predicted from electrophoretic studies, J. Colloid Interface Sci. 218 (1999) 275-281.

[16] P. Chaiyasat, Y. Ogino, T. Suzuki, M. OkuboInfluence of water domain formed in hexadecane core inside cross-linked capsule particle on thermal properties for heat storage application. Colloid Polym. Sci. 286 (2008) 753-759.

[17] H. Zhang, X. D. Wang, D. Wu. Silica encapsulation of n-octadecane via sol-gel process: A novel microencapsulated phase-change material with enhanced thermal conductivity and performance. J. Colloid Interf. Sci. 343 (2010) 246 - 255

[18] C. Alkan, K. Kaya, A. Sara Preparation, Thermal Properties and Thermal Reliability of Form-Stable Paraffin/Polypropylene Composite for Thermal Energy Storage. J. Polym. Environm. 17 (2009) 254-258.

[19] E. Onder, N. Sarier, E. Cimen. Encapsulation of phase change materials by complex coacervation to improve thermal performances of woven fabrics. Thermochim Acta 467 (2008) 63-72.

[20] J. F. Su, L. Ren, L. X. Wang. Preparation and mechanical properties of thermal energy storage microcapsules. Colloid Polym. Sci. 284 (2005) 224-228.

[21] M. You, X. X. Zhang, W. Li, X. C. Wang. Effects of MicroPCMs on the fabrication of MicroPCMs/polyurethane composite foams. Thermochim Acta, 472:(2008) 20-24.

[22] J. F. Su, Z. Huang, L. Ren. High compact melamine-formaldehyde microPCMs containing n-octadecane fabricated by a two-step coacervation method. Colloid Polym. Sci. 285 (2007)1581-1591

[23] F. Salaün, I. Vroman. Influence of core materials on thermal properties of melamine-formaldehyde microcapsules. Eur. Polym. J. 44 (2008) 849-860.

[24] A. Abhat. Low temperature latent heat thermal energy storage. Heat storage materials. Sol. Energy 30 (1983) 313-332.

[25] S. Giraud, S. Bourbigot, M. Rochery, I. Vroman, L. Tighzert, R. Delobel, F. Poutch, Flame retarded polyurea with microencapsulated ammonium phosphate for textile coating, Polym. Degrad. Stab. 88 (2005) 106–113.

[26] K. Hong, S. Park, Characterization of ovalbumin-containing polyurethane microcapsules with different structures, Polym. Test. 19 (2000) 975–984.

[27] O. Maruyama, T. Yamane, N. Tsunemoto, et al. , A preliminary study of microcapsule suspension for hemolysis evaluation of artificial organs, Artif. Organs 23 (1999) 274 - 279.

[28] E. Jabbari, M. Khakpour, Morphology of and release behavior from porous polyurethane microspheres, Biomaterials 21 (2000) 2073–2079.

[29] K. Hong, S. Park, Preparation of polyurea microcapsules with different composition ratios: structures and thermal properties, Mater. Sci. Eng. A 272 (1999) 418 - 421.

[30] K. Mizuno, Y. Taguchi, The effect of the surfactant adsorption layer on the growth rate of the polyurethane capsule shell, J. Chem. Eng. Jpn. 38 (2005) 45 - 48.

[31] L. Torini, J. F. Argillier, N. Zydowicz, Interfacial polycondensation encapsulation in miniemulsion, Macromolecules 38 (2005) 3225-3236.

7

Energy Storage and Transduction in Mitochondria

Bahareh Golfar, Mohsen Nosrati and Seyed Abbas Shojaosadati
Biotechnology Group,
Chemical Engineering Department,
Tarbiat Modares University, Tehran,
Iran

1. Introduction

The processes of energy storage and dissipation in biological systems have been studied during the past few decades in search of alternative energy storage systems to the conventional ones. Based on these studies, living cells have proven to provide appropriate energy storage and consumption patterns for other areas of science and engineering (Alberty, 2003; Lehninger, 1984).

The ability of cells to store energy in an efficient manner and to release it to gain control over the system, has made them an important target for energy related studies and modeling efforts (Qian & Beard, 2006). Since bioenergetics and biochemical thermodynamics specifically deal with energy transductions in biochemical reactions, it would be necessary to investigate these processes from a thermodynamic point of view.

Living organisms usually operate at constant temperature and depend on energy from food consumption or exposure to sunlight for running their vital processes and maintaining their body temperature. Energy transduction takes place in the mitochondrion of animal cells, chloroplast of plant cells and cytoplasm of bacteria. This study focuses on bioenergetics of mitochondria, considering that membranes of mitochondria, chloroplasts and bacteria show many similarities in this regard.

Mitochondria have two types of complexes for obtaining energy from substrates. Complex I includes production of NADH from oxidation of fatty acids, TCA cycle, and glycolysis. Complex II includes $FADH_2$ production from TCA cycle. These complexes vary in different kinds of mitochondria (Cairns et al., 1998). The energy is eventually stored in the body in the form of high-energy molecules such as Adenosine Triphosphate (ATP). ATP molecules have three high-energy bonds which enable them to store energy and then release it as the bonds are broken according to the following equations (Hammes, 2000; Harper et al., 2000):

$$\mathrm{ATP} + \mathrm{H_2O} \rightarrow \mathrm{ADP} + \mathrm{P_i} \quad , \Delta G^{\circ\prime} = (-30.5)\ \mathrm{KJ/mol} \tag{1}$$

$$\mathrm{ADP} + \mathrm{H_2O} \rightarrow \mathrm{AMP} + \mathrm{P_i} \quad , \Delta G^{\circ\prime} = (-27.6)\ \mathrm{KJ/mol} \tag{2}$$

The change in Gibbs free energy for ATP hydrolysis in cells is estimated as follows:

$$\Delta G = \Delta G^{\circ\prime} + R.T.\ln(k) \tag{3}$$

The normal operating conditions of cells would be considered as T= 37°C and k=2300. By replacing these values into equation 3, the total amount of energy released from the hydrolysis of each mole of ATP will be 50 KJ. As a result, an average person with a body mass of 50 K, who needs at least 11700 KJ energy per day, will require over 125 K of ATP. The fact that this amount of energy is produced by only 50 g of this molecule in his body, confirms that ATP is constantly produced and consumed in cycles in the body (Datta, 2002; Haynie, 2003).
Oxidation of different substrates such as 3-hydroxybutyrate, glutamate plus malate (with equal mole fractions), 2-oxoglutarate, and succinate in mitochondria provides the energy that is required to phosphorylate molecules of ADP to form ATP molecules. This process is called "oxidative phosphorylation" and enables the aerobic organisms to obtain more energy from the substrates in comparison to anaerobic organisms (Haynie, 2003; Scheffler, 2000). The overall oxidative phosphorylation process in the mitochondria can be expressed as follows:

$$SH_2 + 0.5O_2 + nADP + nP_i \rightarrow S + H_2O + nATP \quad (4)$$

where S represents the substrate and the stoichiometric coefficient, n, is also determined by the type of substrate (Lemasters et al., 1984). The inner membrane operates very selectively and most of the metabolites and ions such as P_i, ADP, ATP and the respiratory substrates can only cross it through channels or by means of carrier proteins. The transport mechanism of these carriers is usually based on exchanging one substance for the other (Szewczyk & Wojtczak, 2002).
According to the chemiosmotic hypothesis, the electro-chemical driving force for transferring potassium ions into the matrix, leads to the opening of K_{ATP} channel which in turn results in osmotic swelling. In order to maintain electrical balance, the protons released by substrate oxidation are pumped from the matrix to the intermembrane space of mitochondrion. Subsequently, the concentration gradient drives these protons back to the matrix, where they will contribute to the phosphorylation process (Jin & Bethke, 2002; Mitchell, 1961, 1966, 1972). Therefore, proton flux causes a proton motive force (PMF) which is the key factor in energy transduction and ATP production. This process has been schematically shown in figure 1.

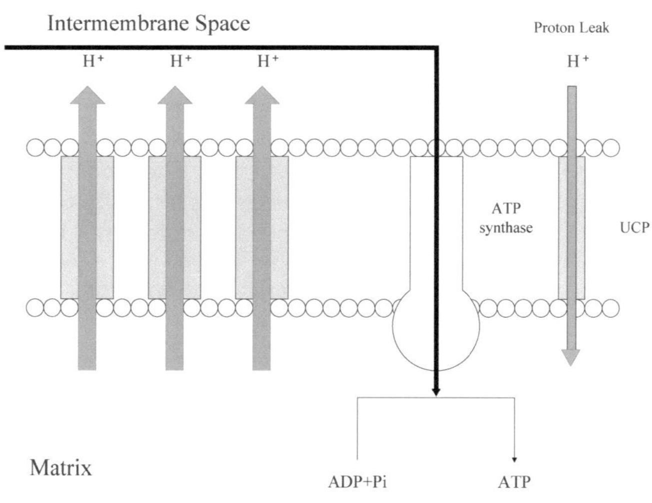

Fig. 1. Proton transport across the inner membrane of mitochondrion

However, coupling of oxidation and phosphorylation processes is not 100 percent complete. As can be observed in figure 1, the uncoupling proteins (UCPs) in the inner membrane let some protons back to the matrix without passing through ATPsynthase (Stuart et al., 2001). This proton leak reduces the coupling of processes and lowers the amount of ATP that is produced (Brand, 2005; Jezek et al., 1998; Jezek, 1999). The cycle of proton pumping and proton leaking across the membrane, also referred to as futile cycle, can release significant amounts of energy (Brand et al., 1999).

The efficiency of oxidative phosphorylation demonstrates the physiological role of mitochondria. In general, mitochondria are capable of taking on different roles to economize energy storage relevant to the current status of the body. These roles can be summarized as (Cairns et al., 1998):

- Maximizing ATP production
- Maximizing P_i production
- Minimizing the costs of energy storage
- A combination of the above

The overall rate of ATP production depends on substrate availability and cellular energy demand. Mitochondria of different tissues have various functions for matching the energy transductions with energy demands; based on these functions, mitochondria will be categorized as either "energetic" or "thermogenic" (Moyes, 2003). Mitochondria with energetic role, e.g. those of liver cells, are designed to maximize ATP production to provide the energy required for vital reactions. On the other hand, mitochondria with thermogenic role, e.g. those of brown adipose tissue (BAT) cells, release a considerable amount of energy as heat to maintain constant body temperature (Porter, 2001; Schrauwen & Hesselink, 2002).

Mitochondria within BAT cells differ from mitochondria of liver cells in that they have a limited rate of ATP synthesis, lower membrane potential, and higher respiratory rates (Kowaltowski, 2000). This is due to varying amounts of proton leak in different mitochondria and also dependents on the body mass (Hulbert, 2003; Else et al., 2004). The major characteristic of the membrane that distinguishes these two types of mitochondria is the property of membrane proton permeability (C_H) which can be used as a criterion for evaluation of proton leak across the membrane. The amounts of C_H are generally much higher in thermogenic mitochondria than the energetic ones (Nicholls, 1997). The main uncoupling protein in BAT is UCP1 which is activated by fatty acids and inhibited by nucleotides (Brand et al., 1999). These remarkable characteristics of BAT cell mitochondria have been a topic of interest among many researchers both from biological and thermodynamic point of views (Matthias et al., 2000).

In order to apply biological energy patterns to current industrial energy systems, an appropriate body of comprehensive models and criteria is required. Unfortunately most of the studies on energetic and thermogenic functions have so far focused on qualitative descriptions (Cadenas et al., 2001; Schrauwen & Hesselink, 2002) and little effort has been made to compare them from a quantitative point of view.

In this study we have proposed a thermodynamic model for ATP synthesis in systems that operate with some distance from equilibrium, by which the energy loss and the efficiency of oxidative phosphorylation can be calculated. Consequently, we have made a quantitative comparison between the rate of energy loss and efficiency of energetic and thermogenic mitochondria by means of this model. This quantitative evaluation of different mitochondria leads to a better understanding of their thermodynamic functions.

2. Non-equilibrium thermodynamics

Although thermodynamics is highly applied in studying biological systems, still many thermodynamic analyses are done based on equilibrium conditions. In other words, it is usually assumed that the systems tend to an equilibrium state after a while. This assumption is not very accurate for biological systems due to the fact that their survival depends on constant mass and energy exchange with their environment. As a result, it would be best to assume that such systems approach a non-equilibrium stationary state (NESS) (Qian & Beard, 2005).

The main difference between equilibrium steady state and non-equilibrium stationary state is that the system needs a constant supply of energy to maintain the latter state, which can be provided by cellular metabolism (Jou & Llebot, 1990). As a result, non-equilibrium thermodynamics (NET) govern such systems.

One of the simplifying assumptions in NET is the local equilibrium assumption, which states that in every small region within the system, thermodynamic properties can be related to state variables by means of equilibrium state equations. A small region is identified by enough number of molecules so that the macroscopic theory can be applied. Therefore, the entropy and specific internal energy can be obtained through the same calculations as in equilibrium state and Gibbs and Gibbs-Duhem relations are also applicable (Demirel & Sandler, 2001, 2004; Hill, 2002; Mazur, 1999).

When applying non-equilibrium thermodynamics to a process, it is important to take into account how far from equilibrium the process is (Demirel & Sandler, 2004). Distance from equilibrium conditions can be determined by the energy dissipation function (Φ) which gives the rate of free energy loss of a system (Caplan & Essig, 1969).

Entropy production in living systems can be viewed from three different aspects (Gnaiger, 1994):

- Stationary low entropy level:
  According to Prigogin, biological systems tend to produce the minimum amount of entropy and maintain almost constant entropy, so that it can be assumed:

$$dS/dt = 0 \tag{5}$$

- Entropy production within the system:
  Energy dissipation from irreversible processes in the system cause an increase in the entropy so that:

$$d_iS/dt > 0 \tag{6}$$

- Stream of negative entropy:
  In order to balance the entropy that is produced in the cell, some entropy is lost through interactions with the surrounding environment. This behavior can be expressed as follows:

$$d_eS/dt < 0 \tag{7}$$

Subsequently, the overall entropy balance of a biological system based on non-equilibrium thermodynamics can be regarded as:

$$dS/dt = d_iS/dt + d_eS/dt \tag{8}$$

Equation 8 suggests that entropy change in cells has two distinct components; d_eS which represents entropy exchange through system boundaries, and d_iS that corresponds to entropy production within the system. For every small volume within the system (dv), energy dissipation function and entropy are related by (Demirel & Sandler, 2001):

$$\phi = T.\, d_iS/dt.\, dv \tag{9}$$

Since energy dissipation is directly proportional to the entropy production, Φ can be used to evaluate the amount of energy released during a process (Demirel & Sandler, 2004). Therefore, one of the main objectives of this study was to determine Φ for different mitochondria in order to compare their functions. However, a prerequisite for determination of Φ is the knowledge of the fluxes and forces in the system.
Based on Linear Non-equilibrium Thermodynamics theory (LNET) for processes with small values of Φ, the relationship between the driving forces (Potential gradients within the system) and thermodynamic fluxes are linear. LNET theory assumes local thermodynamic equilibrium within the system and is valid for many processes in biological systems (Demirel & Sandler, 2001). This can be stated as follows:

$$J_i = \sum_j (L_{ij}.\, X_i) \; ; \; (i,j = 1,2,\ldots,n) \tag{10}$$

In equation 10, J_i represent thermodynamic fluxes, and X_j stand for thermodynamic forces. The L_{ij} coefficients are phenomenological coefficients (PCs) that have the characteristics of conductance and contain some general information on the coupling mechanism of the processes (Aledo & Valle, 2004). According to the Onsager's theory, the matrix of PCs is symmetrical and positive definite. Therefore, the following relations exist (Stucki, 1980):

$$L_{ij} = L_{ji} \tag{11}$$

$$L_{ii} \geq 0 \tag{12}$$

$$L_{ii}.\, L_{jj} \geq L_{ij}^2 \tag{13}$$

Based on the information provided in this section, Φ can be determined for different processes if the relevant fluxes and forces are known. In the next section a thermodynamic model is developed to study oxidative phosphorylation processes and dissipation function.

3. Thermodynamic model

Although many studies have been carried out on ATP production in mitochondria, literature seems to be lacking reliable mathematical models in this area. Such models could be used to provide proper quantitative results on the amounts of energy being stored or released, as well as the entropy production and efficiency of oxidative phosphorylation processes.
In this section a thermodynamic model for determining energy dissipation, ATP production, and efficiency of oxidative phosphorylation processes is presented. This model is based on non-equilibrium thermodynamic equations, and chemiosmotic theory (Golfar et al., 2010).

3.1 Fluxes and forces

The first step is to determine the relationships between thermodynamic forces and fluxes. For oxidative phosphorylation, equation 10 is written as follows:

$$J_{Ox} = L_{OO}.\, A_{Ox} + L_{OH}.\, \Delta\mu_H + L_{OP}.\, A_{Ph} \tag{14}$$

$$J_H = L_{HO}.A_{Ox} + L_{HH}.\Delta\mu_H + L_{HP}.A_{Ph} \quad (15)$$

$$J_{Ph} = L_{PO}.A_{Ox} + L_{PH}.\Delta\mu_H + L_{PP}.A_{Ph} \quad (16)$$

In equations 14, 15 and 16, the subscripts Ox, H, and Ph refer to substrate oxidation, net H^+ flow and ADP phosphorylation. $\Delta\mu_H$ is the electrochemical potential difference across the inner membrane of mitochondrion and can be obtained from the following equation:

$$\Delta\mu_H = F\Delta\psi - 2.3(RT.\Delta pH) \quad (17)$$

where F is the Faraday constant and $\Delta\psi$ is the electrochemical potential difference across the membrane. The values of $\Delta\psi$ vary between 140 to 200 mV for different mitochondria.
A_{Ox} and A_{Ph} are the affinities of oxidation and phosphorylation processes which serve as thermodynamic forces. These affinities can generally be calculated by means of the following equation:

$$A_i = -\sum_j(\nu_{ji}.\mu_j) \quad (18)$$

where ν_{ji} is the stoichiometric coefficient of species j in the ith reaction and μ_j is the electrochemical potential of j (Caplan & Essig, 1969). However, in the case of oxidative phosphorylation, the affinities of processes have been considered equal to Gibbs free energy difference (ΔG) with opposite signs (Lemasters et al., 1984).
In order to establish the phenomenological coefficients, the relationships among them should be verified. The following steps have been taken to determine which coefficients are independent and how to relate them to the dependent ones.
Although J_H can be calculated by equation 15, it can also be written as sum of the flux due to oxidation and the fluxes through ATPsynthase and passive channels (proton leak) as follows:

$$J_H = (-m_O.J_{Ox}) + C_H.\Delta\mu_H + m_P.J_{Ph} \quad (19)$$

where m_O and m_P are the stoichiometric coefficients of the respective pumps and C_H is the proton permeability of the membrane per unit area (Jou & Llebot, 1990). By replacing equations 14 and 16 into equation 19 and its comparison with equation 15 results in the following set of equations:

$$L_{OH} = (-m_O).L_{OO} + m_P.L_{PO} \quad (20)$$

$$L_{PH} = (-m_O).L_{OP} + m_P.L_{PP} \quad (21)$$

$$L_{HH} = (-m_O).L_{OH} + C_H + m_P.L_{PH} \quad (22)$$

Equations 20 to 22 enable us to calculate the desired phenomenological coefficients.
Since mitochondria operate at steady state conditions, all of the protons that are pumped to the intermembrane space will return to the inner membrane either by means of ATPsynthase or other enzymes. Otherwise, the electrochemical potential difference between the two sides of the membrane would increase. Therefore, the total proton flux is equal to zero at steady state ($J_H = 0$). Setting equation 15 equal to zero, $\Delta\mu_H$ will be calculated from the next equation:

$$\Delta\mu_H = ((-L_{PH})/L_{HH}).A_{Ph} + ((-L_{OH})/L_{HH}).A_{Ox} \tag{23}$$

Substituting equation 23 into equations 14 and 16, the fluxes J_{Ox} and J_{Ph} will appear as:

$$J_{Ox} = [L_{OO} - (L_{OH}^2/L_{HH})].A_{Ox} + [L_{OP} - (L_{OH}.L_{PH}/L_{HH})].A_{Ph} \tag{24}$$

$$L_{Ph} = [L_{PO} - (L_{PH}.L_{OH}/L_{HH})].A_{Ox} + [L_{PP} - (L_{PH}^2/L_{HH})].A_{Ph} \tag{25}$$

Now that we are able to determine the values of oxidation and phosphorylation fluxes by means of equations 24 and 25, we can proceed to the next step to evaluate energy dissipation function in mitochondria.

3.2 Energy dissipation function

Under isothermal conditions, Φ can be generally expressed as follows (Caplan & Essig, 1969):

$$\phi = \left[-\textstyle\sum_j\left(\dot{n}_j^{in}.\Delta\mu_j\right) + \sum_j(J_i.A_i^{ex})\right] \tag{26}$$

where n_j represents the number of moles of species j, and $\dot{n}_j$ is defined as:

$$\dot{n}_j = (dn_j)/(dt) \tag{27}$$

The superscripts "in" and "ex" refer to the interior and exterior of the inner membrane of mitochondrion. In case of oxidative phosphorylation the following relation exists for net H^+ flow:

$$\dot{n}_H^{in} = (-)J_H \tag{28}$$

Therefore, the general equation for energy dissipation function (equation 26) takes the following form:

$$\phi = J_{Ox}.A_{Ox} + J_H.\Delta\mu_H + J_{Ph}.A_{Ph} \tag{29}$$

At the steady state the net proton flux is set to zero so that Φ is as follows:

$$\phi = J_{Ox}.A_{Ox} + J_{Ph}.A_{Ph} \tag{30}$$

Substituting equations 24 and 25 into equation 30, the dissipation function for oxidative phosphorylation at steady state will appear as:

$$\phi = \left\{\begin{matrix}[L_{OO} - (L_{OH}^2/L_{HH})] + 2[L_{OP} - (L_{OH}.L_{HP}/L_{HH})].(A_{Ph}/A_{Ox}) \\ +[L_{PP} - (L_{PH}^2/L_{HH})].(A_{Ph}/A_{Ox})^2\end{matrix}\right\}.A_{Ox}^2 \tag{31}$$

In equation 31, L_{OO} (influence of substrate availability on oxygen consumption), L_{PP} (feedback of phosphate potential on ATP production), and C_H (membrane proton permeability) depend on the nature of the inner membrane and are available for various mitochondria. Similarly, the values of m_O and m_P for different substrates are available from the literature. Knowing the amounts of these parameters, L_{OH}, L_{PH}, and L_{HH} can be obtained from equations 20 to 22. L_{OP} (substrate dependency of ATP production) can be determined according to degree of coupling of oxidation and phosphorylation reactions (q) by means of the following relation (Cairns et al., 1998):

$$q = L_{OP}/[(L_{OO}.L_{PP})^{0.5}] \quad (32)$$

q is a dimensionless scale that represents how well the process of oxidation is coupled with phosphorylation. In case of complete coupling, q is equal to one and if the processes are independent from each other, q is equal to zero. For any pair of coupled reactions, q can be viewed as follows:

$$q = \sqrt[N]{\prod L_{ij}/\prod L_{ii}} \quad (33)$$

When the value of q is close to one, the stoichiometric coefficients can be applied with an appropriate precision. As the values of q deviate from one, it would be best to use phenomenological stoichiometric coefficients (Z) that are defined as (Stucki, 1980):

$$Z = \sqrt{L_{PP}/L_{OO}} \quad (34)$$

As q tends to one, values of Z tend to real values of stoichiometric coefficients. The relationship between q and Z is as follows (Lemasters et al., 1984):

$$Z = (-q).(\Delta G_R/\Delta G_P) \quad (35)$$

where ΔG_R and ΔG_P are the Gibbs free energy change for phosphorylation and oxidation reactions respectively.

The degree of coupling has been experimentally determined for some energetic mitochondria (Lemasters et al., 1984; Stucki, 1980), but as for thermogenic ones the data is more limited. Therefore, we have considered the full range of variations of q from 0 to 1. Based on equation 32, for fixed values of L_{OO} and L_{PP}, L_{OP} is minimum at q = 0 and maximum at q = 1. Therefore the range of variations of L_{OP} can be determined for any kind of mitochondrion. After determination of the related parameters, Φ can be calculated for any given A_{Ph}/A_{Ox} using equation 31.

Although Φ is a very useful criterion for comparing different mitochondrial functions, evaluating the efficiency of oxidative phosphorylation processes will provide more insight into these missions and operating regimes.

3.3 Efficiency

The efficiency of oxidative phosphorylation is defined as the percentage of released energy by oxidation process that is consumed by phosphorylation process as follows (Kedem & Caplan, 1965):

$$\eta = (-J_{Ph}.A_{Ph})/(J_{Ox}.A_{Ox}) \quad (36)$$

The J_{Ph}/J_{Ox} ratio (or P/O ratio, in brief) can be theoretically determined from equations 24 and 25 and consequently, the efficiency of oxidative phosphorylation can be easily obtained from equation 36. The J_{Ph}/J_{Ox} ratio is an important criterion in biological systems (Hinkle, 2005) and represents the number of moles of ATP that are produced per consumed moles of oxygen. High values of this ratio imply high efficiency for energy storage processes. Furthermore, the optimum efficiency (η_{opt}) could be determined by means of the following equation (Stucki, 1980):

$$\eta_{opt} = q^2/\left(1+\sqrt{1-q^2}\right)^2 \qquad (37)$$

From equation 37 it can be concluded that optimum efficiency happens when phosphorylation flux is not zero and the priority for the mitochondrion is to maximize ATP production. Clearly, complete coupling of oxidation and phosphorylation processes leads to maximum efficiency.

Such phenomenological thermodynamic models as the present one deal with the role of mitochondria of different organs in utilization and storage of biological energy (or ATP). As a result they could be used to determine energy dissipation function and efficiency of oxidative phosphorylation processes in mitochondria with different thermodynamic functions. The output of these theoretical approaches could be compared with experimental data, if any, to evaluate the model.

4. Results and discussion

In this section, the thermodynamic model is applied to two different types of mitochondria to compare their behaviors based on energy dissipation and efficiency of oxidative phosphorylation processes. We have focused on types of mitochondria for which there is sufficient experimental data available in literature. This will provide the chance to evaluate the theoretical results generated by the model by comparing them against the experimental results.

In order to investigate mitochondria with different thermodynamic roles, rat liver cell mitochondrion with energetic role and BAT cell mitochondrion with thermogenic function have been chosen. Calculations have been carried out for 3-hydroxybutyrate, glutamate plus malate (with equal mole fractions), 2-oxoglutarate and succinate as substrate, all of which have been widely used in previous investigations in this field. The values of different parameters for these two tissues (Jou & Llebot, 1990) and four substrates (Copenhaver & Lardy, 1952; Lee et al., 1996) are listed in tables 1 and 2 respectively. Table 1 includes the parameters related to the structure of the membranes, whereas table 2 contains the parameters corresponding to different substrates.

Parameter	Rat Liver Mitochondrion	Brown adipose tissue Mitochondrion
L_{OO}	1.9 nmolO_2/(mgP.min.mV)	0.5 nmolO_2/(mgP.min.mV)
L_{PP}	7.9 nmolH^+/(mgP.min.mV)	0.4 nmolH^+/(mgP.min.mV)
C_H	3.2 nmolH^+/(mgP.min.mV)	35 nmolH^+/(mgP.min.mV)

Table 1. Parameters related to rat liver and brown adipose tissue mitochondria (Jou & Llebot, 1990).

Energy dissipation function has been calculated in each case by means of equation 31. Since q is approximately 0.98 in rat liver cell mitochondria (Lemasters, 1984), L_{OP} is equal to 3.8 according to equation 32. As for BAT cell mitochondria, data on values of q were not sufficient. As a result, we assigned different values between 0 and 1 to q which lead to values of L_{OP} varying between 0 and 0.4 based on equation 32. The proton gradient across the inner membrane is taken equal to 200 mV in calculations but the results still hold for a large range of affinity ratios for proton gradients from 140 to 200 mV.

Substrate	m_P	m_O	A_O
3-Hydroxybutyrate	4 nmolH+/nmolATP	12 nmolH+/nmolO$_2$	209 KJ/mol
Glutamate+Malate	4 nmolH+/nmolATP	12 nmolH+/nmolO$_2$	220 KJ/mol
2-Oxoglutarate	4 nmolH+/nmolATP	12 nmolH+/nmolO$_2$	307 KJ/mol
Succinate	4 nmolH+/nmolATP	6 nmolH+/nmolO$_2$	151 KJ/mol

Table 2. Parameters related to different substrates (Copenhaver & Lardy, 1952; Lee et al., 1996).

By replacing these values into equation 31, the rate of free energy loss (Φ) has been determined and plotted versus the affinity ratio (A_{Ph}/A_{Ox}) for both energetic and thermogenic mitochondria with 3-hydroxybutyrate, glutamate plus malate, 2-oxoglutarate and succinate respectively. Figures 2 to 5 correspond to these plots.
It is clearly seen that the values of Φ in BAT mitochondria are two to four times greater than rat liver mitochondria, indicating higher amounts of proton leak in BAT mitochondria. These results are in complete agreement with the qualitative descriptions based on biological functions of the two types of tissues and indicate the validity of the proposed model for such calculations.

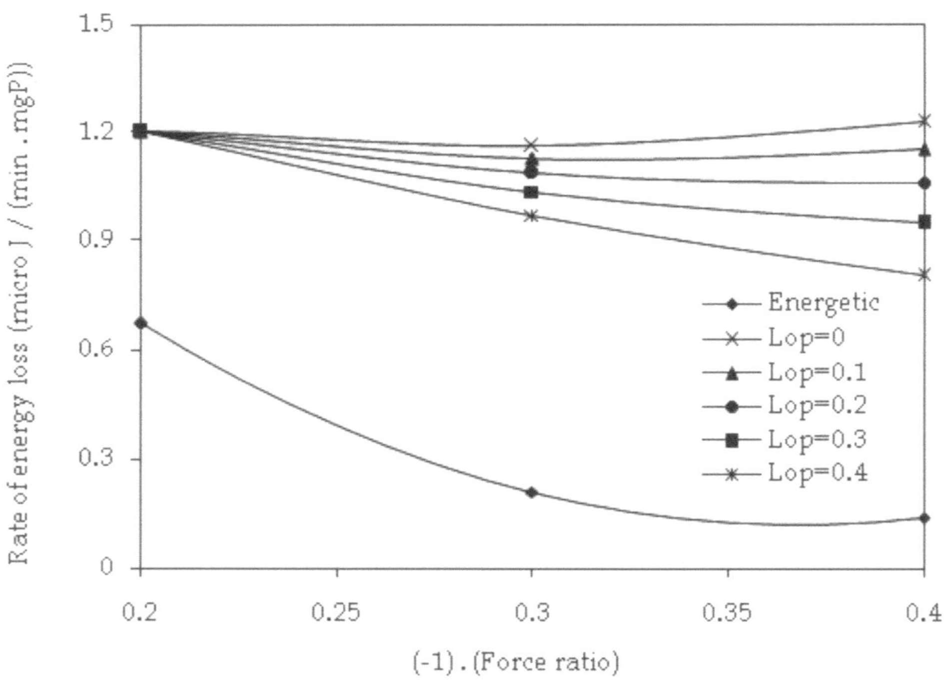

Fig. 2. Rate of energy loss (micro joules/(mgP.min)) vs. force ratio in rat liver and BAT mitochondria with 3-hydroxybutyrate as substrate.

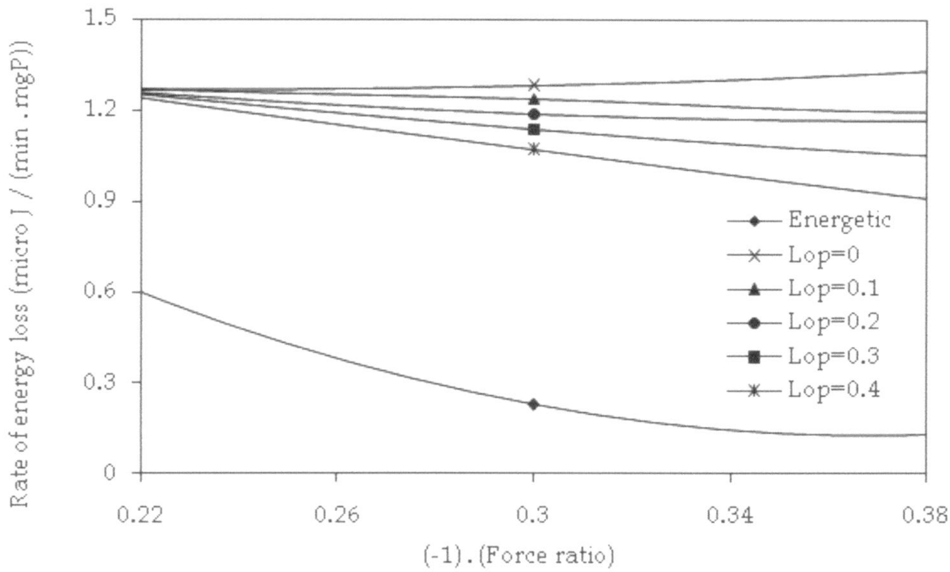

Fig. 3. Rate of energy loss (micro joules/(mgP.min)) vs. force ratio in rat liver and BAT mitochondria with glutamate+malate as substrate.

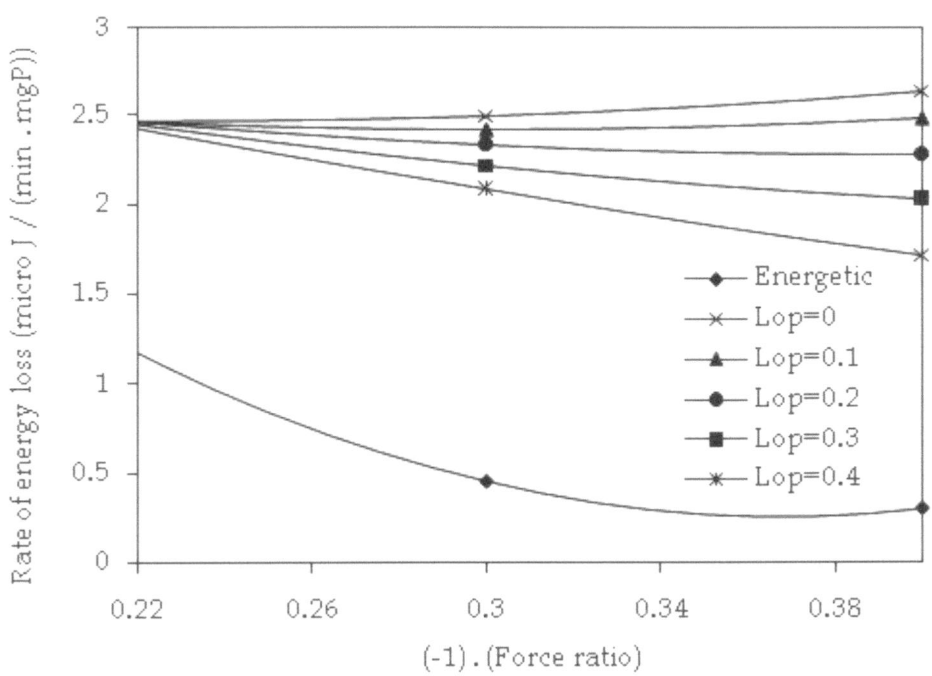

Fig. 4. Rate of energy loss (micro joules/(mgP.min)) vs. force ratio in rat liver and BAT mitochondria with 2-oxoglutarate as substrate.

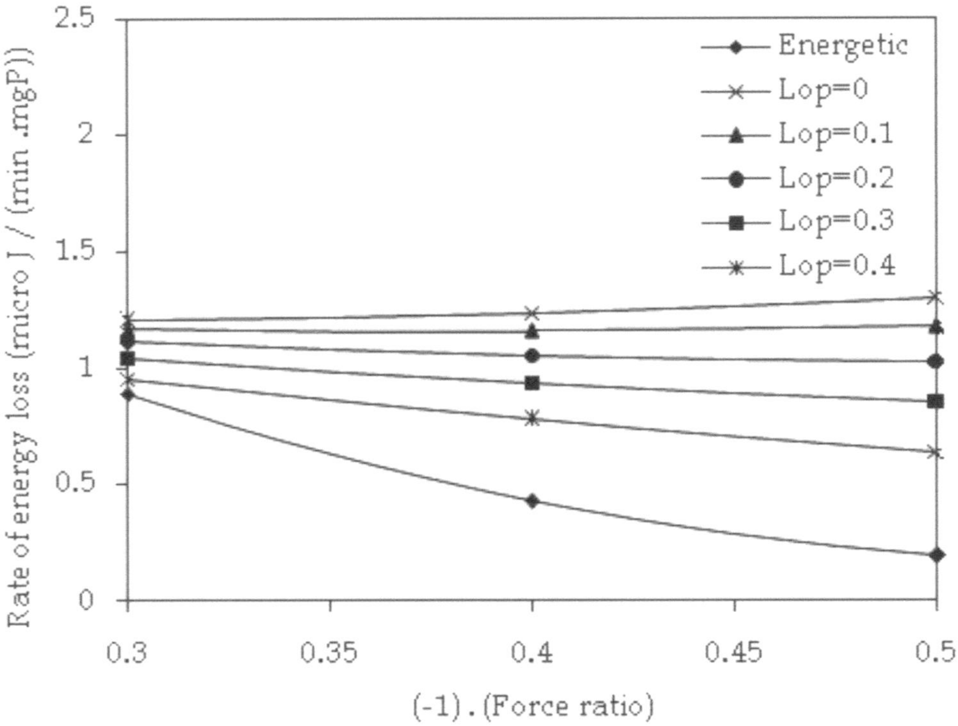

Fig. 5. Rate of energy loss (micro joules/(mgP.min)) vs. force ratio in rat liver and BAT mitochondria with succinate as substrate.

The efficiency of oxidative phosphorylation processes have also been calculated for these mitochondria for different values of L_{OP} with the four selected substrates, and plotted against Φ in figures 6 to 9. The curves in these figures show theoretical results while separate points show some experimental results (Hinkle et al., 1991; Lehninger, 1955; Lemasters, 1984; Nath, 1998; Nicholls, 1974).

From figures 6 to 9 three main points can be made:

- As expected, the efficiency of oxidative phosphorylation is much higher in rat liver than in BAT mitochondria. Lower efficiency is an advantage for BAT mitochondria since it enables them to release heat, conduct thermogenesis and regulate body temperature (Cannon & Nedergaard, 2003).
- In both energetic and thermogenic tissues the values of Φ are low considering the high values of efficiency. Furthermore, in rat liver mitochondria, selection of parameters leads to minimum entropy production with high efficiency. This operating regime in biological systems complies neither with minimum entropy production (MEP) nor maximum power output (MPO) regimes. In fact this conclusion supports the idea that biological systems follow the ecological regime, which involves producing little entropy together with considerable efficiency (Sanitillan et al., 1997).
- Once more the results obtained from the presented model comply with the earlier experimental outcomes. As can be seen in figures, the amounts of efficiency calculated in this model for rat liver mitochondria are close to the experimental results. Therefore,

we are convinced that the predicted values for the efficiency in thermogenic mitochondria will also comply with experimental results. This could be a challenge for further research in order to find proper data for the efficiency in thermogenic mitochondria.

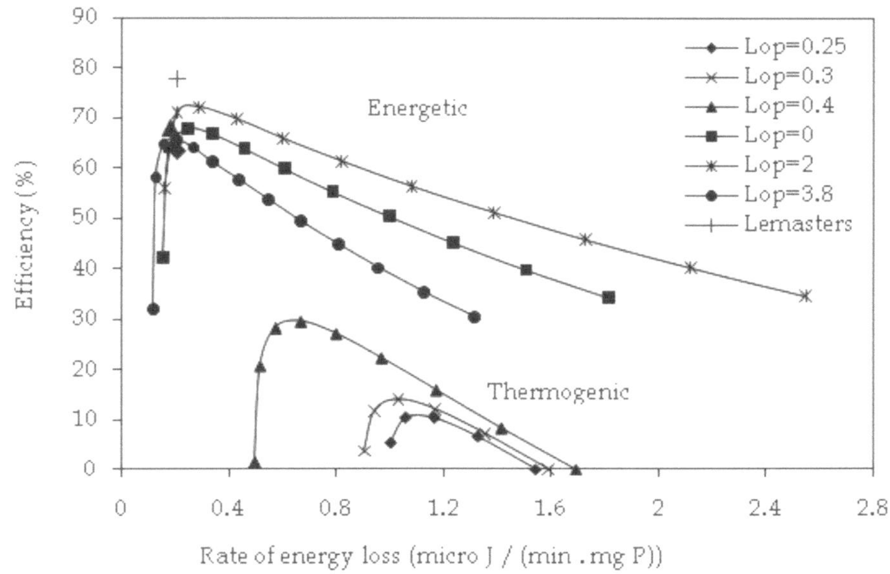

Fig. 6. Efficiency of oxidative phosphorylation vs. rate of energy loss in rat liver and BAT mitochondria with 3-hydroxybutyrate as substrate.

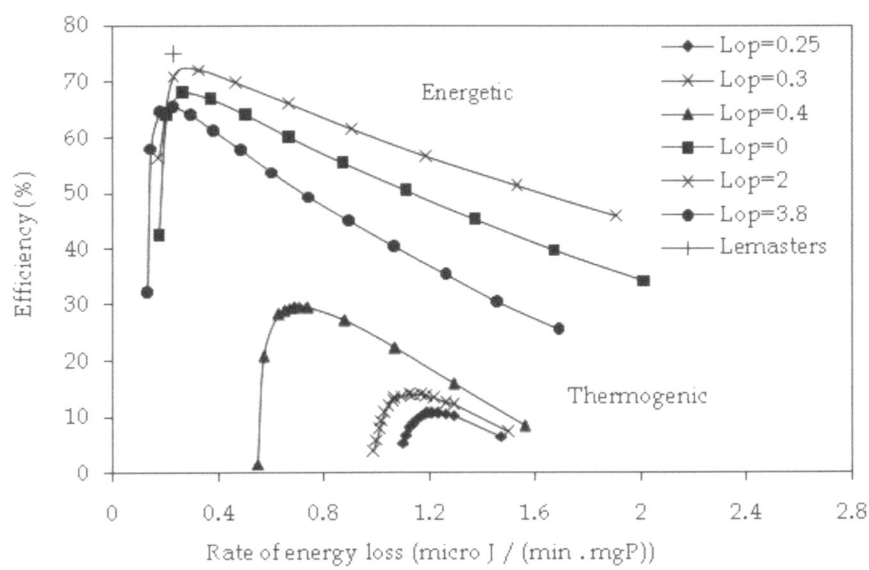

Fig. 7. Efficiency of oxidative phosphorylation vs. rate of energy loss in rat liver and BAT mitochondria with glutamate+malate as substrate.

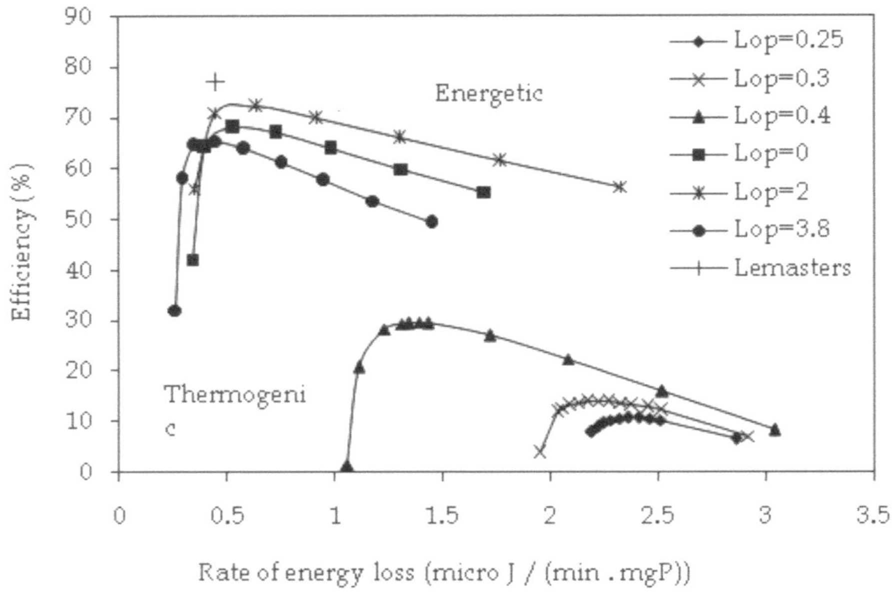

Fig. 8. Efficiency of oxidative phosphorylation vs. rate of energy loss in rat liver and BAT mitochondria with 2-oxoglutarate as substrate.

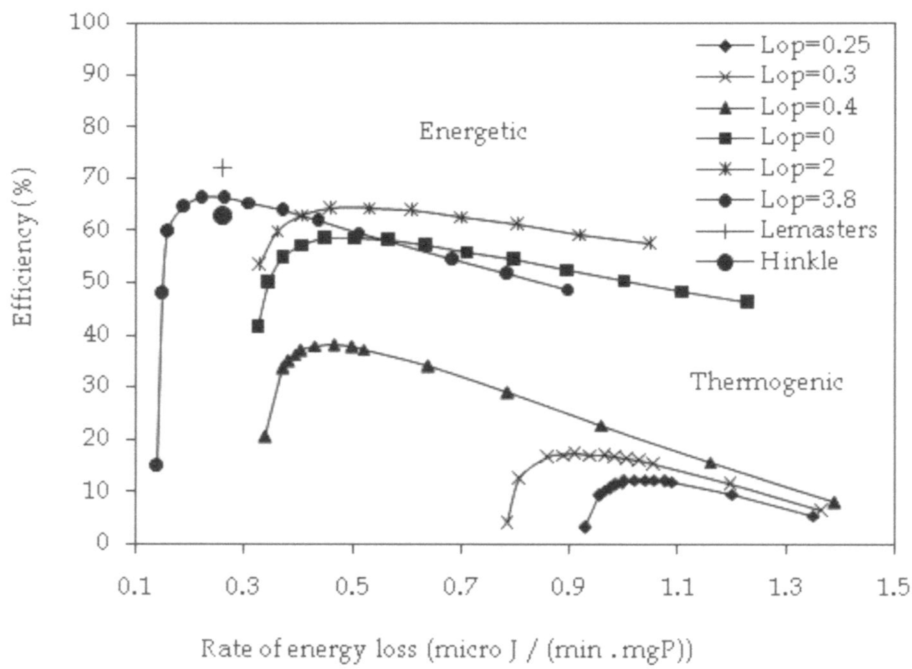

Fig. 9. Efficiency of oxidative phosphorylation vs. rate of energy loss in rat liver and BAT mitochondria with succinate as substrate.

5. Conclusion

Since biological systems are reasonably efficient in energy storage, they can be regarded as appropriate patterns for science and engineering. Thermodynamic models on energy transductions in such systems could play a key role in applying these patterns in industry and other relevant areas. In developing models for energy transductions in biological systems, it is important to apply non-equilibrium thermodynamics since the survival of these systems depend on constant mass and energy exchange with their surroundings, which requires operating at some distance from the equilibrium state.

Energy dissipation function (Φ), along with efficiency of oxidative phosphorylation processes can be viewed as useful criteria in studying the energy storage capabilities of a system and its operating regime. They can also be used to explain different mitochondrial functions.

Mitochondria of various tissues have different functions for matching the energy transductions with energy demands. The rate of energy dissipation and efficiency of energy storage in mitochondria is set according to their roles. In the previous sections, rate of free energy dissipation and efficiency of ATP production were determined for both energetic and thermogenic mitochondria by means of the proposed model and plotted in figures 2 to 9. These plots suggest that mitochondria with energetic function dissipate less energy as heat and store more energy in form of ATP molecules. As a result, the efficiency of oxidative phosphorylation is high in these cases (about 60 to 70 percent in rat liver mitochondria). On the contrary, thermogenic mitochondria release a great deal of energy due to more proton leak across the inner membrane. Therefore, the maximum amount of efficiency in BAT mitochondria is about 30 percent. These theoretical results comply with the experimental results for rat liver mitochondria.

Furthermore, comparison of efficiency values of two types of mitochondria with the rate of their energy dissipation indicates that such systems tend to produce less entropy and store energy in an efficient manner. This conclusion supports the theory of ecological regime in biological systems.

6. Further research

Based on the results of this research as well as the previous works on the subject, developing models for energy transductions in living organisms is of great importance in applying these energy patterns in industry. Therefore, it would be beneficial to figure out such models for other microbial cells such as bacteria and fungi. Studying ATP storage in animals with special characteristics such as hibernating animals or those with high resistance against thirst or starvation might also provide a better insight in this regard.

As mentioned earlier, there is not sufficient experimental data on different kinds of organs and substrates in the literature. Performing such experiments is essential for further research in this field. Furthermore, since Φ and η can be determined for different mitochondria by means of the model, they can be used in diagnosing mitochondrial dysfunctions. Moreover, they could assist in producing therapeutic drugs with mitochondria as their first or secondary target (Szewczyk & Wojtczak, 2002). These amounts can be changed synthetically, to help overcome mitochondrial diseases (Roussel, 2004). But for this to be practically possible, the range of values of Φ and η should be found for different organs.

Such thermodynamic models can also be used in assessing the effect of some drugs used for weight loss or doping. In fact weight loss through reducing the efficiency of ATP production (also reoffered to as "increased metabolic inefficiency") is a topic of interest in nutritional studies (Fine et al., 2004). The underlying mechanism of such drugs is usually based on affecting the mitochondrial membrane and changing the amounts of energy storage or dissipation. Measuring Φ before and after drug injection helps to study the effect of the drug and determine a healthy dosage. Similarly, if the range of Φ is known for an ordinary person, an increase in this function in athletes might be a sign of doping.
We believe the model presented in this chapter has the potential to be applied in various areas of science, pharmaceuticals and industry. Expanding and generalization of this model could be a challenge for further research.

7. Nomenclatures

A affinity of reaction or ΔG of reaction (KJ/mol)
A_i affinity of the ith reaction (KJ/mol)
A_{Ox} affinity of oxidation reaction (KJ/mol)
A_{Ph} affinity of phosphorylation reaction (KJ/mol)
ADP adenosine diphosphate
ATP adenosine triphosphate
BAT brown adipose tissue
C_H membrane proton permeability [nmol H^+/(mg protein. min. mV)]
I subscript for reaction
J subscript for flux
J_H flux of proton transfer [nmol H^+/(mg protein. min)]
J_i thermodynamic flux for i[th] reaction
J_{Ox} flux of oxidation reaction [nmol O_2/(mg protein. min)]
J_{Ph} flux of phosphorylation reaction [nmol ATP/(mg protein. min)]
L_{HH} phenomenological coefficient of proton (H^+) in proton transfer [nmol H^+/(mg protein. min. mV)]
L_{ij} phenomenological coefficient of j species in i[th] reaction [nmol of j species /(mg protein. min. mV)]
L_{OH} phenomenological coefficient of proton (H^+) in oxidation reaction [nmol H^+/(mg protein. min. mV)]
L_{OO} phenomenological coefficient of O_2 in oxidation reaction [nmol O_2/(mg protein. min. mV)]
L_{OP} phenomenological coefficient of ATP in oxidation reaction [nmol ATP/(mg protein. min. mV)]
L_{PH} phenomenological coefficient of proton (H^+) in phosphorylation reaction [nmol H^+/(mg protein. min.mV)]
L_{PP} phenomenological coefficient of ATP in phosphorylation reaction [nmol ATP/(mg protein. min.mV)]
LNET linear non-equilibrium thermodynamics
m_O stoichiometric coefficient of pumps for oxidation reaction (nmol H^+/nmol O_2)
m_P stoichiometric coefficient of pumps for phosphorylation reaction (nmol H^+/nmol ATP)
n_j number of moles of species j

P	protein (ATPsynthase)
PCs	phenomenological coefficients (see L_{ij})
q	degree of coupling of oxidation and phosphorylation reactions (dimensionless)
UCP	uncoupling proteins
$\mathbf{X_j}$	thermodynamic forces for species j (KJ/mol)
$\boldsymbol{\eta}$	overall efficiency of oxidative phosphorylation
$\boldsymbol{\mu}$	electrochemical potential (KJ/mol)
$\boldsymbol{\Delta\mu_H}$	electrochemical potential difference (KJ/mol)
$\mathbf{v_{ji}}$	stoichiometric coefficient of species j in the ith reaction
$\boldsymbol{\Phi}$	energy dissipation function [micro J/(mg protein. min)]

8. References

Alberty, R.A. (2003). *Thermodynamics of biochemical reactions*, Wiley. New Jersey.

Aledo, J.C., & Valle, A.E. (2004). The ATP paradox is the expression of an economizing fuel mechanism, *The Journal of Biological Chemistry*, Vol. 279, No. 53, pp. (55372 - 55375)

Brand, M.D., Brindle, K.M., Buckingham, J.A., Harper, J.A., Rolfe, D.F.S., & Stuart, J.A. (1999). The significance and mechanism of mitochondrial proton conductance. *International Journal of Obesity*, Vol. 23, pp. (4 - 11)

Brand, M.D. (2005). The efficiency and plasticity of mitochondrial energy transduction. *Biochemical Society Transactions*, Vol. 33, No. 5, pp. (897 - 904)

Cadenas, S., Echtay, K.S., Harper, J.A., Jekabsons, M.B., Buckingham, J.A., Grau, E., Abuin, A., Chapman, H., Clapham, J.C., & Brand, M.D. (2001). The basal proton conductance of skeletal muscle mitochondria from transgenic mice overexpressing or lacking uncoupling protein-3. *Journal of Biological Chemistry*, Vol. 277, pp. (2773 - 2778)

Cairns, C.B., Walther, J., Harken, A.H., & Banerjee, A. (1998). Mitochondrial oxidative phosphorylation thermodynamic efficiencies reflect physiological organ roles. *American Journal of Physiology*, Vol. 274, pp. (1376–1383)

Cannon, B., & Nedergaard, J. (2004). Brown adipose tissue: function and physiological significance. *Physiological Reviews*, Vol. 84, pp. (277 - 359)

Caplan, S.R., & Essig, A. (1969). Oxidative phosphorylation: thermodynamic criteria for the chemical and chemiosmotic hypotheses. *Biochemistry*, Vol. 64, pp. (211 - 218)

Copenhaver, J.H., & Lardy, H.A. (1952). Oxidative phosphorylation: pathways and yield in mitochondrial preparations, *Journal of Biological Chemistry*, Vol. 195, pp. (225 - 238)

Datta, A.K. (2002). *Biological and bioenvironmental heat and mass transfer*, Marcel Dekker Inc., New York.

Demirel, Y., & Sandler, S.I. (2001). Linear - nonequilibrium thermodynamics theory for coupled heat and mass transport. *International Journal of Heat and Mass Transfer*, Vol. 44, pp. (2439 - 2451)

Demirel, Y., & Sandler S.I. (2004). Nonequilibrium thermodynamics in engineering and science. *Journal of Physical Chemistry B.*, Vol. 108, pp. (31 - 43)

Else, P.L., Brand, M.D., Turner N., & Hulbert, A.J. (2004). Respiration rate of hepatocytes varies with body mass in birds. *Journal of Experimental Biology*, Vol. 207, pp. (2305 - 2311)

Fine, E.J., Feinman, R.D. (2004). Thermodynamics of weight loss diets. *Nutrition & Metabolism*, pp. (1 - 15)

Gnaiger, E. (1994). Negative entropy for living systems: controversy between nobel laureates schroedinger, pauling and Perutz. *Modern Trends in Biothermokinetics*, Vol. 3, pp. (62 - 70)

Golfar, B., Nosrati, M., Shojaosadati, S.A. (2010). A thermodynamic approach to energy transduction in mitochondria. *Journal of Non-Equilibrium Thermodynamics*, Vol. 35, pp. (15 - 34)

Hammes, G.G. (2000). *Thermodynamics and kinetics for the biological sciences*, Wiley, United States of America

Harper, H.A., Murray, R.K., Granner, D.K., Mayes, P.A., & Rodwell, V.W. (2000). *Harper Biochemistry*, Wiley, New York

Haynie, D.T. (2003). *Biological thermodynamics*, Cambridge University Press, Cambridge

Hill, T.L. (2002). *Thermodynamics of small systems*, Dover publications, New York.

Hinkle, P.C., Arun Kumar, M., Resetar, A., & Harris, D.L. (1991). Mechanistic stoichiometry of mitochondrial oxidative phosphorylation, *Biochemistry*, Vol. 30, pp. (3576 - 3582)

Hinkle, P.C. (2005). P/O ratios of mitochondrial oxidative phosphorylation; a review. *Biochimica et Biophysica Acta*, Vol. 1706, pp. (1 - 11)

Hulbert, A.J. (2003). Life, death and membrane bilayers. *Journal of Experimental Biology*, Vol. 206, pp. (2303 -2311)

Jezek, P., Engstova, H., Zakova, M., Vercesi, A.E., Costa, A.D., Arruda, P., & Garlid, K.D. (1998). Fatty acid cycling mechanism and mitochondrial uncoupling proteins. *Biochim Biophys Acta*, Vol. 1365, No. 1-2, pp. (319 - 327)

Jezek, P. (1999). Fatty acid interaction with mitochondrial uncoupling proteins. *Journal of Bioenergetics and Biomembranes*, Vol. 31, No. 5, pp. (457 - 466)

Jin, Q., & Bethke, C.M. (2002). Kinetics of electron transfer through the respiratory chain. *Biophysical Journal*, Vol. 83, pp. (1797 - 1808)

Jou, D., & Llebot, J.E. (1990). *Introduction to the thermodynamics of biological processes*, Prentice - Hall Inc., New Jersey.

Kedem, O., & Caplan, S.R. (1965). Degree of coupling and its relation to efficiency of energy conversion, *Trans. Faraday Soc.*, Vol. 21, pp. (1897 - 1911)

Kowaltowski, A.J. (2000). Alternative mitochondrial functions in cell physiopathology: beyond ATP production, *Brazilian Journal of Medical and Biological Research*, Vol. 33, pp. (241 - 350)

Lee, C.P., Gu, Q., Xiong, Y., Mitchel, R.A., & Ernster, L. (1996). P/O ratios consistently exceed 1.5 with succinate and 2.5 with NAD-linked substrates, *FASEB Journal*, Vol. 10, pp. (345 - 350)

Lehninger, A.L. (1955). Oxidative phosphorylation. *Harvey Lecture*, Vol. 49, pp. (176 - 215)

Lehninger, A.L. (1984). *Principles of biochemistry*, Worth Publishers Inc., New York.

Lemasters, J.J., Grunwald, R., & Emaus, R.K. (1984). Thermodynamic limits to the ATP/site stoichiometries of oxidative phosphorylation by rat liver mitochondria. *Journal of Biological Chemistry*, Vol. 259, No. 5, pp. (3058 - 3063)

Lemasters, J.J. (1984). The ATP - to - oxygen stoichiometries of oxidative phosphorylation by rat liver mitochondria, *Journal of Biological Chemistry*, Vol. 259, No. 21, pp. (13123 - 13130)

Matthias, A., Ohlson, K.B.E., Fredriksson, J. M., Jacobsson, A., Nedergaard, J., & Cannon, B. (2000). Thermogenic responses in brown fat cells are fully UCP-1 dependent. *The Journal of Biological Chemistry*, Vol. 275, No., 33, pp. (25073 - 25081)

Mazur, P. (1999). Mesoscopic non–equilibrium thermodynamics; irreversible processes and fluctuations. *Physica A*, Vol. 274, pp. (491 - 504)

Mitchell, P. (1961). Coupling of phosphorylation to electron and hydrogen transfer by a chemiosmotic type of mechanism. *Nature*, Vol. 191, pp. (144 - 148)

Mitchell, P. (1966). Chemiosmotic coupling in oxidative and photosynthetic phosphorylation. *Biological Review Cambridge Philosophical Society*, Vol. 41, pp. (445 - 502)

Mitchell, P. (1972). Chemiosmotic coupling in energy transduction: a logical development of biochemical knowledge. *Journal of Bioenergetics*, Vol. 3, pp. (5–24)

Moyes, C.D. (2003). Controlling muscle mitochondrial content. *The Journal of Experimental Biology*, Vol. 206, pp. (4385 - 4391)

Nath, S. (1998). A thermodynamic principle for the coupled bioenergetic processes of ATP synthesis. *Pure & Applied Chemistry*, Vol. 70, No. 3, pp. (639 - 644)

Nicholls, D.G. (1974). The influence of respiration and ATP hydrolysis on the proton electrochemical gradient across the inner membrane of rat liver mitochondria as determined by ion distribution. *Biochimica Biophysica Acta*, Vol. 50, pp. (305–315)

Nicholls, D.G. (1997). The non - ohmic proton leak - 25 years on. *Bioscience Reports*, Vol. 17, No. 3, pp. (251-257)

Porter, R.K. (2001). Mitochondrial proton leak: a role for uncoupling proteins 2 and 3. *Biochimica et Biophysica Acta*, Vol. 1504, pp. (120 - 127)

Qian, H., & Beard, D.A. (2005). Thermodynamics of stoichiometric biochemical networks in living systems far from equilibrium. *Biophysical Chemistry*, Vol. 114, pp. (213 - 220)

Qian, H., & Beard, D.A. (2006). Metabolic futile cycles and their functions: a systems analysis of energy and control. *Systematic Biology*, Vol. 153, No. 4, pp. (192 – 200)

Roussel, D., Dumas, J.F., Simard, G., Malthiery, Y., & Ritz, P. (2004). Kinetics and control of oxidative phosphorylation in rat liver mitochondria after dexamethasone treatment. *Biochemical Journal*, Vol. 382, pp. (491 – 499)

Sanitillan, M., Arias-Hernandez, L.A., & Angulo-Brown, F. (1997). Some optimization criteria for biological systems in linear irreversible thermodynamics. *Il Nuovo Cimento*, Vol. 19, pp. (99 - 109)

Scheffler, I.E. (2000). A century of mitochondrial research: achievements and perspectives. *Mitochondrion*, Vol.1, pp. (3 - 31)

Schrauwen, P., & Hesselink, M. (2002). UCP2 and UCP3 in muscle controlling body metabolism. *The Journal of Experimental Biology*, Vol. 205, pp. (2275 - 2285)

Stuart, J.A., Cadenas, S, Jekabsons, M.B., Roussel, D., & Brand, M.D. (2001). Mitochondrial proton leak and the uncoupling protein 1 homologues. *Biochimica et Biophysica Acta*, Vol. 1504, pp. (144 - 158)

Stucki, J.W. (1980). The optimal efficiency and the economic degrees of coupling of oxidative phosphorylation. *European Journal of Biochemistry*, Vol. 109, pp. (269 - 283)

Szewczyk, A., & Wojtczak, L. (2002). Mitochondria as a pharmacological target. *Pharmacological Reveiws*, Vol. 54, No. 1, pp. (101 - 127)

Part 2

Technologies for Improving Energy Storage Systems

8

Bidirectional DC-DC Converters for Energy Storage Systems

Hamid R. Karshenas[1,2], Hamid Daneshpajooh[2], Alireza Safaee[2], Praveen Jain[2] and Alireza Bakhshai[2]
[1]*Department of Elec. & Computer Eng., Queen's University, Kingston,*
[2]*Isfahan University of Tech., Isfahan,*
[1]*Canada*
[2]*Iran*

1. Introduction

Bidirectional dc-dc converters (BDC) have recently received a lot of attention due to the increasing need to systems with the capability of bidirectional energy transfer between two dc buses. Apart from traditional application in dc motor drives, new applications of BDC include energy storage in renewable energy systems, fuel cell energy systems, hybrid electric vehicles (HEV) and uninterruptible power supplies (UPS).

The fluctuation nature of most renewable energy resources, like wind and solar, makes them unsuitable for standalone operation as the sole source of power. A common solution to overcome this problem is to use an energy storage device besides the renewable energy resource to compensate for these fluctuations and maintain a smooth and continuous power flow to the load. As the most common and economical energy storage devices in medium-power range are batteries and super-capacitors, a dc-dc converter is always required to allow energy exchange between storage device and the rest of system. Such a converter must have bidirectional power flow capability with flexible control in all operating modes.

In HEV applications, BDCs are required to link different dc voltage buses and transfer energy between them. For example, a BDC is used to exchange energy between main batteries (200-300V) and the drive motor with 500V dc link. High efficiency, lightweight, compact size and high reliability are some important requirements for the BDC used in such an application.

BDCs also have applications in line-interactive UPS which do not use double conversion technology and thus can achieve higher efficiency. In a line-interactive UPS, the UPS output terminals are connected to the grid and therefore energy can be fed back to the inverter dc bus and charge the batteries via a BDC during normal mode. In backup mode, the battery feeds the inverter dc bus again via BDC but in reverse power flow direction.

BDCs can be classified into non-isolated and isolated types. Non-isolated BDCs (NBDC) are simpler than isolated BDCs (IBDC) and can achieve better efficiency. However, galvanic isolation is required in many applications and mandated by different standards. The

complexity of IBDCs stems from the fact that an ac link must be present in their structure in order to enable power transfer via a magnetically isolating media, i.e. a transformer.
In this chapter, first some NBDC structures are briefly discussed. As isolation and/or voltage matching is required in many applications, more attention in this chapter is paid on the description of different IBDC configurations. It should be stated that in order to improve the efficiency, almost all recently proposed medium-power IBDC configurations have exploited the benefits of soft-switching or resonant techniques to increase the switching frequency and achieve lower size and weight. In this regard, although a variety of configurations employing soft-switching techniques has been proposed by researchers, they can be divided into a few basic families. In this chapter, IBDCs employing soft-switching techniques are divided into three basic families and investigated in more details. Resonant IBDCs which can be considered as a separate family are not covered in this chapter (Krismer et al., 2005; Li, & Bhat, 2010).

2. Non-isolated BDC

Basic dc-dc converters such as buck and boost converters (and their derivatives) do not have bidirectional power flow capability. This limitation is due to the presence of diodes in their structure which prevents reverse current flow. In general, a unidirectional dc-dc converter can be turned into a bidirectional converter by replacing the diodes with a controllable switch in its structure. As an example, Fig. 1 shows the structure of elementary buck and boost converters and how they can be transformed into bidirectional converters by replacing the diodes in their structure. It is noteworthy that the resulted converter has the same structure in both cases.
Fig. 2 shows the basic waveforms associated with Fig. 1.c. In the buck mode of operation, i.e. when the power is transferred from the high voltage (HV) to the low voltage (LV) side, Q_1 is the active switch while Q_2 is kept off. In the boost mode, i.e. when the power is transferred from LV to HV side, Q_2 acts as a controlled switch and Q_1 is kept off. The switching pattern during power (current) reversal is also shown in Fig. 2. The presence of inductor in the LV side results in lower ripple current which is advantageous in some applications. For example, it is usually preferred to charge/discharge batteries with low ripple current in order to achieve higher efficiency and longer life time.
Some of the major limitations associated with the NBDC shown in Fig. 1.c are:

- It can only operate in buck mode in one direction and boost in the other. In technical terms, this means that the voltage ratio d, which is defined as $d= V_B/V_A$, is either smaller or greater than unity in one direction.
- When the voltage ratio becomes large, this structure becomes impractical.
- The lack of galvanic isolation between two sides.

Many improved structures have been proposed to overcome the first two limitations. When the magnitude of two dc bus voltages is close to each other and the voltage ratio of smaller or greater than unity is required, the buck-boost or Cuk converters are the appropriate choice. Fig. 3.a shows the basic configurations of a NBDC based on buck-boost converter. Note that the polarity of dc buses is reverse with respect to a common ground which is a burden in many applications. This problem can be resolved by adding more switches to this configuration as shown in Fig. 3.b. This new configuration can be envisaged as two back-to-back connected converter of Fig. 1.c.

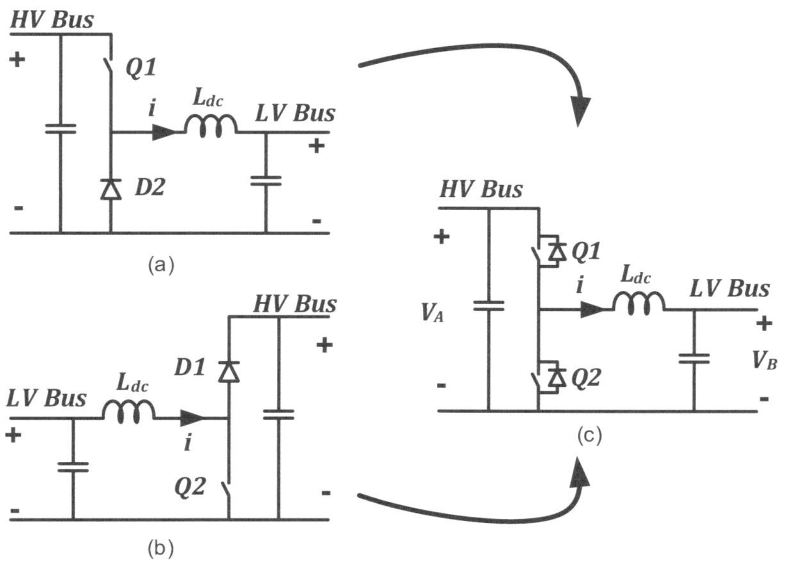

Fig. 1. (a) Elementary unidirectional buck converter, (b) elementary unidirectional boost converter and (c) transformation to bidirectional converter by substituting diodes with a controllable switch.

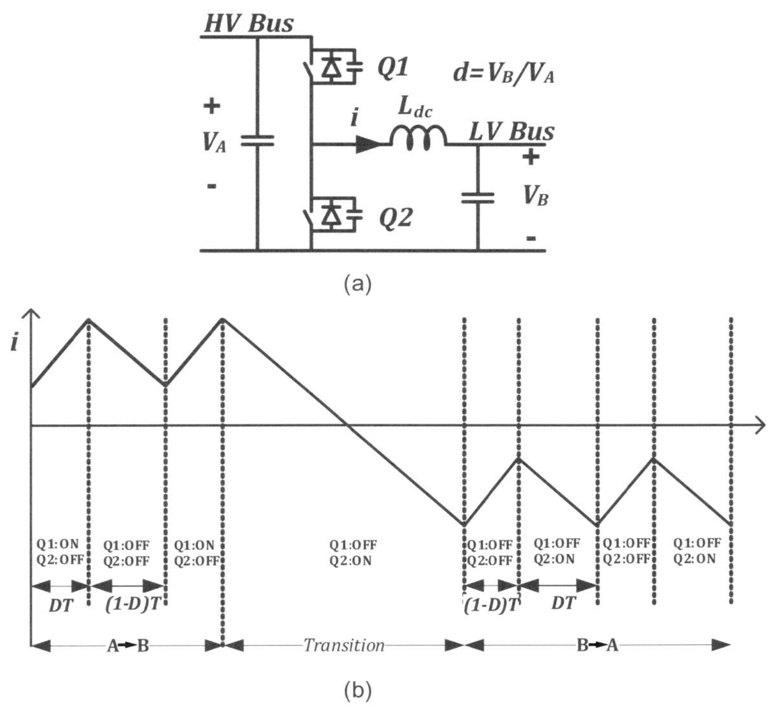

Fig. 2. (a) Basic NBDC and (b) operating waveforms.

The operation of the NBDC of Fig. 3.a is as follows. The inductor is the main energy transfer element in this converter. In each switching cycle it is charged through source side active switch for the duration of $T_{on}=DT$, where $T=1/f_{sw}$ is the switching period and D is the duty-cycle. This energy is then discharged to load during $T_{off}=(1-D)T$. In the four-switch buck-boost converter (Fig. 3.b) the principle of operation is the same. In the left to right power transfer mode, Q_1 and Q_4 act as active switches, while in the right to left power transfer the opposite switches (Q_2 and Q_3) are controlled. Synchronous rectification technique can be employed in this configuration in order to add more features and improve efficiency.

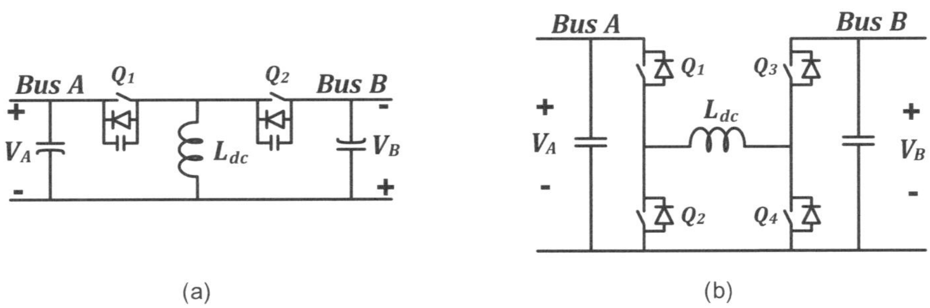

Fig. 3. (a) Bidirectional buck-boost and (b) two back-to-back connected NBDC of Fig. 1.c.

There are other variants of NBDC proposed by researchers. An example is shown in Fig. 4 with the following advantages:

- The structure is symmetrical.
- It inherently has low ripple current on both sides.
- It employs only two switches which simplifies the driver circuitry and decreases the driving power.
- It can work with wide voltage range and different voltage ratios.
- It has intrinsic protection against short circuit.

3. Isolated BDC (IBDC)

Galvanic isolation between multi-source systems is a requirement mandated by many standards. Personnel safety, noise reduction and correct operation of protection systems are the main reasons behind galvanic isolation.

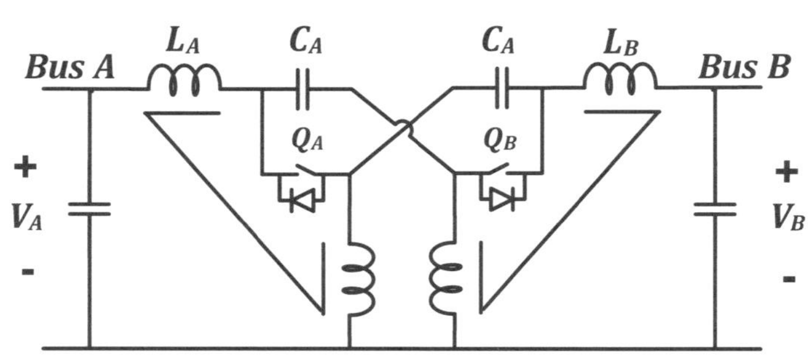

Fig. 4. A different structure for NBDC

Voltage matching is also needed in many applications as it helps in designing and optimizing the voltage rating of different stages in the system. Both galvanic isolation and voltage matching are usually performed by a magnetic transformer in power electronic systems, which calls for an ac link for proper energy transfer. Although this approach is similar to unidirectional dc-dc converters, the need to bidirectional power flow significantly adds to the system complexity. Furthermore, when high efficiency soft-switching techniques are to be applied, this complexity tends to be more.

In this section, the basic structure of common IBDCs is explained. While different terminologies have been proposed and used in the literature, a unified terminology is introduced and used throughout the paper to simplify the comparison between different structures. A classification is provided which helps in understanding the conceptual similarities and differences between different structures.

3.1 IBDC structure

Most, if not all, of medium-power IBDCs have a structure similar to Fig. 5. This structure consists of two high-frequency switching dc-ac converters and a high-frequency transformer which is primarily used to maintain galvanic isolation between two sources. This transformer is also essential for voltage matching in case of large voltage ratio between two sources. The transformer calls for ac quantities at its terminals and thus a dc-ac converter is employed on each side. As energy transfer in either direction is required for the system, each dc-ac converter must also have bidirectional energy transfer capability. With the same token, the dc buses in this structure must also be able to either generate or absorb energy. The dc buses shown in this structure are assumed to have stiff-voltage characteristics, i.e. their Thevenin impedance is negligible. In practice, these buses are connected to a dc source or an active load like battery, ultra-capacitor or dc-link capacitor which resemble an ideal voltage source with stiff voltage characteristics. If the converter is of current-fed type, it is assumed that the required elements to realize stiff current are incorporated inside the converters shown in Fig 5.

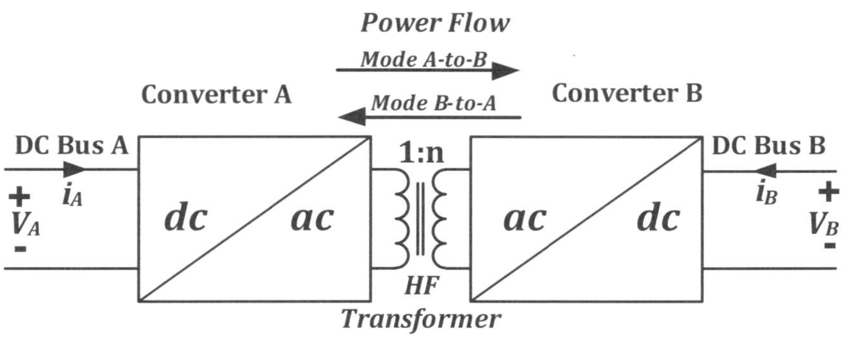

Fig. 5. Basic structure of an IBDC

3.2 Terminology

As the name "bidirectional" implies, there are basically two modes of operation in an IBDC in terms of power transfer. Based on Fig. 5, these modes are denoted by Mode A-to-B and Mode B-to-A in this chapter. It should be noted that other publications in this field may have used different notations. For example, some publications have used the terms "boost" or "step-up" and "buck" or "step-down" modes to describe the two above mentioned modes. This notation usually originates from the fact that the dc voltages at each side have usually different voltage amplitude and thus voltage boosting/bucking takes place along with energy transfer. Other reason behind this notation can be the topology and operation of converters during each mode which resemble conventional buck or boost converters. Some other literature have used "charging" and "discharging" modes, which comes from the fact that at least one of the dc sources in many IBDC applications is a battery, and thus charging and discharging terms become meaningful.

3.3 Classification

Classification of systems with similar functionality but different configurations allows for better comparison among them and helps in understanding the merits and demerits of each of them. The IBDC shown in Fig. 5 can be classified from different viewpoints. The objective of this chapter is not to provide a thorough classification of IBDCs; instead, some basic criteria are presented for better understanding the concepts behind the operation of converters discussed in Sec. 4.

3.3.1 Type of converter

Considering Fig. 5, an important characteristic of an IBDC is the type of converter at each side. Basically, two types of switching converters can be identified. A current-type (or current-fed) structure has an inductor with stiff current characteristic at its terminals which acts like a current source, like conventional boost converter at its input terminals. A voltage-type (or voltage-fed) structure has a capacitor with stiff voltage characteristic at its terminals which acts like a voltage source, like conventional buck converter at its input terminals. The operation, switching strategy and other operational aspects of these converters are different.

3.3.2 Active control in different modes

Bidirectional operation requires both converters in an IBDC to be equipped with controllable switches. Therefore, both converters can be actively controlled in both modes of operation. This capability, however, may or may not be exploited in all proposed IBDCs. In other words, some IBDCs work on the basis of controlling only the source-side converter during each mode, while using the uncontrolled components (i.e. diodes) of the other side for rectification. This, on one hand, reduces the complexity of control, but on the other hand does not allow using the full capabilities and features of the structure. Most recent approaches rely on the active control of both converters irrespective of the direction of power transfer.

4. Common IBDCs

Different configurations have been proposed for IBDC in the literature. Investigation of all these configurations is beyond the scope of this chapter. However, a careful review of various proposed IBDCs shows that they can be categorized into a few basic families into which the majority of configurations fall. Each family can be studied by investigating one of its members that represents the basic operational aspects of that family. To select the representative member of each family, the following criteria have been taken into consideration in this chapter:

- It should be the main member of the family and other schemes have been more or less derived from this configuration.
- It should have been addressed and investigated in more details in the literature.
- Description of its operation should cover the fundamental operational aspects of other members of that family.

Based on the above objectives, the operation of three major IBDC configurations is described in this section by the help of basic illustrative waveforms. Following this description, important characteristics of each configuration are addressed and briefly discussed.

4.1 Configuration 1: (Dual Active Bridge, DAB)

Fig. 6 shows a common IBDC topology which is sometimes called dual active (full) bridge (DAB). The converter is introduced in (De Doncker et al., 1991) and (Kheraluwala et al., 1992). In this configuration, full-bridge voltage-fed converters are used at both sides of the isolation transformer and the control is performed based on soft-switched phase-shift strategy. In its basic form, the diagonal switching pairs in each converter are turned on simultaneously with 50% duty cycle (ignoring the small dead time) and with 180 degrees phase shift between two legs to provide a nearly square wave ac voltage across transformer terminals. The phase shift between two ac voltages, denoted by ϕ, is an important parameter which determines the direction and amount of power transfer between dc buses. By adjusting this phase shift, a fixed frequency operation with full control over the power transfer is possible.

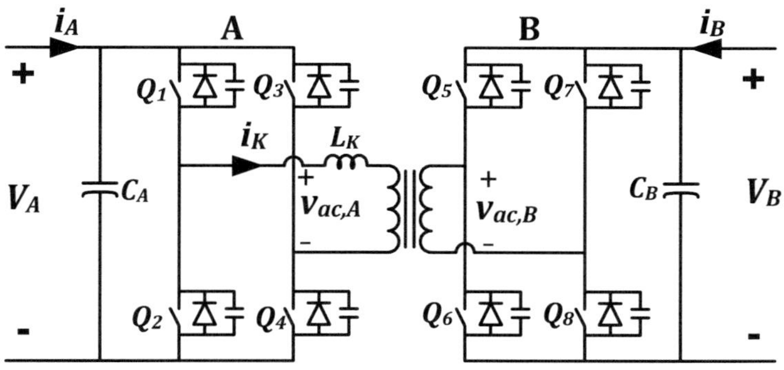

Fig. 6. Circuit diagram of Configuration 1.

Fig. 7 shows the ideal waveforms of A-to-B and B-to-A power transfer modes. The average transferred power can be obtained by calculating the average ac power at the transformer terminals, i.e.

$$P = \frac{1}{2\pi} \int_{0}^{2\pi} (v_{ac,A} \cdot i_K) d(\omega t) \tag{1}$$

which after some mathematical manipulations yields

$$P = \frac{V_A V_B \phi (1 - |\phi|)}{n \pi L_k \omega} \tag{2}$$

where L_K is the transformer leakage inductance (plus any series inductance), n is the transformer turns ratio (Side B to Side A) and ω is the angular frequency. To transfer power from Side A to Side B (A-to-B mode), $v_{ac,A}$ should lead $v_{ac,B}$ and ϕ is considered as positive. In B-to-A mode, $v_{ac,A}$ should lag $v_{ac,B}$ and ϕ is negative. This leading or lagging phase shift is simply implemented by proper timing control of converter switches. L_K is an important element which determines the maximum amount of transferable power with given switching frequency. Therefore, apart from other practical limitations, it is possible to reach a high power density converter with a low leakage transformer.

In most unidirectional dc-dc converters only the input bus has voltage variations and the output voltage is usually regulated. However, in many IBDC applications both dc buses have voltage variations imposed by other sections of the systems. In this regard, an important design parameter in an IBDC which affects soft switching range and other performance characteristics is the voltage ratio defined as

$$d = \frac{V_B}{nV_A} \tag{3}$$

Achieving soft switching over the entire operating range of a converter is always challenging. To have ZVS for the switches in a bridge leg, the current leaving the leg pole (the center of the leg) should lag the pole voltage. In other words, the zero crossing of the leg output current should occur after its voltage zero crossing. It is shown that for the DAB (Fig. 6), soft switching can be achieved in all switches for d = 1 and over the entire control range (De Doncker et al., 1991; Kheraluwala et al., 1992).

Compared to the traditional hard-switched PWM converters, the phase shift converter usually has higher circulating current and thus more conduction losses. However, as the switching frequency increases, the loss reduction caused by soft-switching overweighs the conduction losses and thus the overall efficiency improves.

Some of the advantages of this converter can be listed as below.

i. In this topology, each converter provides an ac waveform with a peak value close to the dc voltage at its terminal, therefore the voltage stress across each switch is limited to the bus voltage level.
ii. The current stresses of all switches on each side are almost equal.
iii. There is no need for additional active or passive elements for having soft switching.
iv. Transformer has a simple structure that simplifies the designing and manufacturing tasks.
v. Another important feature is the fast dynamic behavior due to lack of additional passive components. Note that in practice the soft switching conditions limit the rate of phase shift variation.
vi. Well-known control methods such as average current mode control or peak current mode control are applicable.
vii. Other control techniques that include duty cycle as a second control variable are also possible. This gives another degree of freedom to improve the converter performance (Zhou & Khambadkone, 2009).

Some of the disadvantages are as follows.

i. The currents flowing in dc buses contain high ripple content; therefore appropriate filtering circuits are necessary.
ii. Proper control is required to prevent dc saturation on both sides as there is no inherent dc current blocking capability for transformer windings.

iii. Similar to many other topologies, the converter may lose soft switching in light load conditions.
iv. The control is highly sensitive to slight variations of ϕ, especially when bus voltages are high. Thus if a digital controller is considered, very high resolution phase shift timers are required.
v. Another disadvantage is relatively high component count that leads to larger driver size, higher gate losses and increased cost compared to low switch count topologies.

Recent publications have presented some improvements in the area of duty cycle control, loss reduction and more advanced soft-switching techniques [Zhou & Khambadkone, 2009; Bai & Mi, 2008; Jain & Ayyanar, 2010; Oggier et al., 2009; Krismer et al., 2006).

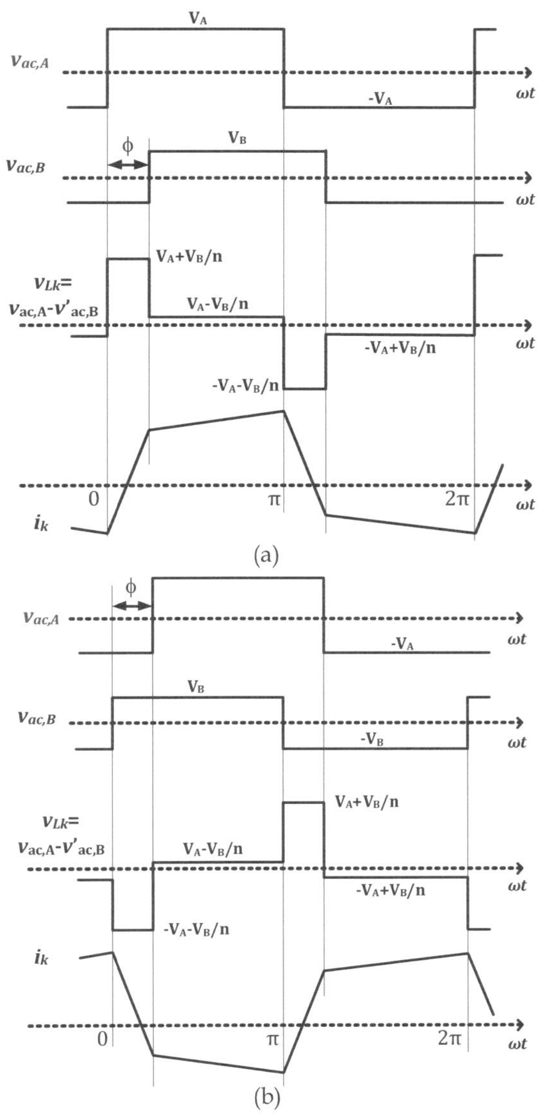

Fig. 7. Operating waveforms of Configuration 1. (a) A-to-B mode and (b) B-to-A mode

4.2 Configuration 2

Fig. 8 shows another configuration proposed in (Wang et al., 1998). This structure consists of a current-fed bridge at Side A and a voltage-fed converter at Side B. The extra transistor Q_C and capacitor C_C at Side A act as an active clamp to limit the overshoots caused by transformer leakage inductance during current commutation (Watson & Lee, 1996; Wang et al., 1998). The operation of this converter is explained as follows.

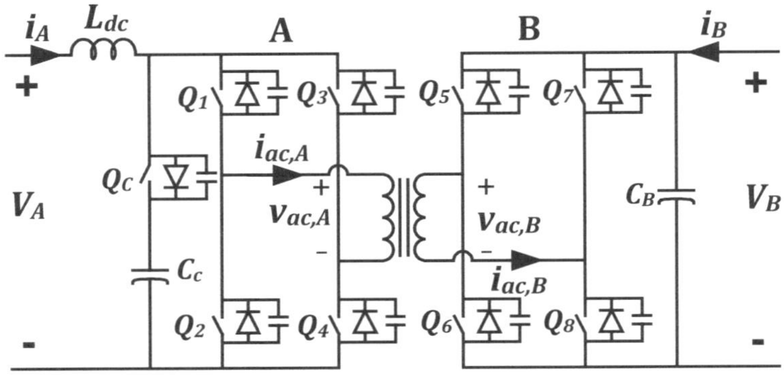

Fig. 8. Circuit diagram of Configuration 2.

4.2.1 Mode A-to-B

Fig. 9.a illustrates the basic idealized waveforms associated with this mode. The circuit operates as an isolated boost full-bridge converter (Wang et al., 1998). Therefore, the reflected output voltage needs to be higher than the input voltage for proper operation. The diagonal switching pairs in the bridge are turned on simultaneously with duty cycle larger than 50% which results in overlapping intervals, as shown in Fig. 5.a. The input inductor is charged during this overlapping interval, and discharged when only one diagonal pair is on. No control has been suggested by (Wang et al., 1998) for the switches on the other side in this mode.

4.2.2 Mode B-to-A

In this mode, the voltage-fed converter (Side B) is the active converter. The control of this converter is performed using conventional phase-shift strategy which enables ZVS operation for Q_5 to Q_8. Furthermore, the active clamp switch Q_C helps in achieving ZCS for one pair of the switches in Converter B. The idealized waveforms are shown in Fig. 5.b. If the switches of Converter A are implemented using MOSFET, they can be controlled to realize synchronous rectification resulting in reduced conduction losses, as shown in Fig. 9.b. More details on soft-switching operation in this configuration along with timing and operation of active clamp circuit can be found in (Wang et al., 1998).

Using current-fed topology in a dc-dc converter constitutes several advantages and disadvantages. Some of the advantages that are beneficial for an IBDC are:

i. Inherent protection against over-current and short-circuit.
ii. Insensitive to transformer saturation in the case of any switching mismatch and even small dc current as the transformer is current driven.
iii. Relatively low-ripple input current which makes it suitable for PFC or battery operated applications.

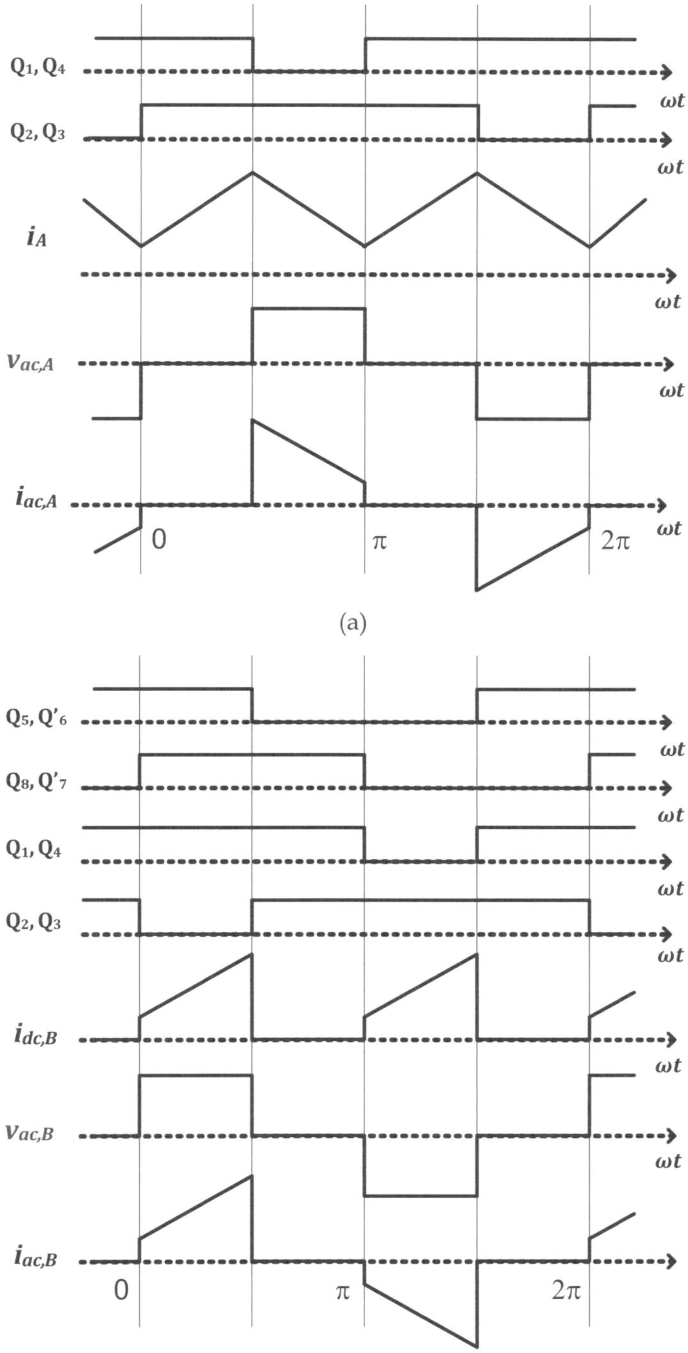

(b)

Fig. 9. Operating waveforms of Configuration 2. (a) A-to-B mode and (b) B-to-A mode

Other advantages of current-fed topologies such as good cross-regulation in multi-output dc-dc converters are not applicable in IBDC.
On the other hand, some of the disadvantages of current-fed topologies are:

i. Difficult start-up procedure which normally calls for extra circuitry.
ii. Voltage spikes due to transformer leakage inductance which could cause high losses in high frequency applications.
iii. Relatively bulky input side inductor.
iv. High ripple output current which calls for high quality capacitors.
v. Need to use semiconductor switches with voltage rating significantly higher than the dc bus voltage (due to both boost operation and voltage spikes).
vi. Susceptibility to loss of gate drive which could lead to current interruption and large voltage spikes.

The characteristics associated with this configuration makes it more suitable where Side B converter is implemented using IGBTs. In such a case, some characteristics like ZCS operation of Side B switches become advantageous. It is worth mentioning that the full benefits of active switches are not exploited in this configuration as Converter B is not actively controlled during A-to-B mode.
Some improvements have been made on this configuration by other researchers. Specifically, more advanced soft-switching techniques have been proposed in (Zhu, 2006; Wu et al., 2010) to reduce the losses associated with active clamp circuit.

4.3 Configuration 3: (Dual Half Bridge, DHB)

Fig. 10 illustrates another IBDC converter introduced in (Peng et al., 2004) for hybrid vehicle applications that is called dual half bridge (DHB). This topology consists of one voltage-fed half bridge converter in Side B (usually higher voltage side) and a modified current-fed half bridge converter (also called boost-half bridge) in Side A. The current-fed side is the lower voltage side because it usually consists of battery or ultra capacitor dc energy sources in which low ripple current is desirable. In practice the voltage amplitude is a few tens of volts for the low voltage side (battery or ultra capacitor) and a few hundreds of volts for the high voltage side. Similar to DAB (Configuration 1) discussed in Sec. III.A, the power regulation is achieved by controlling the phase shift between the voltages applied to two sides of transformer, or equivalently to the leakage inductance of the transformer. The leakage inductance (plus any series inductance) is the energy transfer element like in DAB.
To understand the operation of the converter first note that Q_1 and Q_2 has dual roles in both modes of operations. In the A-to-B mode their first role is acting as a traditional boost converter to produce dc voltage V_M on the auxiliary dc bus from the side A source. At the same time they invert voltage V_M from auxiliary dc bus onto the transformer primary to produce square wave voltage $v_{ac,A}$ as their second role. Q_3 and Q_4 on the other side of transformer rectify the ac current from transformer to transfer power to the side B dc bus. Note that Q_3 and Q_4 rectification is not realized only by acting as diodes or synchronous rectifiers. They have to turn on and off in such a way that the square wave appeared on the transformer secondary, $v_{ac,B}$, has the required phase shift with respect to transformer primary voltage, $v_{ac,A}$. This is also the key to keep the soft switching on Q_3 and Q_4.
In the B-to-A power flow mode Q_3 and Q_4 act as an inverter to produce an ac voltage $v_{ac,B}$ on the transformer secondary . Q_1 and Q_2's first role in this mode is rectifying this ac voltage to produce dc voltage V_M on the auxiliary dc bus. Their second role is acting as a traditional buck converter to send power from auxiliary bus (voltage V_M) to Side A dc bus (voltage V_A).

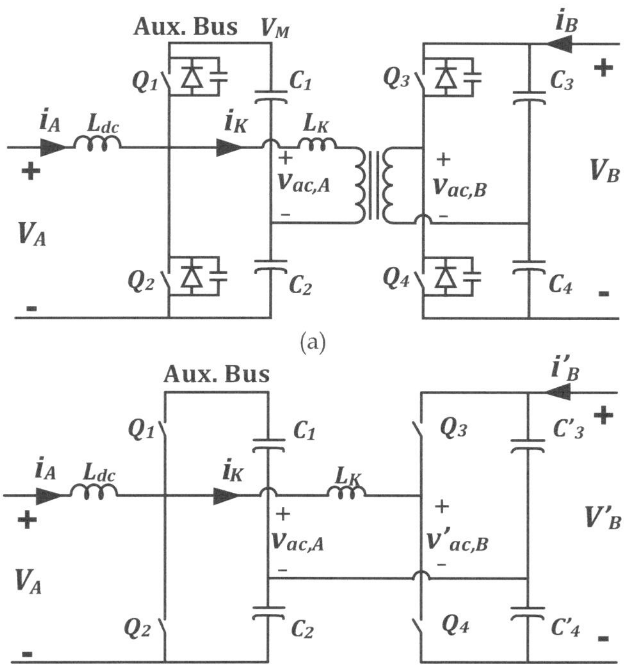

Fig. 10. (a) Circuit diagram of Configuration 3 and (b) idealized model.

Basic waveforms of the converter with 50% duty cycle in both power flow modes are shown in Fig 11 with the help of an idealized model shown in Fig. 10.b. The waveforms of both modes are basically the same, the main difference is negative phase shift and negative dc current in dc input inductor for B-to-A mode. The converter has a unified operation principle and direction of power flow can be changed seamlessly.

The average power transfer can be calculated similar to DAB configuration and based on (1). For DHB configuration it will lead to

$$P = \frac{V_A V_B \phi (1 - |\phi|)}{2n\pi L_k \omega} \tag{4}$$

Similar to DAB configuration, the maximum power transfer is at $|\phi|$=90 degrees. So the converter full range of bidirectional power transfer can be gained by controlling phase shift in -90 to +90 range. To decrease the current stress and increase the efficiency of converter the amount of reactive power transfer through the transformer should be limited. Higher reactive power results in more circulating current and higher conduction losses. Normally higher phase shift angle results in more reactive power similar to sine wave utility systems. From this point of view, it is preferred to design the converter with a lower series inductance (L_K) so that desired power rating of the converter can be reached in lower phase shift values. Control aspects of this configuration can be found in (Li & Peng, 2004; Hui et al. 2005; Ma et al. 2009). Some other IBDCs that are based on this configuration can be found in (Tao et al., 2008; Yu at al., 2010).

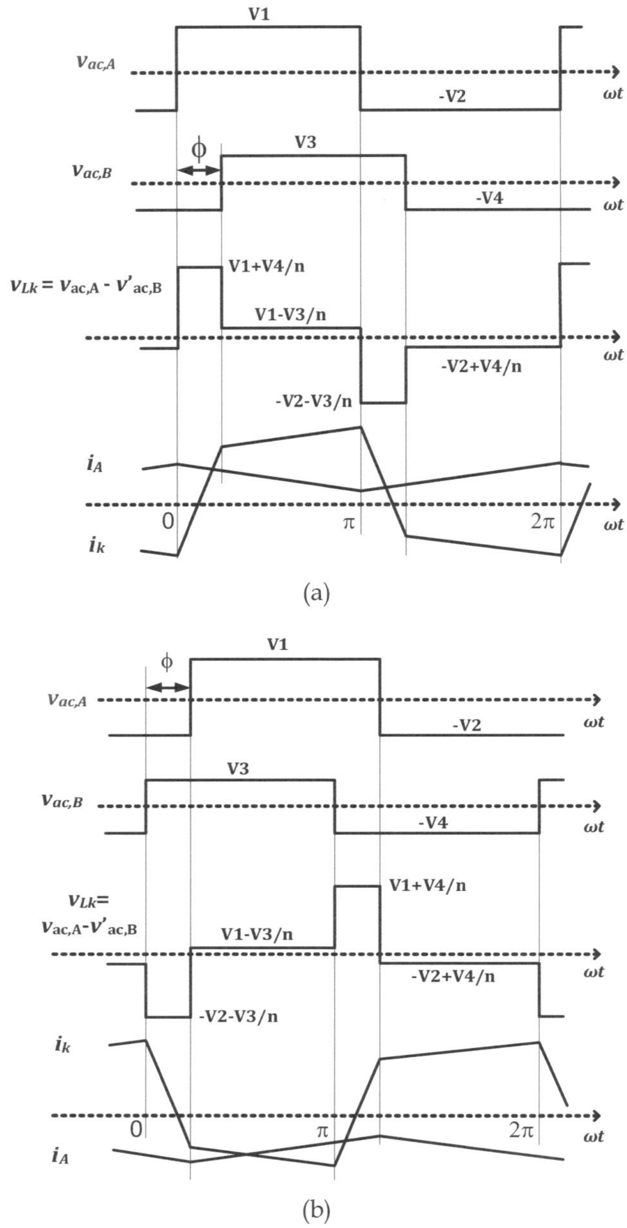

Fig. 11. Operating waveforms of Configuration 3. (a) A-to-B mode and (b) B-to-A mod

With proper design, all converter switches operate in zero voltage switching (ZVS) in a wide range of dc bus voltages or load variations. ZVS for the switches in each leg is achieved when the total current leaving the leg pole is lagging the same point voltage. So in this case, (i_K-i_A) should lag V_{Q2} and (-i_K) should lag V_{Q4}. This way each switch will turn on while its body diode is already conducting (turn on ZVS). Turn off ZVS is also achieved as the device voltage is kept close to zero by snubber capacitor during turn off and the device current is also transferred to snubber capacitors. Snubber capacitors include the device parasitic output capacitors. For more details of ZVS conditions and detailed transition resonance see (Peng et al., 2004).

The main advantages of this configuration can be listed as follows.

i. Low switch count compared to other topologies that normally use full bridge converters.
ii. According to [14] the total device rating of active elements is same as a DAB with the same power.
iii. Relatively wide soft switching (ZVS) range against bus voltages and load variations.
iv. Relatively simple control of the converter based on well known phase shift modulation.
v. Low ripple current at the current fed side that is desirable for batteries and ultra capacitors.

The main drawbacks of the converter are

i. Large ripple current in the splitting capacitors especially in LV side.
ii. Unbalanced current stress between two switches in the LV side.

5. Applications

Minimizing greenhouse gas effects by reducing CO_2 and other emissions is one of the most challenging issues that human is presently facing. As electricity generation is one of the major causes of the pollution, finding alternative clean electricity generation methods is thus becoming attractive. In this regard, renewable energy resources such as wind and solar energy are among the most important substitutes for traditional fuel-based energy production. However, the intermittent nature of most renewable sources does not allow having a reliable and continuous source of energy when these resources are used alone. The fluctuating energy produced by these sources may also cause adverse effects on the power quality of the grid that these resources are connected to. By using energy storage devices, these fluctuations can be absorbed to deliver smooth power to consumers and at the same time maximize the energy output of renewable resources. This reduces the output required from conventional power stations which directly reduces CO_2 emissions.

Besides smoothing the energy output of renewable resources, energy storage systems have other technical applications in the utility grid including grid stabilization, frequency and voltage support, power quality and reliability enhancement and load shifting. Furthermore, restructured electricity markets provide opportunities for exploiting energy arbitrage markets and making revenue by purchaseing the low-cost off-peak electricity and selling the high-cost peak electricity.

In the above mentioned applications and when the storage media is an electrochemical battery, a set of power conversion stages is required to connect the battery to the grid. Fig. 12 shows the structure of a typical energy storage system connected to an ac grid (Inoue & Akagi, 2007). As explained in Sec. 3, isolation and/or voltage matching are usually required in these applications. In the system of Fig. 12, the isolation/voltage matching is done through a line frequency (50 or 60-Hz) transformer. The dc-dc converter shown in this system is usually of non-isolated type, e.g. the converter shown in Fig. 2.

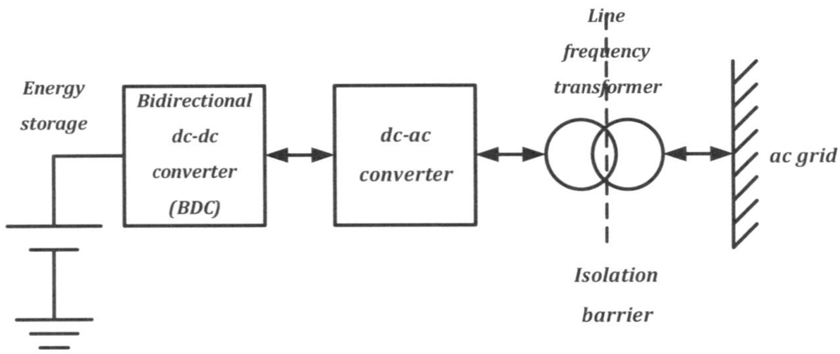

Fig. 12. Basic structure of an energy storage device connected to an ac grid through a line-frequency transformer.

The line frequency transformer in the system of Fig. 12 makes it heavy and bulky. This problem can be largely alleviated by using a high frequency transformer as the isolation stage. Using the existing technologies, the best location to implement such a high frequency transformer is in the dc-dc conversion stage, which results in an IBDC as shown in Fig. 13. In this way, the isolation barrier moves from the low frequency bulky transformer to the high frequency transformer incorporated in the IBDC. Using high frequency transformers lead to more compact and flexible systems. Furthermore, this configuration provides more flexibility in terms of selection of dc voltage amplitude in different stages for optimized operation.

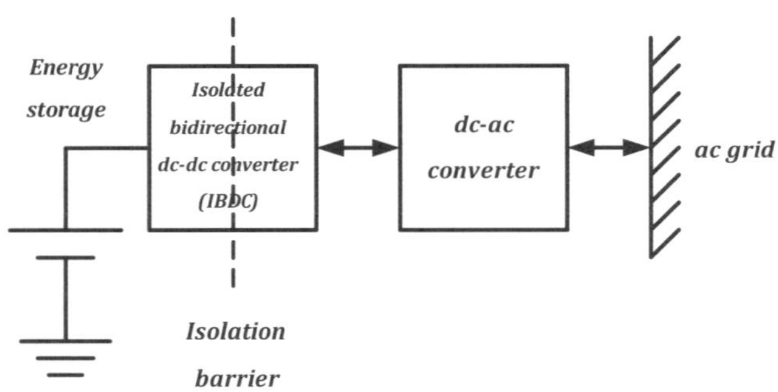

Fig. 13. Basic structure of an energy storage device connected to an ac grid with high frequency isolation barrier inside IBDC.

In (Inoue & Akagi, 2007) an energy storage system based on the structure of Fig. 13 has been discussed. The IBDC structure used in this work is similar to Configuration 1 described in this chapter and rated at 10 kW and 20 kHz with rated dc voltage of 320 V at each side.

6. Conclusion

Bidirectional dc-dc Converters (BDC) are one of the key elements in electrical energy storage systems. They provide a flexible power processing interface between a energy storage device (e.g. battery) and the rest of system. Two main families of BDCs are non-isolated and isolated structures. A review of isolated bidirectional dc-dc converters (IBDC) was presented. The basic structure of these converters along with the terminology used in the literature was described. Despite various configurations proposed for IBDCs in the literature, they can be categorized into a few basic families. A classification was presented to help understand the similarities and differences among the families. The key operational aspects of each family were described by the help of basic waveforms corresponding to a basic member of that family. The advantages and disadvantages of each configuration were briefly stated.

7. References

Bai H. & Mi, C. (2008). Eliminate Reactive Power and Increase System Efficiency of Isolated Bidirectional Dual-Active-Bridge DC–DC Converters Using Novel Dual-Phase-Shift Control. *IEEE Transactions on Power Electronics,* Vol. 23, No. 6, (Nov. 2008), pp. 2905-2914, ISSN 0885-8993

De Doncker, R.W.A.A., Divan, D. M. & Kheraluwala, M. H. (1991). A three-phase soft-switched high-power-density DC/DC converter for high-power applications. *IEEE Transactions on Industry Applications,* Vol. 27, No. 1, (Jan./Feb. 1991), pp. 63-73, ISSN 0093-9994

Inoue, S. & Akagi, H. (2007). A Bidirectional DC-DC converter for an Energy storage system With Galvanic Isolation. *IEEE Transactions on Power Electronics,* Vol. 22, No. 6, (2007), pp. 2299-2306, ISSN 0885-8993

Jain, A. K. & Ayyanar, R. (2010). PWM Control of Dual Active Bridge: Comprehensive Analysis and Experimental Verification. *IEEE Transactions on Power Electronics,* Vol. PP, No. 99, (2008), pp. 1, ISSN 0885-8993

Kheraluwala, M. N., Gascoigne, R. W., Divan, D. M. & Baumann, E. D. (1992). Performance characterization of a high-power dual active bridge DC-to-DC converter," *IEEE Transactions on Industry Applications,* Vol. 28, No. 6, (Nov./Dec. 1992), pp. 1294-1301, ISSN 0093-9994

Krismer, F., Biela J. & Kolar, J.W. (2005). A comparative evaluation of isolated bi-directional DC/DC converters with wide input and output voltage range, *Proceedings of the Fourtieth IAS Annual Meeting,* pp. 599-606, Vol. 1, ISSN 0197-2618, Kowloon, Hong Kong, Oct. 2-6, 2005

Krismer, F., Round, S. & Kolar, J. W. (2006). Performance optimization of a high current dual active bridge with a wide operating voltage range. *Power Electronics Specialists Conference, 2006.* pp. 1-7, ISBN 0-7803-9716-9, Jeju, South Korea, June 18-26, 2006

Li, H. & Peng, F.Z. (2004). Modeling of a new ZVS bi-directional dc-dc converter. *IEEE Transactions on Aerospace and Electronic Systems,* Vol. 40, No. 1, (Jan. 2004), pp. 272-283, ISSN 0018-9251

Li, H., Liu, D., Peng, F.Z. & Su, G.J. (2005). Small Signal Analysis of a Dual Half Bridge Isolated ZVS Bi-directional DC-DC Converter for Electrical Vehicle Applications," *Power Electronics Specialists Conference, 2006.* pp. 2777-2782, ISBN 0-7803-9033-4, Recife, Brazil, June 16, 2006

Li, X. & Bhat, A. K.S. (2010). Analysis and Design of High-Frequency Isolated Dual-Bridge Series Resonant DC/DC Converter. *IEEE Transactions on Power Electronics,* Vol. 25, No. 4, (April 2010), pp. 850-862, ISSN 0885-8993

Ma, G., Qu, Yu, G., Liu, Y., Liang N. & Li, W. (2009). A Zero-Voltage-Switching Bidirectional DC-DC Converter With State Analysis and Soft-Switching-Oriented Design

Consideration. *IEEE Transactions on Industrial Electronics,* Vol. 56, No. 6, (June 2009), pp. 2174-2184, ISSN 0278-0046

Oggier, G.G., Garcia, G. O. & Oliva, A.R. (2009). Switching Control Strategy to Minimize Dual Active Bridge Converter Losses. *IEEE Transactions on Power Electronics,* Vol. 24, No. 7, (July 2009), pp. 1826-1838, ISSN 0885-8993

Peng, F.Z., Li, H., Su, G.J. & Lawler, J.S. (2004). A new ZVS bidirectional DC-DC converter for fuel cell and battery application. *IEEE Transactions on Power Electronics,* Vol. 19, No. 1, (Jan. 2004), pp. 54-65, ISSN 0885-8993

Tao, T., Duarte, J.L. & Hendrix, M.A.M. (2008). Three-Port Triple-Half-Bridge Bidirectional Converter With Zero-Voltage Switching. *IEEE Transactions on Power Electronics,* Vol. 23, No. 2, (March 2008), pp. 782-792, ISSN 0885-8993

Wang, K., Lin, C.Y., Zhu, L., Qu, D., Lee, F.C. & Lai, J.S. (1998). Bi-directional DC to DC Converters for Fuel Cell Systems. *Power Electronics in Transportation,* pp. 47-51, ISBN 0-7803-4398-0, Dearborn, MI , USA, Oct. 23, 1998

Watson, R. & Lee, F.C. (1996). A Soft-switched, Full-bridge Boost Converter Employing an Active-clamp Circuit. *Power Electronics Specialists Conference, 1996.* pp. 1948-1954, ISBN 0-7803-3500-7, Baveno , Italy, June 23-27, 1996

Wu, T.F., Chen, Y.C., Yang, J.G. & Kuo, C.L. (2010). Isolated Bidirectional Full-Bridge DC–DC Converter With a Flyback Snubber. *IEEE Transactions on Power Electronics,* Vol. 25, No. 7, (July 2010), pp. 1915-1922, ISSN 0885-8993

Yu, W., Qian, H. & Lai, J.S. (2010). Design of High-Efficiency Bidirectional DC–DC Converter and High-Precision Efficiency Measurement. *IEEE Transactions on Power Electronics,* Vol. 25, No. 3, (March 2010), pp. 650-658, ISSN 0885-8993

Zhou, H. & Khambadkone, A. M. (2009). Hybrid Modulation for Dual-Active-Bridge Bidirectional Converter With Extended Power Range for Ultracapacitor Application," *IEEE Transactions on Industry Applications,* Vol. 45, No. 4, (July/Aug. 2009), pp. 1434-1442, ISSN 0093-9994

Zhu, L. (2006). A Novel Soft-Commutating Isolated Boost Full-Bridge ZVS-PWM DC–DC Converter for Bidirectional High Power Applications. *IEEE Transactions on Power Electronics,* Vol. 21, No. 2, (March 2006), pp. 422-429, ISSN 0885-8993

9

Bi-Directional DC - DC Converters for Battery Buffers with Supercapacitor

Jan Leuchter
University of Defence,
Czech Republic

1. Introduction

Sources of electrical energy for industry, agriculture or military use are different with respect to the purpose, place, sort of appliance, type of supplied systems. Fixed military facilities, for example, bases and camps are supplied by common electrical network system. Uninterruptible power sources are used to span short network black-outs and to maintain the power quality supplied to computers and other sensitive electronic equipment due to dips, surges and voltage reductions (Kurka & Leuchter, 2008).
In vehicles and aircraft various types of accumulator batteries are used for starting of engines, on-board generators as main source of electrical energy are driven by operating engines. Stable electrical power generating sets are used for power supply in military bases and camps, where the connection to network is not possible. Mobile generating sets are also used for general use to supply various appliances, facilities, systems, for heating, illumination, and other purposes in industry, agriculture or military units. Most of up to date systems are equipped by specialized built-in generating sets. In studies focused on this problem the sources based on the small nuclear generators, sun and wind energy are speculated, but no one from these technologies is suitable from the mobility, safety and operativity of corresponding application point of view. The majority of above mentioned electrical energy sources use some means of energy accumulation (electrical energy buffers or accumulators) to secure the reliable operation under all possible circumstances and conditions. It is considered as self-evident that fuel cells become another mobile power sources and electrical energy buffers (batteries). Fuel cells were marked as one of the new energetic source alternatives for military applications. In the comparison with classical conversion of fuel (where the efficiency reaches 15 to 35%, and with the gas turbine with max. 40%), fuel cells based on the direct conversion of chemical energy to the electric one reach efficiency 50 to 70%, according to the type, power, used chemicals and design. In the combination with semiconductor converters, systems based on the fuel cells become practically the universal source of electric energy (Kurka & Leuchter, 2008).
With the development of new technologies in transportation, vehicles, renewable energy sources, UPS, mobile electrical energy generating sets and in other branches the accumulation of electrical energy, its transformation and transportation represents one common problem. For this purposes various types of electrical energy buffering methods and converters including buffers are used. As will be shown, supercapacitors are well suited to replace classical batteries and conventional capacitors in many applications. Electrical

energy generating sets (EGS) still are and will probably be in near future based mainly on generators driven by combustion engines using fossil fuels. EGS initially developed and produced mainly for military purposes because EGS enable the independence on the common electrical power network. They are used in ground and air transport, in health service and in other military branches. The EGS are quite indispensable in civil defence, crisis management forces, and naturally in security forces. Sophisticated weapon systems, including aircraft and air defence, artillery systems, transport means, logistical structure and training systems based on computer simulation and virtual reality concepts, require also modern and reliable EGS, corresponding to new conditions and requirements. In order to increase efficiency, decrease the fuel consumption and optimize the operational conditions of mobile electrical power generating sets, the VSCF (variable speed-constant frequency) technology is used. Variable output voltage and frequency are transformed to constant values by means of power electronic converters. Due to unconvenient dynamical properties of VSCF based EGS the electrical energy buffers (accumulators) create the essential part of the system securing its reliable operation and can help improve a dynamic behaviour of diesel engine which is limited mainly by fuel injecting. (Kurka & Leuchter, 2008 and 2000) and (Leuchter et al., 2009).

In the following pages we shall propose the energy buffer of EGS with VSCF technology to illustrate the feature, requirements and advantages of systems with energy buffer. EGS with optimum variable speed is really fine example, where power buffer e.g. can improve dynamic behaviour, improve efficiency, and reduce volume.

2. Electrical power buffer

A simplified block diagram of an electrical generator sets (EGS) with variable speed control can be seen in Figure 1, where ω_a represent actual engine speed and ω_o optimum engine speed. As a consequence of varying the engine speed when using the optimum variable speed control, both the output voltage and the output frequency of the generator vary and must be regulated to a constant value as required by the load. Therefore, a power electronic converter is required to regulate the output voltage and frequency. The real drawback of the concept with optimum variable speed is the inferior engine-generator dynamics. In case of sudden power output increase, the engine can not deliver the requested torque and the result is further decrease of the speed and torque of the engine until the undesirable stop. The diesel engine has namely a time constant of few seconds, which is further limited by fuel injection limitation. Therefore at high change of the speed the engine undesirably stops. This problem of the EGS dynamic behaviour can be improved by means of electrical power buffer. The engine-generator dynamics, poses during sudden transients from low-load to high-load conditions, still poses a challenge in this regard. (Leuchter et al., 2010)

The dynamic behavior analyses of 4 kW EGS with 7.5 kW diesel engine can be summarized to the figure 2. Thus, we see that output power of engine depends on engine speed. We now can describe three types of power for every engine speed. The first power can be denoted as P_{opt}, which represents the power provided by engine operating at the optimal angular speed to achieve the minimum fuel consumption. The second P_{Lmax} represents a maximum power that can be obtained for every speed. If the load power is higher then P_{Lmax} then the engine cannot deliver the requested torque and the result is further decrease of the speed and torque of the engine until the undesirable stop. For carrying out the successful speed change, P_{Lmax} must be higher than power required by the load i.e. the condition $P_{L2} < P_{Lmax}$

must be fulfilled. The third curve P_{rez} is the difference between P_{Lmax} and P_{opt} and it indicates the reserve of power of the system operating at optimal angular speed. For example: if the engine operates with an angular speed of 150 rad$\cdot$s^{-1} then the output power of engine is 1900 W and the power margin of engine is as high as 1700 W. If the power increase is higher, then the diesel engine cannot develop enough power required by the load. (Leuchter et al., 2009).

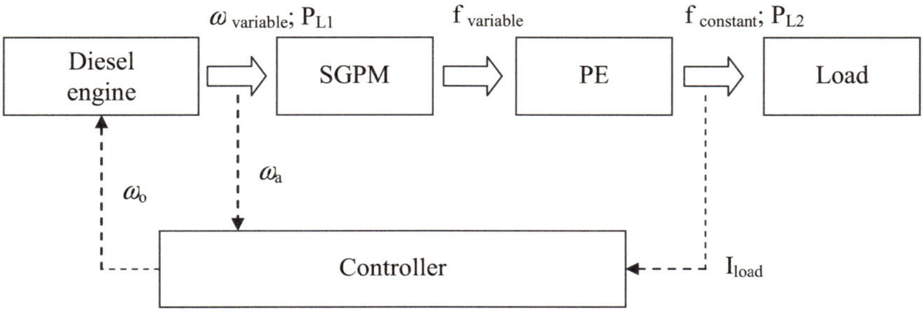

Fig. 1. Block diagram of EGS system with VSCF technology (SGPM-synchronous generator with PM; PE-power electronics)

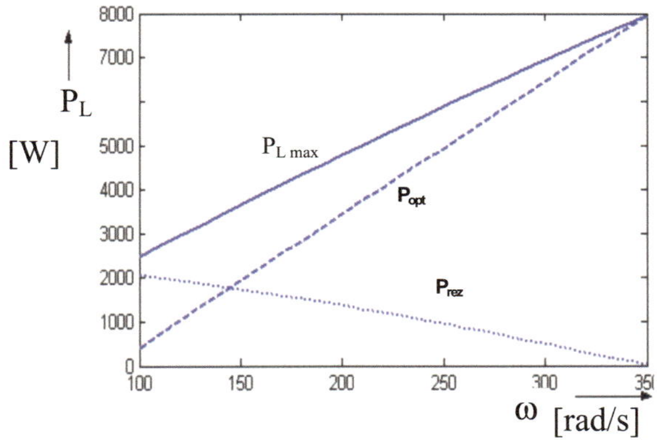

Fig. 2. Identification of the power margin (Diesel engine Hatz 1D40; 7.5 kW)

A power buffer, connected via an electronic converter, can improve the dynamic behaviour of the system with diesel engine by means of injecting stored energy into the dc-link by the dc-dc converter, see Fig. 3. This concept is based on the delivery of peak power from the energy storage to the link capacitor of the dc-dc converter during the low to high speed transition of the diesel engine. The requested energy W is given by the maximal required power P and the average time of the regulation T_R, as given in the following equation

$$W = P \cdot T_R \tag{1}$$

The topology of EGS with battery can achieve some operating advantages. The main advantage of using batteries to cover high power requirements to supply load during short time can be permitted. Or if the batteries are designed to supply half of maximum power while the diesel engine supplies the other half. Such design strategy can reduce the size and cost of the EGS. From it should be apparent that the EGS from figure 3 can bring profitable results, where is not necessary to note that battery is required by diesel engine for start at any case. Such an approach requires a battery interface by DC bus and bi-directional dc-dc converter. The dc-dc converter makes voltage step up of the low voltage of batteries to the constant dc value e.g. 600 V required by three phase inverter. This chapter has only been able to touch on the most general features of the bi-directional dc-dc converter (blue one) that is shown in figure 3 with battery. Battery and bi-directional converter make power buffer together and our intention here is to highlight to the implementation of converter and battery to the system of EGS and the development in power electronics will be briefed. Also some control method will be discussed regarding the different characteristics of the various EGS systems. (Leuchter et al., 2009).

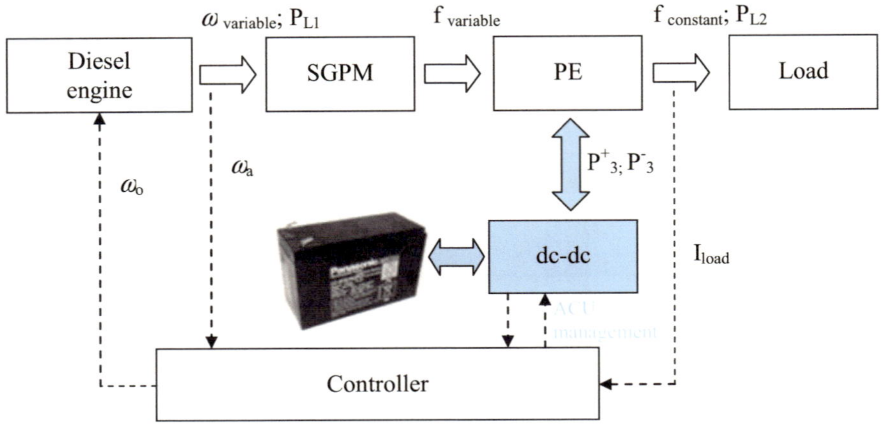

Fig. 3. The bi-directional dc-dc converters

In the following pages main power electronic configurations will be presented and explained with focusing mainly for the dc-to-dc converters, which make interface between battery and loads that is supported by power buffer. Our intention here is to highlights into most important converter topologies, namely bi-directional. Looking ahead to the application of this, we find that these converters operate according quadrants operations.

3. Review of basic electrical multi-quadrant operations

Many types of applications, such as variable speed of diesel engines or wind turbines, use power electronics systems as interface. Power electronics has changed rapidly during the last 20 - 30 years and number applications have been increasing, mainly due to IGBTs (Insulate Gate Bipolar Transistors) devices. Power electronics converters are constructed by power electronics devices, driving, protection and control circuits. A converter, depending on the topology and application, may allow both direction of power flow and can interface between the load and generator sets. There are two different types of converter systems: thyristor converters and pulse modulated (PWM) converters. The high frequency switching of a PWM-converter may produce converters with better power density in comparison with thyristor converters. Due to the high frequencies, the harmonics are relatively easier to be removed what leads to use smaller size filters, especially inductors. On the other hand, the thyristor converters have three important issues in using a power electronic system. These are reliability, efficiency and cost. This part of chapter discusses the modern power

electronics topologies, which play an important role in the area of modern energy source. (Blaabjerg & Chen, 2006)
This leads to some basic principles to show in the following pages. The most important goal of all efforts in developing the product range of power devices and converters is to reach minimum power losses to achieve the maximum efficiency.
Depending on the application, the output to the load may have two main forms: dc and ac. The power converters usually consist of more then one power conversion stage. Converters can be divided into the following categories: ac-to-dc, dc-to-dc, dc-to-ac and ac-to-ac. Our intention here is to highlight and briefly review some of the basic concepts of dc-to-dc conversions. The dc-to-dc converters are widely used in regulated switch-mode dc power sources and dc motor drive applications, where circuits convert fixed dc voltage to variable dc voltage. Such dc converters are very often called as choppers. We can define the multiple - quadrant operation. As shown in Fig. 4, the quadrant I (I-Q) operates with positive voltage and positive current and quadrant II (II-Q) operates with positive voltage and negative current. Quadrant III (III-Q) operates with negative voltage and negative current and quadrant IV (IV-Q) operates with negative voltage and positive current.

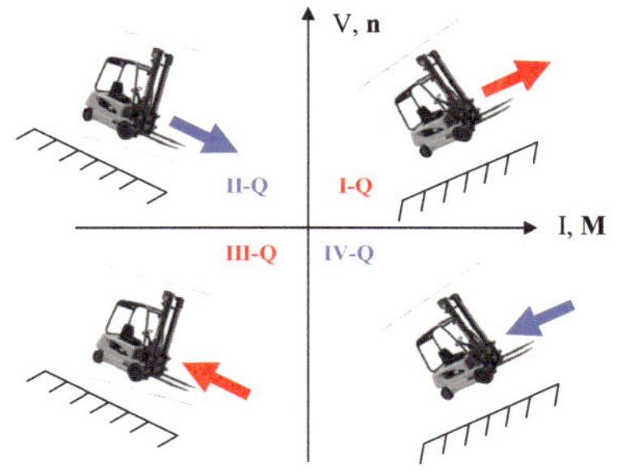

Fig. 4. Four-quadrant operation

We begin our study with a variable speed drive for a DC motor to understand what four-quadrant operation is. We assume that its operation is restricted to I-Q. Machines are seldom DC used as generators (II-Q and IV-Q). However, they operate as a generator while braking, where their speed is being reduced. During the braking operation, the polarity of armature voltage (V_L) does not change, since the direction of the rotation has not changed. If the terminal voltage (V) polarity is also reversed, the direction of the rotation of the motor will reverse. Therefore, a DC motor can operate in either direction and its electromagnetic torque can be reversed for braking, as shown by the four quadrant of the torque/speed plane in Fig. 4. (Mohan et al., 2002)

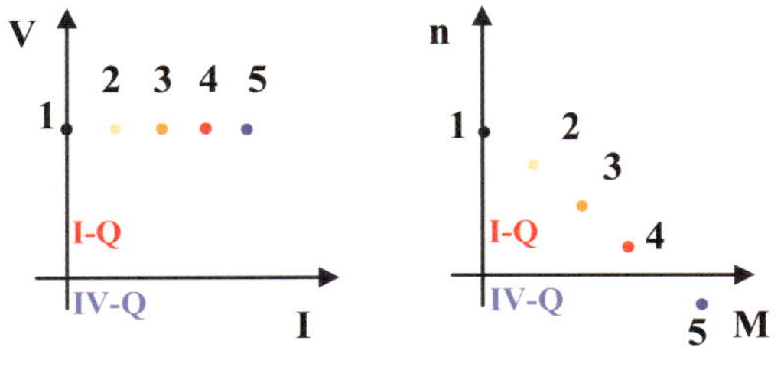

Fig. 5. Operating modes of and DC electric drive in I-Q and IV-Q of the current/voltage plane of the source and torque/speed plane of the drive

The field excitation is fixed and the speed is varied by the armature voltage. A pulse converter is connected between the armature and DC source. In addition, the converter processor can be set for any desired motor speed (n) and torque (M). Using the analogy between electric circuits and car behaviour, we can obtain results as follows. The slope of the street has effect on the results in a change of the load torque (M) of DC drive. In the Fig. 5 the change of the slope can be seen, where the point 1 represents no-load and next points make higher slope of the street (2<3<4<5). The higher loads, in this case the higher load represents higher slope of the street, produce higher load torque of the drive and their speed is being reduced. The case of point 5 represents a generator mode, where DC drive was reversed for braking, which was achieved by load up of motor mode. Therefore, higher loads produce higher load torque and higher current until the power is possible to produce. If the required power by the load is higher then power produce by source, then DC drive cannot deliver power to go car up and operate in I-Q, see Fig. 4 again. In this case, the direction of the drive is changed and car goes down and I-Q move to IV-Q.
A DC drive can run in forward or reverse running. The forward process when armature voltage (V_L) and current (I_L) are both positive. Using previous analogy with DC drive, we can draw down the next figures with a pulse converter, which is connected between the armature voltage and DC source. The I-Q of converter and DC motor can be seen in Fig. 6a. The output voltage of forward motoring operation (I-Q) is calculated by Eq. (2), where T is the repeating period, V_{IN} is the input voltage, t_{on} is the switch-on time.

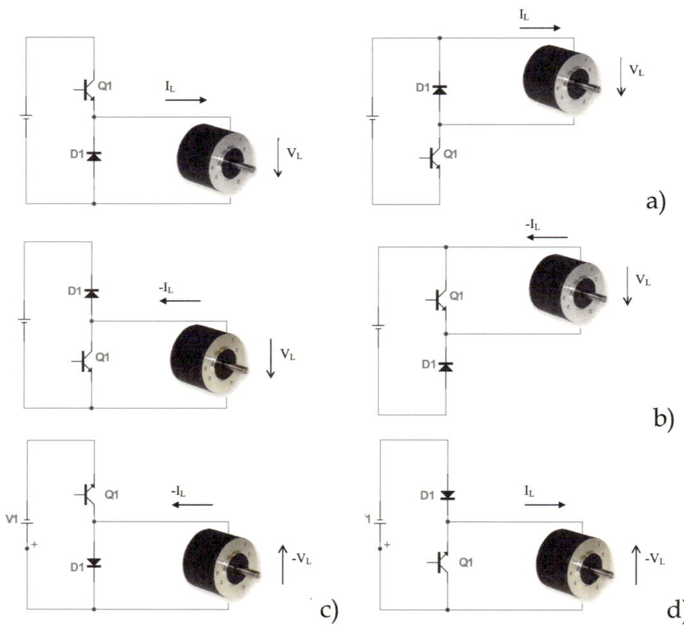

Fig. 6. a) I-Q (motor); b) II-Q (generator); III-Q (motor); IV-Q (generator)

$$V_L = \frac{t_{on}}{T} \cdot V_{IN} , \tag{2}$$

$$V_L = \frac{t_{off}}{T} \cdot V_{IN} = \left(1 - \frac{t_{on}}{T}\right) \cdot V_{IN}, \quad (3)$$

During the forward braking process its armature voltage is still positive and its armature current is negative. This state can be called as the forward generating operation (II-Q), where the output load voltage are as defined in Eq. (3), where t_{off} is the switch-off time. During the reverse process the DC motor armature voltage and current are both negative (III-Q). The output voltage can be calculated by the formula (2): During the reverse braking process its armature voltage negative and its armature current is positive (IV-Q). The output voltage can be calculated by the formula (3). (Luo & Ye, 2000)
Two-quadrant control is shown in Fig. 7 for I-Q and II-Q. Dual quadrant operation is usually requed in the system with two voltage sources.

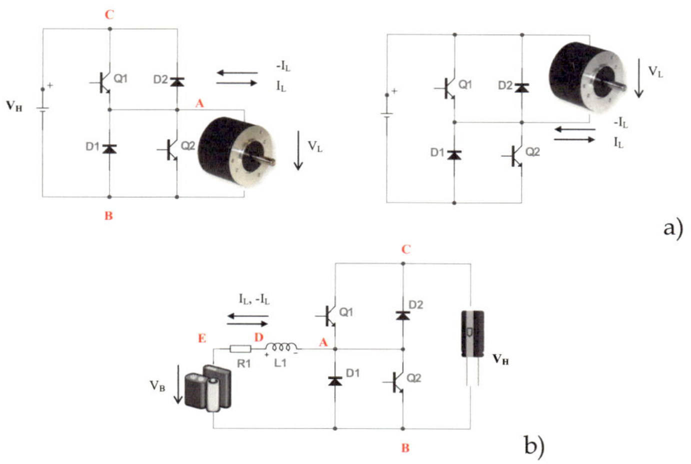

Fig. 7. Two-quadrant convertrs (I-Q, II-Q)

Consider Fig. 7 in which two switches Q1 and Q2 are connected across a dc voltage source V_H. The switched open and close alternately in such a way that when Q1 is switch off, Q2 is switch on and vice versa. The output voltage for a period T oscillates and its average value is given by:

$$V_L = \frac{t_{on}}{T} \cdot V_H = D \cdot V_H, \quad (4)$$

where D is the duty cycle, V_H is the positive voltage during a period T_α. (Wildi, 1997). It is apparent that the circuit between point A and B is never open. If current I_L happens to flow into terminal A, it can find its way back to terminal B either via Q2 (if Q2 is closed) or via Q1 (if Q2 is open). Because one of the switches is always closed and it is evident that current I_L can always circulate, which is important feature of this converter. It can be called a two-quadrant converter because the current I_L can flow in either direction, but the polarity of the dc voltage is fixed, where voltage between points A–B (V_{AB}) is always positive.
In point of fact, from this is evident that concept of two-quadrant converter is a variant, which can be used for our application of bi-directional power delivering between battery and dc-dc line. Suppose we want to transfer dc power from terminal A-B to a load such as a

battery, whose dc voltage between E-B (V_{EB}) has value V_B, which is constant while V_{AB} is fluctuating. Despite this we need to apply the buffer between both sides. We could use a resistor R_1, but the efficiency of converter is reduce or inductance L_1, as shown in Fig. 8. The inductor L_1 is an ideal component. During I-Q operation, Q1 and D2 works, and Q2 and D1 are idle. Vice versa, during II-Q operation, Q2 and D1 work, and Q1 and D2 are idle. Consequently, the both voltage (V_B and V_H) can be fixed by duty cycle D and relation between the two voltage sources can be calculated by the formula:

$$V_B = D \cdot V_H \; ... \; I-Q; \quad V_B = (1-D) \cdot V_H \; ... \; II-Q. \tag{5}$$

If V_B is exactly equal to V_{AB}, no dc current will flow and no dc power exchange. Whereas if is V_B is less than V_{AB}, a dc current I_L will flow from terminal A into terminal E and average value is given by:

$$I_L = \frac{(V_{AB} - V_{EB})}{R_1}. \tag{6}$$

This dc power can only come from the higher voltage source V_H. In this mode, if V_{EB} is less then V_L, the converter works like step-down (buck), which will be discussed in the below in details. (Wildi, 1997)
On the other hand, if V_B is greater than V_{AB}, a dc current I_L will flow out of terminal E and into terminal A, where I_L is also given by (6). Power now flows from the low-voltage battery side V_B to the higher voltage side V_H. In this mode, with V_B greater than V_L, the converter operates like a step-up (boost) converter. Therefore, system of converter from Fig. 7b is able to transfer dc power in both directions by means of changing of current flow and again, such two-quadrant converter operates either in I-Q or II-Q. The detail of system operation can be seen in Fig. 8.

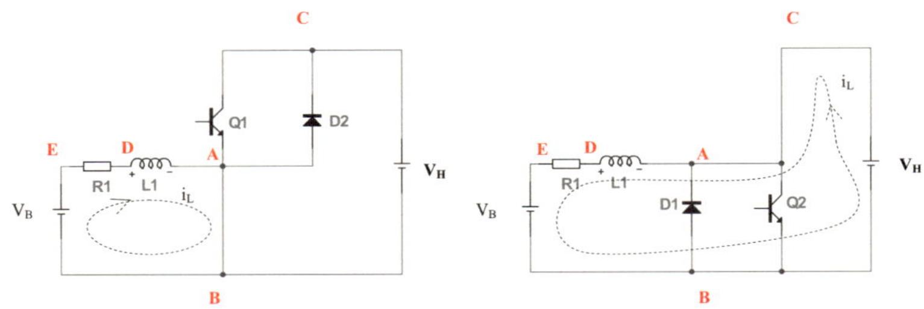

Fig. 8. Two-quadrant dc-to-dc converter (principle of operation of I-Q and II-Q)

If the switch Q2 is closed, the current i_L is given by the voltage V_{DB} and by resistor R_1 as:

$$i_L = \frac{(V_B - V_{DB})}{R_1} \tag{7}$$

$$V_{DB} = V_H - V_B - (R_1 \cdot i_L). \tag{8}$$

And inductor accumulates volt-second during time, when Q2 is closed, and then the voltage across the inductor is given by (8). Terminal voltage V_{DA} is negative and therefore the current i_c is decreasing. The volt-seconds discharging during the time of the switch Q1 is closed together with previous volt-second during time (when Q2 is closed) gives the current change (peak-to peak ripple). The two-quadrant converter is the basic building block for most switch bi-directional application. (Wildi, 1997)

The multi-quadrant converter for the III-Q and IV-Q operation is shown in Fig. 9a. Both voltage polarities are defined in the figure. Suppose that the $V_H > V_B$ and L is ideal inductor, than III-Q operation, Q1 and D2 work, and Q1 and D2 are idle. The relation between the two voltage sources can be calculated by:

$$V_B = D \cdot V_H \quad ... \quad III-Q; \quad V_B = (1-D) \cdot V_H \quad ... \quad IV-Q. \tag{9}$$

The next two-quadrant converter can be seen in Fig. 9b, which operates in I-Q and IV-Q.

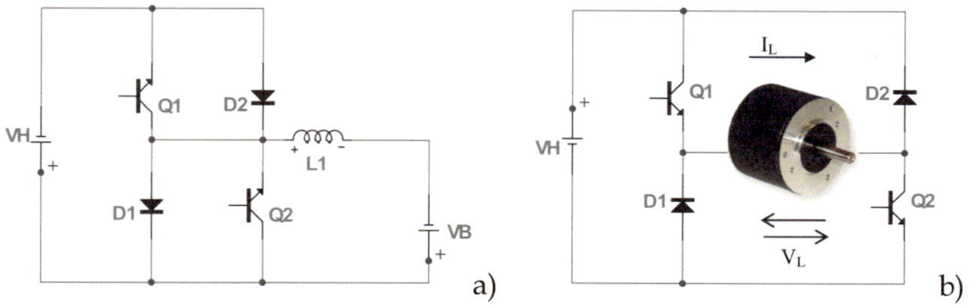

Fig. 9a,b. Two-quadrant dc-to-dc converter a) III-Q and IV-Q) b) I-Q and IV-Q

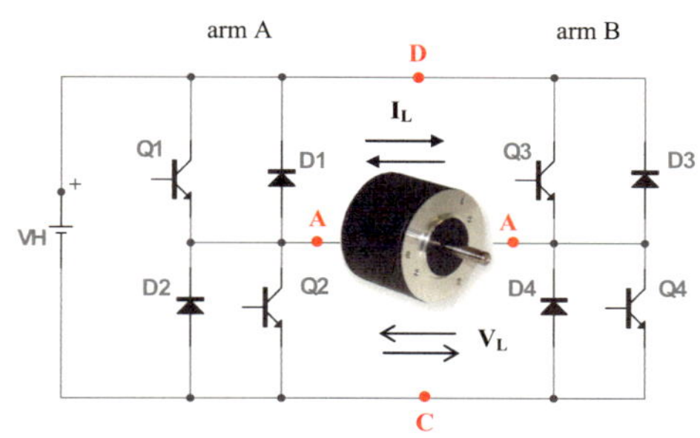

Fig. 9c. Four-quadrant dc-to-dc converter (I-Q, II-Q, III-Q, IV-Q).

As shown in Fig. 4 and 5, the load of these converters is often variable, and therefore it will fluctuate due to changes in the voltage polarities of the load. Switch-mode converters are used to convert the unregulated dc (or ac) input into a controlled dc output at a voltage level. Looking ahead to the application of these converters, we find that these converters are very often used in switch-mode dc power supplies and uninterruptible power sources, energy conversion, transportation, traction control of electrical vehicles, and battery charges. The four-quadrant is shown in Fig. 9c. The input voltage is positive; output voltage can be either positive or negative. The output voltage is given by (10). Anyway, the fourth-quadrant converter consists of two identical two-quadrant converters. Switches Q1 and Q2 in converter (arm A) open and close alternately, as do switches Q3 and Q4 in converter arm B with the same frequency as switches in arm A. The switching sequences are such that Q1 and Q4 open and close simultaneously; similarly, Q2 and Q3 also open and close simultaneously. Consequently, if the duty cycle for Q1 is D, it will also be D for Q4 and than duty cycle for Q2 and Q3 is (1-D). (Wildi, 1997), (Luo & Ye, 2000)

$$\begin{aligned} &V_L = D \cdot V_H \quad ...I-Q; \qquad V_L = (1-D) \cdot V_H \quad ... \quad II-Q \\ &V_L = -D \cdot V_H \quad ...III-Q; \quad V_L = -(1-D) \cdot V_H \quad ... \quad IV-Q \end{aligned} \tag{10}$$

The dc voltage between terminals A nd C (V_{AC}) is given by (11) and the dc voltage between terminals B nd C (V_{BC}) is given by (12). The dc voltage V_L, which is terminal voltage between A and B is difference between V_{AC} and V_{BC} is given by (13).

$$V_A = D \cdot V_H \tag{11}$$

$$V_B = (1-D) \cdot V_H \tag{12}$$

$$V_L = V_A - V_B = D \cdot V_H - (1-D) \cdot V_H = V_H \cdot (2 \cdot D - 1) \tag{13}$$

From (13) we can obtaine that the dc voltage is zero when D=0.5, the voltage changes linearly with D, becoming $+V_H$ when D=1 and $-V_H$ when D=0. The polarity of the output voltage can be either positive or negative. Moreover, the dc current flow can be also either from A to B or from B to A. Therefore, the voltage V_L between the output terminals A and B oscillates between $+V_H$ and $-V_H$. Therefore we can deliver power from V_H to V_L, or vice versa, by simply adjusting the duty cycle D.
In the following pages we shall consider in power electronics dc-to-dc circuits. The review of power electronics dc-to-dc converters will be consecutively briefed.

4. Review of voltage dc-to-dc circuits

The dc-to-dc converter is one of the most widely frequent components of all electronic circuits. Looking ahead to the application of this, we find that these converters are very often used with an electrical isolation transformer in the switch-mode dc power converters or without an isolation transformer. They can be sorted into following groups: Step-down (buck); Step-up (boost); Step-down and Step-up (buck-boost). All of them consist of a switch, a diode and an inductor and represent the basic converter topologies. These three switch conditions permit the formation of the three converter topologies, each of which has two switch conditions, diode and switch. Therefore, for only two switch conditions, we have

$$V_{L1} \cdot D + V_{L2} \cdot (1-D) = 0, \tag{14}$$

where

$$V_L = V_{L1} \quad ...Q1 = 1; Q2 = 0 \quad \text{or} \quad V_L = V_{L2} \quad ...Q1 = 0; Q2 = 1. \tag{15}$$

As well now, the fundamental converters listed are shortly discussed in the following and can be seen in Fig. 10.
The output voltage of Buck converter from Fig. 10a is given by (16) and then it is possible write equation (17). The output voltage of Boost converter from Fig. 10b can be written by (18) and for converter from Fig. 10c by (19).

$$V_{out} = (V_{in} - V_{out}) \cdot D - V_{out}(1-D) = 0, \tag{16}$$

$$V_{out} = D \cdot V_{in} \tag{17}$$

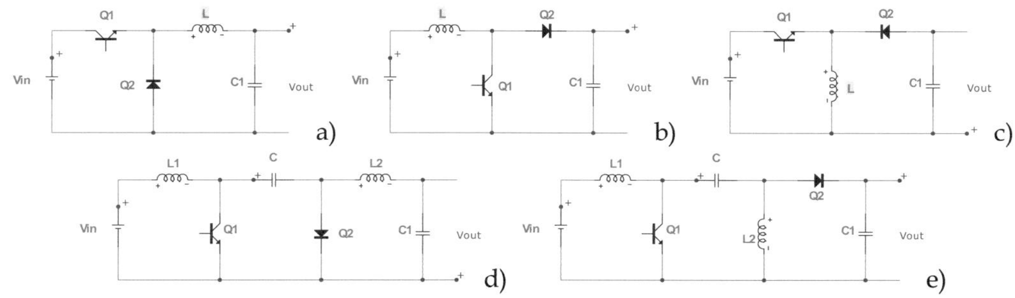

Fig. 10. Dc-to-dc converter a) Buck; b) Boost; c) Buck-Boost; d) Cúk; e) SEPIC

$$V_{out} = \frac{V_{in}}{1-D} \tag{18}$$

$$V_{out} = -\frac{D \cdot V_{in}}{1-D} \quad . \tag{19}$$

These equations and fundamentals topologies from Fig. 10 are well now, but we need show these to explain modes of operations of the following more complicated dc-to-dc converters.

From previous is evident that inductance (L) is one of the most important components of these circuits. For i_L>0, the voltage waveforms are rectangular, being either of amplitude V_{L1} or V_{L2} in accordance with Eq. (15). Therefore, the inductor current is ac (triangular). For very large L, the ac components of the inductor current may be considered negligible (continuous conduction). As L becomes smaller, for a given value of the load (R_L), the ratio of peak inductor current to average inductor current becomes larger. The peak-to-peak sawtooth amplitude becomes large enough that i_L would down below zero (discontinuous conduction). The value of L is defined as a critical inductance L_C for which i_L=0. The following questions show the results of critical inductance criterion. Equations (20) and (21) are for Buck and Boost converter, respectively. (Mitchell, 1998)

$$L_{C\,buck} = \frac{R_L \cdot (1-D)}{2 \cdot f_s} \tag{20}$$

$$L_{C\,boost} = \frac{R_L \cdot (1-D)^2 \cdot D}{2 \cdot f_s} \tag{21}$$

Two very important configurations of dc-to-dc converter using the capacitor as the primary element for storing and transferring energy from input to the output are shown in Fig. 10d and 10e. The advantage of these configurations can be found as the possibility of transformer saturation due to dc offset is precluded by the series capacitor. For converter Cúk and SEPIC, we have

$$V_{out} = \frac{D \cdot V_{in}}{1-D} \tag{22}$$

Cúk converter is obtained by using the similarity with buck-boost. Similar to the buck-boost, Cúk converter provides a negative output voltage polarity. An advantage of this Cúk circuit is that the input current are ripple free and it is possible eliminate the ripples completely by means of external inductor filter. Compared SEPIC to Cúk converter, output voltage is preserved and V_C is smaller, what is main reason why use a SEPIC for design with tantalum electrolytic capacitors. The following figures represent test results which were made with focusing on the voltage and current ripple of previous converters.

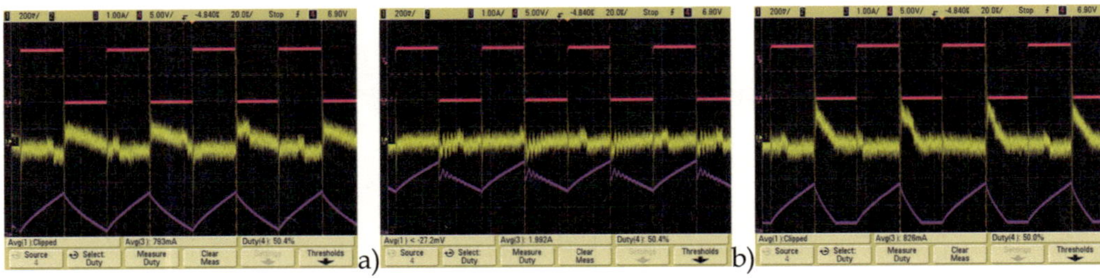

Fig. 11. Converter waveforms a) Boost; b) Buck-Boost; c) SEPIC

(U_{in}=10 V, f_s=20 kHz, L=64 uH, C= 330 uF, R=158 Ω, D=0.5)

All converters shown above in their basic forms, off the concept from Fig. 10, are capable of transferring energy only in one direction in I-Q. The concept of two-quadrant of dc-to-dc converter operating in I-Q and also in II-Q was shown just in Fig. 7b. This converter topology is capable of a bi-directional power flow and provides the basic topology for a design of bi-directional converters. A full-bridge converter topology from Fig. 9c is also capable of a bi-directional power flow. This capability operates in four quadrants and provides a good system topology, because the output current through these PWM full-bridge dc-dc converter does not become discontinuous. In Fig. 12, the switch utilization factor P_{out}/P_T is shown for the previously considered converters. The switch peak voltage rating V_T and the peak current rating are calculated as $P_T=V_T \cdot I_T$. (Mohan et al., 2002)

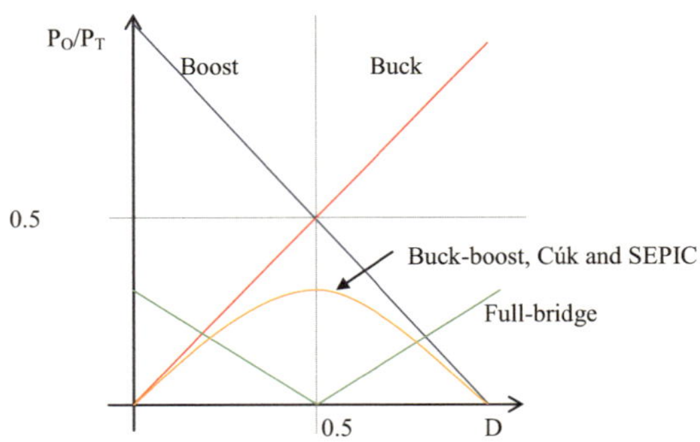

Fig. 12. Switch utilization in dc-dc converters (Mohan et al., 2002)

The new developed converters are created from the dc-to-dc converters e.g. positive Luo-pump, modified Buck-Boost and many others, like switched components SI/SC, soft-switching ZCS, ZVS, ZT. There are more than 500 existing prototypes of these and can provide the some advantages like lower output voltage ripple in comparison with fundamentals. The output voltage ripple of all developed types is usually lower then 2%.

The positive Luo-pump converter is shown in Fig. 13a. It can be derived from the buck-boost converter. The output voltage is calculated by (22). They work in the I-Q with large voltage gain. The inductor L1 transfers the energy from source to capacitor C during switch-off time, and then the stored energy on C is delivered to the load during switch-on.
Cascade concept of converter can help to implement the output voltage increasing with simpler structure. For example: the basic form of Boost converter was shown in Fig. 10b and output voltage can be calculated by (18). The two-stage boost circuit is set up from boost converter and adding the parts L2, D2, D3 and C2. Output voltage of the first-stage (V1) is also given by (18) and the voltage across capacitor C_2 is charged to V_{out} by (23).

$$V_{out} = \frac{1}{1-D} V_1 = \left(\frac{1}{1-D} \right)^2 V_{in} \tag{23}$$

$$V_{out} = \left(\frac{1}{1-D} \right)^3 V_{in} \tag{24}$$

The three-stage and higher-stage connections are derived from the two-stage boost circuit by adding the parts L3, D5, D4 and C3 like in previous case. Then the voltage across capacitor C_3 is charged to V_{out} by (24).

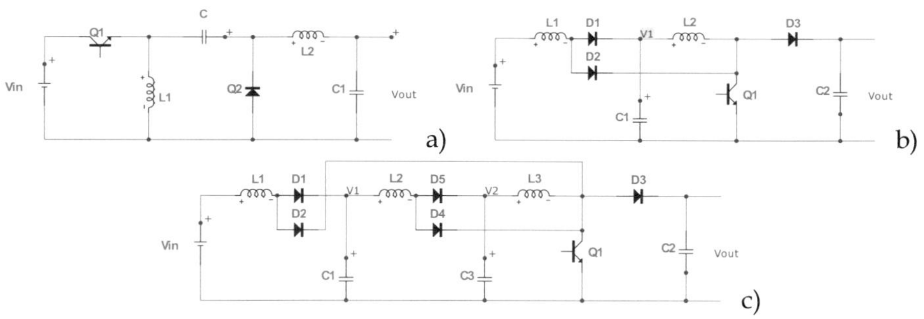

Fig. 13. Developed dc-to-dc converter a) positive Luo-converter; b) Two-stage Boost; c) Three-stage Boost (Luo & Ye, 2000).

Higher stage can be designed by just multiple repeating of parts. Many other circuits can be derived from these baseline topologies using fundamentals converters. Using the analogy of cascade stage converters is evident that these concepts can operate with higher voltage gain, but in the other hand cascade concept require higher number of components such as diodes, inductors and mainly capacitors, what produce lower efficiency and lower reliability. Voltage converters with transformer isolation can provide voltage increasing as well with high efficiency. Fig. 14 shows the most common Buck-transformer isolated voltage converters, called Forward.
The forward converter is a single-ended topology using only one switch Q1. During Q1 (on-time) is Q3 conducted and during Q1 is Q2 (off-time). The use of the forward converter is generally confined to low voltage and power applications. The output voltage can be calculated by

$$V_{out} = D \cdot N \cdot V_{in} , \tag{25}$$

where N is transformer turn radio. Some of the common modifications of the forward converter topology is shown in Fig. 14b. The two switches Forward converter can operate with voltage rating of the each of the switches are one-half of that in a single-switch topology. The paralleling Forward converters are also common used, where the advantage can be found at high power levels.

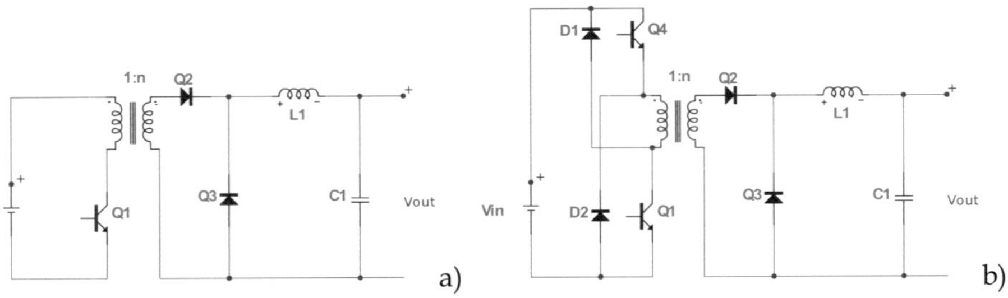

Fig. 14. Buck-derived dc-to-dc converters with transformer isolation (Forward)

The Push-Pull converter topology, which is also based on the Buck topology, can be seen in Fig. 15. These can operate in I-Q and III-Q of the transformer, therefore their are smaller. For a well designed tightly coupled transformer, there is not necessary to use an addition free-wheeling diode for the inductor. The output voltage is equal

$$V_{out} = 2 \cdot D \cdot N \cdot V_{in} \qquad 0 < D < 0.5\,. \tag{26}$$

However, they are still confined to relatively low input voltage application. (Mitchell, 1998) Fig. 15b shows a push-pull arrangement of the boost converter called Current-fed converter. This converter is particularly useful in high output voltage applications where the inductor prevents what otherwise might be damaging switching current spikes due to the transformer and winding capacitance. The output voltage is equal

$$V_{out} = \frac{N \cdot V_{in}}{1 - D}\,. \tag{27}$$

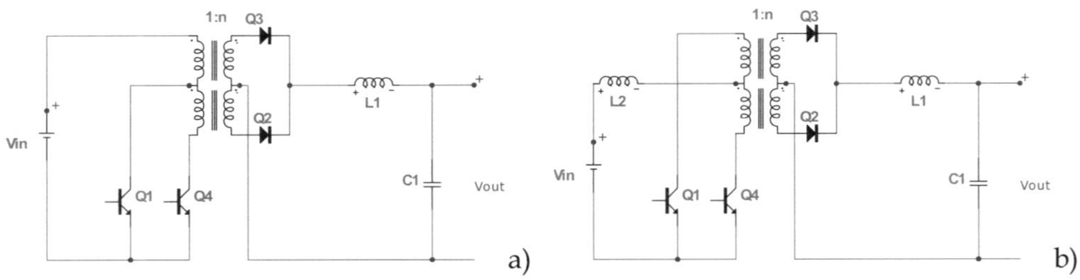

Fig. 15. Buck-derived converters with transformer (Push-Pull); b) Current-fed converter

Fig. 16a and 16b show the Full-bridge and Half-bridge converters which are derived also from Buck converter. Bridge converters are often used for higher input voltage application. The output voltage of Full-bridge is calculated by (26) as well and output voltage of half-bridge is half.

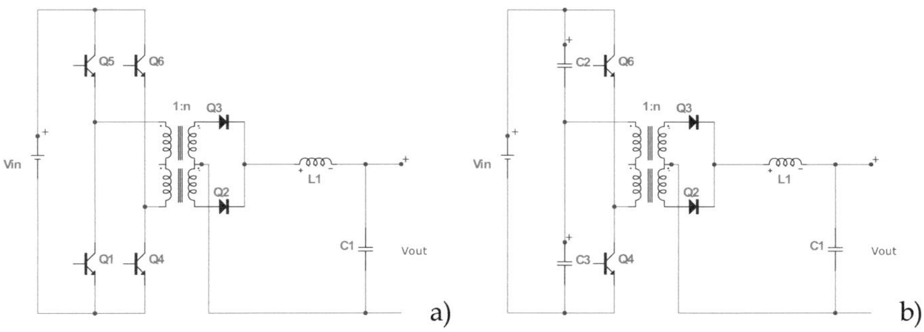

Fig. 16. Dc-to-dc converters with transformer isolation a) Full-bridge; b) Half-bridge

These previous topologies of Boost dc-to-dc concept of converter with transformer isolation can be also build with the bridge configuration. The most common switching converter of dc-to-dc converter with transformer isolation – isolated buck-boost converter, called Flyback and ZETA. The Flyback converter topologies are shown in Fig. 17. The output voltage is calculated of these is by (28). Two transistor concept has advantage of such that the voltage rating of the switches is a one-half of the single-transistor version. The version of paralleling Flyback converters can be found at high power levels, where can be connected more Flyback converts in parallel. The ZETA converters are also a transformer type converter with a low-pass filter. The output voltage ripple is small and output voltage can be calculated also by (28). (Mitchell, 1998).

$$V_{out} = \frac{N \cdot D \cdot V_{in}}{1 - D} \tag{28}$$

The concept of Cúk and SEPIC derived dc-to-dc converters with transformer isolation is shown in Fig. 18a and Fig. 18b.

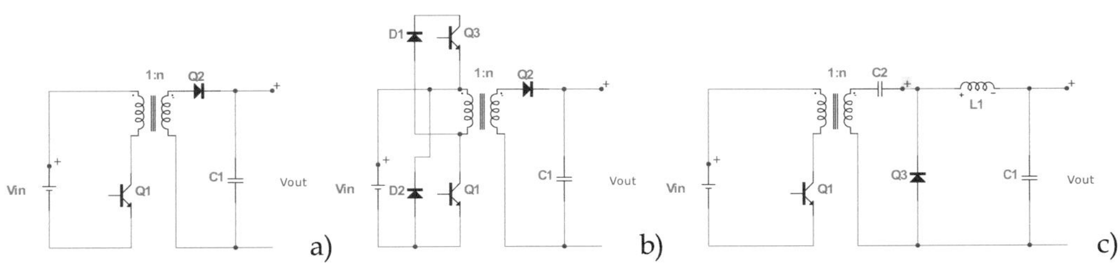

Fig. 17. Buck-Boost derived dc-to-dc converters with transformer isolation (Flyback) a) basic; b) two transistor Flyback; c) ZETA converter

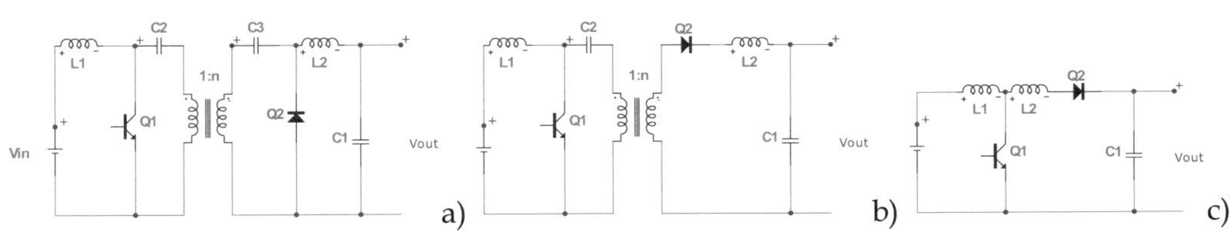

Fig. 18. a) Cúk; b) SEPIC derived dc-to-dc converters with transformer isolation; c) Tapped-Inductor Boost converter

In order to show complete family dc-to-dc converters to achieved higher voltage transformer gain, we need to add next developed converters shortly. For instance, tapped inductor converters are the next most common converters of dc-to-dc converters based on the buck, boost or buck-boost topologies. The example of these converters is shown in Fig. 18c and (Luo & Ye, 2000). Here the tapped inductor ratio is

$$n = \frac{n_1}{n_1 + n_2} \quad . \tag{29}$$

The other kinds of converters are based on the technology of Switched-capacitor. These dc-to-dc converters consist of only capacitors without inductors. They have advantages such as lower power losses. Zero-current or voltage switching (ZVS, ZCS) and zero-transition (ZT) converters can performed in two and four-quadrant operation with high power range. Although the treatment of the resonant converters is beyond the scope of this chapter, it should be observed that they are becoming increasingly popular for bi-directional applications (Luo & Ye, 2000).

The final analyses of previous concepts will be discussed in the following. The Buck-Boost dc-to-dc converter (Fig. 10c) is simple concept with low gain of transfer voltage. This topology of converters can be used for high switching frequency, what is good from the view of power density. The disadvantages are no galvanic isolation between input and output and high level of disturbances. Whereas, the various types of dc-to-dc with isolation usually have high transfer voltage gain and high isolation between both sides what is advantage in the comparison with previous fundamentals dc-to-dc converters. Their gain depends on the transformer turn radio N, which can be 1000. The Flyback converters (Fig. 17) use the demagnetizing effect and works in I-Q. The advantages of this Flyback concept are also simple topology, possibility to use more then one inputs and high efficiency. The disadvantages are design of transformer and also high level of disturbances.

The Boost dc-to-dc converter (Fig. 10b) can be described as well as Buck-Boost like simple circuit for high switching frequency, nevertheless with low level of disturbances and possibilities to operate with continuous current. The disadvantage is non-galvanic isolation and regulation feed-back control is non-stabile. The Buck (Fig. 10a) is also simple and for high switching and operates with high efficiency, on the other hand the concept is with non-galvanic isolation and operates with high efficiency. The concept with Tapped-Inductor Boost converter (Fig. 18c) can achieved higher transfer voltage gain in comparison with Boost concept (Fig. 10b). The other parameters are similar as a Boost.

The Cúk concept (Fig. 10d) operates with high efficiency, continuous input and output current, low level of disturbances, with high switching frequency and with low voltage and current ripple. The disadvantage is also no galvanic isolation between input and output, the output voltage is inverted. The concept of Cúk with transformer (Fig. 18a) has a galvanic isolation and other features are the same as a Cúk concept (Fig. 10d). SEPIC concept (Fig. 10e) are resistant for short-current, wild range of input voltage and both inductances can be at the same core. The disadvantage can be found in higher ripple and these concepts are good only for low power application.

Forward concept (Fig. 14) is also simple circuit with galvanic isolation by transformer and with low level of voltage ripple. The design of transformer is slightly complicated and converter has bad response of load changing. The concept with two-switch has lower request for switch and lower output ripple. The concept with paralleling Forward

converters can be useful for high power. The Push-Pull (Fig. 15) includes a good efficiency and ZETA converter (Fig. 17c) includes a low voltage ripple. The Bridge converters (Fig. 16) are often used for higher input voltage application.
What was shown above, a forward converter is a transformer type buck converter and t works in I-Q. Push-pull converter works in push-pull state, which effectively avoids the iron core saturation. The converters which can operate in I-Q and III-Q are Push-pull, Half-bridge and Full-bridge. Therefore, these will be discussed in the following, because these can be used as bi-directional converters.

5. Bi-directional dc-to-dc converter topology for low power application

This part presents the main purpose of this chapter. Bi-directional converters allow transfer of power in either direction according previous part of converter quadrants, where was shown the I-Q and II-Q are necessary for these, or the variant of all quadrants converters. Due to ability of reversing, current direction of flow and voltage polarity must be unchanged; they are being used in application of sources with battery or supercapacitors. Achievable implementation of bi-directional converters can be with hard-switching technology, or with resonant or soft switching technology. All of these lead to an increase components which make decreasing total efficiency and lower reliability. The main points of our focus on these dc-to-dc converters are: high efficiency; high switching frequency; high voltage transfer gain between input and output; low current ripple; to allow bi-directional power control flow; galvanic isolation between input and output; low level of EMI; low stress on the switches.
Therefore, the main goal is finding the middle ground of these previous points. The proposed topologies are a fundamental Buck-Boost concept from Fig. 7, bridge-converters from Fig. 16, a current push-pull (Fig. 15), and tapped inductor (Boost or Buck) converter (Fig. 18c). As was mentioned above, the proposed converter from Fig. 7, is a combination of two well-known topologies, Buck and Boost. The converters provide the bi-directional flow of power for battery charging and discharging using only just one inductor. On the other hand, they require a bulky input inductor L to limit the current ripple, especially when high voltage transfer gain is required. This concept can be used for kW what is main advantage besides simplicity. Other advantages of this concept include low stress on the switches and high switching frequency, low level of disturbances. The both modes (Buck and Boost) operate with high efficiency. The main disadvantage of this is no galvanic isolation between input and output. (Profumo et al., 2006)
To minimise the input inductor size and the current ripple, as well as to reduce the single switch current stress, the previous concept of dc-to-dc converter can be designed with multiple legs interleaving each other by means of an input coupling inductor. The concept is based on the controlling the two identical legs with a phase shifting equal to one-half of the switching period. In this, the current ripple can be reduced and the size of inductor then can be lower then concept with just one inductor (Fig. 7c). Such concept can reduce also switch current stress. Additional advantage of multiple legs of bi-directional Buck and Boost converter is also high power using such as in hybrid electrical vehicles or fuel cell application. To increase its power density, the design can take on a small inductor with multiphase interleaving to operate in discontinuous conduction mode, nevertheless EMI noises increasing.

The first chosen topology was taken from Buck and Boost topology (Fig. 10a and 10b). The second topology being considered is bi-directional concept derived from Buck-Boost (Fig. 10c), which is shown in Fig. 19b. Can be obvious that for I-Q operation is Q1 and D2 switching and Q2 and D1 switching during battery charging (II-Q). The third bi-directional concept of dc-to-dc topology can be seen in Fig. 19c. A Cascade Buck-Boost converter allows controlling in a separate way a switch for each particular condition of operation. The main advantage of this concept is an accurate control of the output voltage by duty cycle. The duty cycle can allow setting from 0 to 1 for both way of current flow. The drawback of cascade concept is double switches and diodes, which are duplicated with respects to the converter arrangement.
The proposed concepts from Fig. 7c, 19b and 19c require just one inductor and two electrolytic capacitors. The current ripple of Cascade Buck-Boost is lower in comparison with a Conventional bi-directional Buck-Boost topology, and requires an inductor with reduced size, what can be beneficial. (Caricchi et al., 1998)

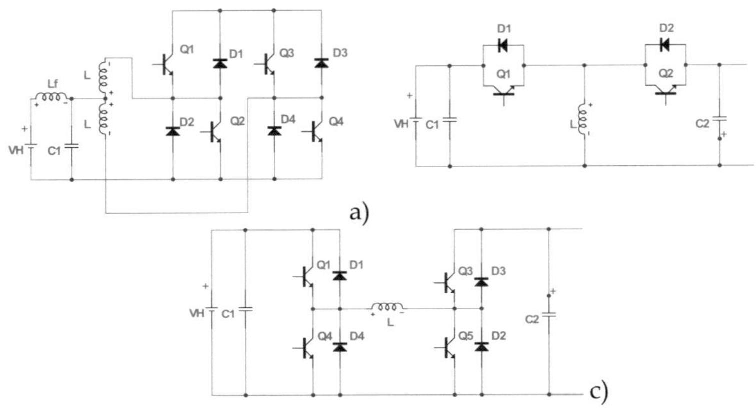

Fig. 19. Bi-directional dc-to-dc converter a) multiple legs Buck and Boost converter); b) Conventional bi-directional Buck-Boost dc-to-dc converter (delivered from Fig. 20); c) Cascade Buck-Boost converter)

The next considered topologies are Cúk, SEPIC and Luo converter. All was shown above for I-Q operation in Fig. 10d, 10e and 13. Circuits show converters these converters, which can operate also in II-Q and therefore are the bi-directional, there are shown in Fig. 20. The multiple-quadrant operation Luo converters are shown in Fig. 20c.

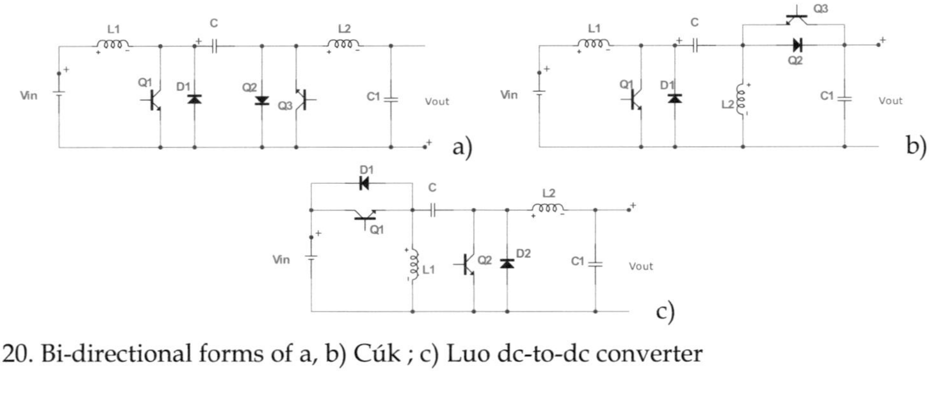

Fig. 20. Bi-directional forms of a, b) Cúk ; c) Luo dc-to-dc converter

Each consists of two switches, two diodes and two inductors. The main advantage is its reduced input and output current ripples with possibility of ripple-free current when integrating the inductances. Certainly, the advantage is also easy system to make isolation between input and output. Nevertheless, the main drawback is requirement of two large inductors and also need a large output capacitor. (Schupbach & Balda, 2003) The next drawback is transfer capacitor, for high power application must be rated at i.e. 600 V, which is actually expensive technology and this concept is available just only for low power application.

Anyway, the main disadvantage of these previous circuits is that the voltage transfer gain is still limited. The concept of dc-to-dc converter with Two-stage according Fig. 12 or Three-stage boost can desire gain improve, but the combination of this with concept from Fig. 19a together is so complicated and the output efficiency of this is out of the ordinary. However, when a very high dc voltage transfer gain is required, a high frequency transformer must be used. As a possible concepts of the most used topologies are the push-pull and Bridge converter. The first one is derived from Buck concept (Fig. 10a) and there were shown in Fig. 15 with diode rectifier on the output. The concept of un-bi-directional can be improved by converter which can also operate in I-Q and II-Q to achieve a bi-directional form of dc-to-dc converter. Therefore, converter consists of two active full-bridge units linked by high-frequency transformer, called dual Full-bridge topology, shown in Fig. 21a.

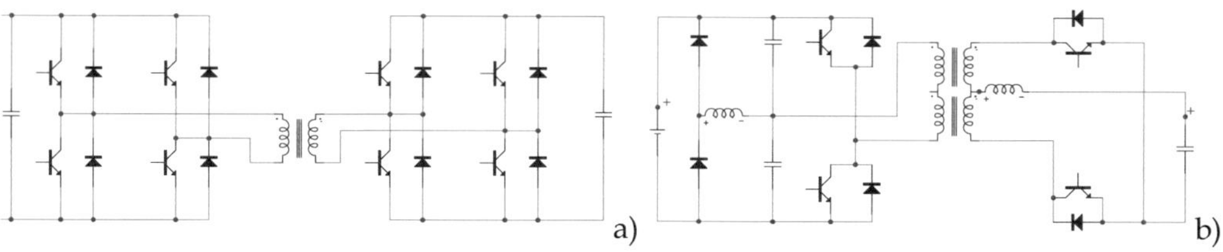

Fig. 21. a) Dual Full-bridge topology; b) Bi-directional dc-to-dc converter of half-bridge + current fed push-pull (Jain et al., 2000)

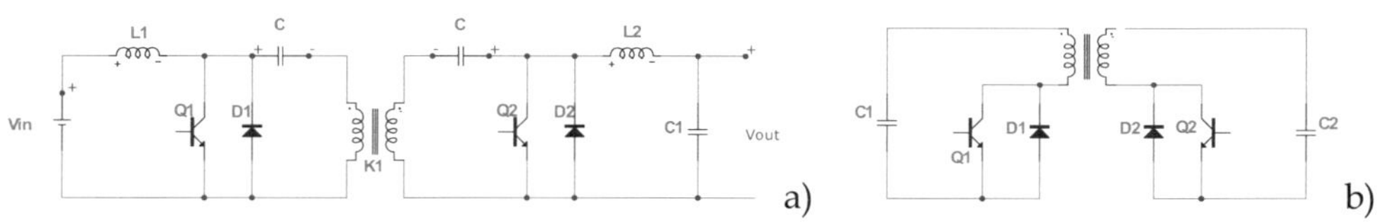

Fig. 22. Bi-directional form of a) Cúk dc-to-dc converter with transformer (Aboulnaga & Emadi, 2004); b) Bi-directional form of Flyback

An additional solution can be proposed by reducing of number of the power switches by means of half-bridge form of dc-to-dc converter. The concepts of bi-directional converters based on the push-pull technology are shown in the following figure (Fig. 21b). The bi-directional converter is set into one by two well-know concept, namely the half-bridge is putted in the primary side and a current fed push-pull (Fig. 15) in the secondary of a high frequency transformer. Such simple concept of galvanic isolated topology provides a good solution of these bi-directional circuits for battery charger circuits of UPS sources. The advantage is a low ripple charging current of battery, low stresses across switches and minimal number of active switches (efficiency). (Jain et al., 2000)

The analyses of current bi-direction dc-to dc converter for low power application, namely for telecommunication system use the concept of Cúk isolated bi-directional converter with integrated magnetics. (Aboulnaga & Emadi, 2004). The converters provide high efficiency, low noise and low EMI. In addition, the converter features high power density, low cost for low power and electric isolation. These are achieved mainly by integration of magnetic components (input and ouput inductor). The isolated bi-directional Cúk dc-to-dc converter is shown in Fig. 22a.

A bi-directional Flyback converter concept is shown in Fig. 22b. Advantage of the Flyback concept includes its simplicity, its known dynamic behaviour and its capability of stepping the dc voltage up and down. The drawback is low power using. The push-pull converter is a cost-effective solution for power up-to kWs and the control a strategy is simple. Anyway, the Bridge concepts can get more wide power ranging and so is used.

The major disadvantages of previous circuits are that the transformer leakage inductance causes high transient voltage across the bridge switches, which increases the switching rating and decreasing the reliability. For closing of this part of this chapter we can affirm that the Cúk, SEPIC and Luo converter are capable for low power application and require two inductors instead one in comparison with basic converters Buck-Boost (Fig. 7). The main advantage of these is its reduced input and output current ripples. The Bridge converters can be proposed with inductor lower in comparison in previous Cúk, SEPIC and Luo converter. Also from above analyses of topology is evident that converters Buck-Boost and Bridge concepts are capable for middle and high power application. In addition, potentially, both can be achieved the highest efficiencies. The main drawback of the Bridge converters is its discontinuous output current, when operating as a boost converter.

The Buck-Boost converter is simple and provides the bi-directional flow of power for battery charging and discharging using just by one inductor. On the other hand, it requires a bulky input inductor L to limit the current ripple, especially when high voltage transfer gain is required. Consequently, the Bridge concept is more frequent for this case of high voltage transfer gain request.

Both (Buck-Boost and Bridge converter) can be used for kW what is our use. Other advantages of the Buck-Boost concept include low stress on the switches and high switching frequency, low level of EMI. They can operate with high efficiency. The main disadvantage of this is no galvanic isolation between input and output. Therefore, the Buck-Boost or Bridge concepts are good candidate for middle power application. (Schupbach & Balda, 2003).

6. A design of bi-directional dc-to-dc converter for EGS

A simplified block diagram of an EGS with variable speed control was shown in Fig. 5. As a consequence of varying the engine speed when using the optimum variable speed control, both the output voltage and the output frequency of the generator vary and must be regulated to a constant value as required by the load. Therefore, a power electronic converter is required to regulate the output voltage and frequency. As mentioned above, the real drawback of EGS with optimum variable speed is the engine-generator dynamics at sudden transients from low load to high load. In case of sudden power output increase, the engine can not deliver the requested torque and the result is further decrease of the speed and torque of the engine until the undesirable stop. The diesel engine has namely a time constant of few second, which is further limited by fuel injection limitation. Therefore at high change of the speed the engine undesirably stops.

Diesel engine coupled with synchronous generator with permanent magnets has been widely use on many power related systems. The output voltage of the generator has an unregulated frequency as it is generated by a varying engine speed. However, the unregulated voltage and frequency of the generator must be changed by power electronics to obtain constant e.g. three-phase ac 400 V/ 50 Hz. Fig. 23 shows the block diagram of the power conversion steps including an ac-dc, dc-dc and dc-ac stage. As can be seen, all power flows from generator to the load through the converter. The fact that power flow through the converter is always makes intention to highlight into efficiency. Furthermore, the output alternating voltage of the EGS must be independent of its load and engine speed.

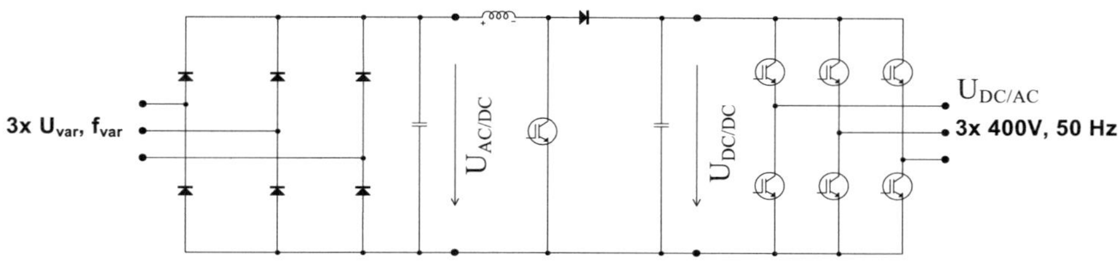

Fig. 23. The power electronics of an experimental model of EGS

As mentioned above and e.g. (Leuchter et al., 2009), the engine rotational speed is to be controlled optimally by adapting the operating strategy for every load to the required output power and torque, while minimizing the fuel consumption. The engine-generator dynamics, during sudden transients from low-load to high-load conditions, still poses a challenge in this regard. The following experimental results (Fig. 24) show the oscilloscope reading of the EGS dynamic behaviour. The measurements were performed in open-loop mode to show the dynamic behaviour of only the generator set with its static speed error e_s (Δ_n) and also to determine the generator set time constant T_C. The yellow curve (CH 1) shows the output current of ac-dc rectifier of power electronics (PE) and the red curve (CH 4) presents the engine speed. The output power is calculated from the output dc-voltage and dc-current. The result of this calculation is shown in Fig. 24 as the violet curve (MATH). The exact point where the load is changed from low load to high load is indicated by arrow A. During time interval t_1 the system was loaded with 640 W at 1560 rpm. During time interval t_2 (580 ms) the load was changed to 2480 W and the transient occurs. Time interval t_3 shows the system operating in steady-state condition. In Fig. 24a drop in speed Δ_n = 160 rpm is shown. This drop in speed corresponds to the static error e_s of the diesel engine inter-regulator. The time constant of a diesel engine T_C is approximately 0.6 s.

Fig. 24b shows a load change, which can cause to stop the diesel engine during time interval t_2. The engine is unable to sustain the required torque under these conditions, and consequently the EGS is unable to deliver sufficient power to the load. The previous measurements indicate the serious nature of the dynamic behaviour problem associated with the EGS with variable speed: its unsatisfactory reliability during sudden load changes as a result of the inferior dynamic behaviour of the EGS. (Leuchter et al., 2009)

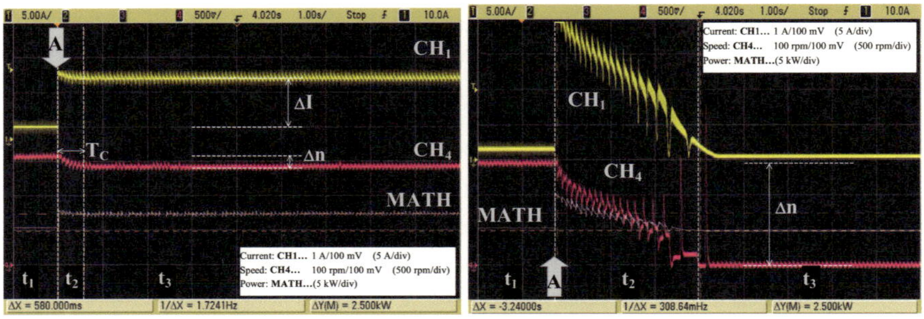

Fig. 24. The oscilloscope record of the load change from a) 0.64 kW to 2.480 kW at engine speed of 1560 rpm b) 0.6 kW to 6 kW at 1440 rpm

An energy buffer, connected via an electronic converter, can improve the dynamic behavior of the EGS system by injecting stored energy into the dc-link by more quadrant dc-dc converter. This concept is based on the delivery of peak power from the energy storage to the link capacitor of the dc-dc converter during the low to high speed transition of the diesel engine. The requested energy W is given by the maximal required power P and the average time of the regulation T_R. For providing 6 kW of power during approximately 4 s with small buffer overlarge to have enough energy for change a load during a short time:

$$W = P \cdot T_R \doteq 6000 \cdot 4 = 24 \text{ kJ} \tag{30}$$

$$C = \frac{\Delta t}{\Delta U} \cdot I_d = \frac{\Delta t}{\Delta U} \cdot \frac{P}{U_{DC}} \tag{31}$$

The EGS's energy buffer must therefore provide 24 kJ in order to satisfy the peak energy demand during a speed change. A storage bank, consisting of electrolytic capacitors, is feasible but the energy level of 24 kJ is very high. The electrolytic capacitor can accumulate a low amount of energy in comparison with convenient batteries or supercapacitors. The capacity which is required for 6 kW system is given by Eq. 31, where ΔU is the enabled voltage drop and U_{DC} is the nominal voltage of the dc-link.
Capacity request according Eq. 31 can be implemented by Supercapacitors. The electrical parameters are comparable with batteries. Very low internal resistance can achieve really high discharge currents. Present supercapacitors are readily available with capacity 3500 F. The advantage of supercapacitors is that their electrical characteristics and parameters are keep at -40°C. Interesting result, why use superacapacitors, can be seen in the following Fig. 25. The first reason is a lifetime, second is continues power and temperature range of using. The next advantages of these are good electrical behavior and relatively low initial cost in comparison with modern batteries, such as Li-Pol, Li-Fe (LiFePo4), Li-Ion and so on. Therefore, supercapacitors are one of the latest innovations in terms of electrical energy storage and its efficiency during charging and discharging is higher than with batteries. The energy stored within the electrical field of a capacitor by reducing its voltage to the half of its initial value can be expressed by the capacitor's state equation:

$$W = \frac{1}{2} \cdot C \cdot \Delta U^2 = \frac{1}{2} \cdot C \cdot \left(U_{nom}^2 - \left(\frac{U_{nom}}{2} \right)^2 \right) \tag{32}$$

The value of capacitor is given by:

$$C = \frac{2 \cdot W}{U_{nom}^2 - \left(\frac{U_{nom}}{2}\right)^2} = \frac{2 \cdot 24\,000}{50^2 - \left(\frac{50}{2}\right)^2} \approx 26\ \text{F} \tag{33}$$

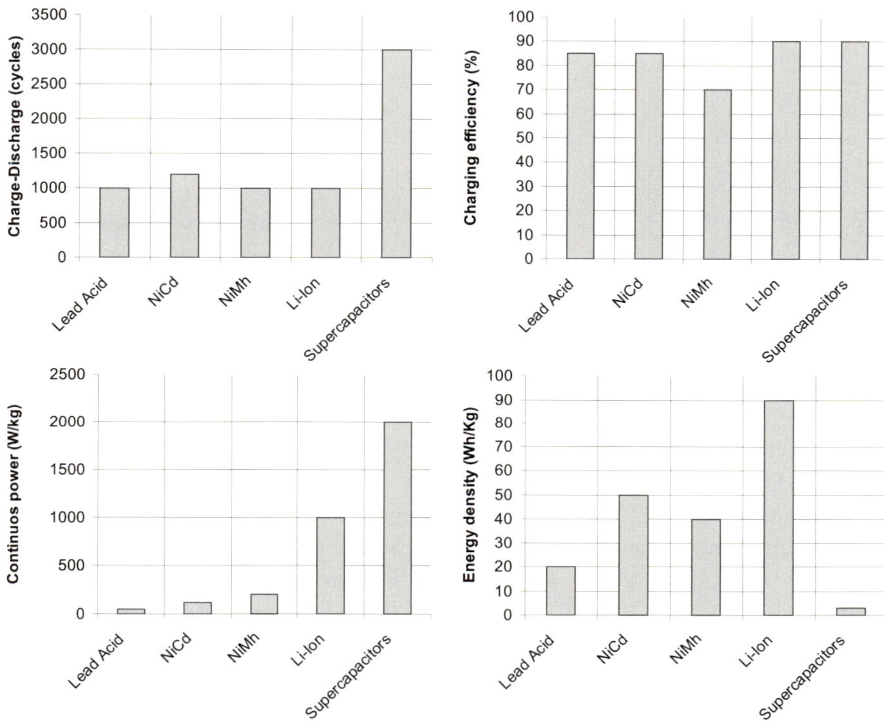

Fig. 25. Battery comparing (Leuchter et al., 2010)

How to set up such power buffer with supercapacitor technology is shown in Fig. 26. Figure shows two main concepts of bi-directional converters, which were described above.

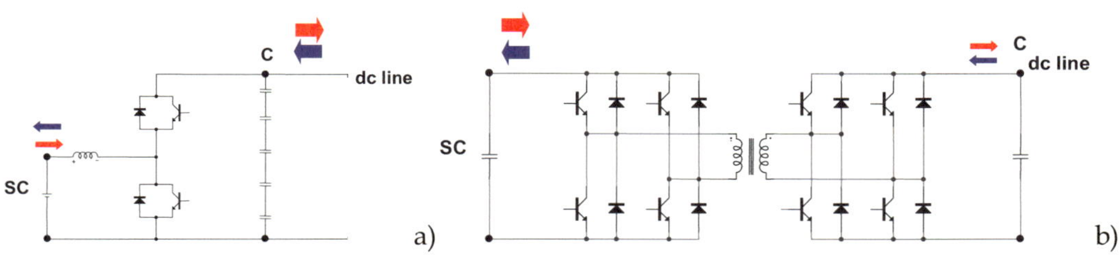

Fig. 26. Concepts of bi-directional converters with supercapacitors

To verify the steady stated and dynamic states performances of basic bi-directional dc-to-dc converter, a small-scale laboratory model where voltage was reduced is built it and set-up. The following block diagram of our experimental set-up of EGS with supercapacitor and bi-directional converter is shown in Fig. 27.

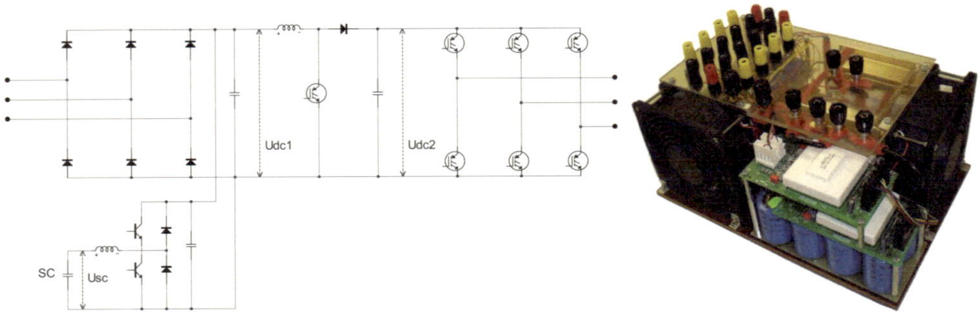

Fig. 27. Concepts of bi-directional converters with supercapacitors

In the experimental set-up, IGBT (SKM100GB128DN, Semikron) are used as switches. At the same time, one power diode bridge (SKD30/12, Semikron) in input are applied. In the following pages we shall discussed some performance evaluations. The value of supercapacitor capacity is equal to the reciprocal of the per-unit of W as defined in (31) and (33). Thus, if the per-unit value of W is 24 kJ and nominal voltage 50 V, the output capacity is 26 F, which is given by (33). Actually, we have selected more over-dimension superacacitor with 84 F which is directed into the automotive and transportation sector, where the wider range of power delivery requirements. In the experimental set-up, BMOD0083E48 is used. The important parameters of these supercapacitors are: capacitance 83 F; tolerance capacitance -5%/+20%; rated voltage 48.6 Vdc; ESRdc 10.3 m Ohm; ESRac 9.8 m Ohm; ,aximum continuous current 61 A; maximum peak current (1 s) 1 090 A; leakage current 3 mA; usable power density 2700 W/Kg; energy available 27 Wh; operating temperature range -40°C to +65°C. Charge and discharge characteristics by 10 A during supercapacitor testing are shown in 28.

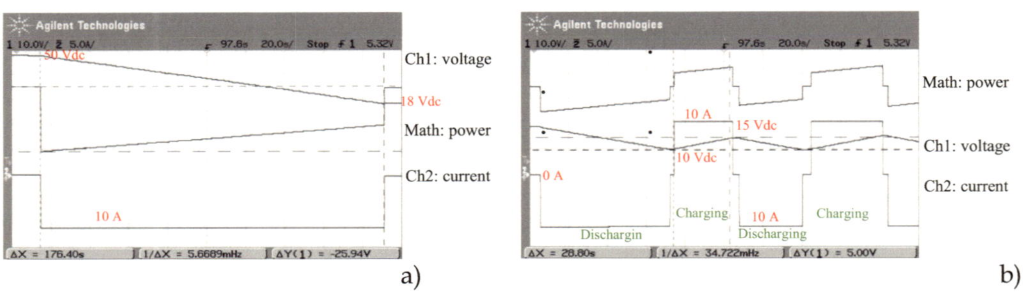

Fig. 28. Charging and discharging of supercapacitor (BMOD0083E48)

Bi-directional dc-to-dc converter according circuit topology from fig. 26a is set-up. The flow of power from or to supercapacitor is controlled according requirements of power management. The Boost mode is given by minimum input voltage 10 V and maximum input voltage 45V. The output voltage is controlled by PID regulator to achieve 100 Vdc voltage output of bi-directional converter. The input and output voltage is shown in Fig. 29. In this figure can be seen the current flowing through L. From these it can be evident that converter conduction operates in discontinuous modes.

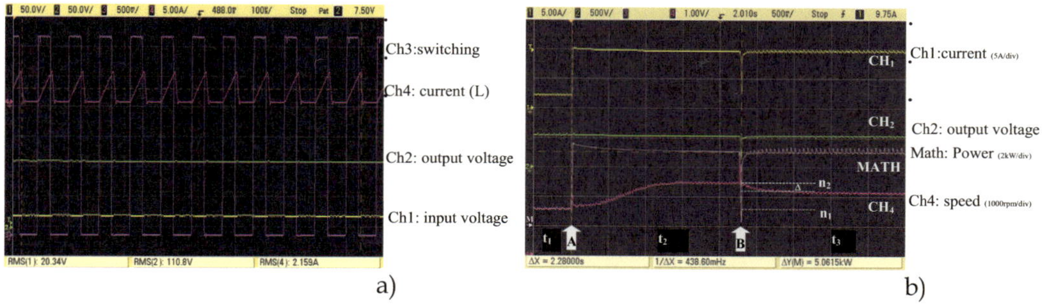

Fig. 29. a) The characteristic of Bi-directional converter with supercapacitor b) The verification of the EGS concept with supercapacitor and bi-directional converter

In accordance with these modes, the outputs voltages are given by (17) and (18). The critical inductance criterions were given in eq. (20) and (21). For very large L, the ac components of the inductor current may be considered negligible with respects to the dc components for the purposes of designing the inductor. As L becomes smaller, for given value of the load resistance, the radio of peak inductor current to average inductor current becomes larger. However, the peak-to-peak sawtooth amplitude becomes large enough that i_L would dip below zero by increasing R (Mitchell, 1998). In our case, converter can operate in discontinuous conduction modes and output voltage is set by PID regulator. The design specifications of the circuit are shown next. Power of converter is 1 kW; output maximum current is 20 A; maximum Δi_L= 2 A and switching frequency is selected 20 kHz. The output inductance of inductor can be design about 100 to 500 uH. Magnetic circuit which was chosen is given: U core; 1924 mm^2 (37x52 mm) and profile a=19 mm; B_{Max}=0.3 T; permeability μ_{rFE}=1000 a path length l_{FE}=0.28m. Then number of inductor winding is N=200 and the airgap are l_v= 0.001 (m). Therefore, the output L is around 250 uH. The following figures Fig. 30. shows the results of inductance (L) test versus frequency. From this analys is evident that output L a constant, whereas the output resistance are variable, what have effect on the output losses. The following figures show test results of converter as a function of output characteristics versus duty of Boost modes. Fig. 31 also shows the test results of efficiency during Boost mode for bi-directional dc-to-dc converters according Fig. 26a with inductor and Fig.26b with transformer. It can be seen that our efficiency of converter based on the Buck and Boost topology is actually very good.

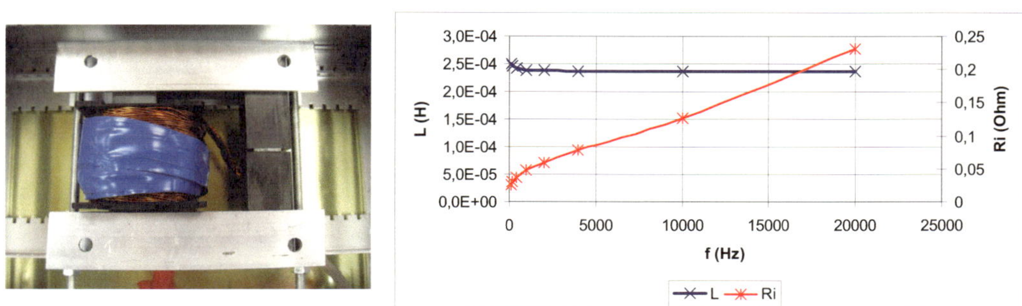

Fig. 30. Inductance versus frequency

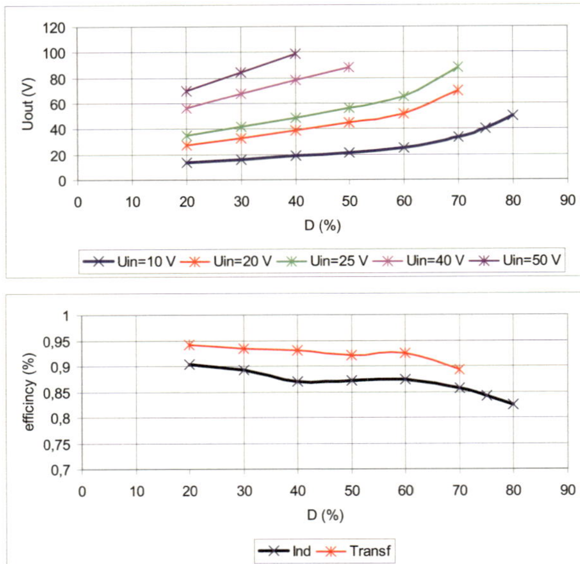

a)

b)

Fig. 31. a) Output characteristics of Boost mode b) Output efficiency

The experimental verification of the EGS concept with supercapacitor and bi-directional converter is shown in Fig. 29b. Channel 1 curve (CH 1) represents the same output current of dc line as was shown in Fig. 24. The red curve (CH 4) shows the engine speed. The output dc voltage is shown by the green curve (CH 2), while the output power is shown as the violet curve (MATH), which is calculated from the output dc voltage and dc current. In time interval T_1 the system was loaded with 1 kW at 1500 rpm. In time interval T_2 the power load was changed from 1 kW to 5.06 kW. During this time (2.28 s) the transition takes place and the power is extracted initially from the energy storage element. Time interval T_3 shows the steady-state condition of the system. It is possible to see that the system is now capable of handling the power load change combined with the feed-back control of the engine speed (set at an optimal speed of the diesel engine). This proves the feasibility of the EGS with variable speed of engine concept.

7. Conclusion

The review of dc-to-dc converter topologies has been shown and dynamic behavior of EGS with power buffer shown that power buffer with supercapacitor and bi-directional dc-to-dc converter can improve dynamic characteristics. This chapter includes detailed bi-directional converter topologies analysis as well as identification of its performance for power buffers. The real drawback of the proposed bi-directional dc-to-dc converters is higher cost because supercapacitor technology is still expensive.

8. Acknowledgment

The research work is supported by the Grant Agency of the Czech Republic (project no. 102/09/0013)

9. References

Aboulnaga, A.A. & Emadi, A. (2004). High performance bidirectional Cuk converter for telecommunication systems, *Proceedings of the INTELEC*, pp. 182-189, ISBN 0-7803-8458-X, 2004

Blaabjerg, F. & Chen, Z. (2006). *Power Electronics for Modern Wind Turbines*, Morgan and Claypool Publishers, ISBN 1-59829-032-0, USA

Boeke, U. (2007). High Efficiency Flyback Converter Technology, *Proceedings of the Power Conversion Conference*, pp. 1268-1273, ISBN 1-4244-0844-X, Nagoya, Japan, 2007

Caricchi, F. et al. (1998). Study of Bi-directional Buck-Boost Converter Topologies for Application in Electrical Vehicle Motor Drives, *Proceedings of the APEC*, pp. 287-293, ISBN 0-7803-4340-9 , Anaheim, CA, 1998

Jain, M.; Jain, P.K. & Daniele, M. (1997). Analysis of a bi-directional DC-DC converter topology for low power application, *IEEE Electrical and Computer Engineering*, Vol. 2, No. 1, (May 1997), pp. 25-28.

Jain, M.; Daniele, M. & Jain, P.K. (2000). A bidirectional DC-DC converter topology for low power application, *IEEE Transactions on Power Electronics*, Vol. 15, No. 4, (Jul 2000), pp. 595-606.

Kurka, O. & Leuchter, J. (2008). Super-capacitor as Electrical Power Buffers in Military Use, *Proceedings of the 4th International Scientific Conference of The Military Technical Collage (ICEENG)*, 13 p., Cairo, Egypt, 2008

Luo, F.L. & Ye, H. (2004). *Advanced DC/DC Converters*, CRC Press, ISBN 0-8493-1956-0, Florida, USA

Leuchter, J. et al. (2007). Bidirectional DC-DC Converters for Supercapacitor Based Energy Buffer for EGS, *Proceedings of the EPE Conference*, 10 p., Aalborg, Denmark, 2007

Leuchter, J. (2008). New Generation of Mobile Electrical Power Sources, *Proceedings of the the International Conference on Electrical Machines (ICEM)*, pp. 1366-1369, ISBN 951-22-5097-7, Helsinki, Finland, 2000

Leuchter et al. (2009). Dynamic Behavior Modeling and Verification of Advanced Electrical-Generator Set Concept, *IEEE Transactions on Industrial Electronics*, Vol. 56, No. 1, (January 2009), pp. 266-279, ISSN 0278-0046.

Leuchter, J.; Bauer, P. & Stekly, V. (2010). Battery-supercapacitors mixed as electrical power buffers, *Proceedings of the 5th IET International Conference on Power Electronics, Machines and Drives (PEMD 2010)*, 6 p., ISBN 978-184919231-6, Brighton, UK, 2010

Leuchter, J.; Bauer, P. & Zobaa, A., F. (2010). Power electronics and energy management of hybrid power sources with Supercapacitors, *Proceedings of the 5th IET International Conference on Power Electronics, Machines and Drives (PEMD 2010)*, 6 p., ISBN 978-184919231-6, Brighton, UK, 2010

Mitchell, D.M. (1998). *DC-DC Switching Regulator Analysis*, McGraw-Hill, ISBN 0070425973, USA

Mohan, N.; Undeland, T.M. & Robbins, W. (2003). *Power Electronics, Converters, applications and Design*, John Wiley & Sons, Inc., ISBN 0-471-22693-9, USA

Profomo, F. et al. (2006). Fuel cells for Electric Power Generation: Peculiarities and Dedicated Solutions for Power Electronic Conditions systems, *EPE Journal,* Vol. 16 No. 1 (Februarry2006). Pp. 44-49.

Schupbach, R.M. & Balda, J.C. (2003). Comparing DC-DC converters for power management in hybrid electric vehicles, *Proceedings of the IEMDC'03,* pp. 1369- 1374, ISBN 0-7803-7817-2, 2003

Schupbach, R.M. & Balda, J.C. (2004). 35 KW ultracapacitor unit for power management of hybrid electric vehicles: bidirectional DC-DC converter design, *Proceedings of the PESC,* pp. 2157- 2163, ISBN 0-7803-8399-0, 2004

Wildi, T. (1997). *Electrical Machines, Drives, and Power Systems,* Prentice Hall Publishers, ISBN 0-13-367889-X, USA

10

Bio-Inspired Synthesis of Electrode Materials for Lithium Rechargeable Batteries

Kisuk Kang and Sung-Wook Kim
Seoul National University, Republic of Korea

1. Introduction

Human history has been made through endless challenges, searching for universal truths of nature. Sometimes, nature becomes a crucial barrier that human beings should overcome, however, repeatedly, it inspires us to make progress in science and results in a better life. Nature always provides pointers in developing technologies; emulating nature serves as a very helpful methodology for such development (Bensaude-Vincent et al., 2002). Figure 1 shows some examples of creations that were invented through the emulation of nature. Especially, living organisms are excellent teachers whose metabolism, vital activity, and growth present novel synthetic routes for the formation of organic (or inorganic) biomaterials (Sanchez et al., 2005). The study of on the biomaterials, highly ordered forms of molecules in a biological system with complex nanostructures, has opened up a new era for fabricating nanomaterials through the emulation of biological processes (Dickerson et al., 2008).

This chapter briefly introduces the bio-inspired synthetic routes of nanostructured electrode materials for lithium (Li) rechargeable batteries using biomaterials as structural templates. Various biomaterials have been synthesized both naturally, *i.e.*, inside living bodies (*in vivo)*, and intentionally in the laboratory (*in vitro)*, (Sanchez et al., 2005; Dickerson et al., 2008). One can synthesize biomaterials that possess unique nanostructures without much difficulty. By controlling the synthesis conditions, the nanostructure of biomaterials can be varied from a simple 0-D structure to complex 3-D structures (Lv et al., 2008). The unique nanostructures of the biomaterials can be applied to various research fields, including not only bio-applications but also non-bio-applications such as semiconductors, display devices, catalysts, and energy conversion/storage devices, by hybridizing them with various functional materials at the nanoscale (Katz et al., 2004; Su et al., 2008; Li et al., 2009). As the minimizing of a material's dimension in a certain shape often provides distinctive material properties due to a large surface-to-volume ratio, geometry, and/or quantum effects, This could lead to breakthroughs in overcoming the limitations of conventional bulk materials (Moriarty, 2001). Thus, the hybridization of nanostructured biomaterials with functional materials frequently offers improved material properties under simple nanofabrication principles.

Fig. 1. Photographs of creations inspired by nature: (a) thistles (left) and Velcro® hooks and loops (right), (b) eye of fishes (left) and a fisheye lens (right), (c) kingfisher (left) and bullet trains (right), and (d) bumble bee (left) and micro air vehicle (right). All images were obtained from http://en.wikipedia.org/.

The (Li) rechargeable battery is the leading candidate for large scale energy storage devices due to its high specific capacity, high operation voltage, and thus, high energy density (Tarascon & Armand, 2001). Although the (Li) rechargeable battery has been used most widely as an energy storage system for small portable devices such as lap-top computers and mobile phones, its electrochemical performance is not sufficient to power larger scale energy storage systems such as electric vehicles and load-leveling systems (Kang et al., 2007). In this respect, investigating nanostructured electrode materials has become essential because improvements in electrochemical performance, such as higher specific capacity, higher rate capability, and better cyclability, are expected in this dimension. The nanoscale dimension offers some advantages to the electrochemical performance because of the large surface area contacting electrolyte, short (Li) ion diffusion length, and facile strain accommodation induced by volume change (Bruce et al., 2008).

Various synthetic routes have been investigated extensively to synthesize novel nanostructured electrode materials. The use of nanostructured templates is one of the most promising approaches because target nanostructures can be obtained simply from the structural duplication of the nanostructured templates (Cheng et al., 2008). Biomaterials, whose varieties of nanostructures are easily obtained by simple control of the synthesis conditions, are considered useful structural templates for nanofabrication (Cui et al., 2010). Also, their surface groups can offer possible nucleation sites for the growth of electrode materials (Ryu et al., 2010a). Nanostructured electrode materials based on biomaterial templates can show improved electrochemical performance compared with that of bulk materials, and their fabrication processes are often more environmentally friendly compared with other methods of preparing nanomaterials.

2. Biomaterials

In living organisms, biomaterials are produced from interpreting the genetic information in nucleic acids such as deoxyribonucleic acids (DNAs) and ribonucleic acids (RNAs). Genetic expression produces proteins, cells, tissues, organs, and finally, bodies. One interesting feature of biomaterials is that they are constructed spontaneously by self-assembly. The term 'self-assembly' refers to the organization of highly ordered nanostructures from disordered components through spontaneous non-covalent interactions between the components under specific conditions without any external driving forces (Reches & Gazit, 2006). When specific condition for self-assembly are satisfied, the components, or building blocks, start to create the nanostructure by themselves, forming various complex nanoarchitectures. Thus, self-assembly is a useful way to fabricate nanostructured materials. The self-assembly Phenomenon frequently occurs naturally in biological systems, forming complex nano-patterned structures (Fraden & Kamien, 2000).

For the past several decades, many research efforts have been focused intensively on the synthesis of biomaterials in the laboratory and on using them to develop conventional bio-applications such as tissue regeneration and artificial organs (Geise et al., 2006). Figure 2 illustrates various shapes of self-assembled biomaterials formed naturally and artificially. They exhibit exclusive structures from nanoscale to macroscale (Brachmann & Cagan, 2003; Ryu & Park, 2008; Xia et al., 2004).

Recently, it has been widely demonstrated that self-assembled biomaterials can serve as structural motifs for non-biological nanostructured materials (Sanchez et al., 2005). Biomaterials are composed of organic materials that contain various functional groups, so their acidities and polarities at the surface can provide possible adsorption sites for precursors of the target nanostructured materials. When the adsorbed precursors react exclusively on the surface to form the target materials, the complex shapes of the biomaterials are duplicated generating isostructural target materials. Some examples of duplicated nanomaterials are presented in Figure 3 (Zhou et al., 2007; Zhang et al., 2010).

Nanomaterials can fully or partially cover the surface of the biomaterials, forming nanoarchitecture similar to that of the template. The duplications of the nanostructure of the biomaterials offers new possibilities for biomaterials in a broad range of research fields beyond the conventional bio-applications by enabling the synthesis of various functional materials in complex forms.

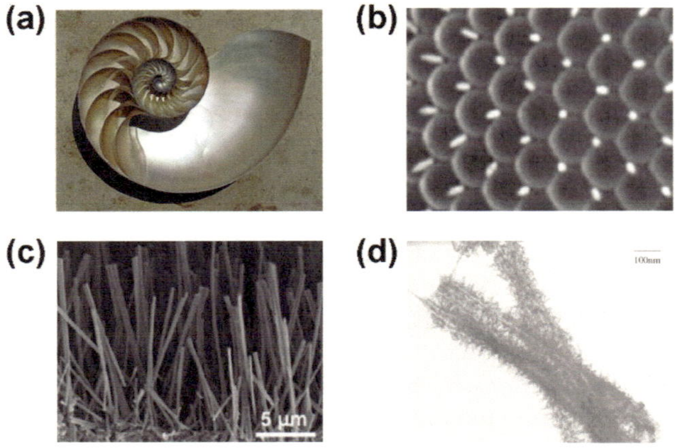

Fig. 2. Photographs of naturally (a-b) and artificially (c-d) self-assembled biomaterials: (a) shell of nautilus (http://en.wikipedia.org/), (b) hexagonal array of eye of *drosophilia* (Brachmann & Cagan, 2003), (c) well-aligned peptide nanowires (Ryu & Park, 2008), and (d) polyaniline-naphthol blue black nanotubes (Xia et al., 2004).

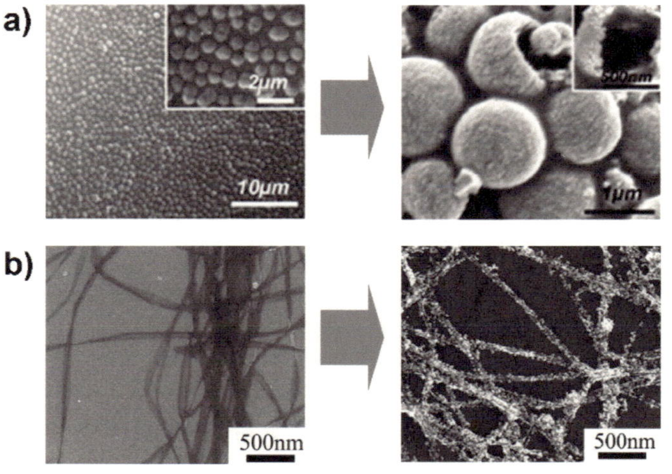

Fig. 3. Nanomaterials fabricated using biomaterials as structural templates: (a) *Str. theromophilus* (left) and ZnO hollow nanospheres fabricated using *Str. theromophilus* as the structural template (right) (Zhou et al., 2007) and (b) bacteria-cellulose nanofibers (left) and Au-bacterial-cellulose nanocomposite (right) (Zhang et al., 2010).

In this respect, the structural control of the biomaterial itself becomes an important technical issue. Artificially self-assembled biomaterials can display various nanostructures depending on the self-assembly conditions. Because the self-assembly is derived from the complicated combination of non-covalent interactions including hydrogen bonds, electrostatic interactions, hydrophobic interactions, and van der Waals interactions between the building blocks and environment, the morphology of biomaterials is significantly affected by the local environment. For example, Figure 4 illustrates a series of nanostructures of a self-assembled aromatic dipeptide, which were produced by controlling the dissolving solvents (Han et al., 2008). During the process of dissolving and cooling diphenylalanine (NH_2-Phe-Phe-COOH) in H_2O, CH_3OH, C_2H_5OH, or CH_2Cl_2, the diphenylalanine self-assembled into a nanotube shape in H_2O (Figure 4(a)), nanoribbon shape in CH_3OH and C_2H_5OH (Figure 4(b-c)), and a nanoribbon/nanowire shape in CH_2Cl_2 (Figure 4(d)). The solvent polarity affects the force balance between the non-covalent interaction, resulting in various nanostructures obtained with different solvents. If one can prepare the self-assembled biomaterial with precise control of the specific nanostructure, from a simple 0-D to a complex 3-D structure, it would be a very attractive template for the nanostructured functional material. So far, naturally structured biomaterials have been used frequently as templates, and thus, their structural duplications have been studied extensively in research fields where nanostructured materials are required. However, despite the extensive research efforts to control the morphology of biomaterials, further studies are needed to fabricate self-assembled biomaterials with specific shapes.

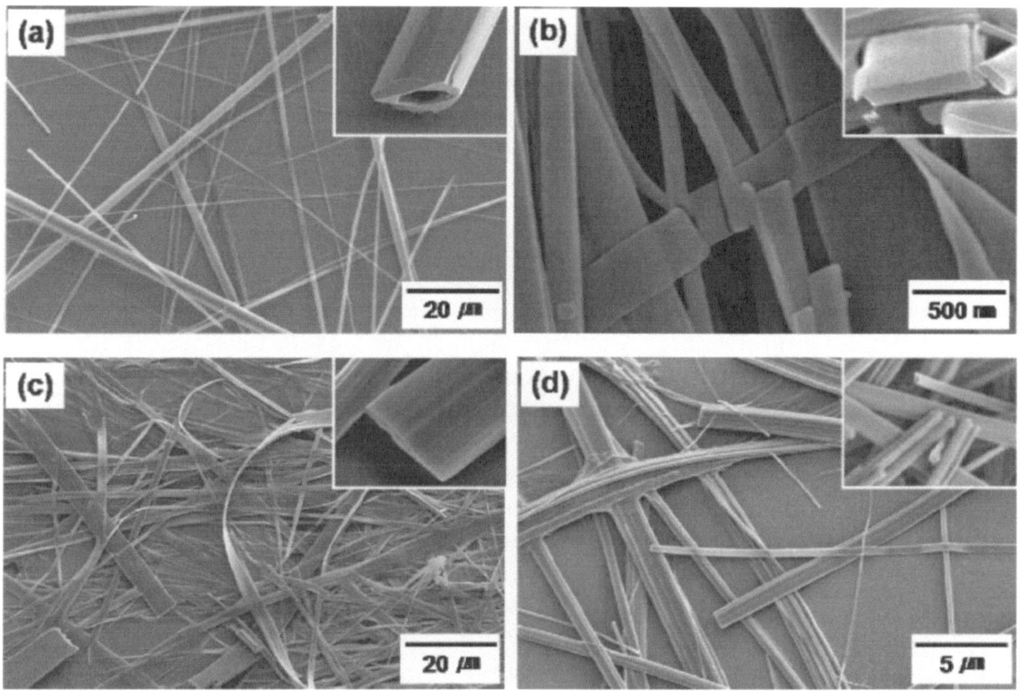

Fig. 4. Morphology of diverse nanostructures of the peptide formed in various solvents: (a) nanotubes formed in H_2O, (b) nanoribbons in CH_3OH, (c) nanoribbons in C_2H_5OH, and (d) nanoribbons and nanowires formed in CH_2Cl_2 (Han et al., 2008).

3. Bio-inspired synthesis of materials for lithium rechargeable batteries

Recently, researchers have tried to find a way to combine biomaterials as a nano-sized structural template with a functional material with the hope that the nanoscale dimension and morphology can improve performance of the material. Among various functional materials under consideration, materials for energy storage/conversion devices attract much interest because energy problems, such as exhaust and CO_2 emissions of fossil fuels, have become severe (Tarascon et al., 2001). Nanostructured electrode materials for Li rechargeable batteries offer improved electrochemical activity, resulting in enhanced battery performance. However, the realization of the expected nanostructure, *i.e.* morphology and dimension, requires further investigation of the synthetic routes. The bio-inspired synthesis of nanostructured electrode materials can be a useful approach because surface coating onto the nanostructure biomaterials produces materials conformally (Sanchez et al, 2005). When one can control the nanostructure of biomaterials by adjusting the self-assembly conditions, it is easy to synthesize the nanostructured electrode materials with the intended morphology and dimension.

The various synthetic routes for the bio-inspired synthesis of nanostructured electrode materials are described in this section. Potential electrode materials, such as the Co_3O_4 anode, TiO_2 anode, and amorphous $FePO_4$ cathode, could be fabricated onto biomaterials by wet-chemistry or vapor deposition processes. Because the biomaterials generally possess polarity or various functional groups on the surface, many chemical species such as ions and molecules can be adsorbed on the surface, reacting into the electrode materials. Organic-inorganic hybrid materials, ensembles of the organic biomaterial templates and the inorganic electrode materials, can be adopted as electrode materials. Additionally, removing the biomaterial template in the hybrid materials leaves hollow-structured materials with superior electrochemical performances as the electrode.

3.1 Virus-based hybrid electrode materials

Enhanced electrochemical activity in the nanoscale dimension is an important reason for challenges to apply bio-inspired synthesis to Li rechargeable battery materials, due to its precise structure controllability. Another reason is the soft, flexible, and self-standing properties of the biomaterials. Increasing demands for portable, wearable, and stretchable electronic devices have created a need for flexible batteries, which can only be realized by using flexible electrode materials. As such, biomaterials are considered as excellent supporting materials for electrode materials. Biomaterials such as viruses and peptides are promising templates for the nanostructured electrode materials due to their capabilities to form unique nanostructures uniformly over a large area. When the electrode materials precipitate onto the surface of the biomaterials, forming organic-inorganic hybrid materials, it is expected that the hybrid materials can exhibit electrochemical activity combined with flexibility.

Viruses are kind of parasitic pathogens inside living organisms that replicate themselves. Their genetic information is stored in either DNA or RNA. They can replicate only inside the infected cells of the hosts because they do not have any organs for metabolism and energy production. Since the first observation of tobacco mosaic virus in 1892, various viruses have been reported inside all types of living organisms, from small protozoa to large mammals. Among them, the M13 virus is one of the most frequently investigated viruses for nanotechnology (Lee et al., 2003). The M13 virus, shown in Figure 5, possesses a wire-like anisotropic structure approximately 6.5 nm in diameter and 880 nm in length (Nam et al., 2004). The M13 virus particle is composed of circular single-stranded DNA encapsulated by a major coat protein (p8) and capped by minor coat proteins (p3, p6, p7, and p9) at the end of the virus.

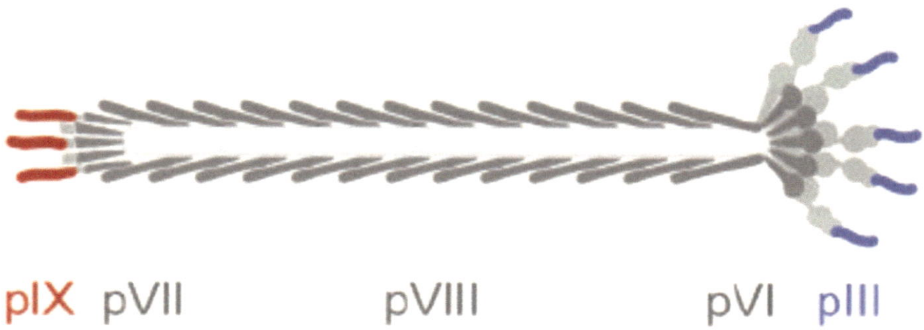

Fig. 5. Schematic representation of a genetically modified M13 virus (Nam et al., 2004).

Hybridizing the M13 virus with functional inorganic materials is possible through modification of the major and/or minor coat proteins because the modified proteins provide nucleation sites for the inorganic materials. The configuration of the proteins is determined by the genetic information. Hence, modifying the genetic information in DNA creates modified proteins and thereby the modified M13 viruses. Figure 6(a) briefly describes the hybridization process of ZnS-M13 virus with genetic modification. The major coat protein, p8, is modified to induce the nucleation of ZnS nanoparticles. The configuration of the protein is determined by considering the compatibility with ZnS nanoparticle growth. In this case, the A7 (Cys-Asn-Asn-Pro-Met-His-Gln-Asn-Cys) peptide strongly adsorbs Zn^{2+} ions. The modified M13 virus is stored in $ZnCl_2$ solution, and then Na_2S solution is introduced to synthesize the ZnS-virus hybrid material. As a result, ZnS nanoparticles cover the surface of the virus conformally (Figure 6(b-c)).

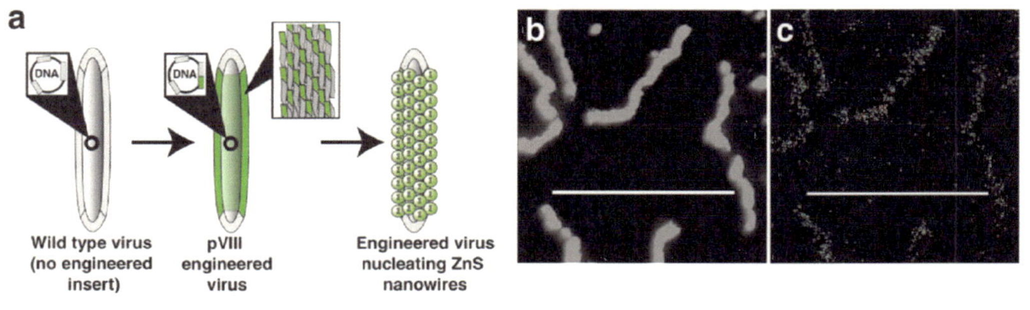

Fig. 6. (a) Schematic diagram of the hybridization process of ZnS-virus nanowire through the genetic modification of p7 proteins, (b) STEM images of the synthesized ZnS-virus nanowire, and (c) EDS mapping of S of the same sample. (Scale bars: 800 nm) (Mao et al., 2003).

As the M13 viruses are able to bind inorganic materials through the genetic modification, it is possible to use the modified M13 viruses as biotemplates for the nanostructured materials used in Li rechargeable battery electrodes. Figure 7 represents the M13 virus genetically modified to bind amorphous $FePO_4$ cathode material and highly conductive carbon nanotubes (CNTs) (Lee et al., 2009). Amorphous $FePO_4$ is an attractive cathode material due to its high specific capacity and safety originating from the strong $(PO_4)^{3-}$ covalent bond, but its poor electronic/Li-ionic conduction precludes further investigation

(Okada et al., 2005). Therefore, developing nano-sized amorphous $FePO_4$ with a highly conductive agent is essential for the high performance Li rechargeable battery. The modified M13 virus enables the formation of amorphous $FePO_4$ nanoparticles at the p8 proteins and, at the same time, combines with the CNT at the p3 proteins as the result of genetic modification of the p8 and p3 genes.

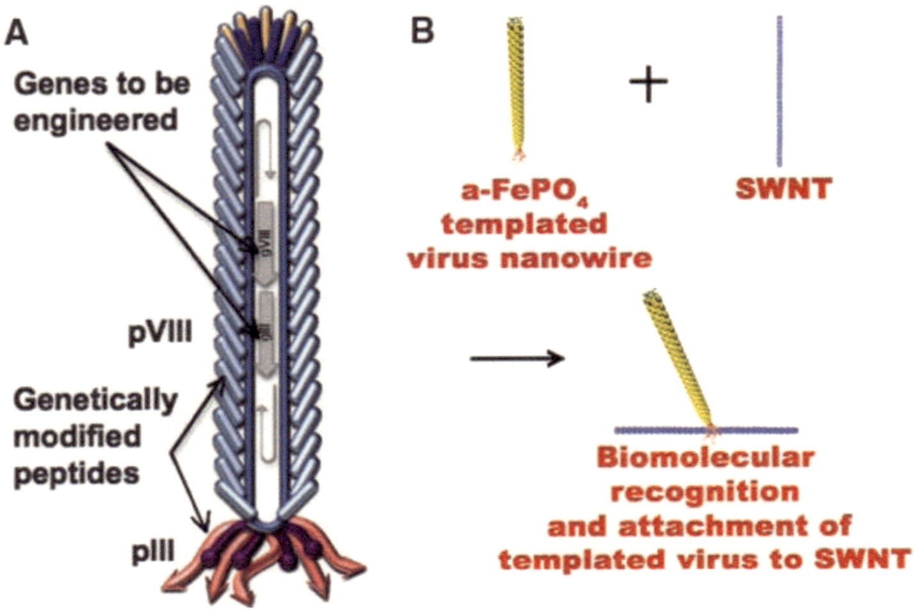

Fig. 7. Schematic illustration presenting the process of the virus-amorphous $FePO_4$-CNT hybrid material for a high-power Li rechargeable battery: (a) multifunctional M13 virus obtained through genetic modification and (b) the hybridization process with the modified M13 virus (Lee et al., 2009).

The hybrid nanowire is synthesized by the following sequence: (i) genetic modification of the M13 virus to bind amorphous $FePO_4$ at p8 proteins and CNTs at p3 proteins, (ii) Au nanoparticle loading onto the virus, (iii) formation of amorphous $FePO_4$ onto the Au-loaded virus template to enhance electrical conduction, and finally, (iv) reacting the virus-amorphous FePO4 nanowires with CNT suspensions with surfactant. Because the modified p8 proteins contain extra carboxyl acid groups, metallic cations can interact readily with the virus. The modified p3 proteins have a high affinity for CNTs; thus, the virus-amorphous $FePO_4$ nanowire can bind CNTs to form the hybrid nanowire. The synthesized hybrid material composed of virus template-amorphous $FePO_4$ electrode materials-CNT conductive agent is shown in Figure 8 (Lee et al., 2009). The hybrid material possesses a complex network structure of individual nanowires, which has the same morphology as that of the virus template without any hybridization process. The diameter of amorphous $FePO_4$ nanoparticles is approximately 10-20 nm, and that of CNTs is approximately 4-5 nm. Amorphous $FePO_4$ and CNTs are well connected each other, as shown Figure 8(e), indicating that electrons are rapidly supplied to the amorphous $FePO_4$ cathode through the CNT networks.

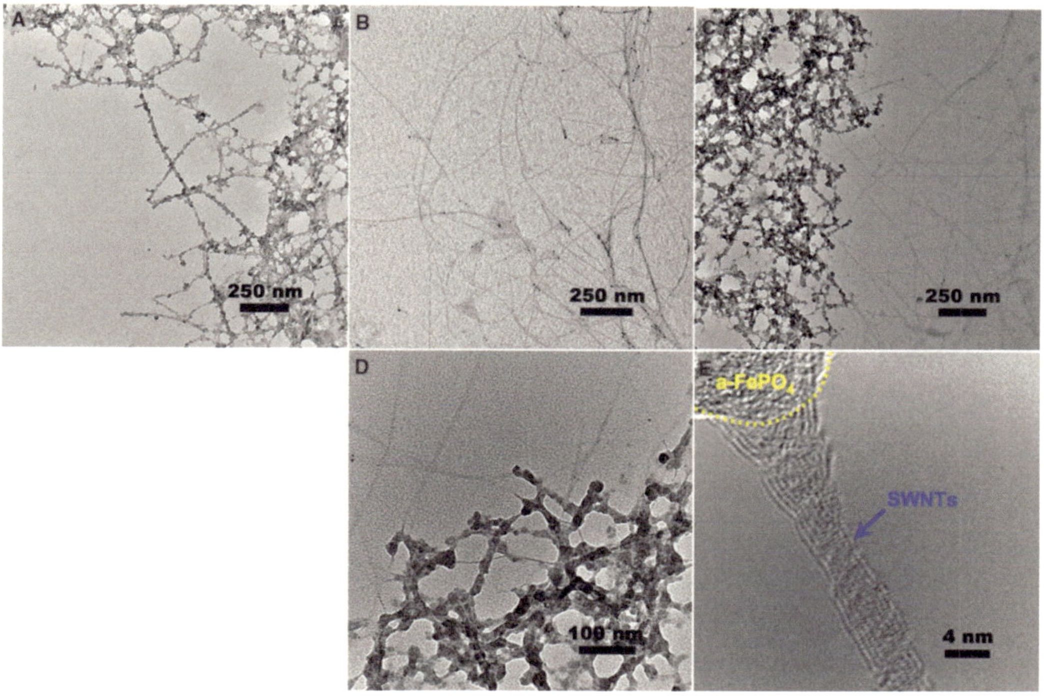

Fig. 8. Nanostructure of synthesized hybrid nanowires: (a) virus-amorphous $FePO_4$ hybrid nanowires without CNTs, (b) bare CNTs, and (c-e) virus-amorphous $FePO_4$-CNT hybrid nanowires at different resolutions (Lee et al., 2009).

The simultaneous supply of Li ions and electrons in/outside of active electrode materials is essential for the operation of the Li rechargeable battery. It is obvious that a faster supply results in higher performance during operation. The dimension shrinkage of electrochemically active material, *i.e.*, amorphous $FePO_4$, reduces the distance for Li-ion penetration. The CNT networks continuously supply electrons to the active material. Hence, superior electrochemical performance is accomplished in the virus-amorphous $FePO_4$-CNT hybrid nanowires, as depicted in Figure 9 (Lee et al., 2009). Figure 9 compares the electrochemical properties of three types of hybrid nanowires; which are virus-amorphous $FePO_4$ hybrid nanowires without CNTs (E4), virus-amorphous $FePO_4$-CNT hybrid nanowires with moderate binding affinity to CNTs (EC#1), and those with strong binding affinity to CNTs (EC#2). EC#1 and EC#2 display higher specific capacity at every current rate compared with E4, indicating that the CNT networks sufficiently provide electrons during the electrochemical reaction. Especially, EC#2 exhibits superb electrochemical performance. The specific capacity of EC#2 at a current of C/10 (1C = 178 mA g^{-1}) is approximately 170 mAh g^{-1}, which is comparable to that of a theoretical one. Its high capacity is maintained at higher current rates, indicating that the hybrid nanowires are promising candidates for high-power applications. The specific capacity of EC#2 at 10C is as high as 130 mAh g^{-1}. Additionally, all samples show excellent cycle retention up to 50 cycles, indicating that the hybrid nanowires remain stable upon electrochemical cycling.

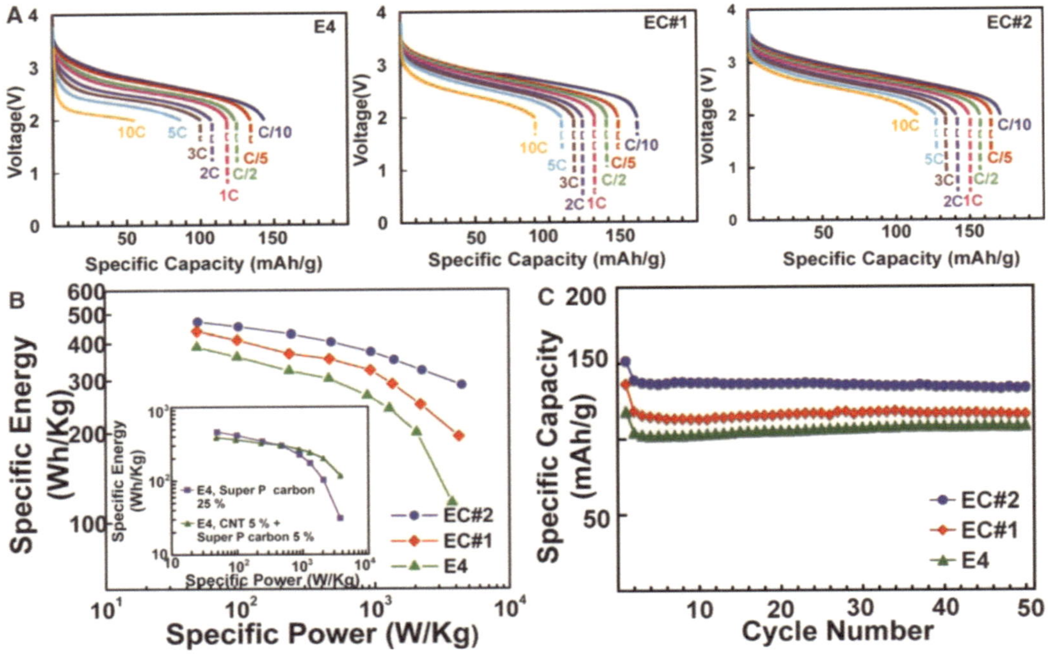

Fig. 9. Electrochemical performance of hybrid nanowires based on multifunctional M13 virus templates: (a) discharge profiles of E4 (virus-amorphous $FePO_4$ nanowires), EC#1 (virus-amorphous $FePO_4$-CNT nanowires with moderate affinity for CNTs), and EC#2 (virus-amorphous $FePO_4$-CNT nanowires with strong affinity for CNTs) at current rates from C/10 to 10C, (b) Ragone plots of the hybrid nanowires (inset: Ragone plot of E4 as a function of carton contents), and (c) capacity depending on the number of cycles at a current rate of 1C for 50 cycles (Lee et al., 2009).

The same strategy of a modified virus as a structural template could be adoptable for other kinds of electrode materials. Co_3O_4 is one of the most promising anode materials for the Li rechargeable battery due to its high specific capacity (~890 mAh g^{-1}) through a conversion reaction (Kang et al., 2005). Although the Li_2O phase, a product of the conversion reaction of Co_3O_4, is electrochemically inactive, it becomes electrochemically active when the dimension of Li_2O is reduced to nanoscale. Thus, the nanofabrication of Co_3O_4, the mother phase of Li_2O, is essential for the highly active Co_3O_4 anode (Li et al., 2008). The genetically modified M13 virus could define the nanostructure of Co_3O_4. Figure 10(a) simply illustrates the modification of the M13 virus (Nam et al., 2006). The surrounding p8 proteins are modified to two types: one contains a Co nucleating motif, and the other contains Au binding motif. The virus-Au-Co_3O_4 hybrid nanowires are fabricated through an Au-binding followed by Co_3O_4 nucleation and growth. Viruses that do not contain the Au binding motif cannot bind Au nanoparticles at all. Uniform Au nanoparticles (~5 nm in diameter) are finely dispersed in the hybrid nanowires, as shown in Figure 10(b-c) (Nam et al., 2006). The hybrid nanowires show excellent electrochemical properties due to the improved electrochemical activity of the conversion reaction in the nanoscale dimension. The Au-containing nanowires exhibit better performance, indicating that Au nanoparticles further improve the electrochemical activity. This is due to the increased electronic conduction, or catalytic effect of the Au nanoparticles. The increase in the current density at CV measurement shown in Figure 10(e) indicates that the incorporating of Au nanoparticles increases the reaction rate of the Co_3O_4 anode (Nam et al., 2006).

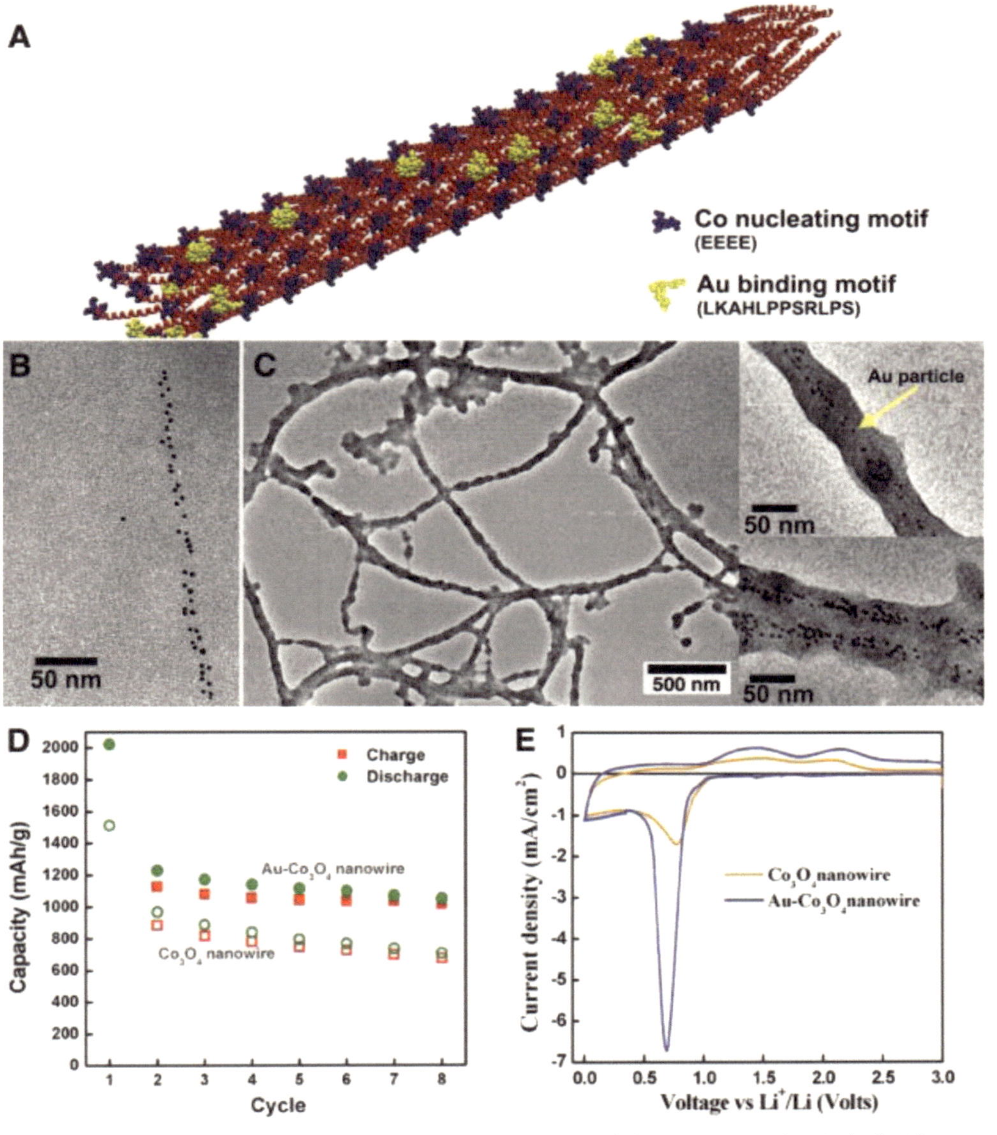

Fig. 10. Nanostructures and electrochemical properties of the virus-Au-Co_3O_4 hybrid nanowires: (a) schematic visualization of the modified M13 virus with both a Co-nucleating motif (blue) and an Au-binding motif (yellow), (b) the virus-Au hybrid nanowire before Co_3O_4 growth, (c) the virus-Au-Co_3O_4 hybrid nanowire, (d) specific capacity depending on the number of cycles of the hybrid nanowires with and without Au at a current of C/26.5 for 8 cycles, and (e) CV curves of the nanowires with and without Au at a scan rate of 0.3 mV s^{-1} (Nam et al., 2006).

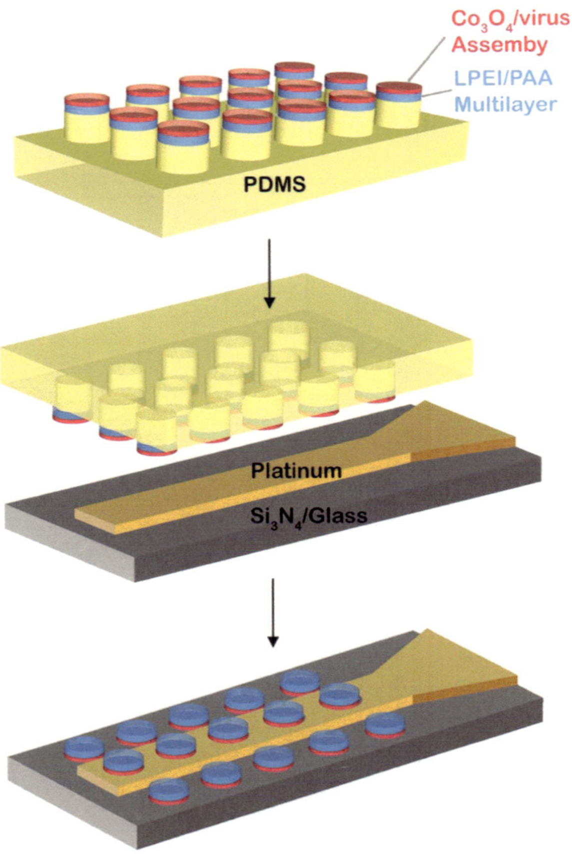

Fig. 11. Schematic diagram of the fabrication procedure for a virus based microbattery electrode (Nam et al., 2008).

When a virus is genetically modified to have high affinity for a certain substrate, the virus can be assembled with the substrate without difficulty. This enables the growth of the virus in any place with the expected shapes by surface modification using a thin layer of the substrate having high affinity to the virus. This means that the virus-based synthesis could be applied widely in micro/nano-patterned devices. In this respect, microbattery fabrication with modified virus has been investigated, as illustrated in Figure 11 (Nam et al., 2008). As the modified M13 virus is known to be easily assembled onto polyelectrolyte multilayers, a linear-polyethylenimine (LPEI)/polyacrylic acid (PAA) multilayer is prepared as a solid electrolyte as well as a separator on the patterned polydimethylsiloxane (PDMS) substrate. The positive LPEI and negative PAA layers electrostatically combine to form a multilayer film through a layer-by-layer deposition technique. The modified virus is grown onto the film by dropcasting, and then Co_3O_4 is nucleated on the virus template. Figure 12 shows AFM images of the fabricated virus-Co_3O_4 anode onto the multilayer film (Nam et al., 2008). The virus tends to form a 2-D liquid crystalline assembly as a result of the interaction between the viruses and LPEI/PAA multilayers. Such a 2-D nanostructure confers an advantage upon the microbattery due to its high packing density. Finally, the virus-Co_3O_4 anode/multilayer film structure is transferred to Pt current collectors by a stamping process, and the electrode is electrochemically characterized, as shown in Figure 13 (Nam et al., 2008). The electrode can store and release Li-ions reversibly, demonstrating that the virus assembly represents an adequate process for fabricating microbattery electrodes.

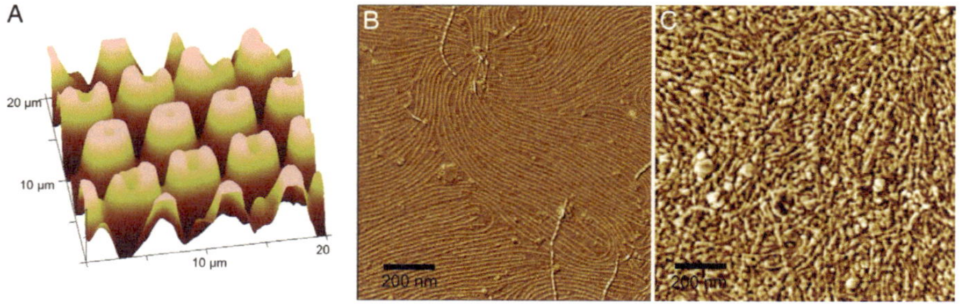

Fig. 12. AFM images of the virus-based microbattery: (a) height image of stacks of the virus-Co_3O_4/(LPEI/PAA) multilayer/PDMS electrode, and phase image of the virus assembly onto the multilayer (b) before and (c) after the nucleation of Co_3O_4 (Nam et al., 2008).

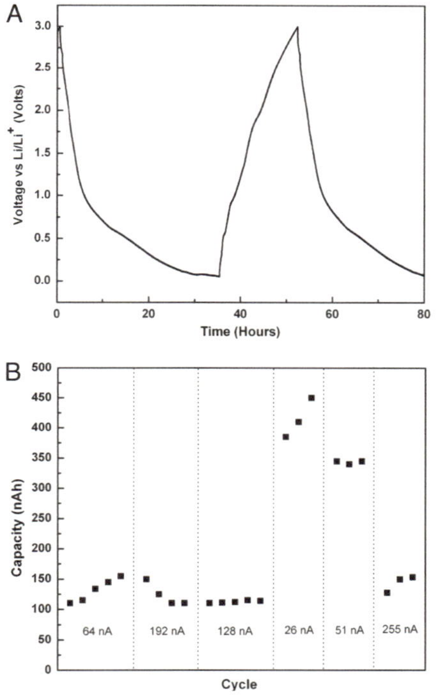

Fig. 13. Electrochemical properties of the virus based microbattery electrode: (a) charge-discharge profiles at a current rate of 26 nA in the voltage range of 0.01-3 V and (b) capacity depending on current rate, from 26 nA to 255 nA (Nam et al., 2008).

The self-assembled biomaterials are flexible and self-standing, even in the macroscopic dimension. LPEI/PAA multilayers (just described above) could be fabricated as flexible, transparent, and self-standing films. When the modified M13 virus is assembled onto the surface of the film (Figure 14(a-b)), Co_3O_4 anode material can be grown through the virus template (Nam et al., 2006). The self-standing film with the anode material can be rolled and bent without any crack propagations as shown in Figure 14(c) (Nam et al., 2006). Figure 14(d) shows that the virus-Co_3O_4 anode assembled on the multilayer film exhibits high specific capacity, even at high current rates (Nam et al., 2006). This result indicates that the genetic modification of the virus to assemble onto the self-standing polymer film is a promising strategy for investigating flexible Li rechargeable battery electrodes.

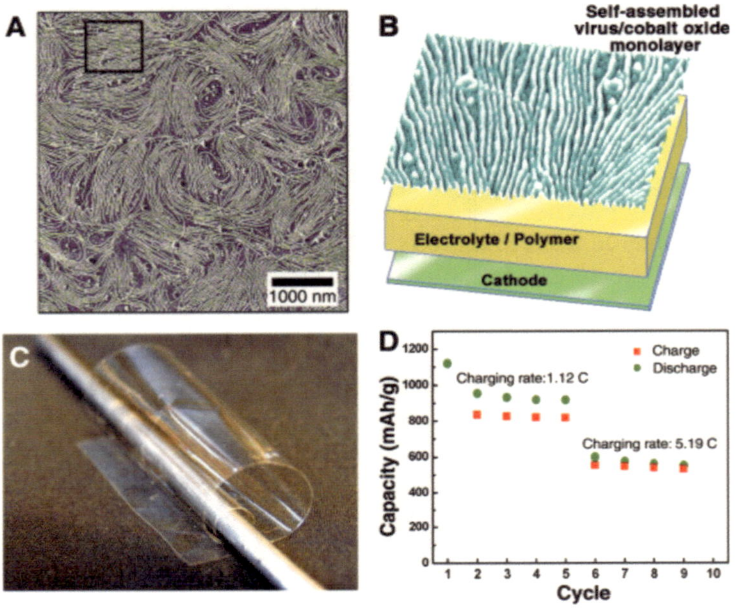

Fig. 14. Self-standing and flexible electrode film based on the virus-Co_3O_4 nanowires: (a-b) phase-mode AFM image of the virus-Co_3O_4 nanowires, (c) photograph of the electrode film composed of virus-Co_3O_4 on a LPEI/PAA multilayer with excellent flexibility, and (d) electrochemical property of the virus-Co_3O_4 anode at current rates of 1.12C and 5.19C. (Nam et al., 2006).

The virus, which can be genetically modified to change the proteins to control its affinity for other materials, provides a promising approach to fabricating the nanostructured electrode materials for Li rechargeable batteries. The improved affinity offers potential nucleation sites for not only various electrode materials (*e.g.*, Co_3O_4 and amorphous $FePO_4$) but also highly conductive materials (*e.g.*, CNT and noble metals such as Au, Ag, and etc). The virus-based hybrid materials, which contain the electrochemically active nanomaterial with the conductive agent, exhibit superior electrochemical performance for high-power Li rechargeable batteries due to the enhanced Li-ionic and electronic conduction. More interestingly, the genetic modification can control the nanostructures of the virus templates, as depicted in Figure 15 (Nam et al., 2005; Huang et al., 2004). Although virus assembly to simple wire-like virus template is well established, further studies should be undertaken to investigate the control of the shapes of the virus assembly, from simple 0-D to complex 3-D architectures, and their use as structural templates because the electrode performance could be strongly influenced by the nanostructure. When genetic technologies to precisely control the virus assembly and the affinity for the electrode materials are achieved, the barriers limiting the conventional Li rechargeable battery could be overcome.

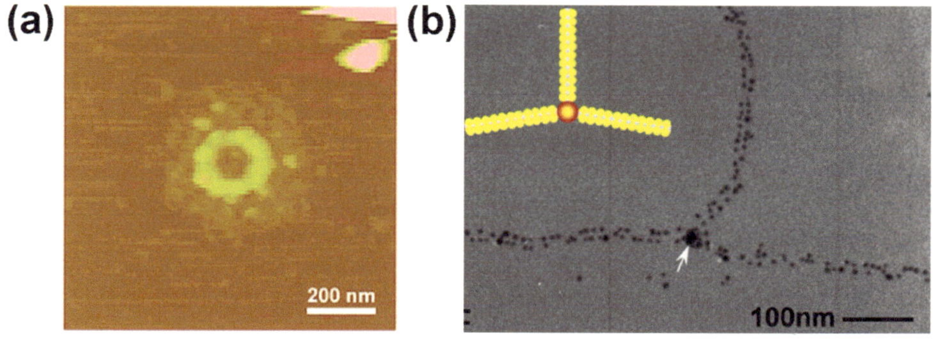

Fig. 15. Various nanostructures of modified viruses: (a) ring-shape (Nam et al., 2005) and (b) Y-shape (Huang et al., 2004).

3.2 Nanotubular electrode materials based on peptide assembly

When a carboxyl group (-COOH) of one amino acid meets an amino group (-NH_2) of another amino acid, the functional groups of each molecule can interact to form a covalent bond (-C(=O)NH-) through a dehydration reaction, as illustrated in Figure 16. This covalent bond is called a 'peptide bond' (Stoker, 2010). The formation of peptide bonds occurs continuously in living organisms to form proteins. Proteins are biological molecules composed of one or more polypeptides, which are long chains linked through number of peptide bonds. Commonly, a peptide is discriminated from a protein in terms of the chain length, *i.e.*, the number of peptide bonds. When the chains are short enough to be synthesized *in vitro*, the molecule is generally called a peptide, but this classification is not always consistent. Peptides are classified by the number of peptide bonds, such as dipeptide, tripeptide, tetrapeptide, and so on. The proteins in the genetically modified viruses possess a high affinity for the electrode materials of Li rechargeable batteries, as described above. Because peptides and proteins are identical in terms of their component species, peptides are also able to provide nucleation and growth sites for various electrode materials. The surface functional groups of the peptides, such as carboxyl groups, are beneficial for coating the electrode materials onto the peptides.

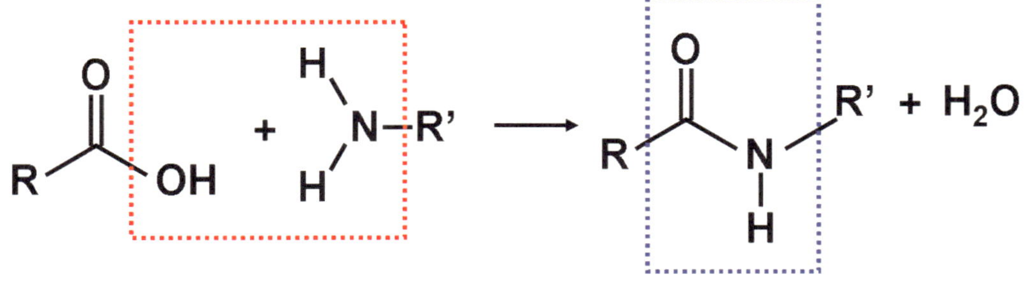

Fig. 16. Formation of a peptide bond between a carboxyl group and an amino group through a dehydration reaction.

The nanostructures of peptides are easily varied by controlling the self-assembly conditions, as shown in Figure 4 (Han et al., 2008). The structural duplication with the electrode materials is achieved by various surface-coating methods, including conventional aqueous solution-based synthesis as well as vacuum vapor deposition . For example, Figure 17 shows hybrid nanowires of diphenylalanine peptide nanowires and Co_3O_4 nanoparticles (Ryu et al., 2010b). Co_3O_4 nanoparticles are formed on the peptide surface by a simple reduction and oxidation process. Although a higher specific capacity is attributed after Co_3O_4 coating (Figure 18), the electrochemical performance of the hybrid nanowires is not sufficiently high, despite the nanoscale nature of Co_3O_4 (Ryu et al., 2010b). The reversible specific capacity of the pure peptide nanowires and the hybrid nanowires is approximately 25 and 80 mAh g^{-1}, respectively. The specific capacity of the hybrid nanowire is very low compared to that of pure Co_3O_4 (890 mAh g^{-1}). This is because the portion of Co_3O_4 nanoparticles in the hybrid is too limited.

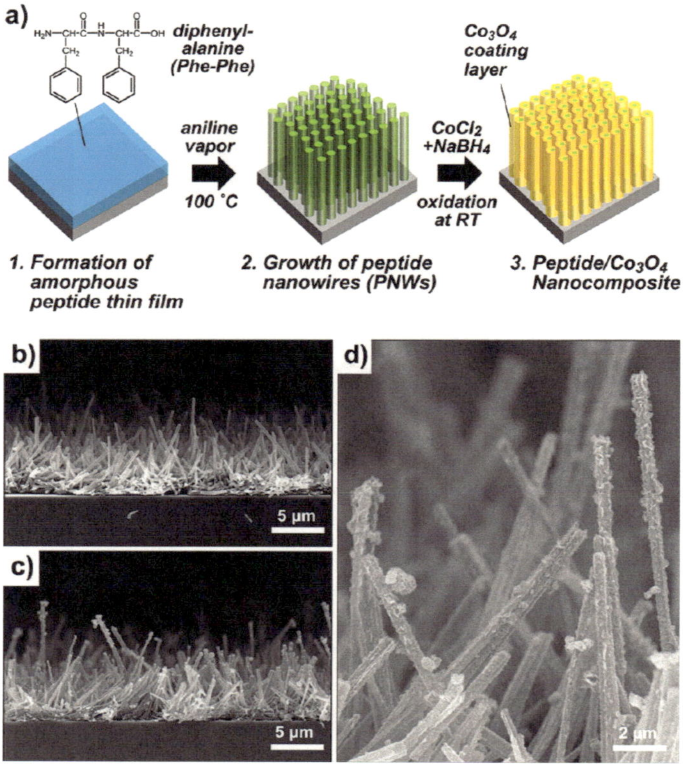

Fig. 17. Diphenlyalanine-Co_3O_4 hybrid nanowires: (a) schematic illustration of the fabrication process of hybrid nanowires and SEM images of (b) the pure peptide nanowires and (c-d) the hybrid nanowires (Ryu et al., 2010b).

When hybrid electrode materials contain electrochemically inactive or less active species to Li ions, their gravimetric and volumetric energy densities decline, as in the case of peptide-Co_3O_4 hybrid nanowires. It is essential to increase the portion of active species, *i.e.*, electrode materials, in the hybrid materials for higher energy-delivery capability. Thus, controlling the portion of the peptide template and the electrode material becomes very important in the fabrication of peptide-based nanostructured electrode materials. In this respect, removing the peptide template has been suggested to increase the energy-delivery capability (Kim et al., 2009; Ryu et al., 2010b).

The minimized portion of the peptide of the hybrid materials can exhibit a capacity approaching the theoretical capacity of the electrode material. Additionally, the template removal leaves a nanotubular structure of the electrode material, which has advantages over simple nanostructures for battery performance.The unique property of the nanotube is that the structure has surfaces both inside and outside of the tube. The dual-surface system is beneficial for the electrode material because it allows two-directional Li-ion diffusion, twice the contact area to electrolyte, and stress relaxation by the hollow during battery operation.

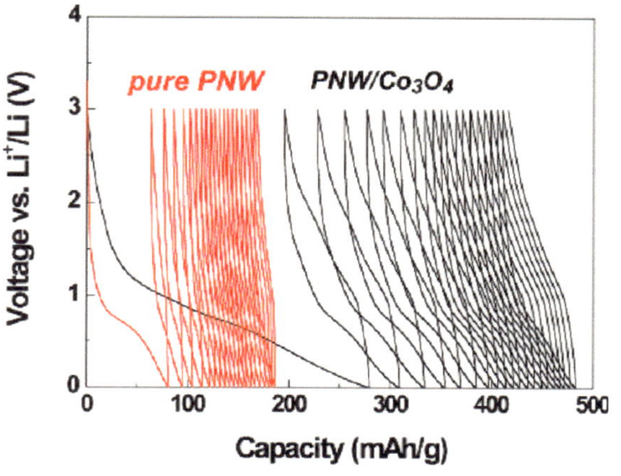

Fig. 18. Discharge-charge profiles of pure peptide nanowires (red) and peptide-Co_3O_4 hybrid nanowires (black) (Ryu et al., 2010b).

The fabrication process for the nanotubular electrode material based on the peptide template is briefly explained in Figure 19 (Kim et al., 2009). After self-assembly of the peptide molecules, the electrode material is coated onto the peptide surface. When the electrode materials fully cover the peptide surface, the peptide template is removed by adequate heat treatment. As a result, only the nanotubular electrode material remains. In this case, a TiO_2 anode is coated onto the peptide template by atomic layer deposition (ALD). ALD is a kind of vapor deposition method that is generally used in fabricating thin films of semiconductor devices (Kim et al., 2008). The self-limited monolayer adsorption, *i.e.*, only chemisorption and not physisorption, of ALD is adoptable for coating the peptide template because the self-limited reaction takes place only at the surface where chemical species for the ALD precursor adsorption site exist (Zhao et al., 2009). The various surface functional groups of the peptide template provide the adsorption sites. This is similar to the adsorption of metal ions in aqueous solution. In ALD, thin-film thickness is controlled digitally by controlling the number of cycles. One cycle is composed of four steps: (i) injection of metal precursor, (ii) Ar purge to remove the residual precursor, (iii) injection of reactant gas, and (iv) another purge (George, 2010). The amount of adsorbed precursors is not varied in ALD because the number of chemisorption sites is identical after the substrate surface is fully covered with the deposited thin film, hence the thin-film thickness is determined by the number of cycles. The digital thickness controllability is beneficial for fabricating a thin film of a specific thickness.

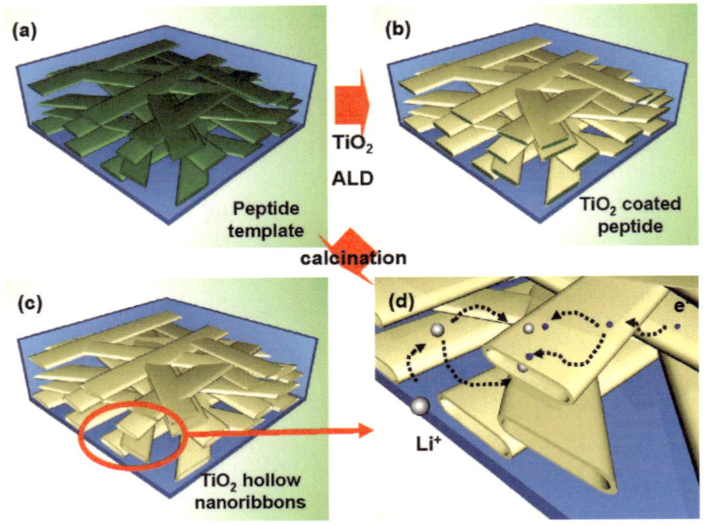

Fig. 19. Schematic illustration of the synthesis process for a network structure of TiO_2 nanotubes based on peptide assembly and ALD: (a) peptide network template, (b) TiO_2 coated peptide network by ALD, (c) TiO_2 nanotube network after removing the peptide template, and (d) Li-ion and electron transports expected in TiO_2 nanotube network structure (Kim et al., 2009).

Fig. 20. SEM images of (a) the peptide template and (b) the TiO_2 nanotube network (inset: magnified cross-sectional image of TiO_2 nanotube) (Kim et al., 2009).

Figure 20 represents SEM images of the peptide template and the TiO_2 nanotubes after coating TiO_2 and removing the peptide (Kim et al., 2009). The dimension of the mother phase, the peptide template, is successfully duplicated with slightly larger dimensions forming a continuous thin film on the peptide through ALD. The inset in Figure 20(b) clearly indicates that fabricated TiO_2 is the nanotubular structure, as expected. Such a network structure of TiO_2 nanotubes is expected to exhibit improved kinetics. Facile Li-ionic conduction through the inside and outside of the nanotube and electronic conduction through the interconnected TiO_2 nanotubes seem to be promising for high-power operation.
TiO_2 is known to be an anode material with high safety because the formation of a solid-electrolyte-interphase (SEI) layer, which could trigger heat generation during malfunctioning of the battery, is suppressed due to the relatively high operation voltage window (>1.0 V) (Xu et al., 2008). Nevertheless, its relatively low practical capacity (167.7 mAh g^{-1}) is a significant drawback. The strong Li-Li repulsion restricts Li-ion intercalation more than $Li_{0.5}TiO_2$. It is reported that this is possible in nanoscale due to the surface layer of TiO_2 (<7 nm) (Wagemaker et al., 2007). The discharge-charge profiles of the fabricated nanostructured TiO_2 provided in Figure 21(a) show that Li-ion intercalation of more than $Li_{0.5}TiO_2$ is achieved (Kim et al., 2009). The reversible capacity is approximately 198.3 mAh g^{-1} ($Li_{0.59}TiO_2$) after the initial cycle. The slight initial capacity loss might be attributed to surface defects, which are observed frequently in nanostructured materials. The relatively high first-cycle irreversible capacity can be a general drawback associated with nanostructure electrode materials.
The high specific capacity of the fabricated TiO_2 nanotube network and that of the conventional spherical TiO_2 nanoparticle are compared in Figure 21(b) (Kim et al., 2009). Although the specific capacity of 20-nm TiO_2 nanoparticles is comparable to that of the TiO_2 nanotube network at a low current rate, the TiO_2 nanotube network shows much better high-rate capability due to the improved Li-ionic and electronic conduction, as illustrated in Figure 19(d) (Kim et al., 2009). Furthermore, the TiO_2 nanotube network shows excellent cyclability, as shown in Figure 21(c) (Kim et al., 2009). When Li ions are inserted (or extracted) into (or out of) the host material, a volume change generally occurs, inducing strain on the host material. Repeated Li-ion insertions and extractions can break out a fracture of the host material, resulting in poor cycle stability. Nanotubular structures, which contain hollow space, are believed to accommodate this strain because the volume change is bidirectional, *i.e.*, inward and outward relative to nanotubes. Additionally, the highly ordered network structure enhances the structural stability upon electrochemical cycling. As a result, such a network structure of nanotubes based on peptide assembly is considered to represent a promising strategy for investigating electrode materials with superior electrochemical performance.
Although the nanotube network structures based on ALD onto the peptide assembly are expected to possess excellent electrochemical properties, the relatively high production cost and low precursor efficiency of ALD are demerits for commercialization. In this respect, an aqueous solution-based mineralization technique can be advantageous in fabricating the nanotube network structure. Figure 22(a) simply describes the process for the synthesis of the network structure based on the sequential adsorption method using aqueous solution (Ryu et al. 2010a). An amorphous $FePO_4$ cathode is chosen as a coating material because Fe^{3+} and PO_4^{3-} ions are dissolved readily in H_2O. The conformal thin-film coating in ALD is achieved by separate injections of the metal precursor and the reactant gas. Similarly, separate injections of an aqueous solution of Fe^{3+} and another solution of PO_4^{3-} are believed

to result in conformal coating of amorphous $FePO_4$. Peptide molecules (Fmoc-diphenylalanine; Figure 22(b)), are assembled into the network in H_2O, as depicted in Figure 22(c-d) (Ryu et al. 2010a). The peptide hydrogel is formed in a filtration apparatus, and then the solution of Fe^{3+} is injected and stored inside the apparatus for adsorption of Fe^{3+} ions onto the peptide surface. The solution containing Fe^{3+} is removed by filtration, and then the solution containing PO_4^{3-} is injected, stored, and filtered. The sequential injection of two solutions is repeated to obtain a suitable amount of coated $FePO_4$. This process is very similar to that of the ALD discussed with regard to the TiO_2 nanotube network.

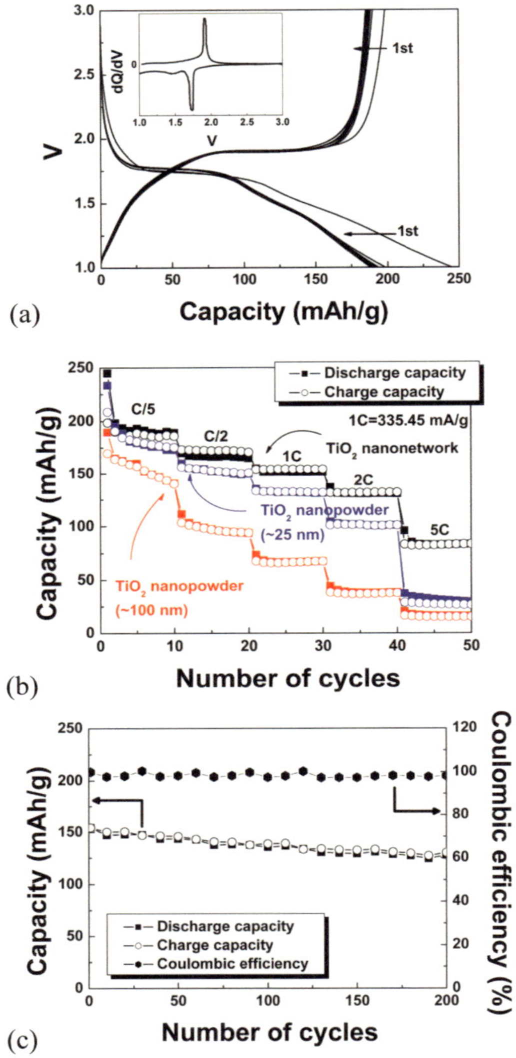

Fig. 21. Electrochemical properties of a TiO_2 nanotube network: (a) discharge-charge profiles at a current rate of C/5 (inset: differential capacity curve at the first cycle), (b) high rate capability compared with conventional TiO_2 nanoparticles, and (c) capacity retention and Coulombic efficiency at a current rate of 1C (1C = 335.45 mAh g-1) (Kim et al., 2009).

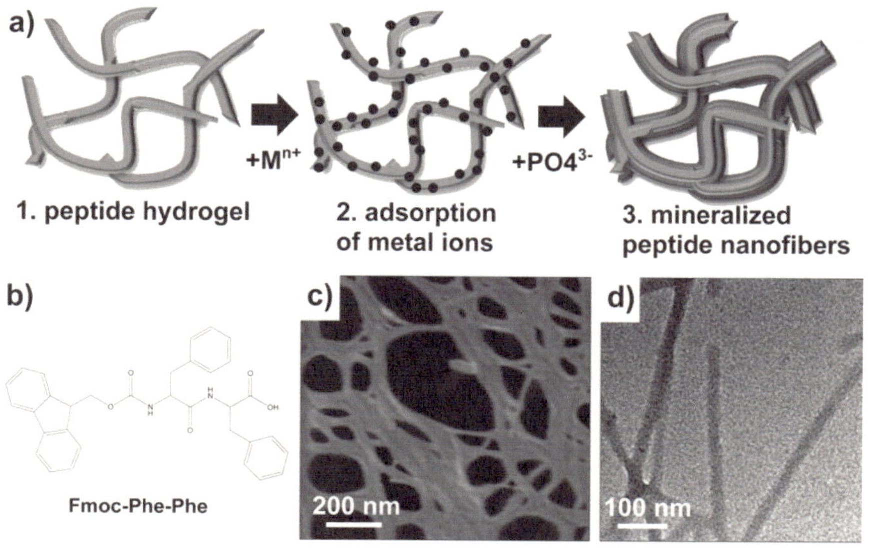

Fig. 22. Fabrication of peptide-amorphous $FePO_4$ hybrid material using the sequential injection method: (a) schematic diagram describing the fabrication process, (b) molecular structure of Fmoc-diphenylalaine, (c) SEM image and (d) TEM image of the peptide network (Ryu et al., 2010a).

Fe^{3+} ions, adsorbed during the Fe^{3+}-containing solution treatment, react with PO_4^{3-} during the injection of solution containing PO_4^{3-} to form a hybrid material of peptide template and electrochemically active amorphous $FePO_4$, as shown in Figure 23(a-b) (Ryu et al. 2010a). Interestingly, other transition metal phosphates also could be coated onto the peptide assembly by the same method as shown in Figure 23 (Ryu et al. 2010a).

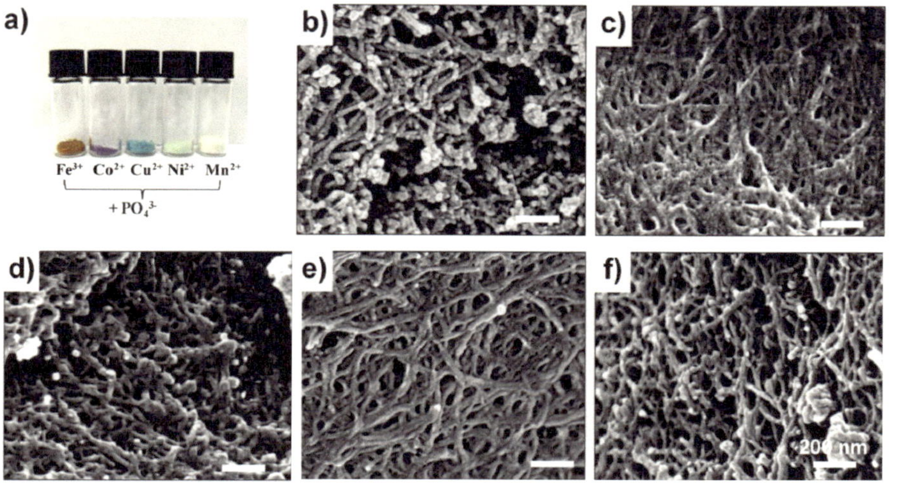

Fig. 23. (a) Photographs and (b-f) SEM images of various peptide-metal phosphate hybrid materials: injected metal ions are (b) Fe^{3+}, (c) Co^{2+}, (d) Cu^{2+}, (e) Ni^{2+} and (f) Mn^{2+} ions, respectively. (Ryu et al., 2010a).

The fabricated peptide-amorphous $FePO_4$ is annealed in adequate condition to remove the inactive peptide template, forming the amorphous $FePO_4$ nanotube, as illustrated in Figure 24(a) (Ryu et al. 2010a). Hollow space is formed inside the amorphous $FePO_4$, resulting in a nanotubular structure, as shown in Figure 24(c-d) (Ryu et al. 2010a). Another purpose for the heat treatment is to remove crystal water because it tends to form a hydrated phase of $FePO_4 \cdot 2H_2O$ when the $FePO_4$ phase is synthesized from aqueous solution. Heat treatment at too high temperature induces the crystallization of the amorphous phase, lowering the electrochemical activity (Okada et al., 2005). Thus, controlling the temperature is an important factor for the heat treatment. The individual nanotubes are interwoven with one another to form a complex network structure, as expected. The peptide leaves a thin amorphous carbon layer inside the tube after heat treatment, as which is identified by the FT-IR spectra shown in Figure 24(e) (Ryu et al. 2010a). Additionally, the FT-IR spectra confirm that the $FePO_4$ phase is successfully formed. The residual carbon layer is expected to provide a facile pathway for electron transport.

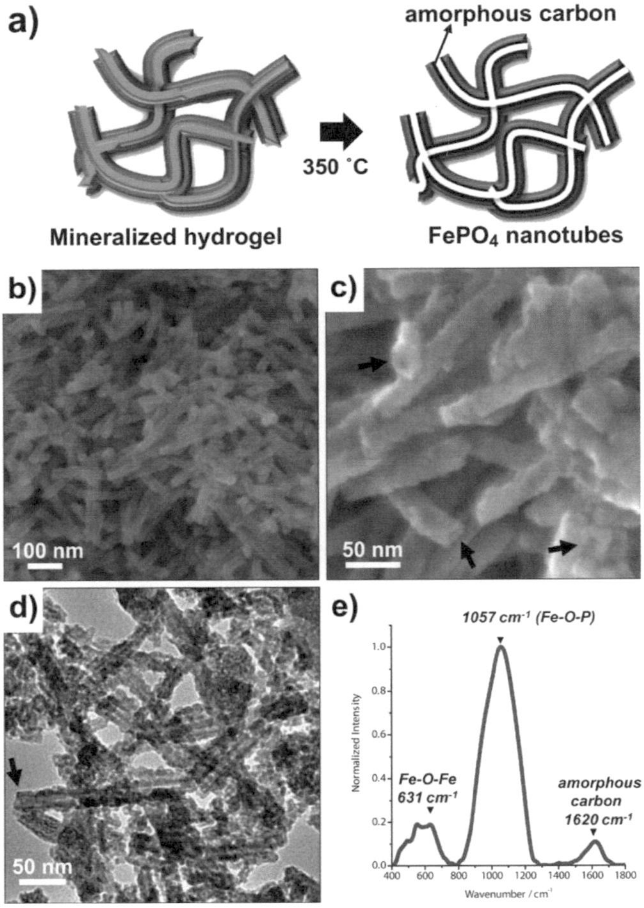

Fig. 24. Amorphous $FePO_4$ nanotube synthesized by a peptide template: (a) schematic diagram for synthesis procedure, (b-c) SEM images, (d) TEM image, and (e) FI-IR spectra (Ryu et al., 2010a).

As a cathode material, the amorphous $FePO_4$ nanotube coated by a carbon layer exhibits a high specific capacity, as depicted in Figure 25 (Ryu et al. 2010a). Its first discharge capacity is approximately 170 mAh g^{-1}, which is comparable to the theoretical capacity (178 mAh g^{-1}). Then, the specific capacity is saturated to 150 mAh g^{-1}, with excellent reversibility. Similar to the TiO_2 nanotube network, it is believed that the nanoscale dimension and unique nanostructure improve the electrochemical activity of amorphous $FePO_4$. The carbon layer inside the nanotube, which sufficiently supplies electrons for the electrochemical reaction, is also impressive.

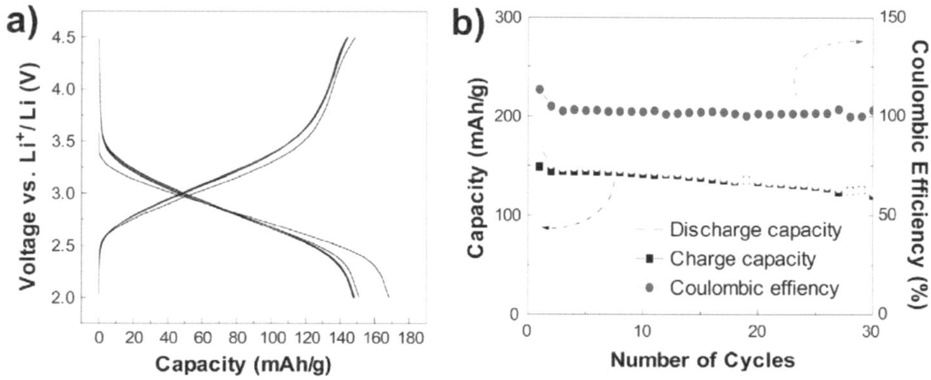

Fig. 25. (a) Discharge-charge profiles and (b) capacity retention of the amorphous $FePO_4$ nanotube at a current rate of 10 mA g^{-1} (Ryu et al., 2010a).

The morphology of the peptides is easily controllable because the self-assembly of the peptides is influenced simply by the assembly condition. Generally, peptides display numerous acidic and polar moieties on their surface. These moieties provide adsorption and nucleation sites for the electrode materials because they have high affinities for the ions or molecules forming the electrode materials. Various nanostructured electrode materials can be fabricated based on peptide assemblies. Among these, the network structure of hollow nanotubes is considered a high-performance structure. The unique nanostructure has advantages for battery performance, especially high-rate performance and capacity retention upon cycling. The two-directional Li-ion transport leads to improved Li-ion supply for the electrochemical reaction and the highly ordered nanotube network enhances cycle stability due to stress relaxation through the hollow and improved structural stability by interweaving. Furthermore, the anisotropic nanostructure is speculated to show faster electronic transport than the conventional isotropic nanostructure through one long dimension (Bruce et al., 2008). Although further investigation of the precise control of the peptide morphology and simple coating techniques is warranted, the strategy using the peptide assembly as the structural template is expected to be applied universally to fabricate various electrode materials with various nanostructures, and thus, to fabricate high performing Li rechargeable batteries.

4. Bio-inspired materials for other energy conversion devices

The synthesis and electrochemical characterization of the electrode materials based on virus and peptide templates for Li rechargeable batteries have been described in the above sections. In this section, the nanostructured materials based on the biomaterial templates for other energy-conversion devices, such as solar cells and fuel cells, are discussed briefly. Solar cells and fuel cells generate electricity from sunlight and fuels such as H_2, respectively. As they generate the electricity whereas Li rechargeable batteries store the electricity, their combination can give rise to a new concept of energy device.

When a solar cell is exposed to sunlight, an electron-hole pair is generated by photons and simultaneously voltage and current are created. The solar cell is regarded as a most promising renewable green energy device due to the permanence of the sun. Among many types of solar cells, the Si-based quantum-dot solar cell is usually fabricated in the form of thin film, which typically requires a lithography process for structure determination. Here, biomaterials can be applied to the lithography process as a mask layer for etching (Huang et al., 2010). Ferritin, an Fe-containing protein, is assembled into a 2-D array on the SiO_2 surface. Shells of the protein are removed by heat treatment, leaving iron oxide cores for the mask. Following an adequate etching process, a 2-D array of the quantum-dot solar cells is successfully fabricated with 10-nm dimensions as shown in Figure 26 (Huang et al., 2010).

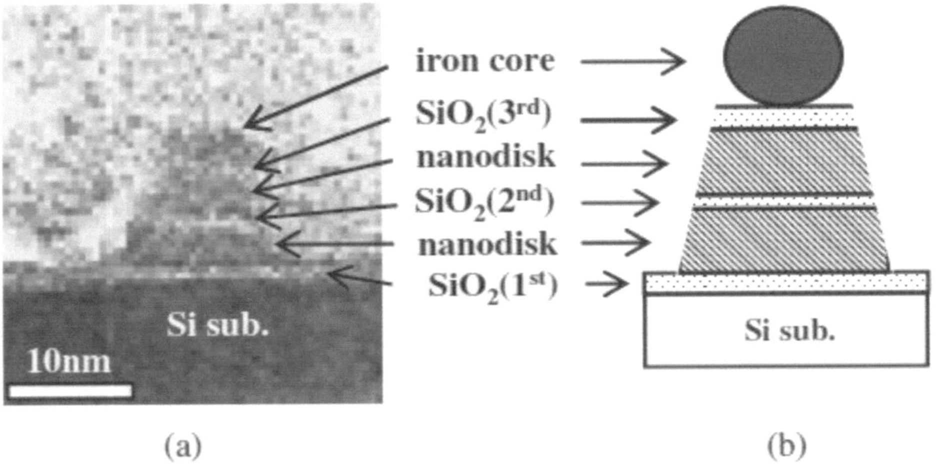

Fig. 26. (a) STEM image and (b) schematic diagram of a Si-based solar cell using ferritin protein as a mask layer (Huang et al., 2010).

Recently, dye-sensitized solar cells have attracted much interest due to their potentially lower production cost (Grätzel, 2005). Dye adsorbed onto a photo-anode generates electrons. The electrons are transferred to the photo-anode, generally TiO_2. Because the amount of dye adsorbed determines the solar-cell efficiency, a nanostructured photo-anode is required for a sufficiently high surface area to adsorb the dye. Figure 27 depicts the process for a nanostructured TiO_2 photo-anode based on a butterfly wing (Zhang et al., 2009). The butterfly-wing template is immersed in $Ti(SO_4)_2$ solution to sock the precursor solution. The socked template is placed on the TiO_2/FTO/glass substrate and then clamped tightly. The structurally duplicated TiO_2 photo-anode is prepared after a final heat treatment, as shown in Figure 28 (Zhang et al., 2009). The original honeycomb structure of the wing is well duplicated in the TiO_2 photo-anode. Its high surface area increases the amount of dye adsorbed. Furthermore, its unique nanostructure is advantageous for light absorption. As a result, the honeycomb structured TiO_2 photo-anode enhances the solar cell efficiency compared with that of conventional TiO_2 photo-anode without the butterfly-wing template.

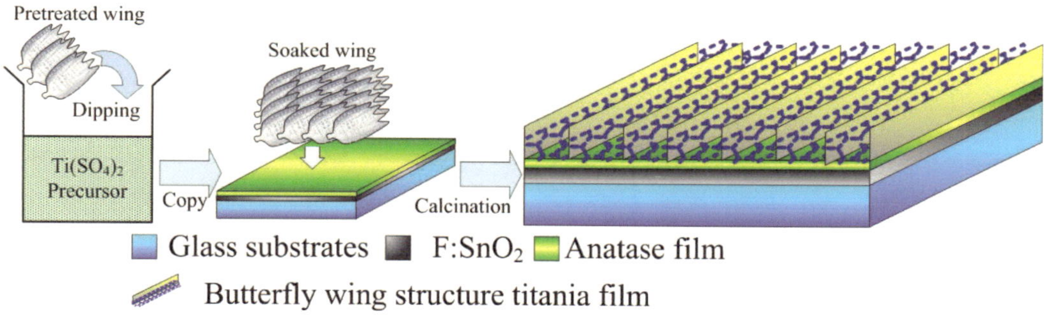

Fig. 27. Schematic diagram of the fabrication of a TiO_2 photo-anode based on butterfly-wing assembly (Zhang et al., 2009).

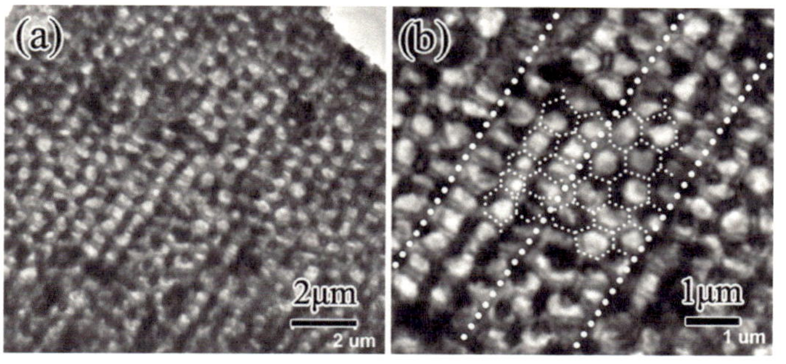

Fig. 28. TEM images of the butterfly wing-shaped TiO_2 photo-anode: (a) Medium resolution image and (b) high resolution image (Zhang et al., 2009).

Fuel cells generate electric energy from the oxidation of the fuel. For example, oxidation of H_2 generates the electric energy with the simultaneous reduction of O_2, leaving H_2O in a H_2-based fuel cell. When the fuel cell operates, a noble metal catalyst, such as Pt and Pd, is used to improve the oxidation efficiency, thereby increasing the output-energy density (Liu et al., 2006). Because the size and shape of the catalyst affect the oxidation efficiency significantly, nanostructure control of the catalyst is an important issue for a high-performance fuel cell. Bacteria are also adoptable as the template for the noble metal nucleation, as depicted Figure 29 (Yong et al., 2010). Pd nanoparticles are formed on the bacteria surface by sequential adsorption and reduction of Pd^{2+} ions. The particle size and homogeneity are influenced by the type of bacteria. The *D. desulfuricans*-Pd hybrid material shown in Figure 29(b) contains homogeneous small nanoparticles. The bacteria-Pd hybrid materials are sintered to serve as proton exchange membrane fuel cell catalysts. Furthermore, the fabrication of nanocatalysts with intermetallic compounds of noble metals is also possible. Compared to the commercial Pd catalyst, the catalyst fabricated based on the bacteria template shows enhanced power output. A similar outcome is achieved using tobacco mosaic virus (TMV) as the template, as shown in Figure 30 (Yong et al., 2010). TMV-Pt hybrid nanowires have been synthesized successfully with almost uniform individual size. The precursor concentration has a significant relationship with the size of the synthesized nanoparticles. As the catalyst for a fuel cell, the TMV-Pt hybrid nanowires exhibit better efficiency compared with simple Pt nanoparticles due to the increased electrochemically active surface area.

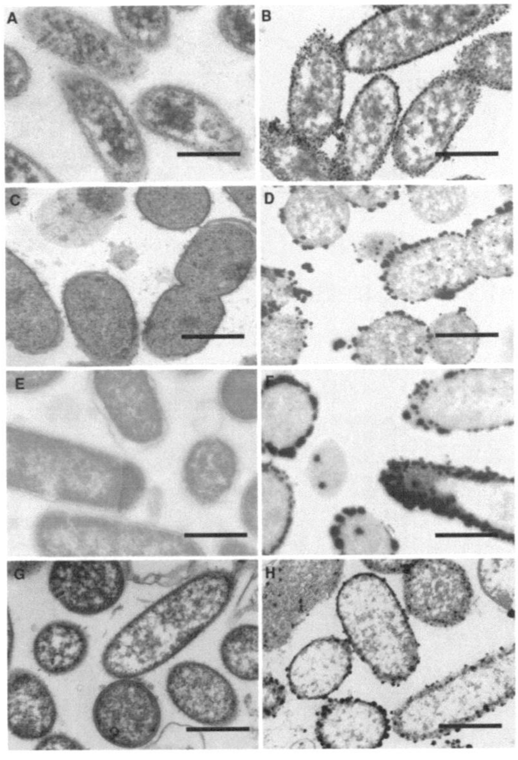

Fig. 29. TEM images of bacteria templates before (a, c, e, and g) and after (b, d, f, and h) the formation of Pd nanocatalyst: (a-b) *D. desulfuricans*, (c-d) *E. coli* MC4100, (e-f) *E. coli* IC007, and (g-h) *C. metallidurans* (Scale bar: 500 nm) (Yong et al., 2010).

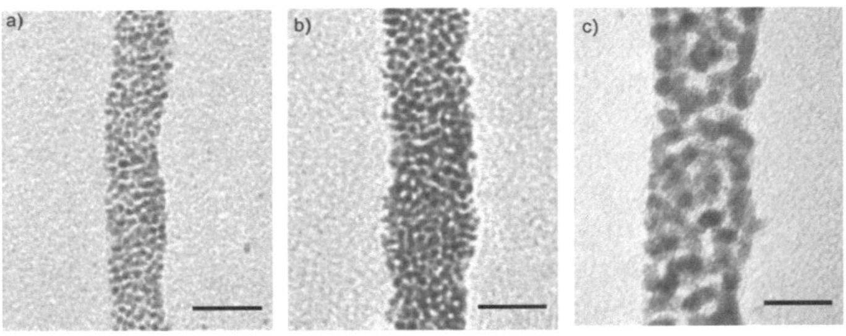

Fig. 30. TEM images of Pt-TMV hybrid nanowires prepared at various Pt precursor concentration. The concentrations increases from (a) through (c). (Scale bar: 20 nm) (Górzny et al., 2010).

5. Summary and perspective

In this chapter, the bio-inspired synthesis of electrode materials was briefly introduced for the application of next-generation Li rechargeable batteries. The use of self-assembled viruses and peptides as the templates for nanostructured electrode materials has been studied extensively. The electrode materials are coated onto the surface of the biomaterial templates, making easily obtained isostructured electrode materials. The numbers of acidic and polar moieties at the surface of the biomaterials lead to high affinity of the coating electrode materials. The nanoarchitectures of biomaterials are easily controllable, and thus, the electrode materials can be varied by applying different structured biomaterial templates. The nanostructured electrode materials based on the biomaterial templates exhibit superior electrochemical performance, such as specific capacity, rate capability, and cyclability, due to the improved Li-ion and electron supply and the strain accommodation upon cycling.

Despite the potential of such bio-inspired syntheses, further research is required to realize their application in the Li rechargeable batteries. The electrode materials introduced in this chapter include the Co_3O_4 anode, TiO_2 anode, and amorphous $FePO_4$ cathode. More promising electrode materials should be tested. For example, amorphous $FePO_4$ has a severe drawback in that a Li-deficient phase is not preferred in the cathode. Conventional Li rechargeable batteries are composed of a Li-deficient anode and a Li-containing cathode. Therefore, investigations of the coating techniques for Li-containing cathode materials are needed. Additionally, the production cost of biomaterials such as peptides is relatively high for commercialization, and nanostructure formation with large-area uniformity is questionable from the perspective of mass production. Thus, it is necessary to investigate the technologies for controlling the morphology of biomaterial over a large area with reduced production costs.

Bio-inspired synthesis can generate various classes of materials, from metals to complex chemical compounds such as metal phosphates; thus, it represents a promising way to fabricate various kinds of devices, that require nanostructure materials for improved performance, as well as Li rechargeable batteries. Nature is the oracle that always inspires new developments by human beings for the a better life. Nature generates innumerable materials on the earth, and we only use some of them. Deep insight for material generation, not just from living organisms but from all products in nature, will present novel strategies with the opportunity to investigate materials for a better life.

6. Acknowledgment

This work was supported by a grant from the Fundamental R&D Program for Technology of World Premier Materials and Energy Resource Technology R&D Program (20092020100040) funded by the Ministry of Knowledge Economy, Republic of Korea. This work was also supported by the National Research Foundation of Korea Grant (R11-2008-058-01003-0, NRF-2009-0082069, NRF-2009-0094219) and Converging Research Center Program (2010K001088) funded by the Ministry of Education, Science, and Technology, Republic of Korea.

7. References

Bensaude-Vincent, B. ; Arribart, H. ; Bouligand, Y. & Sanchez, C. (2002). Chemists and The School of Nature, *New Journal of Chemistry*, Vol.26, No.1, (January 2002), pp. 1-5, ISSN 1144-0546

Sanchez, C.; Arribart, H. & Guille, M. M. G. (2005). Biomimetism and Bioinspiration as Tools for The Design of Innovative Materials and Systems, *Nature Materials*, Vol.4, No.4, (April 2005), pp.277-288, ISSN 1476-1122

Dickerson, M. B.; Sandhage, K. H. & Naik R. R. (2008). Protein- and Peptide-Directed Snytheses of Inorganic Materials, *Chemical Review*, Vol.108, No.11, (November 2008), pp. 4935-4978, ISSN 0009-2665

Lv, J.; Liu, H. & Li, Y. (2008). Self-Assembly and Properties of Low-Dimensional Nanomaterials Based on π-Conjugated Organic Molecules, *Pure and Applied Chemistry*, Vol.80, No.3, (March 2008), pp. 639-658, ISSN 0033-4545

Katz, E.; Willner, I. & Wang, J. (2004). Electroanalytical and Bioelectroanalytical Systems Based on Metal and Semiconductor Nanoparticles, *Electroanalysis*, Vol.16, No.1-2, (January 2004), pp. 19-44, ISSN 1040-0397

Su, H.; Han, J.; Dong, Q.; Zhang, D. & Guo, Q. (2008) *In Situ* Synthesis and Photoluminescence of QD-CdS on Silk Fibroin Fibers at Room Temperature, *Nanotechnology*, Vol.19, No.2, (January 2008), pp. 025601 , ISSN 0957-4484

Li, X.; Fan, T.; Zhou, H.; Chow, S.-K.; Zhang, W.; Zhang, D.; Guo, Q. & Ogawa, H. (2009) Enhanced Light-Harvesting and Photocatalytic Properties of *Morph*-TiO_2 from Green-Leaf Biotemplates, *Advanced Functional Materials*, Vol.19, No.1, (January 2009), pp. 45-56, ISSN 1616-301X

Moriarty, P. (2001), Nanostructured Materials, *Reports on Progress in Physics*, Vol.64, No.3, (March 2001), pp. 297-381, ISSN 0034-4885

Tarascon, J.-M. & Armand, M. (2001). Issues and Challenges Facing Rechargeable Lithium Batteries, *Nature*, Vol.414, No.6861, (November 2001), pp. 368-377, ISSN 0028-0836

Bruce, P. G.; Scrosati, B. & Tarascon, J.-M. (2008), Nanomaterials for Rechargeable Lithium Batteries, *Angewante Chemi International Edition*,Vol.47, No.16, (April 2008), pp. 2930-2946, ISSN 1433-7851

Kang, K.; Meng, Y. S.; Brе́ger, J.; Grey, C. P. & Ceder, G. (2006). Electrodes with High Power and High Capacity of Rechargeable Lithium Batteries, *Science*, Vol.311, No.5763, (February 2006), pp. 977-980, ISSN 0036-8075

Cheng, F.; Tao, Z.; Liang, J. & Chen. J. (2008). Template-Directed Materials for Rechargeable Lithium-Ion Batteries, *Chemistry of Materials*, Vol.20, No.3, (Febrary 2008), pp. 667-681, ISSN 0897-4756

Cui, H.; Webber, M. J. & Stupp, S. I. (2010). Self-Assembly of Peptide Amphiphiles: From Molecules to Nanostructures to Biomaterial, *Peptide Science*, Vol.94, No.1, (January 2010), pp. 1-18, ISSN 0006-3525

Ryu, J.; Kim, S.-W.; Kang, K. & Park, C. B. (2010a). Mineralization of Self-Assembled Peptide Nanofibers for Rechargeable Lithium Ion Batteries, *Advanced Materials*, Vol.22, No.48, (December 2010), pp. 5537-5541, ISSN 0935-9648

Reches, M. & Gazit, E. (2006). Molecular Self-Assembly of Peptide Nanostructures: Mechanism of Association and Potential Uses, *Current Nanoscience*, Vol.2, No.2, (May 2006), pp. 1573-4137, ISSN 1573-4137

Fraden, S. & Kamien, R. D. (2000). Self-Assembly in Vivo, *Biophysical Journal*, Vol.78, No.5, (May 2000), pp. 2189-2190, ISSN 0006-3495

Giese, C.; Demmler, C. D.; Ammer, R.; Hartmann, S.; Lubitz, A.; Muiiler, L.; Müller, R. & Marx, U. (2006). A Human Lymph Node in Vitro – Challenges and Progress, *Artificial Organs*, Vol.30, No.10, (October 2006), pp. 803-808, ISSN

Brachmann, C. B. & Cagan, R. L. (2003). Patterning the Fly Eye: The Role of Apotosis, *Trends in Genetics*, Vol.19, No.2, (February 2003), pp. 91-96, ISSN 0168-9525

Ryu, J. & Park, C. B. (2008). High-Temperature Self-Assembly of Peptides into Vertically Well-Aligned Nanowires by Aniline Vapor, *Advanced Materials*, Vol.20, No.19, (October 2008), pp. 3754-3758, ISSN 0935-9648

Xia, H.; Chan, H. S. O.; Xiao, C. & Cheng, D. (2004). Self-Assembled Oriented Conducting Polyaniline Nanotubes, *Nanotechnolgy*, Vol.15, No.12, (December 2004), pp. 1807-1811, ISSN 0957-4484

Zhou, H.; Fan, T. & Zhang, D. (2007). Hydrothermal Synthesis of ZnO Hollow Spheres Using Spherobacterium as Biotemplates, *Microporous and Mesoporous Materials*, Vol.100, No.1-3, (March 2007), pp. 322-327, ISSN 1387-1811

Zhang, T.; Wang, W.; Zhang, D.; Ma, Y.; Zhou, Y. & Qi, L. (2010). Biotemplated Synthesis of Gold Nanoparticle-Bacetria Cellulose Nanocomposites and Their Application in Biosensing, *Advanced Materials*, Vol.20, No.7, (April 2010), pp. 1152-1160, ISSN 0935-9648

Han, T. H.; Park, J. S.; Oh, J. K. & Kim, S. O. (2008). Morphology Control of One-Dimensional Peptide Nanostructures, *Journal of Nanoscience and Nanotechnology*, Vol.8, No.10, (October 2010), pp. 5547-5550, ISSN 1550-7033

Lee, S.-W.; Lee, S. K. & Belcher, A. M. (2003). Virus-Based Alignment of Inorganic, Organic, and Biological Nanosized Materials, *Advanced Materials*, Vol.15, No.9, (May 2003), pp. 689-692, ISSN 0935-9648

Nam, K. T.; Peele, B. R.; Lee, S.-W. & Belcher, A. M. (2004). Genetically Driven Assembly of Nanorings Based on the M13 Virus, *Nano Letters*, Vol.4, No.1, (January 2004), pp. 23-27, ISSN 1530-6984

Mao, C.; Flynn, C. E.; Hayhurst, A.; Sweeney, R.; Georgiou, G.; Iverson, B. & Belcher, A. M. (2003). Viral Assembly of Oriented Quantum Dot Nanowires, *Proceedings of the National Academy of Sciences*, Vol.100, No.12, (June 2003), pp. 6946-6951, ISSN 0027-8424

Lee, Y. J.; Yi, H.; Kim, W.-J.; Kang, K.; Yun, D. S.; Strano, M. S.; Ceder, G. & Belcher, A. M. (2009). Fabricating Genetically Engineered High-Power Lithium-Ion Batteries Using Multiple Virus Genes, *Science*, Vol.324, No.5390, (May 2009), pp. 1051-1055, ISSN 0036-8075

Okada, S.; Tamamoto, T.; Okazaki, Y.; Yamiaki, J.; Tokunaga, M. & Nishida, T. (2005) Cathode Properties of Amorphous and Crystalline $FePO_4$, *Journal of Power Sources*, Vol.146, No.1-2, (August 2005), pp. 570-574, ISSN 0378-7753

Kang, Y.-M.; Song, M.-S.; Kim, J.-H.; Kim, H.-S.; Park, M.-S.; Lee, J.-Y.; Liu, H. K. & Dou, S. X. (2005). A Study on The Charge-Discharge Mechanism of Co_3O_4 as an Anode for The Li Ion Secondary Battery, *Electrochimica Acta*, Vol.50, No.18, (June 2005), pp. 3667-3673

Li, Y.; Tan, B. & Wu, Y. (2008), Mesoporous Co_3O_4 Nanowire Array for Lithium Ion Battery with High Capacity and Rate Capability, *Nano Letters*, Vol.8, No.1, (January 2008), pp. 265-270, ISSN 1530-6984

Nam, K. T.; Kim, D.-W.; Yoo, P. J.; Chiang, C.-Y.; Meethong, N.; Hammond, P. T.; Chiang, Y.-M. & Belcher, A. M. (2008). Virus-Enabled Synthesis and Assembly of Nanowires for Lithium Ion Battery Electrodes, *Science*, Vol.312, No.5775, (May 2006), pp. 885-888, ISSN 0036-8075

Nam, K. T.; Wartena, R.; Yoo, P. J.; Liau, F. W.; Chiang, Y.-M.; Hammond, P. T. & Belcher, A. M. (2008). Stamped Microbattery Electrode Based on Self-Assembled M13 Viruses, *Proceedings of the National Academy of Sciences*, Vol.105, No.45, (November 2008), pp. 17227-17231, ISSN 0027-8424

Huang, Y.; Chiang, C.-Y.; Lee, S. K.; Gao, Y.; Hu, E. L.; Yoreo, J. D. & Belcher, A. M. (2005). Programmable Assembly of Nanoarchitectures Using Genetically Engineered Viruses, *Nano Letters*, Vol.5, No.7, (July 2005), pp. 1429-1434, ISSN 1530-6984

Stoker, H. S. (2010). Proteins, In: *General, Organic, and Biological Chemistry*, Kilean Kennedy, (Ed.), pp. 655-697, Brooks Cole, ISBN 978-054-7152-81-3, California, USA

Ryu, J.; Kim, S.-W.; Kang, K. & Park, C. B. (2010b). Synthesis of Diphenylalanine/Cobalt Oxide Hybrid Nanowires and Their Application to Energy Storage, *ACS Nano*, Vol.4, No.1, (January 2010), pp. 159-164, ISSN 1936-0851

Kim, S.-W.; Han, T. H.; Kim, J.; Gwon, H.; Moon, H.-S.; Kang, S.-W.; Kim, S. O. & Kang, K. (2009). Fabrication and Electrochemical Characterization of TiO_2 Three-Dimensional Nanonetwork Based on Peptide Assembly, *ACS Nano*, Vol.3, No.5, (May 2009), pp. 159-164, ISSN 1936-0851

Kim, S.-W.; Kwon, S.-H.; Jeong, S.-J. & Kang, S.-W. (2008). Improvement of Copper Diffusion Barrier Properties of Tantalum Nitride Films by Incorporating Ruthenium using PEALD, *Journal of The Electrochemical Society*, Vol.155, No.11, (September 2008), pp. H885-H888, ISSN 0013-4651

Zhao, Y.; Wei, M.; Lu, Jun.; Wang, Z. L. & Duan, X. (2009). Biotemplated Hierarchical Nanostructure of Layered Double Hydroxides with Improved Photocatalysis Performance, *Nano Letters*, Vol.3, No.12, (December 2009), pp. 4009-4016, ISSN 1530-6984

Geroge, S. M. (2010). Atomic Layer Deposition: An Overview, *Chemical Reviews*, Vol.110, No.1, (January 2010), pp. 111-131, ISSN 0009-2665

Xu, J.; Wang, Y.; Li, Z. & Zhang, W. F. (2008). Preparation and Electrochemical Properties of Carbon-Doped TiO_2 Nanotubes as an Anode Material for Lithium-Ion Batteries, *Journal of Power Sources*, Vol.175, No.2, (January 2008), pp. 903-908, ISSN 0378-7753

Wagemaker, M.; Borghols, W. J. H. & Mulder, F. M. (2007). Large Impact of Particle Size on Insertion Reactions. A Case for Anatase Li_xTiO_2, *Journal of The American Chemical Society*, Vol.129, No.14, (April 2007), pp. 4323-4327, ISSN 0002-7863

Huang, C.-H.; Igarashi, M.; Horita, S.; Takeguchi, M.; Uraoka, Y.; Fuyuki, T.; Yamashita, I. & Samukawa, S. (2010). Novel Si Nanodisk Fabricated by Biotemplate and Defect-Free Neutral Beam Etching for Solar Cell Application, *Japanese Journal of Appliced Physics*, Vol.49, No.4, (April 2010), pp. 04DL16, ISSN 0021-4922

Grätzel, M. (2005). Solar Energy Conversion by Dye-Sensitized Photovoltaic Cells, *Inorganic Chemistry*, Vol.44, No.20 (October 2005), pp. 6841-6851, ISSN 0020-1669

Zhang, W.; Zhang, D.; Fan, T.; Gu, J.; Ding, J.; Wang, H.; Guo, Q. & Ogawa, H. (2009). Novel Photoanode Structure Templated from Butterfly Wing Scales, *Chemistry of Materials*, Vol.21, No.1, (January 2009), pp. 33-40, ISSN 0897-4756

Liu, H.; Song, C.; Zhang, L.; Zhang, J. ; Wang, H. & Wilkinson, D. P. (2006). A Review of Anode Catalysis in The Direct Methanol Fuel Cell, *Journal of Power Sources*, Vol.155, No.2, (April 2006), pp. 95-110, ISSN 0378-7753

Yong, P.; Mikheenko, I. P.; Deplanche, K.; Redwood, M. D. & Macaskie, L. E. (2010). Biorefining of Precious Metals from Wastes: An Answer to Manufacturing of Cheap Nanocatalyst for Fuel Cells and Power Generation via and Integrated Biorefinery?, *Biotechnology Letters*, Vol.32, No.12, (December 2010), pp. 1821-1828, ISSN 0141-5492

Górzny, M. L.; Walton, A. S. & Evans, S. D. (2010). Synthesis of High-Surface-Area Platinum Nanotubes Using a Viral Template, *Advanced Functional Materials*, Vol.20, No.8, (April 2010), pp. 1295-1300, ISSN 1616-301X

11

Thioether Bond Containing Polymers as Novel Cathode Active Materials for Rechargeable Lithium Batteries

Zhang J.Y., Zhan H., Tang J., Zhan L.Z., Song Z.P., Zhou Y.H. and Zhan C.M.
College of Chemistry & Molecular Sciences, Wuhan University, PR China

1. Introduction

The development of lightweight and high energy-density rechargeable batteries is of importance with the increasing demand for mobile power and the extensive use of portable devices. To build high energy batteries, both the anode (negative electrode in batteries) and cathode (positive electrode in batteries) must have a high specific capacity. Lithium, with a theoretical specific capacity of 3860 mA h g^{-1} at the anode, may offer a high-energy density, if the cathode material also has a high capacity. Most of the cathode materials currently in use are inorganic materials, such as transition metal oxide, the specific capacity of which is very low(less than 200 $mAhg^{-1}$) compared with lithium and limited the specific capacity of lithium secondary battery.

Sulfur is of high specific capacity, cheap, and environmentally benign as cathode material for rechargeable lithium batteries. It has attracted great interest and has being extensively studied. The shortcomings are its electric insulation, dissolution in solvents, and migration of S^{2-} from cathode.

Since Liu and Visco et al. (M. Liu et al., 1989; Visco et al., 1989) found that organic disulfide compounds could be used as cathode materials for rechargeable lithium battery in late 1980s, many researches have been focused on organic disulfide materials. These disulfide compounds could be divided into five categories (Scheme 1): (1) Small organic compounds (Scheme 1a), from which small molecular anions (−SRS−) were formed in the reductive state (Z. J. Liu et al., 2005; Maddanimath et al., 2003; Tsutsumi et al., 1996, 1997, 2001). Therefore this kind of organic disulfide compounds has the similar disadvantage to elemental sulfur. (2) Disulfide ladder polymers (Scheme 1b) have polymeric backbone linking disulfide bonds and do not dissolve in solvents during discharge process. But it is difficult to recombine perfectly during charging process, so that the cycling property is poor (Huang et al., 2003; Naoi et al, 1998; Su et al., 2004). (3) Disulfide polymers have disulfide bonds as side-chains (Scheme 1c), in which the two S atoms of one disulfide bond (–S–S–) are linked to the same polymeric main chain. Hence the recombination of disulfide bonds could be improved (Amaike & Iihama, 2006; Deng et al., 2006; J. X. Li et al., 2004; Uemachi et al., 2001; Xu et al., 2006; Xue et al., 2003). (4) Polysulfide compounds (Myachina et al., 2006; Trofimov et al.,

2002, 2006) (Scheme 1d) have high specific capacity in the first cycle and serious cycling fading due to its polysulfide structure. (5) Sulfurized polymers were synthesized through heating the mixture of organic polymer compounds and sulfur. Their structure is unclear; the electrochemical behaviors relied on the heating conditions. Some of them have high specific capacity and good cycling properties (Wang et al., 2002, 2003; Trofimov et al., 2002, 2003). The electrode reactions of all these compounds are based on the same mechanism: the cleavage-recombination of disulfide (S–S) bonds.

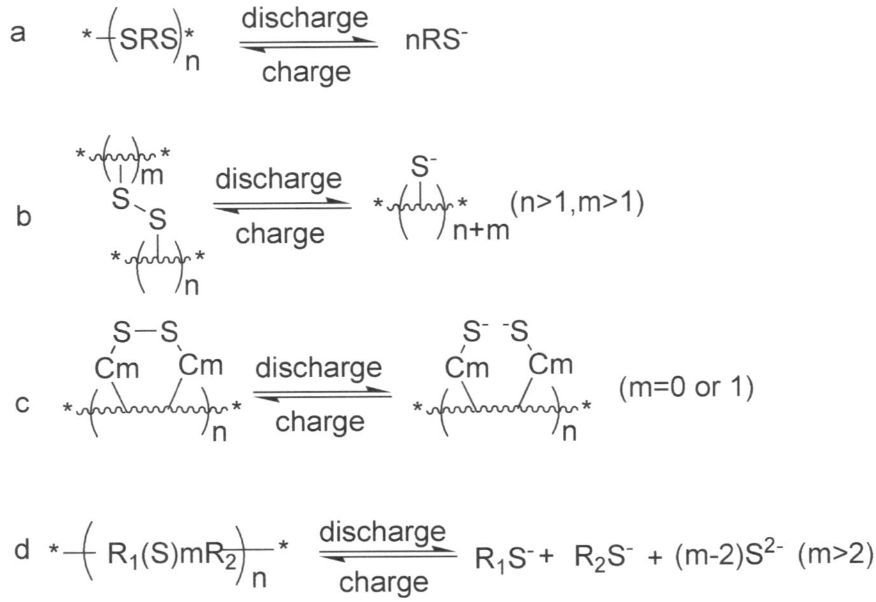

Scheme 1. Organic disulfide compounds for cathode active material in rechargeable lithium batteries

We found that thioether bond containing polymers (thioether polymers) other than conventional disulfides could also storage energy as cathode active materials and reported the research results in series papers since 2007 (Tang et al., 2008; Zhan et al., 2008, 2009; Zhang et al., 2007, 2008, 2009). The thioether polymers contain thioether (C–S–C) bonds instead of disulfide (-C–S–S–C–) bonds, in which the thioether bonds are the function groups to undergo electrode redox reaction. The redox reactions of thioether compounds (Scheme 3) do not involve bond cleavage, thus it may offer better stability as cathode active material. Furthermore, thioether bond may afford multi electron redox cathode reaction and higher specific capacity. As a novel structure, it showed some advantage and need further study and exploration. In this chapter, some thioether polymers as cathode active materials in lithium rechargeable battery are reviewed.

2. Structure, syntheses, and characterization of the thioether polymers

2.1 Structure and syntheses routes of the thioether polymers

The structure and synthesis routes of thioether polymers investigated as cathode active material in our project are summarized in Scheme 2.

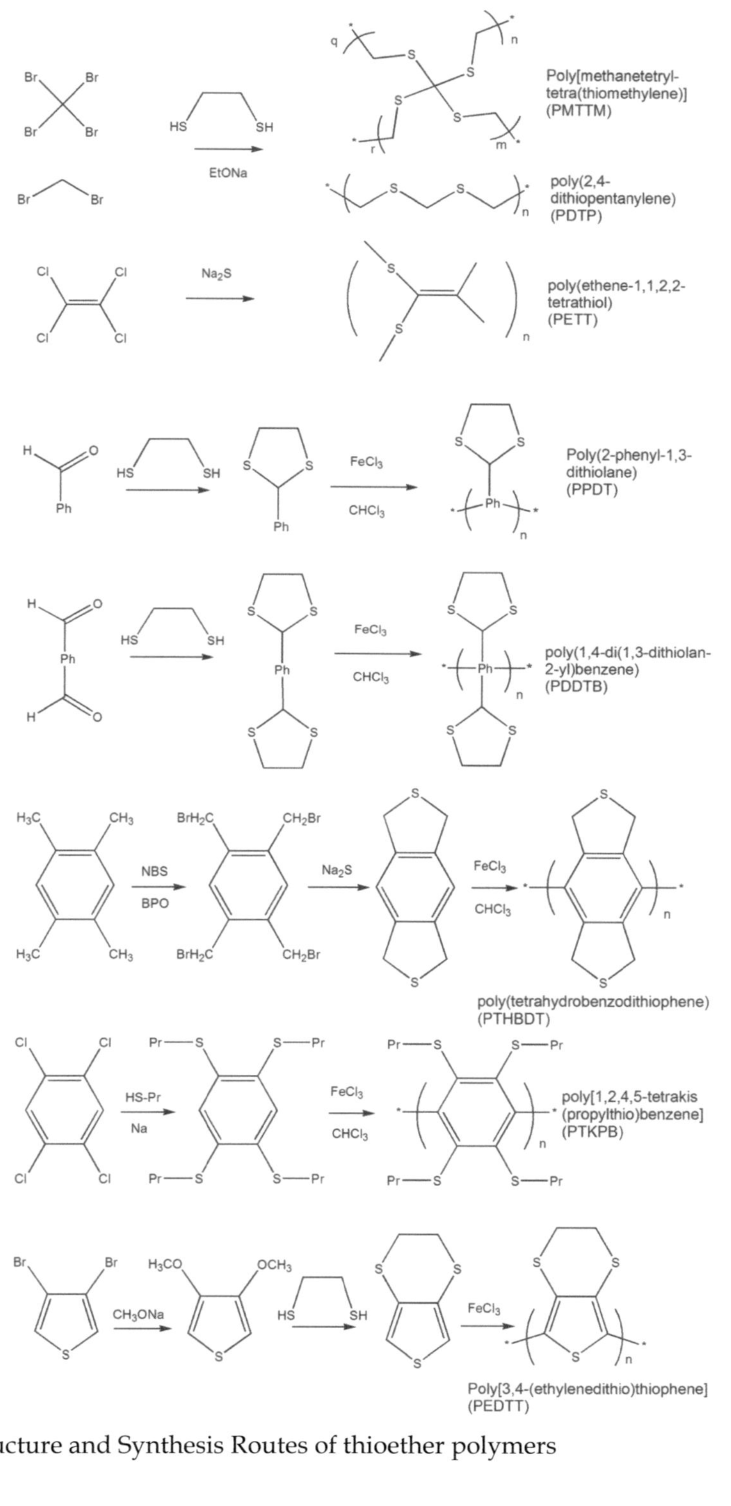

Scheme 2. Structure and Synthesis Routes of thioether polymers

Here Poly[methanetetryl-tetra(thiomethylene)] PMTTM, Poly(2,4-dithiopentanylene) PDTP, and Poly(ethene-1,1,2,2-tetrathiol) PETT are mainchain thioether polymers, in which sulfur atoms link carbon atoms to form polymeric backbones. The other polymers are side chain thioether polymers, those polymeric mainchains are consisted of conjugating aromatic moieties, with thioether side chains as pendants. Among them Poly(2-phenyl-1,3-dithiolane) PPDT, Poly(1,4-di(1,3-dithiolan-2-yl)benzene) PDDTB, poly(tetrahydrobenzodithiophene) PTHBDT, and poly[1,2,4,5-tetrakis(propylthio)benzene] PTKPTB have polyphenylene main chain, linking thiolane on benzene moieties as pendants. Similarly, Poly[3,4-(ethylenedithio)thiophene] PEDTT has polythiophene backbone, linking cyclo-thiolane on the 3,4- position on thiophene ring.

The syntheses of thioether polymers are facile and listed in Scheme 2. The mainchain thioether polymers were prepared by condensation polymerization. The conjugated polymers were prepared by facile oxidation coupling polymerization.

Most of thioether polymers were insoluble in organic solvents and were characterized by FT-IR, FT-Raman, XPS (X-ray photoelectron spectroscopy), and elemental analysis.

2.2 Materials characterization

In FT-IR spectra, the absorption peaks of C-S stretching vibration in C-S-C bonds usually appear around 700 cm^{-1}. In FT-Raman spectra, the peaks of C-S stretching vibration are between 700-600 cm^{-1}, and the peaks of C-S-C skeletal deformation are located at 270 to 380 cm^{-1}. In XPS spectra, the binding energy peak of S2p in thioether is at 163.6 eV or 164.2 eV. Most of the thioether polymers show amorphous in XRD curves.

3. Electrochemical properties

3.1 Cyclic voltammetry measurement

The cyclic voltammetry (CV) tests were carried out at a scan rate of 50 μVs^{-1} between 1.5 and 4.4 V at room temperature. The cathode was prepared by mixing polymer powder with carbon black and Teflon in a weight ratio of 4:4:2, the Li foil was used as the reference electrode and counter electrode with 1M $LiN(CF_3SO_2)_2$ (LiTFSI) in 1,2-dimethoxyethane (DME)/1,3-dioxolane(DOL) (1:2, by weight) as electrolyte. The cyclic voltammograms (CVs) results of the first cycle of various thioether polymers are summarized in Table 1.

Thioether Polymer	Oxidation Peaks/V[b]	Reduction Peaks/V[b]
poly[methanetetryl-tetra(thiomethylene)] PMTTM	2.20, 3.9	1.89, 2.00
poly(2,4-dithiopentanylene) PDTP	2.16, 3.9	1.72, 1.85
poly(ethene-1,1,2,2-tetrathiol) PETT	2.23, 4.03	2.00, 2.38
poly(2-phenyl-1,3-dithiolane)[a] PPDT	2.71, 3.60	1.55, 1.85
poly(1,4-di(1,3-dithiolan-2-yl)benzene)[a] PDDTB	1.95, 3.60	1.65, 1.75
poly(tetrahydrobenzodithiophene) PTHBDT	3.4-4.2	1.7-2.2, 2.5-2.8

[a] Cyclic voltammograms were tests through powder microelectrodes packed with only the pure thioether polymers in 1M $LiN(CF_3SO_2)_2$, 1,2-dimethoxyethane/1,3-dioxolane (1:2, by weight) solution at room temperature; scan rate, 5mVs^{-1}.

[b] Li foil was used as the reference electrode

Table 1. CV redox peaks of the thioether polymers

There are two oxidation peaks and two reduction peaks for all of the thioether polymers. The two oxidation peaks usually appear around 2V and 4V, while the reduction peaks are between 1.5-2.5V. The high oxidative peak indicates that the polymer may be charged at higher voltage as cathode material. But the reductive peaks are in lower voltage, which imply a lower discharge voltage.

3.2 Charge-discharge tests

The charge-discharge tests were performed on Land Battery test system. Working cathodes was prepared by mixing specific polymer powder with carbon black and Teflon in a weight ratio of 4:4:2. Lithium foil was used as the anode and the tests was processed at a rate of 50 mA/g in 1M $LiN(CF_3SO_2)_2$ (LiTFSI) in 1,2-dimethoxyethane (DME)/1,3-dioxolane(DOL) (1:2, by weight) solution. The cut-off voltages were 1.5V and 4.4V.

3.2.1 The cycling specific capacity and proposed redox reaction

The cycling specific capacity of the thioether polymers is summarized. In Fig.1 Poly[methanetetryl-tetra(thiomethylene)] (PMTTM) displays specific capacity of 504 $mAhg^{-1}$ at the third cycle, which decreases rapidly and maintains the value around 200 $mAhg^{-1}$ at the 16th cycle. Similarly, Poly(2,4-dithiopentanylene) (PDTP) exhibits a maximal specific capacity of 174 $mAhg^{-1}$ and a stable specific capacity about 100 $mAhg^{-1}$ between 16th and 50th cycles. The cycling capacity results of aliphatic thioether polymers suggest that polymers with thioether groups alone can offer energy storage without disulfide bond or conjugated polymeric main chain. There are only three kinds of bonds in PMTTM and PDTP: C-C, C-H, and C-S-C bonds. The saturated C-C and C-H bonds cannot afford redox reaction, hence it can be deduced that thioether bond is the only electrochemical active function group to offering energy storage. The oxidation of thioether had been studied and reported in literatures (Gilbert et al., 1973; Glass et al., 1977, 1990; Momose et al, 1987; Musker & Roush, 1978; Musker et al., 1978; Musker, 1980; Symons, 1974; Werst, 1991; Wilson, 1979), which illuminated the formation and existence of thioether cation.

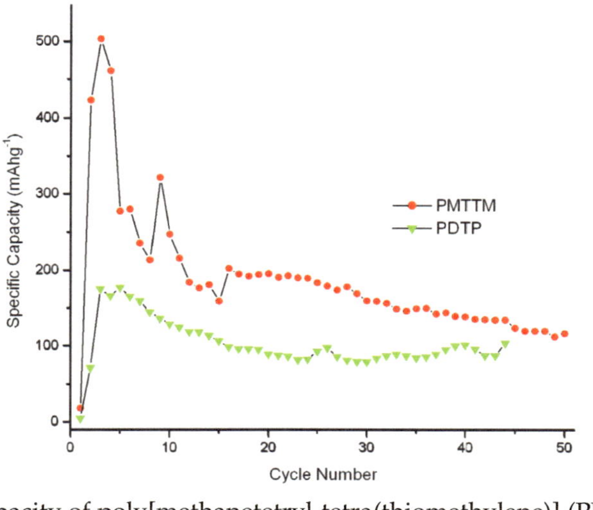

Fig. 1. Cycling capacity of poly[methanetetryl-tetra(thiomethylene)] (PMTTM) and poly(2,4-dithiopentanylene) (PDTP)

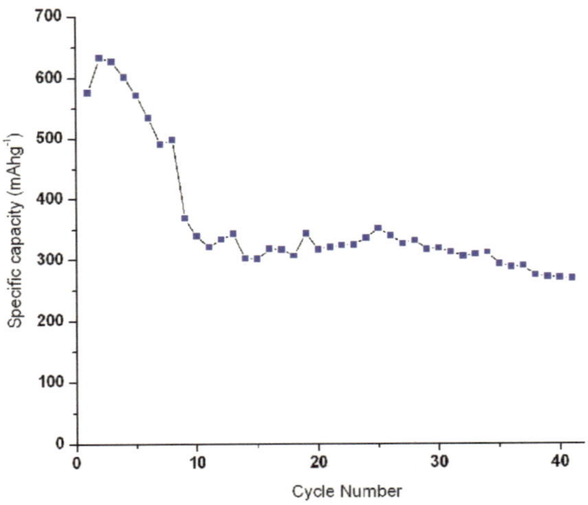

Fig. 2. Cycling capacity of Poly(ethene-1,1,2,2-tetrathiol) PETT In Fig.2 Poly(ethene-1,1,2,2-tetrathiol) PETT displays a stable discharge specific capacity value of ca. 300 $mAhg^{-1}$ from the 10th cycle to the 40th cycle and shows maximal specific capacity of 632 $mAhg^{-1}$ at the second cycle. The coulomb efficiency is at high level of 90% from the 10th cycle to the 40th cycle. It shows more stable discharge specific capacity compare to saturate aliphatic thioether polymers; indicating vinyl structure benefits the electrode reaction.

In Fig. 3 Poly(2-phenyl-1,3-dithiolane) (PPDT) displays a specific capacity of 129 $mAhg^{-1}$ at the first cycle and a specific capacity of 238 $mAhg^{-1}$ at the second cycle, and then it decreases and maintains the value around 100 $mAhg^{-1}$. But Poly(1,4-di(1,3-dithiolan-2-yl)benzene) (PDDTB) exhibits a maximal specific capacity of 378 $mAhg^{-1}$, and a stabilized specific capacity of about 300 $mAhg^{-1}$. The two polymers have same mainchain and pendant but different thiolane contents; hence the specific capacity value is determined by its dithiolane contents.

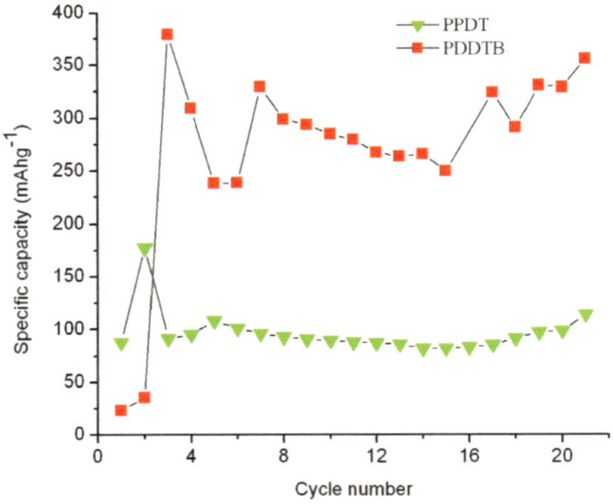

Fig. 3. Cycling capacity of Poly(2-phenyl-1,3-dithiolane) (PPDT) and Poly(1,4-di(1,3-dithiolan-2-yl)benzene) (PDDTB)

The initial discharge specific capacity of Poly[1,2,4,5-tetrakis(propylthio)benzene] PTKPB is around 200 $mAhg^{-1}$. From the second cycle on, the specific capacity decreases dramatically and becomes stable after the tenth cycle (Fig. 4).

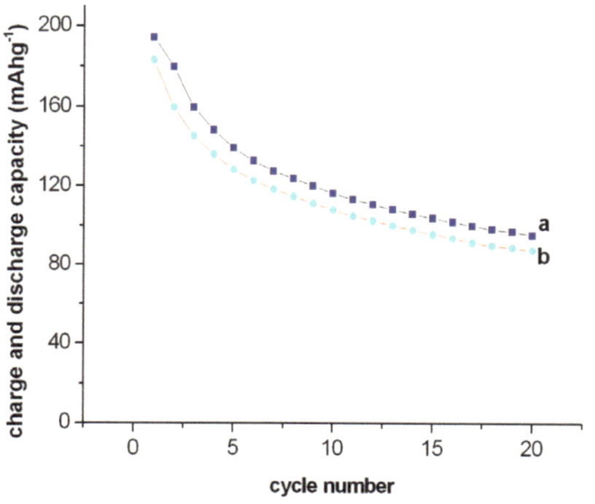

Fig. 4. Cycle life of Poly[1,2,4,5-tetrakis(propylthio)benzene] PTKPB (a) charge and (b) discharge

The discharge specific capacity of Poly(tetrahydrobenzodithiophene) PTHBDT was above 550$mAhg^{-1}$. The maximal specific capacity of 820 $mAhg^{-1}$ was obtained at the 9th cycle. After 16 cycles, the specific capacity was above 550 $mAhg^{-1}$ (Fig.5). Similarly, the polythiophene mainchain thioether polymer, Poly[3,4-(ethylenedithio)thiophene] PEDTT showed discharge specific capacity of 600 $mAhg^{-1}$ at 3rd cycle and ca. 400 $mAhg^{-1}$ from 19th cycle to 34th cycle (Fig. 6) [29].

Generally, the specific capacity of polyphenylene and polythiophene mainchain thioether polymers, PPDT, PDDTB, and poly[1,2,4,5-tetrakis(propylthio)benzene], looked more stable than that of aliphatic thioether polymers, PDTP and PMTTM. It means that electric conductive mainchain is benefit to the charge-discharge behavior of the thioether polymers.

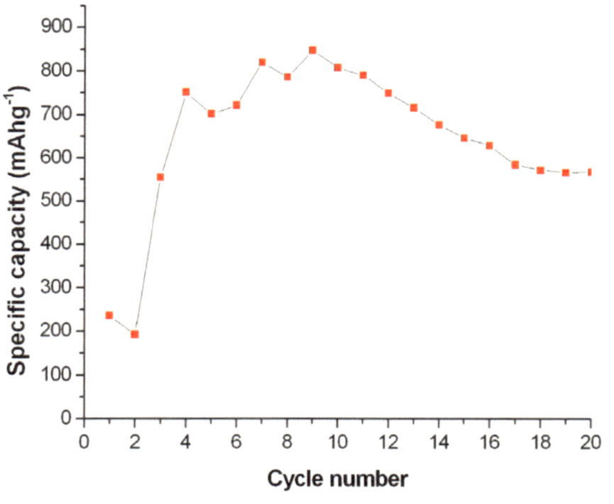

Fig. 5. Specific capacity ($mAhg^{-1}$) of poly(tetrahydrobenzodithiophene) vs. cycle numbers.

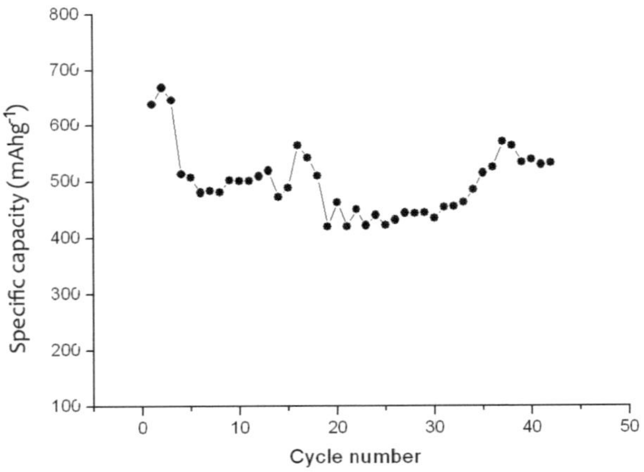

Fig. 6. Specific capacity of poly[3,4-(ethylenedithio)thiophene]

The measured maximal specific capacity of the thioether polymers are listed in Table 2. According to proposed redox reaction on S atom of thioether bond (Scheme 3), the theoretic specific capacity of the thioether polymers was calculated and listed in Table 2. It is very know that conjugated electro conductive polymers have discharge specific capacity come from "doping" mechanism as cathode active materials in lithium batteries, but the values are very lower hence are neglected in this paper. It can be seen that for most of thioether polymers their maximal specific capacity value is roughly agree with the theoretical specific capacity. One exception is Poly[3,4-(ethylenedithio)thiophene] PEDTT, its maximal specific capacity is 650 mAh g^{-1} , higher than its theoretic value. But from the 19th cycle to the 34th cycle, the discharge specific capacity is ca. 400-450 mAh g^{-1} (Fig.6), still agrees with the theoretic value. Generally, the results give support to our proposed mechanism.

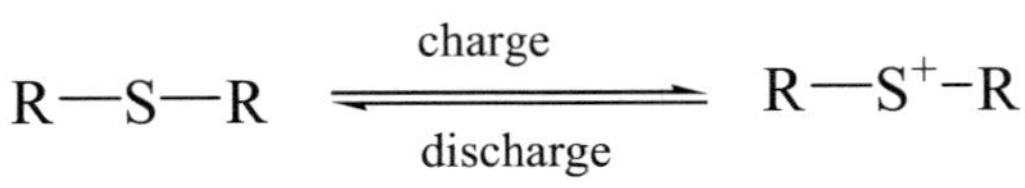

Scheme 3. Single electron cathode redox of the thioether polymers

Thioether Polymer	theoretic value	Practical maximal value
poly[methanetetryl-tetra(thiomethylene)] PMTTM	570	504
poly(2,4-dithiopentanylene) PDTP	505	200
poly(ethene-1,1,2,2-tetrathiol) PETT	595	632
poly(2-phenyl-1,3-dithiolane) PPDT	294	238
poly(1,4-di(1,3-dithiolan-2-yl)benzene) PDDTB	374	378
poly[1,2,4,5-tetrakis(propylthio)benzene] PTKPB	288	200
poly(tetrahydrobenzodithiophene) PTHBDT	279	820
Poly[3,4-(ethylenedithio)thiophene] PEDTT	467	650

Table 2. The theoretic and practical maximal specific capacity of thioether polymers ($mAhg^{-1}$)

But Poly(tetrahydrobenzodithiophene) PTHBDT has much higher specific capacity than its theoretic specific values. The theoretic specific capacity of PTHBDT is only 279 $mAhg^{-1}$, but its practical discharge maximal specific capacity is of 820 $mAhg^{-1}$ at the 9th cycle and at most cycle the discharge specific capacity is above 500 $mAhg^{-1}$ (Fig 5). New redox mechanism can be deduced from the rare high specific capacity: multi electrons are transferred on one sulfur atom in this thioether polymer [28]. To represent multi-electron redox reaction, a general reaction formula is proposed in Scheme 4. Here n is the electron number involved in the electrode redox. Corresponding to the specific capacity of 500 $mAhg^{-1}$ and 820 $mAhg^{-1}$, the electron number is 2 and 3 respectively.

$$R-S-R \underset{+\,ne}{\overset{-\,ne}{\rightleftharpoons}} R-S^{\pm n}-R$$

Scheme 4. Multi-electron Redox reaction of thioether polymer in lithium battery

Poly[3,4-(ethylenedithio)thiophene] PEDTT showed high discharge specific capacity above 650 $mAhg^{-1}$ (Fig. 6) at 3rd, the maximal value is also higher than that of theoretic value of single electron redox, which may due to the multi-electron redox reaction too.
In addition, solvents are important to the cathode redox reaction. The oxidation of thioether can be facilitated by electro-donating group (Glass, 1977), and thioether cation can be stabilized by electron donation (Wilson et al., 1979). The specific capacity results of thioether polymers were tested with 1, 2-dimethoxyethane and 1, 3-dioxolane mixture as electrolyte. Both two electrolytes are electron donors; the electrolyte may be helpful to the redox of thioether group.

3.2.2 The charge-discharge curves

The charge-discharge curves of various thioether polymers are summarized and shown in Fig. 7 to Fig.12.
The charge–discharge curves of poly[methanetetryl-tetra(thiomethylene)] PMTTM are shown in Fig.7. It can be seen that the capacity plateaus of the two polymers are at ca. 2.2V on discharge and 2.8 and 3.8V on charge, corresponding to the oxidative potential and reduction potential in CV (Table 2).
The charge-discharge curves of poly(ethene-1,1,2,2-tetrathiol) PETT are shown in Fig.8. The charge curve contains two plateaus: the low plateaus are at 2.6–2.8V and the high plateaus are between 3.3 and 3.8 V. But the discharge voltage is at 2.7–2.0 V. It showed higher discharge voltage than saturate aliphatic thioether polymer, poly[methanetetryl-tetra(thiomethylene)]. The results suggest that molecular structure played important role in the electrochemical behavior of thioether polymer. However, the discharge voltage fading was dramatic for the two mainchain thioether polymers. Fig. 9 to Fig. 11 show charge - discharge curves of the three polyphenylene mainchain thioether polymers —Poly(1,4-di(1,3-dithiolan-2-yl)benzene) PDDTB, Poly[1,2,4,5-tetrakis(propylthio)benzene] PTKPB, and poly(tetrahydrobenzodithiophene) PTHBDT. There are well-defined plateau for charge and discharge curves, the voltages are very close to each other : 2.4 V on charge and 2.3 V on discharge. These results indicated that high voltage may come from the polyphenylene backbone, those high oxidation potential afforded high discharge voltage. In addition, their charge-discharge voltages are very close to each other, which mean higher voltage efficiency than the mainchain thioether polymers.

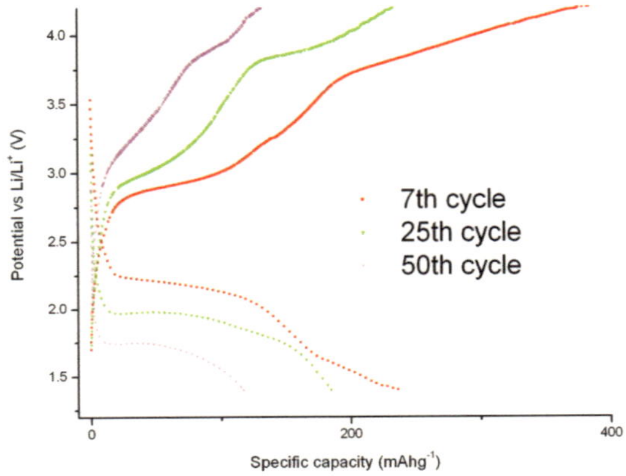

Fig. 7. The charge–discharge curves of poly[methanetetryl-tetra(thiomethylene)]

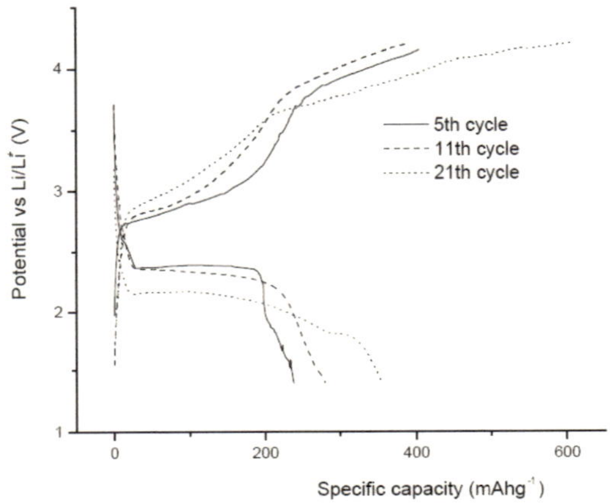

Fig. 8. The charge–discharge of poly(ethene-1,1,2,2-tetrathiol) PETT

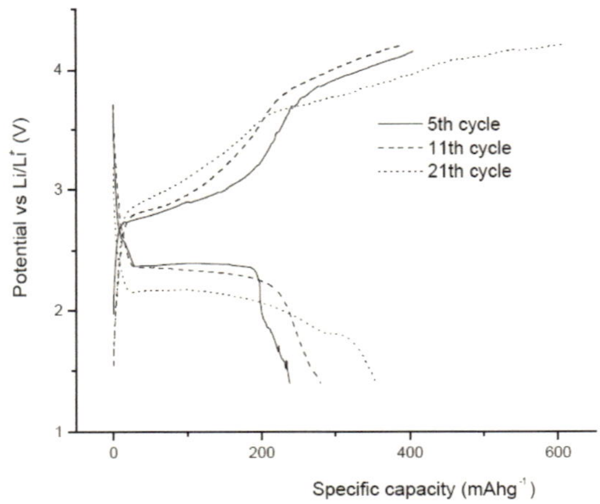

Fig. 9. Charge–discharge curves of poly(1,4-di(1,3-dithiolan-2-yl)benzene) PDDTB

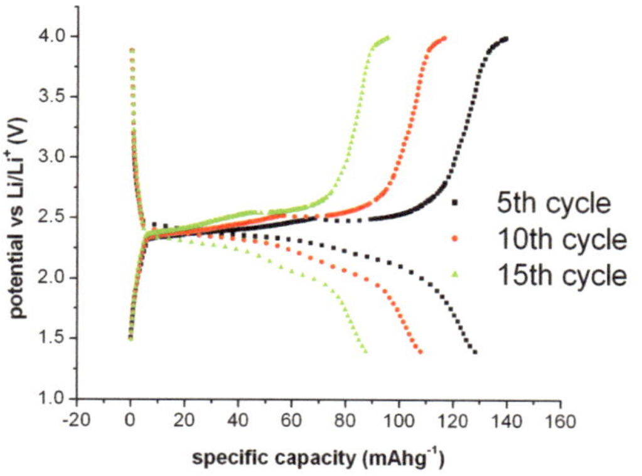

Fig. 10. Charge-discharge curves of poly[1,2,4,5-tetrakis(propylthio)benzene] PTKPB

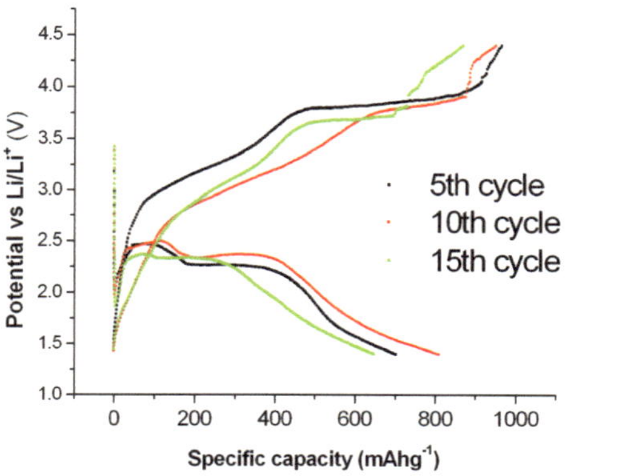

Fig. 11. The charge–discharge curves of Poly(tetrahydrobenzodithiophene) PTKP

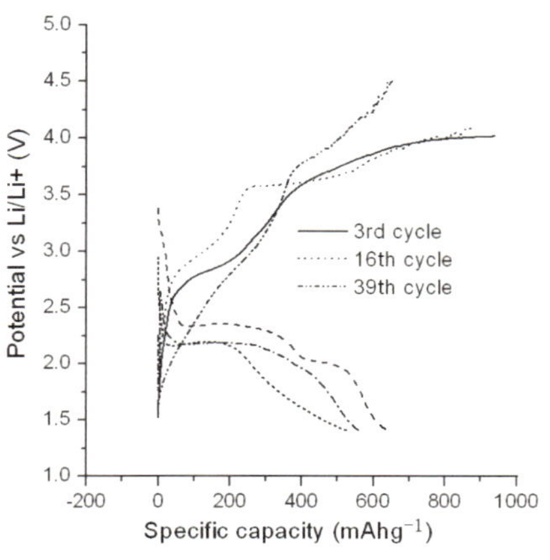

Fig. 12. Charge–discharge curves of Poly[3,4-(ethylenedithio)thiophene] PEDTT

Further, the voltage fading is slighter than those mainchain thioether polymers. The results mean that electric conductive mainchains improved the charge-discharge behavior of the thioether polymers.

The charge - discharge curves of Poly[3,4-(ethylenedithio)thiophene] PEDTT (Fig. 12) show two charge plateaus at ca2.8V and 4.0 V and discharge plateaus at ca. 2.2 V, respectively. The charge voltage and discharge voltage correspond to the oxidative potentials and reductive potentials in the CV test, as shown in Table 1. The gap between charge voltage and discharge voltage indicates the lower voltage efficiency.

4. Conclusion

Thioether polymers were prepared by facile methods and tested as novel cathode active materials in lithium secondary batteries. The thioether polymers showed high discharge specific capacity up to 800 $mAhg^{-1}$, discharge voltage above 2V, and good cycling stability. The results demonstrated that thioether polymers are cathode active material and thioether bond(C–S–C) is the function group to offer cathode redox reaction. It is proposed that S cation formed during charge-discharge process and somehow multi-electron redox reaction took placed, which imparted high specific capacity. Electric conductive mainchain can improve the charge –discharge performance of the thioether polymers.

5. References

Amaike, M. Iihama, T. (2006). Chemical polymerization of pyrrole with disulfide structure and the application to lithium secondary batteries. Synth. Met., Vol.156, No. 2-4, pp.239-243, ISSN: 0379-6779

Deng, S. R. Kong, L. B. Hu, G. Q. Wu, T. Li, D. Zhou, Y. H. Li, Z. Y. (2006). Benzene-based polyorganodisulfide cathode materials for secondary lithium batteries. Electrochim. Acta, Vol.51, No.13, pp. 2589-2593, ISSN: 0013-4686

Gilbert, B. C. Hodgeman, D. K. C. Norman, R. O. C. (1973). Electron spin resonance. XXXVIII. Formation of dimeric radical-cations in the one-electron oxidation of sulfides. J. Chem. Soc., Perkin Trans. 2(1972-1999), Vol.13, pp.1748-1752, ISSN: 1472-779X

Glass, R. S. Broeker, J. L. Firouzabadi, H. (1990). Stereochemistry of the formation and hydrolysis of a dithioether dication, J. Org. Chem., Vol.55, No.22, pp. 5739-5746, ISSN:0022-3263

Glass, R. S. Duchek, J. R. Klug, J. T. Wilson, G. S. (1977). Facilitation of dialkyl sulfide oxidation by neighboring groups. J. Am. Chem. Soc., Vol.99, No.22, pp.7349-7350, ISSN: 0002-7863

Huang, L. M. Wen, T. C. Yang, C. H. (2003). Electrochemical copolymerization of aniline and 2,2 '-dithiodianiline. Mater. Chem. Phys., Vol.77, No.2, pp. 434-441, ISSN: 0254-0584

Li, J. X. Zhan, H. Zhou, L. Deng, S. R. Li, Z. Y. Zhou, Y. H. (2004). Aniline-based polyorganodisulfide redox system of high energy for secondary lithium batteries. Electrochem. Commun., Vol.6, No. 6, pp.515-519, ISSN:1388-2481

Liu, M. Visco, S.J. De Jonghe, L.C. (1989). Electrochemical properties of organic disulfide/thiolate redox couples. J. Electrochem. Soc., Vol.136, No. 9, pp. 2570-2575, ISSN: 0013-4651

Liu, Z. J. Kong, L. B. Wang, L. M. Zhou, Y. H. Zhan, C. M. (2005). Poly (tetrathiooxalic acid)-Novel cathode active material for secondary lithium batteries. Journal of New Materials for Electrochemical Systems, Vol. 8, No.4, pp. 269-272, ISSN: 0370-047X

Maddanimath, T. Khollam, Y. B. Aslam, A. Mulla, I.S. Vijayamohanan, K. (2003). Self-assembled monolayers of diphenyl disulphide: a novel cathode material for rechargeable lithium batteries. J. Power Sources, Vol.124, No.1, pp. 133-142, ISSN: 0378-7753

Momose, T. Tanimura, R. Ushida, K. Shida, T. (1987). Comparative analysis of electronic states of saturated dioxa and dithia heterocycles and their radical cations. J. Phys. Chem., Vol.91, No.22, pp. 5582-5586, ISSN: 0022-3654

Musker, W. K. (1980). Chemistry of aliphatic thioether cation radicals and dications. Acc. Chem. Res., Vol.13, No.7, pp.200-206, ISSN: 0001-4842

Musker, W. K. Roush, P. B. (1976). Preparation and reactivity of dithioether dications [R2S+-+SR2]: a new functional group. J. Am. Chem. Soc., Vol.98, No.21, pp. 6745-6746, ISSN: 0002-7863

Musker, W. K. Wolford, T. L. Roush, P. B. (1978). An investigation of mesocyclic and acyclic dithioether cation radicals and dications. J. Am. Chem. Soc., Vol.100, No.20, pp. 6416-6421, ISSN: 0002-7863

Myachina, G. F. Korzhova, S. A. Papernaya, L. K. Deryagina, E. N. Skotheim, T. A. Trofimov, B. A. (2006). A polysulfide based on bis(thien-2-yl) sulfide: Synthesis and testing as an active cathode material. Phosphorus, Sulfur Silicon Relat. Elem., Vol. 181, No.6, pp. 1241-1247, ISSN: 1042-6507

Naoi, K. Kawase, K. I. Mori, M. Komiyama, M. (1998). A new polymer cathode -conducting polymer interconnected with sulfur-sulfur bonds: poly(2, 2'-dithiodianiline). Mater. Res. Soc. Symp. Proc., 496, pp.309-314.

Su, Y. Z. Niu, Y. P. Xiao, Y. Z. Xiao, M. Liang, Z. X. Gong, K. C. (2004). Novel conducting polymer poly [bis(phenylamino)disulfide]: Synthesis, characterization, and properties. J. Polym. Sci., Part A: Polym. Chem., Vol.42, No.10, pp.2329-2339, ISSN:0887-624X

Symons, M. C. R. (1974). Electron spin resonance detection of alkylthio radicals in irradiated solids. Radicals of type RSSR-, RSSR2, and R2SSR2+, J. Chem. Soc., Perkin Trans. 2 (1972-1999), Vol.13, pp.1618-1620, ISSN: 1472-779X

Tang, J. Song, Z. P. Shang, N. Zhan, L. Z. Zhang, J. Y. Zhan, H. Zhou, Y. H. Zhan, C. M. (2008). Poly[3,4-(ethylenedithio)thiophene]: High specific capacity cathode active material for lithium rechargeable batteries. J. Power Sources. Vol.185, No. 2, pp. 1434-1438, ISSN: 0378-7753

Trofimov, B. A. (2003). Sulfurization of polymers: a novel access to electroactive and conducting materials. Sulfur Reports, Vol.24, No.3, pp.283-305, ISSN:0196-1772

Trofimov, B. A. Deryagina, E. N. Sukhomazova, E. N. Levanova, E. P. Myachina, G. F. Korzhova, S. A. Mikhaylik, Y. V. Skotheim, T. A. (2002). Poly(methylene diselenide) and poly(methylene seleno-polysulfido-selenides), -[-CH2Se2-]n-, -[-CH2SeSmSe-]n, Sulfur Lett., Vol.25, No.5, pp.229-234, ISSN:0278-6117

Trofimov, B. A. Morozova, L. V. Markova, M. V. Mikhaleva, A. I. Myachina, G. F. Tatarinova, I. V. Skotheim, T. A. (2006). Vinyl ethers with polysulfide and hydroxyl functions and polymers there from as binders for lithium-sulfur batteries. J. Appl. Polym. Sci., Vol.101, No. 6, pp. 4051-4055, ISSN: 0021-8995

Trofimov, B. A. Vasil'tsov, A. M. Petrova, O. V. Mikhaleva, A. I. Myachina, G. F. Korzhova, S. A. Skotheim, T. A. Mikhailik, Y. V. Vakul'skaya, T. I. (2002). Sulfurization of polymers 6. Poly(vinylene polysulfide), poly(thienothiophene), and related structures from polyacetylene and elemental sulfur. Russ. Chem. Bull., Vol.51, No. 9, pp.1709-1714, ISSN:1066-5285

Tsutsumi, H. Okada, K. Fujita, K. Oishi, T. (1997). New type polyamides containing disulfide bonds for positive active material of lithium secondary batteries. J. Power Sources, Vol.68, No.2, pp. 735-738, ISSN: 0378-7753

Tsutsumi, H. Okada, K. Oishi, T. (1996). Electrochemical behavior of new polyamides containing disulfide bonds and pyridine rings in organic electrolyte solution. Electrochim. Acta, Vol.41, No.16, pp.2657-2659, ISSN:0013-4686

Tsutsumi, H. Oyari, Y. Onimura, K. Oishi, T. (2001). Electrochemical behavior of polyamides with cyclic disulfide structure and their application to positive active material for lithium secondary battery. J. Power Sources, Vol.92, No.1-2, pp.228-233, ISSN:0378-7753

Uemachi, H. Iwasa, Y. Mitani, T. (2001). Poly(1,4-phenylene-1,2,4-dithiazol-3',5'-yl): the new redox system for lithium secondary batteries. Electrochim. Acta, Vol.46, No.15, pp. 2305-2312, ISSN: 0013-4686

Visco, S. J. Mailhe, C. C. De Jonghe, L. C. Armand, M. B. (1989). A novel class of organosulfur electrodes for energy storage. J. Electrochem. Soc., Vol.136, No.3, pp. 661-664, ISSN:0013-4651

Wang, J. L. Yang, J. Wan, C. R. Du, K. Xie, J. Y. Xu, N. X. (2003). Sulfur composite cathode materials for rechargeable lithium batteries, Adv. Funct. Mater., Vol.13, No.6, pp.487-492,ISSN:1616-301X

Wang, J. L. Yang, J. Xie, J. Y. Xu, N. X. (2002). A novel conductive polymer-sulfur composite cathode material for rechargeable lithium batteries. Advanced Materials, Vol.14, No.13-14, pp.963, ISSN: 0935-9648

Werst, D. W. (1991). Radical cation-arene p-molecular complexes: Thioether radical cations in aromatic solvents. J. Am. Chem. Soc., Vol.113, No.11, pp. 4345-4346, ISSN: 0002-7863

Wilson, G. S. Swanson, D. D. Klug, J. T. Glass, R. S. Ryan, M. D. Musker, W. K. (1979). Electrochemical oxidation of some mesocyclic dithioethers and related compounds. J. Am. Chem. Soc., Vol.101, No.4, pp.1040-1042, ISSN: 0002-7863

Xu, G. X. Qi, L. Yu, B. T. Wen, L. (2006). PVC disulfide as cathode materials for secondary lithium batteries. Chin. J. Polym. Sci., Vol.24, No.3, pp.307-313 , ISSN : 0256-7679

Xue, L. J. Li, J. X. Hu, S. Q. Zhang, M. X. Zhou, Y. H. Zhan, C. M. (2003). Anthracene based organodisulfide positive active materials for lithium secondary battery. Electrochem. Commun., Vol.5, No.10, pp.903-906, ISSN:1388-2481

Zhan, L. Z. Song, Z. P. Zhang, J. Y. Tang, J. Zhan, H. Zhou, Y. H. Zhan, C. M. (2008). Synthesis and properties of novel organic thiolane polymer as cathode material for rechargeable lithium batteries. J Appl. Electrochem., Vol.38, No.12 , pp.1691-1694, ISN: 0021-891X

Zhan, L. Z. Song, Z. P. Zhang, J. Y. Tang, J. Zhan, H. Zhou, Y. H. Zhan, C. M. (2009) Poly(tetrahydrobenzodithiophene): High discharge specific capacity as cathode material for lithium batteries. J. Power Sources, Vol.193, No.2, pp. 859-863, ISSN: 0378-7753

Zhang, J. Y. Kong, L. B. Zhan, L. Z. Tang, J. Zhan, H. Zhou, Y. H. Zhan, C. M. (2007). Sulfides organic polymer: Novel cathode active material for rechargeable lithium batteries. J. Power Sources, Vol.168, No.1, pp. 278-281, ISSN:0378-7753

Zhang, J. Y. Kong, L. B. Zhan, L. Z. Tang, J. Zhan, H. Zhou, Y. H. Zhan, C. M. (2008). Aliphatic thioether polymers as novel cathode active materials for rechargeable lithium battery. Electrochem. Comm., Vol.10, No.10, pp. 1551-1554, ISSN: 1388-2481

Zhang, J. Y. Song, Z. P. Zhan, L. Z. Tang, J. Zhan, H. Zhou, Y. H. Zhan, C. M. (2009). Poly(ethene-1,1,2,2-tetrathiol): Novel cathode material with high specific capacity for rechargeable lithium batteries. J. Power Sources, Vol.186, No.2, pp. 496-499, ISSN: 0378-7753

12

Nanostructured MnO_2 for Electrochemical Capacitor

Mao-wen Xu[1] and Shu-Juan Bao[2]
[1]*Laboratory of New Energy Material Chemistry, College of Chemistry & Chemical Engineering, Xinjiang Normal University,*
[2]*Institute of Applied Chemistry, Xinjiang University, P.R. China*

1. Introduction

Energy is always a priority issue for human beings. In the 21st century, pollution due to combustion of fossil fuel, which has triggered the biggest environmental issue "global warming", has become a worldwide problem. The renewable and clean energy sources as well as efficient use of energy are highly necessary to make our economy, environment, society, and human species sustainable (Simon & Gogotsi, 2008; Xu et al., 2010). Energy storage, an intermediate step to energy, creates a new approach to use energy versatility, cleanly, and efficiently. Electrochemical capacitors (ECs) or supercapacitors (SCs), as a new energy storage/conversion device, have gained enormous attention owing to their higher power density and longer cycle life compared to secondary batteries and higher energy density than conventional electrical double-layer capacitors (EDLC). It is also characterized by environmental friendliness, high safety, and good efficiency and can be operated in a wide temperature range with a nearly infinitely long cycle life. Therefore, supercapacitors have been applied in and are showing potential application in communications, transportation, consumer electronics, aviation, and related technologies (Siegwart, 2001; Burke, 2000; Yoda & Lshihara, 1997; Becker, 1957; Yoshino et al., 2004).

Supercapacitors or ultracapacitors, ECs can be fully charged or discharged in seconds. Their energy density (about 5 Wh/kg) is lower than in batteries, but a much higher power delivery or uptake (10 kW/kg) can be achieved for shorter times (a few seconds) (Simon & Gogotsi, 2008; Siegwart, 2001). They have had an important role in complementing or replacing batteries in the energy storage field, such as for uninterruptible power supplies and load-levelling. A recent report by the US Department of Energy assigns equal importance to supercapacitors and batteries for future energy storage systems and articles on supercapacitors appearing in business and popular magazines show increasing interest by the general public in this topic (Simon & Gogotsi, 2008; Whittingham et al., 2004).

Depending on the charge storage mechanism as well as the active materials used, several types of ECs can be distinguished. One group, Electric double layer capacitor (EDLC) uses carbon as the electrodes and stores charge in the electric field at the interface. A second group, known as pseudo-capacitors or redox supercapacitors, uses fast and reversible surface or near-surface reactions for charge storage, which use transition metal oxides and conducting polymers as active materials. Hybrid capacitors, combining a capacitive or pseudo-capacitive electrode with a battery electrode, are the latest kind of EC, which benefit from both the capacitor and the battery properties.

As pseudocapacitors electrode materials, metal oxides, which store energy through highly reversible surface redox (faradic) reactions in addition to the electric double-layer storage, attracted increasing attention in recent years. Among the available metal oxides materials, RuO_2 shows the best performance with high specific capacitance (720 F/g) and 1.4 V potential window, but it is very expensive and toxic, which greatly limits its commercialization (Toupin et al., 2002; Wang et al., 2008; Reddy et al., 2009). Other metal oxides have also been tested as possible candidates for electrochemical supercapacitor devices. Interesting capacitance values have been reported for IrO_2 or CoOx electrodes but they are still expensive compounds (Conway 1997; Lin, 1998). On the other hand, NiO, $Ni(OH)_2$, MnO_2 systems seem more promising due to their lower cost. However, NiO and $Ni(OH)_2$ cannot be used at voltage windows above 0.6 V (Xing et al., 2004). Extensive studies have been conducted to explore alternative economic supercapacitor materials with good performance. Among of the being studied metal oxides, manganese oxide is an attractive candidate for supercapacitor designs due to its availability, low cost, low toxicity and wide voltage windows (Jeong & Manthiram, 2002; Reddy & Reddy, 2003; Reddy & Reddy, 2004). Then, one potential disadvantage of MnOx relative to RuO_2 as a psuedocapacitive material is its lack of metallic conductivity (Long et al., 2003).

It is well known that the theoretical value of specific capacitance of MnO_2 is ~1380 F/g. At present, only 30% or even lower of theoretical value can be obtained (Xu et al., 2010). To further improve the performance of MnO_2-based supercapacitor, it is necessary to design new MnO_2 materials with excellent performance and further understand its charge storage mechanism as electrode of Ecs. In this chapter, the physicochemical features, synthesis methods, and charge storage mechanism of MnO_2 as well as the current status of MnO_2-based supercapacitors are summarized and discussed in detail. The future opportunities and challenges related to MnO_2-based supercapacitors have also been proposed.

2. Crystalline structures of different MnO_2

As an important functional metal oxide, manganese dioxide is one of the most attractive inorganic materials because of its physical and chemical properties and wide applications in catalysis, ion exchange, molecular adsorption, biosensor, and particularly, energy storage (Qi et al.1999; Shen et al., 1993; Cao & Suib,1994). MnO_2 is a best representation of a general class of materials exhibiting a rich chemistry. MnO_2 is a very interesting and attractive material because it is diverse in crystalline structure and rich in Mn valence. Normally, MnO_2 is a complex and nonstoichiometric oxide and often contains foreign cations, physisorbed and structural water molecular, and structural vacancies. Because of the presence of foreign species, the average valence of Mn in MnO_2 generally locates between 3 and 4. However, the variety of MnO_2 in structure and valence comes from only one basic structural unit, MnO_6 octahedron. It acts likes string in a string theory world. In a MnO_2 world, this little tiny MnO_6 octahedron enables the buildup of a colorful and diverse world, in which every version is just one combination of MnO_6 octahedra (Qi et al., 1999; Tsuj & Abe, 1985). The combination of MnO_6 octahedra provides a veritable "toolbox" , from which to design, optimize, and synthesize specific MnO_2 for a specific purpose. Therefore, it is necessary to give a brief review on crystalline structures of MnO_2.

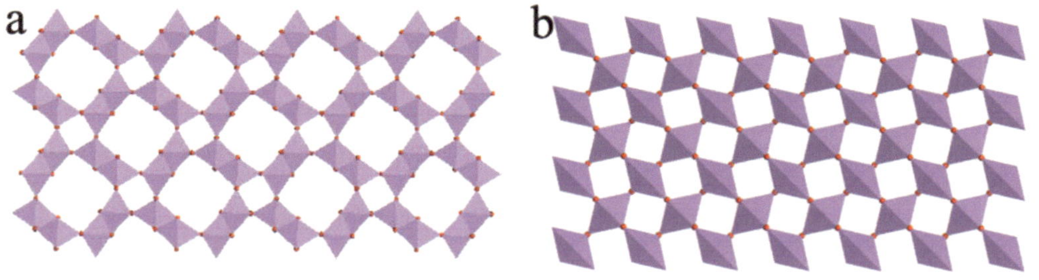

Fig. 1. The structure of α-MnO_2 with double chains of [MnO_6] octahedron (a) and β- MnO_2 with single chains of [MnO_6] octahedron (b).

Fig. 1 illustrates the schematic structure of a-MnO_2 and β-MnO_2. As shown in Fig. 1a, a-MnO_2 is consist of interlinking double chains of octahedral MnO_6 and an interstitial space comprised of one-dimensional channels of relative dimensions (2×2) and (1×1) that extend in a direction parallel to the c axis of a tetragonal unit cell (Wang & Li, 2002). The schematic structure of β-MnO_2 is different to that of a-MnO_2, which is composed of single chains of the octahedral [MnO_6] (Fig. 1b).

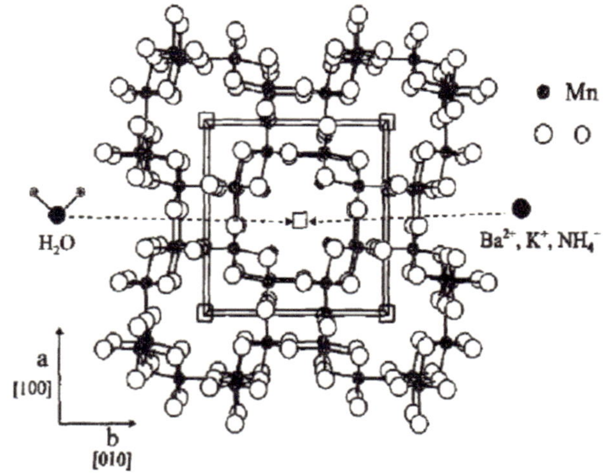

Fig. 2. Schematic structure of a-MnO_2 structure as viewed down the c-unit cell axis (Johnson et al., 1997).

In order further understand the structure of a-MnO_2 more clearly, a [001] projection of the a-MnO_2 framework structure is given in Fig. 2. The structure has tetragonal symmetry with space group I4/m. The framework has an interstitial space consisting of unidimensional channels of relative size (1 ×1) and (2 ×2). The (1 ×1) channels represent the interstitial space that is found in β-MnO_2 (rutile-type structure). Cations such as Ba^{2+} (hollandite), and K^+ (cryptomelane) and NH_4^+ or O^{2-} anions [from H_2O (or H_3O^+) and Li_2O] that stabilize the a-MnO_2 framework partially occupy sites at the center of the (2×2) channels at a special position (0, 0, z), usually close to (0, 0, 1/2) (Johnson et al., 1997; Rossouw et al., 1992). If oxygen occupies this site, the structure adopts a distorted close-packed oxygen array, with the close-packed oxygen layers parallel to the (110) planes. The α-MnO_2 framework structure in natural (mineral) form is stabilized by large cations such as Ba^{2+} (hollandite) and

K^+ (cryptomelane) or by NH_4^+ ions located within the large (2×2) channels of the structure (Brock et al.,1998).
The intergrowth of two or more tunnel phases occurs also in the manganese oxides. An irregular intergrowth of (1×1) tunnels (pyrolusite) and (1×2) tunnels (ramsdellite) in the structure of γ-MnO_2 (nsutite) is well known to electrochemists (Devaraj & Munichandraiah, 2008). Fig. 3 is the structure of γ-MnO_2. This intergrowth structure can be described in terms of De Wolff disorder and microtwinning. Ramsdellite is composed of double MnO_6 chains (Fig. 3b) linked together to form tunnels with a (1×2) octahedra cross-section. Ramsdellite is only observed in nature while the synthetic product γ-MnO_2 contains pyrolusite (1×1) tunnels intergrowths (de Wolff defects) and microtwinnings (Johnson et al., 1997; Portehault et al., 2009; Hill &Verbaere, 2004;. Wolff & P, 1959).

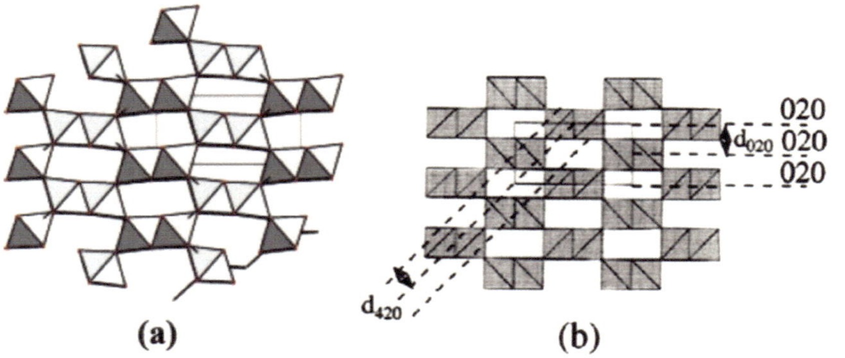

Fig. 3. a. The structure of γ-MnO_2 with irregular intergrowth of (1×1) tunnels (pyrolusite) and (1×2) tunnels (ramsdellite), b, ramsdellite-MnO_2 (Johnson et al., . 1997; Devaraj & Munichandraiah, 2008).

δ- MnO_2 (Fig. 4) is a 2D layered structure with an interlayer separation of ~7 Å. It has a significant amount of water and stabilizing cations such as Na^+ or K^+ between the sheets of MnO_6 octahedra (Devaraj & Munichandraiah, 2008; Ma et al., 2004).

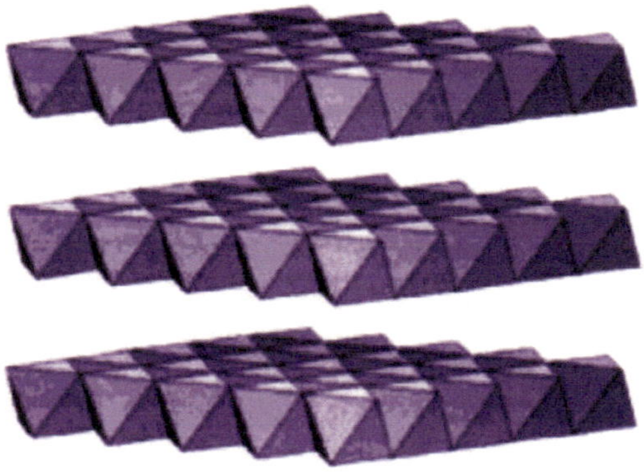

Fig. 4. The structure of δ- MnO_2 with 2D layered structure.

In the spinel structure of λ-MnO_2, the Mn ions occupy the 16d sites in *Fd*3*m* and form a three-dimensional (3D) array of corner-sharing tetrahedra as shown in Fig. 5.

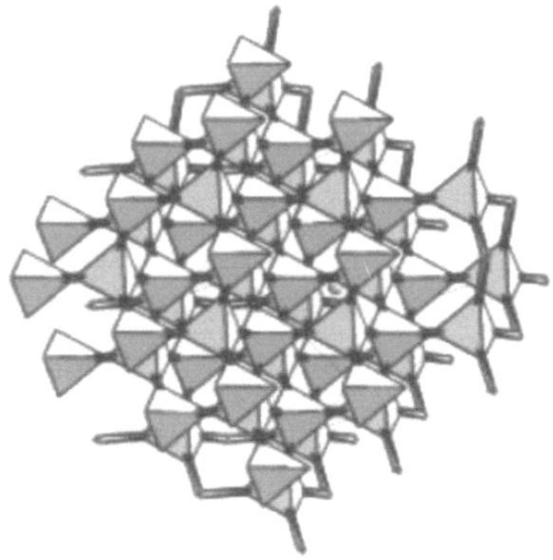

Fig. 5. The structure of λ- MnO_2 with spinel structure (Devaraj & Munichandraiah, 2008).

XRD patterns of MnO_2 listed above are shown in Fig. 6. The curves marked α and α(m) exhibit the patterns of α-crystallographic form (JCPDS no. 44-0141). The curves marked β and γ displayed the XRD patterns of β- (JCPDS no. 24-0735) and γ- (JCPDS no. 14-0644) crystallographic forms of MnO_2, respectively. Broad peaks at 2θ = 12.2, 24.8, 37.0, and 65.4° in the pattern marked δ correspond to δ-MnO_2 (JCPDS no. 18-0802). This pattern of the sample marked λ was indexed to cubic symmetry with space group $Fd3m$ of λ-MnO_2 (JCPDS no. 44-0992, $a = b = c = 8.03$ Å).

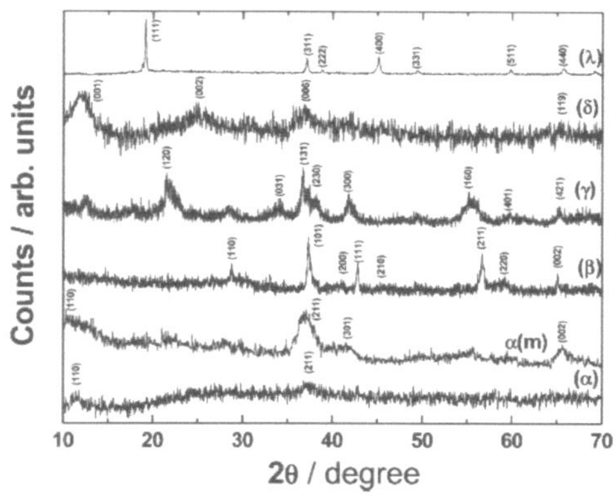

Fig. 6. XRD pattern of α-, β-, γ-, δ- and λ-MnO_2. The data were recorded at a sweep rate of 0.5° min^{-1} using Cu Kα source (Devaraj & Munichandraiah, 2008).

Complex intergrowths of the (2×2) tunnels (hollandite) and the (2×3) tunnel (romanechite) in fibrous manganese oxide minerals and intergrowth of (3×2) tunnels to (3×7) tunnels in natural todorokite have been detected by HRTEM measurements (Qi et al.,1999). However, almost all the intergrowths are random, so that regular periodicity or superstructure may not be apparent. In here, we don't review it again.

3. Synthesis methods of MnO_2

Because of the variety in structures, MnO_2 can be synthesized in dozens of crystalline and disordered forms, each with unique physicochemical properties, which are determined by the synthesis methods and post treatment procedures. The physicochemical properties in turn influence its electrochemical performance. Therefore, it is necessary to give a brief review on the synthesis of MnO_2.

3.1 Hydrothermal process

Hydrothermal process is a very useful and unique method for the preparation of different structured manganese oxides. By controlling the synthesis process, treatment temperature, pH value, post synthesis procedures, etc., different micro-structured MnO_2 could be obtained. Recently, many MnO_2 with different crystal structure and morphology have been successfully prepared by hydrothermal method (Zhou et al., 2011; Zhang et al., 2011; Jiang et al., 2011; Tang et al., 2011; Song et al., 2010; Liu et al., 2006; Yang et al., 2010).

In our previous work (Xu et al., 2007), α-MnO_2 hollow spheres and hollow urchins are synthesized via a simple hydrothermal process without using any template or organic surfactant. Further changing the treatment temperature, polyhedron structured β- MnO_2 be prepared.

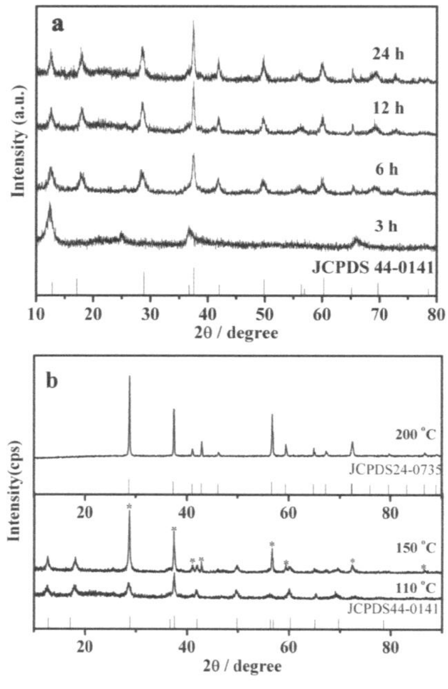

Fig. 7. a. XRD patterns for the standard values and the samples obtained at 110 °C for different reaction time; b, XRD patterns for the standard values and the samples obtained at different reaction temperature for 12 h.

It can be seen from Figure 7a, at 110 °C, after 3 h hydrothermal reaction, only four very weak peaks are observed, and the main peaks can be indexed to α-MnO_2 phase (JCPDS, card no: 44-0141), suggesting the α-MnO_2 forms in the process of hydrothermal treatment for 3h. With increasing of hydrothermal reaction time, all of these peaks intensities increase significantly, with lattice constants of a = 4.399 Å and c = 2.874 Å, which match very well with the standard XRD pattern (Figure 7a, bottom). However, in Figure 7b, for the XRD patterns of the sample obtained at 150 °C for 12 h, the intensive diffraction peaks appeared at 12.68, 18.06, 28.68, 37.36, 49.88, and 60.16 °, respectively, are assigned to the characteristic peaks for α-MnO_2, and the peaks occurred at 28.68, 37.36, 41.04, 42.82, 46.02, 56.65, 59.37, 72.38 and 86.18 °, respectively, should be ascribed to the characteristic peaks for β-MnO_2. Hence, the sample should be composed of α-MnO_2 and β-MnO_2. When the reaction temperature enhanced to 200 °C, the XRD pattern of the obtained sample shows highly crystalline β-MnO_2, all of the diffraction peaks can be indexed to β-MnO_2 (JCPDS 24-0735), which demonstrating the high purity of the β-MnO_2 may obtain in the process of hydrothermal treatment at 200 °C.

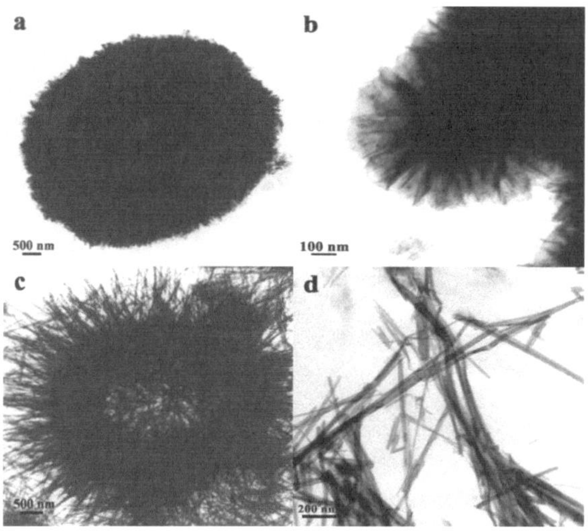

Fig. 8. TEM of α-MnO_2 obtained at 110 °C for different reaction time (a, 3 h; b, 6 h; c, 12 h; d, 24 h).

As shown in Fig. 8, several obvious evolution stages could be clearly observed. In the initial stage (shorter reaction time, 3 h), only a close-grained sphere is observed; after hydrothermal reaction for 6 h, the surface of α-MnO_2 sphere has changed to flower-like nanostructure which consists of nanoflakes and nanowires. When the reaction time was prolonged to 12 h, an interior cavity sphere is easily observed; after reaction for 24 h, the sphere structures disappear completely, only nanorods can be observed.

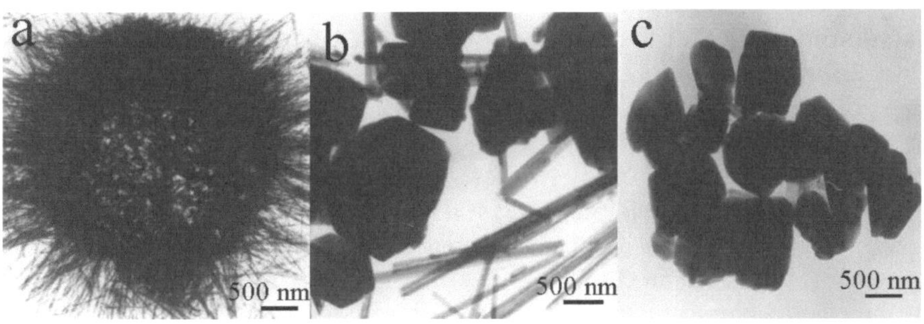

Fig. 9. TEM of MnO_2 obtained at different reaction temperature for 12 h (a, 110 °C; b, 150 °C; c, 200 °C).

The morphology of the products obtained at different temperature for 12 h were also observed by using TEM. Fig. 9a demonstrates the interior cavity spheres of the sample obtained by hydrothermal reaction at 110 °C. Upon increasing the reaction temperature to 150 °C, the sphere structure of the products disappeared completely, as shown in Fig. 9b, a mixture of nanorods and blocks were observed. When the reaction temperature enhanced to 200 °C, the polyhedron structure develops further and becomes the dominant product with good crystallization and regular morphology (in Fig. 9c).

3.2 Template directed synthesis

Since electrolyte diffusion within the bulk electrode materials is a rate-limiting step, a crucial issue to improve the rate capacity of ECs is to optimize the electrolyte transport paths without sacrificing electron transport. Hence, the development of novel synthesis routes to low dimensional and porous manganese oxides attracted increasing attention of scientists, since these compounds offer promising electrochemical properties and a rich application in many field (Wei et al., 2011). Host–guest compounds represent a new and promising class of material that can be used for the controlled preparation of complex organized structures or composites in the nanoscale regime.

3.2.1 Carbon template

CNTs are the most representative nanostructured carbons with one dimensional tubular structure and exhibit outstanding physicochemical properties such as high electrical conductivity, high mechanical strength, high chemical stability, and high activated surface areas. By using CNTs as template and reducing agent, heterogeneous nucleation of MnO_2 were deposited on CNTs, and MnO_2–CNT composite was obtained in literatures (Subramanian et al., 2006; Yan et al., 2009; Ma et al., 2008; Jiang et al., 2009; Xue et al., 2009).

Ordered mesoporous carbon materials are aother attractive type with a nanostructured hierarchy with desirable electrolyte transpnort routes. Dong et al. presented a novel MnO_2/mesoporous carbon composite structure, synthesized by embedding MnO_2 into the mesoporous carbon walls through the redox reaction between permanganate ions and carbons (Dong et al., 2006). A similar process was applied to obtain Mn_2O_3-templated mesoporous carbon composite (Zhang et al., 2009). A kind of MnO_2/mesoporous carbon composite was prepared by Zhu S. M. et al. through incorporating MnO_2 nanoparticles inside the pore channels of CMK-3 ordered mesoporous carbon under sonochemical process (Zhu et al., 2005).

Three dimensional (3D)-assemblies of silica spheres were used as a hard template to synthesize porous carbon materials with large mesopores (~100 nm) and large surface areas reaching up to 900 m^2/ g. Birnessite-type MnO_2 was deposited by a chemical co-precipitation method in the porous network (Lei et al., 2008).

In our past work (Xu et al., 2010), porous MnO_2 was synthesized via a simple and efficient in situ reduction process by using different carbon materials as sacrificed template and reducing agent. It is discovered that the microstructure of the samples has a remarkable effect on the electrochemical capacitive behaviors of the samples, of which the mesoprous MnO_2 prepared by using mesoporous carbon provides good conductivity and high capacitance.

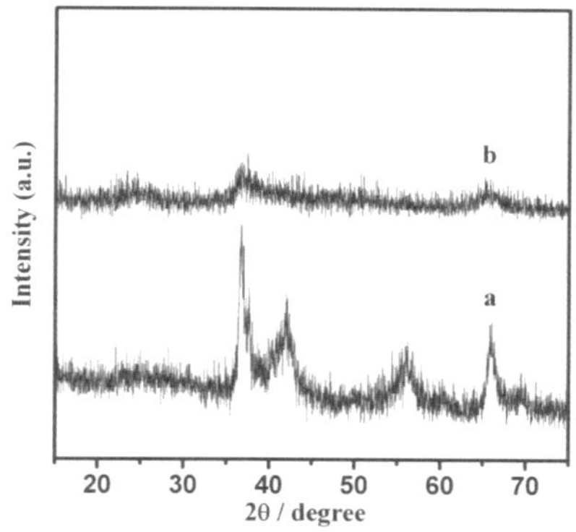

Fig. 10. XRD patterns of the samples obtained by using different carbon sources (a, CNTs; b, Mesoporous carbon)

As shown in Fig. 10, in which all of the diffraction peaks for the sample obtained from CNTs can be indexed to δ-MnO_2. Compared with the XRD peaks of the sample obtained by using CNTs, those of the sample obtained by mesoporous carbon as carbon sources are weak and wide, presents amorphous MnO_2 type, which may be relevant to different raw materials.

Fig. 11 displays the image of the MnO_2 prepared by using CNTs as carbon source is composed of uniform spheres and its built-up interleaving sheets or flakes (Fig. 11b). Interestingly, the MnO_2 prepared by using mesoporous carbon tends to form porous clusters (Fig. 11c). To reveal the actual structure of the cluster, high magnification FESEM were recorded. Fig. 11d clearly shows the surface structure of MnO_2 cluster, which consists of very small nanoparticle and nanowires. The detail structure of the MnO_2 fabricated by using mesoporous carbon was further investigated by TEM.

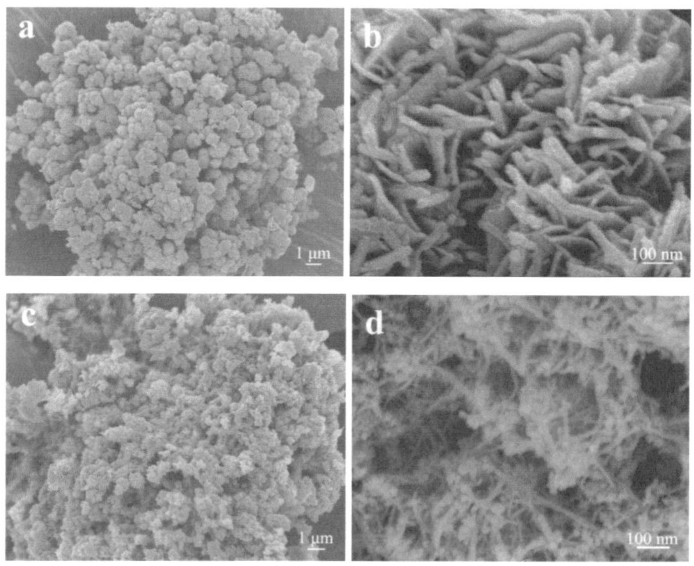

Fig. 11. SEM images of carbon sources and products. (a and b, MnO_2 obtained using CNTs; c and d, MnO_2 obtained using mesoporous carbon.)

A panoramic TEM image of the MnO_2 fabricated by using mesoporous carbon (Fig. 12a) gives more detail structure and morphology, in which lots of nanoflakelets and nanowires intercross with each other to form a slack MnO_2 cluster for a high specific surface area. A high-magnification TEM image (Fig. 11b) further illustrates that the flake has many small pores.

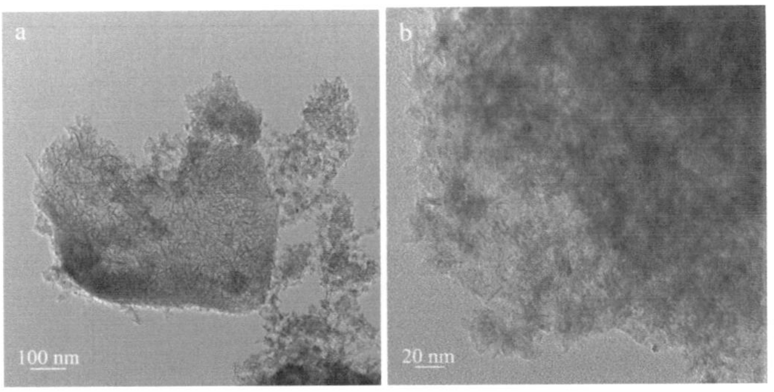

Fig. 12. TEM images of the MnO_2 obtained using mesoporous carbon.

3.2.2 Supramolecular template

In more recently, the discovery of M41S materials by the supramolecular templating mechanism ushered in a new era in synthesis chemistry (Wang et al., 2001; Lee et al., 2002). By using supramolecular template, some materials with high surface area, narrow pore size distribution and large pore volume could be prepared easily.
In our previous work (Xu et al.,2007), a kind of very slack mesoporous amorphous MnO_2 was prepared by using polyacry-lamide (PAM) and polyvinyl-alcohol (PVA) as template. The as synthesized material has large surface area and uniformed pore distribution is expected to favor ion transfer in the pore system and increase the MnO_2-electrolyte interfacial area, respectively.

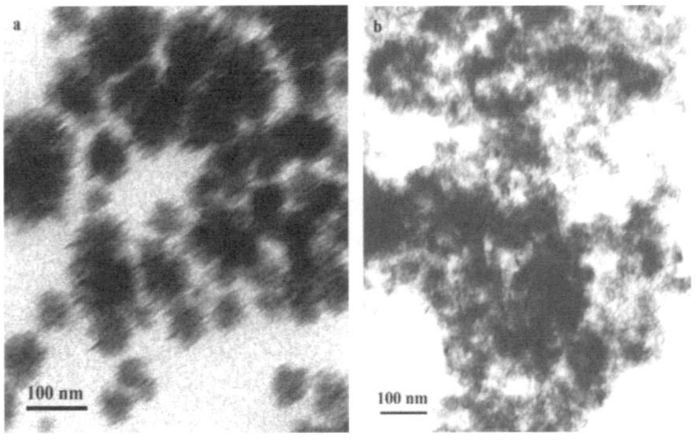

Fig. 13. TEM of MnO_2 obtained by without using supramolecular template (a) and using supramolecular template (b).

The morphology of the samples prepared under different condition was observed by transmission electron microscope (TEM). Figure 13a presents the morphology of the MnO_2 powders obtained without using supramolecular as template. It can be clearly seen that the sample was actually made up of small lamellar nanoparticles that agglomerate with each other to form the clusters. Figure 13b display the morphology of MnO_2 obtained using supramolecular as template. The MnO_2 is very loose and consisted of many small particles. Comparing the morphology of MnO_2 obtained under different conditions, great distinctions could be clearly seen, which indicates that the supramolecular template has played an important role in controlling the morphology of samples. The BET surface areas, calculated from adsorption isotherms, show that the surface area of the MnO_2 obtained using supramolecular template is 171 m^2/g, which increases significantly compared with that of the MnO_2 prepared without using supramolecular template (85 m^2/g).

3.2.3 Modified anodic aluminum oxide (AAO) template

Anodic aluminum oxide (AAO) template offers a promising route to synthesize a high surface area, ordered nanowire electrodes because of its advantages (Zhang et al., 2011; Zhao et al., 2011; Pan et al., 2004; Sui et al., 2001). Recently, amorphous manganese oxide nanowire arrays for high energy and power density electrodes were prepared by the AAO template method (West et al., 2004). But this method is difficult to be used in practical purpose due to the fragility of the AAO template. In our previous work (Xu et al., 2006), AAO films are successfully grown on Ti/Si substrate and is used as a template to synthesize high surface area and ordered MnO_2 nanowire array electrode for electrochemical capacitors. The experimental results indicated that this kind of template has unique electrodeposition properties and can bond well with the deposited materials.

From Fig. 14, we can find that many clusters protrude from the Ti/Si substrate which provide high surface area electrode. The clusters could result from the situation in which the nanowires are uncovered from the framework of the porous anodic alumina template but freestanding incompletely. When the porous anodic alumina template was dissolved away, the nanowires embedded in the template were released gradually and inclined to agglutinate together to minimize the system free energy. Fig. 14 also shows that the nanowires are abundant, uniform and well ordered in the large area.

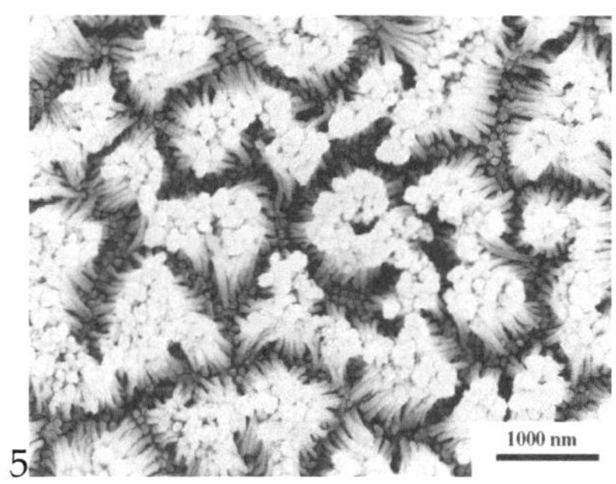

Fig. 14. FESEM image of MnO_2 nanowire arrays grown on AAO/Ti/Si substrate

Since lyotropic liquid crystals (LLCs) have unchanged topology of the phase throughout the process of the reaction and the calcination, it offers the ability to fabricate mesoporous materials with high specific surface areas (Elliott et al., 1999; Yamauchi et al., 2008). In the further work of Xu C. L. et al., lyotropic liquid crystals were introduced as a assisted template to prepared mesoporous MnO_2 (Xu et al., 2009). The mesoporous MnO_2 nanowire array architecture exhibits enhanced capacitance and charge/discharge performance, which is attributed to its intriguing architectures consisting of mesopores and nanowire arrays.

This approach will have potential applications in the fabrication of a wide range of mesoporous nanowire array materials. As can be seen from Fig. 15, the MnO_2 nanowires have uniform mesoporous structures and a continuous porous network is formed. The pores are patterned after the removal of the lyotropic liquid crystals template. The electron diffraction pattern (inset of Fig. 15(B)) demonstrates that the as-synthesized samples are amorphous.

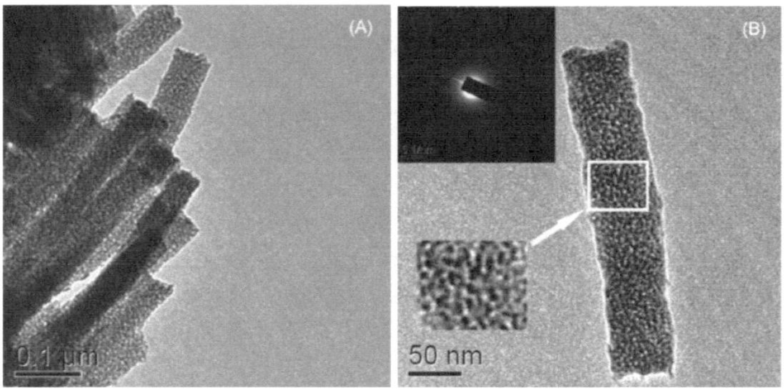

Fig. 15. TEM images of the mesoporous MnO_2 nanowires: (A) low magnification image, and (B) high magnification image. Inset is the electron diffraction pattern of nanowires in TEM.

3.3 Sol-gel process

The sol-gel process offers a number of potential advantages over traditional synthetic procedures. Sol-gel chemistry provides homogeneous mixing of reactants on the molecular level and can also be used to control shape, morphology, and particle size in the resulting products (Choy et al., 1999; Wu et al., 2004. Pang et al. conducted the first research on sol-gel processing of thin film MnO_2 electrodes for EC application in 2000 (Wei et al.2011; Pang et al., 2000; Lide et al., 2009). Stable colloidal MnO_2 was prepared by reducing tetrapropylammonium permanganate with 2-butanol or adding solid fumaric acid to 0.2 M $NaMnO_4$, mixing manganese acetate with a citric acid containing n-propyl alcohol at room temperature, reacting $KMnO_4$ with H_2SO_4 solutions (Wei et al., 2011; Chin et al.2002; Long et al., 2003; Lin et al., 2007; Lin et al., 2009; Lin et al., 2009). Sol-gel-derived nanoparticle MnO_2 thin films were then formed by either dip-coating or ''drop-coating'' colloidal MnO_2 directly onto conductive substrates, followed by calcination at various temperatures. The calcination temperature was found to have significant influence on the surface morphology, specific surface area, and specific capacitance of sol–gel derived MnO_2 thin films, may because the calcination at proper temperatures can generate high porosity and a well-defined pore size distribution through evaporation of the adsorbed water, solvent, and organic molecule (Wei et al., 2011).

4. Charge storage mechanism

The history of MnO_2 for electrochemical energy storage applications can be tracked to more than 100 years ago in a primary Zn/MnO_2 cell, which dominated in primary battery chemistry for centuries (Xu et al., 2010.) In such cells, MnO_2 is used as a cathode in the aqueous electrolytes and stores charge by a so-called double-injection process, which involves the insertion of protons from the aqueous solutions and the reduction of Mn in oxides by electrons from external circuit (Xu et al., 2010; Be´langer et al., 2008; Trasatti, 1991). This double-injection process could be expressed as:

$$MnO_2 + H^+ + e^- \rightarrow MnOOH \tag{1}$$

If the electrode continues to discharge, MnOOH could be further reduced at the second step:

$$MnOOH + 3H^+ + e^- \rightarrow Mn^{2+} + 2H_2O \tag{2}$$

As the ever-increasing needs in energy density of advanced consumer electronics and related technologies, high-voltage secondary lithium ion batteries have attracted more concerns. MnO_2 materials have been studied as cathode materials for lithium ion batteries, and spinel $LiMn_2O_4$ has even been commercialized for mass applications. In this case, the charge storage of MnO_2 in lithium ion batteries is considered to be: (Wu et al., 2004; Armand et al., 1985; Bao et al., 2005).

$$\frac{1}{2}Li^+ + \frac{1}{2}e^- + 2\lambda\text{-}MnO_2 \Leftrightarrow Li_{0.5}Mn_2O_4 \tag{3}$$

$$\frac{1}{2}Li^+ + \frac{1}{2}e^- + Li_{0.5}Mn_2O_4 \Leftrightarrow LiMn_2O_4 \tag{4}$$

Lithium cations from the electrolyte diffuses and stores in the tunnels of MnO_2; meanwhile, electrons travel to the neighboring Mn(IV) sites to balance the charge (Xu et al., 2010; Kadoma et al., 2007; Johnson et al., 2007). Correspondingly, Mn(IV) ions become Mn (III) and electrons are stored. The insertion process of lithium ions and variation of Mn valence between Mn(IV) and Mn(III) are totally reversible.

As soon as capacitive behavior of MnO_2 in the mild aqueous electrolytes was discovered, the researchers began to seriously consider the capacitive charge storage mechanism. An intercalation/chemisorption of proton into the solid phase MnO_2 mechanism was first proposed as following (Pang et al., 2000; Wu et al., 2004; Hu & Tsou, 2002; Bao et al., 2005).

$$MnO_2 + H^+ + e^- \leftrightarrow MnOOH \tag{5}$$

In the process of reversible insertion/desertion of protons (H^+) in MnO_2, Mn valence vary between Mn(IV) and Mn(III).

However, some phenomena, in which the specific capacitance of MnO_2 was directly affected by the species and concentrations of the alkaline metal cations with the same pH value, indicated that the charge storage mechanism is not as simple as expressed by Eq. (5).

It is found that the specific capacitance of MnO_2 electrode were affected by the nature of Li^+, Na^+, or K^+, significantly (Xu et al., 2008; Wen et al., 2004). Hence, a mechanism based on the surface adsorption of electrolyte cations (C^+) on MnO_2 has been proposed subsequently and be described as following: (Xu et al., 2010)

$$(MnO_2)_{surface} + C^+ + e^- \leftrightarrow (MnO_2^-C^+)_{surface} \tag{6}$$

Where $C^+ = H^+, Na^+, K^+, Li^+$.

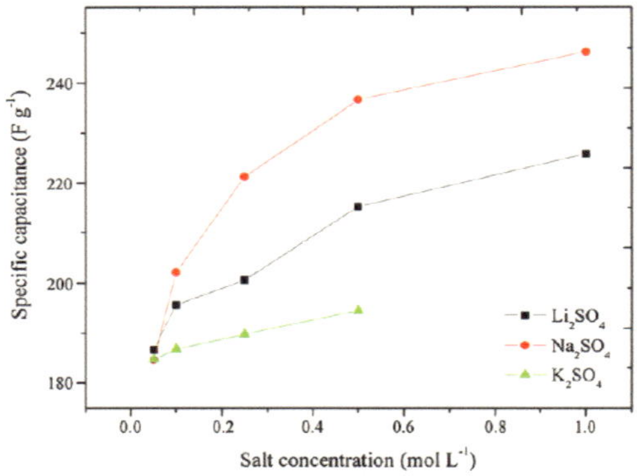

Fig. 16. Dependence of SC values on different electrolytes at sweep rate of 2 mV/s (Xu et al., 2008).

In order to further understand the charge storage mechanism of MnO_2 electrode, Toupin et al. investigated the variation of Mn valence in MnO_2 electrode during charge and discharge process by cyclic voltammetry and X-ray photoelectron spectroscopy (Toupin et al., 2004). A thin MnO_2 films deposited on a platinum substrate and thicker MnO_2 composite electrodes were used. X-ray photoelectron spectroscopy (XPS) measurements (Mn 3s and O 1s) with the thick composite electrodes did not reveal any change that could be assigned to a variation of the manganese valence, and, at this point, the charge storage mechanism could be based on electrostatic effects only. In fact, the charge storage would be similar to that observed for carbon electrodes (Toupin et al., 2004). On the other hand, a completely different XPS behavior was noticed for the thin film electrodes. Both the Mn 3s and O 1s spectra were consistent with manganese oxidation states of +3 and +4 for the reduced and oxidized forms, respectively. The XPS data also show that Na^+ cations from the electrolyte are involved in the charge storage process of MnO_2 thin film electrodes. The Na/Mn ratio for the reduced electrode is much lower than what is anticipated for charge compensation dominated by Na^+ and suggests the involvement of protons. The apparent discrepancy between the XPS data (Mn 3s and O 1s spectra) indicate that only a thin layer of MnO_2 is involved in the redox process and is electrochemically active. The authors think that this thin surface layer cannot be probed for the composite electrode may because this region is brought back to the chemical (oxidation) state of the bulk by internal redox interconversion (Xu et al., 2010; Toupin et al., 2004; Kuo et al., 2006).

In the same paper, a reversible expansion and shrinkage in lattice spacing of the oxide during charge transfer at manganese sites upon reduction/oxidation of MnO_2 was also demonstrated by in situ synchrotron x-ray diffraction. A similar result was also obtained for a layered MnO_2. An increase of the interlayer spacing from 0.70 to 0.72 nm upon electrochemical oxidation indicated that alkaline metal cations, Na^+ ions, intercalated in the 2D tunnels of MnO_2 (Xu et al.,2010; Athouel et al., 2008). The facts of lattice expansion and shrinkage during redox process indicated that the insertion of cations in the electrolytes predominates in the charge storage process of MnO_2.

Hence, based on surface adsorption of electrolyte cations C^+ (K^+, Na^+...) as well as proton incorporation, a more reasonable charge storage mechanism of MnO_2 electrode was proposed (Simon & Gogotsi, 2008).

$$MnO_2 + xC^+ + yH^+ + (x+y)e^- \leftrightarrow MnOOC_xH_y \quad (7)$$

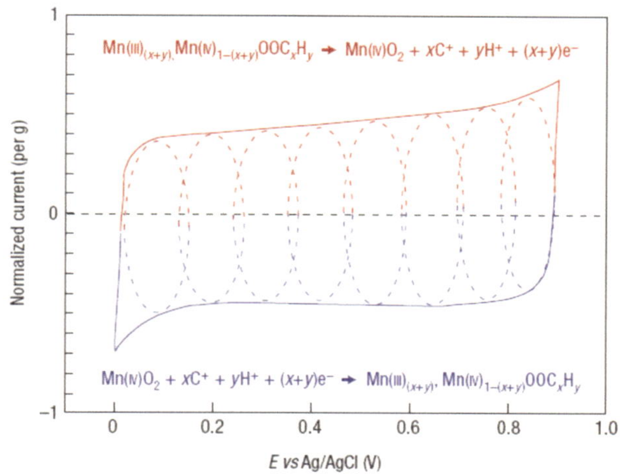

Fig. 17. Cyclic voltammetry of MnO_2-electrode cell in mild aqueous electrolyte (0.1 M K_2SO_4). This schematic of cyclic voltammetry for a MnO_2-electrode cell shows the successive multiple surface redox reactions leading to the pseudo-capacitive charge storage mechanism. The red (upper) part is related to the oxidation from Mn(III) to Mn(IV) and the blue (lower) part refers to the reduction from Mn(IV) to Mn(III) (Simon & Gogotsi, 2008).

As seen in Fig. 17, a cyclic voltammogram of a single MnO_2 electrode in mild aqueous electrolyte, the fast, reversible successive surface redox reactions define the behaviour of the voltammogram, whose shape is close to that of the EDLC.

It is really that the charge storage process of MnO_2 electrode is consists of double layer charging and pseudocapacitive surface redox process according to the research results of many scientists. However, as for the MnO_2 electrode, which process, double layer charging and pseudocapacitive surface redox, give more contribution to its capacitance.

In our previous work (Xu et al., 2007), we found although some double layer charging can significantly contribute to the measured capacitance due to the high surface area of the material, it cannot solely explain the whole capacitance of the electrode. The calculation of the pure double layer capacitance using the BET surface area is far lower than that measured in our work. So it is believed that the main part of the capacitance comes from the pseudocapacitive surface redox process. It may because when the specific surface area of MnO_2 electrode material increases, not only the double layer capacitance will increase, the redox active sites will increase subsequently, so the pseudocapacitance will increase significantly.

The capacitance of MnO_2 electrode also is affected by their microstructure. Ouassim G. et al., study the electrochemical performance of different structured MnO_2 in 0.5 M K_2SO_4 electrolyte (Ghodbane et al., 2009). The various curve shapes in Fig. 18 demonstrate the electrochemical response dependence on the MnO_2 microstructure and indicate that cooperative charge-storage mechanisms may exist. In fact, the typical rectangular shape of pseudocapacitive behaviors is observed for the pyrolusite, Ni-doped todorokitr (Ni-todorokite), ramsdellite, and spinel forms only. For the birnessite, cryptomelane, and octahedral molecular sieves (OMS-5) compounds, the shapes of the CV curves exhibit more or less pronounced redox waves. This behavior suggests that faradic phenomena occur during the charge-storage mechanism. The presence of redox waves during the charge/discharge process was already reported for several MnO_2-based electrodes (Hu & Tsou, 2002; Brousse et al., 2006; Devaraj & Munichandraiah, 2008; Ghodbane & Favier 2009; Chang et al., 2009; Lee et al., 2010). The electrochemical experiments in their work demonstrate that the crystallographic form of MnO_2 influences the electrochemical performance, and the small tunnel of β-MnO_2 was not suitable to store cations, while the large tunnel size of α-MnO_2 favors the storage of cations Devaraj & Munichandraiah, 2008; Cheng et al., 2010). In addition, the presence of other metal cations in the tunnel in advance hinders the diffusion and storage of the electrolyte cations leading to a decrease in capacitance. More recently, further research on the tunnel storage of MnO_2 even revealed that the aprotic ionic liquids anions with the far bigger diameters than that of alkaline metal ion, could also be stored in the tunnels of MnO_2 (Xu et al. 2010 ; Chang et al., 2009; Lee et al., 2010).

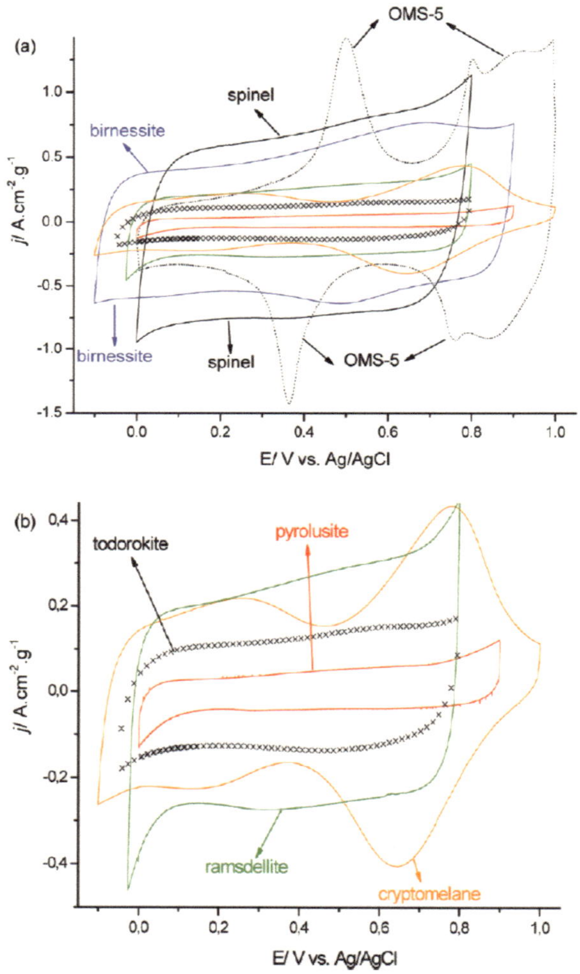

Fig. 18. (a) CV curves (fifth cycle) of different MnO_2 forms recorded in aqueous 0.5 M K_2SO_4 at 5 mV/s. (b) Enlargement of the CV curves of pyrolusite, Ni-todorokite, ramsdellite, and cryptomelane phases (Ghodbane et al., 2009).

More recently, a multivalent cation storage mechanism was proposed by Xu et al., they think not only univalent alkaline metal cations could store in the tunnels of MnO_2, bivalent alkaline-earth metal cations also are present in it (Xu et al., 2009; Xu et al., 2009). The realization of the theoretical capacity of a host material offered by its redox levels is determined by the number of intercalated ions concurrent with the charge transfer of the required number of electrons. From a vacancy viewpoint, polyvalent cations may open up the possibility of electrodes, which exhibit higher gravimetric capacity and energy density, because each multivalent intercalated ions cation will force multiple Mn^{4+} ions to become Mn^{3+} ion and store multiple electrons (Xu et al., 2010). This improvement method may be the most convenient and low cost for enhancing the capacitance of MnO_2-based supercapacitors.

5. Electrolytes system for MnO_2-based supercapacitors

5.1 Aqueous-based electrolytes

For aqueous electrolytes, the maximum operating voltage is theoretically limited by the electrolysis of water to 1.229 V (at 25 ^oC). Aqueous electrolytes tend to produce faster rates of charge/discharge due to the relatively high conductivity and low viscosity of concentrated solutions. Further advantages of aqueous electrolytes are their low cost and ease of manufacture in comparison with hygroscopic organic based and ionic liquid electrolytes (Hall et al., 2010).

The most commonly employed aqueous electrolytes for MnO_2-based supercapacitors are Na_2SO_4 and KOH. Other some neutral electrolytes, such as NaCl, KCl, Li_2SO_4, K_2SO_4 also been used as electrolytes. More recently, alkaline-earth metal cations (Mg^{2+}, Ca^{2+}, and Ba^{2+}) are considered for use as charge storage media of MnO_2-based supercapacitors to replace conventional alkaline cations (Li^+, Na^+, and K^+) (Xu et al., 2009). As for electrolytes, relatively concentrated electrolytes are required to minimize the equivalent series resistance and maximize power output. However, the use of concentrated electrolytes also increases the rate of self discharge displayed by the capacitor.

5.2 Organic-based electrolytes

Organic electrolytes allow for increased operating voltages of up to 3 V and a consequent increase in energy density. Typically, the operating voltage is set to 2.5 V to prevent oxidation of the electrolyte through over-charging. To ensure that these electrolytes can operate at the higher voltages they must be handled in an atmosphere free of water and oxygen. This ensures that the evolution of H_2 and O_2 gases at potential differences above 1.23 V does not occur (Hall et al., 2010; K€otz et al., 2000).

The two most common organic solvents used are propylene carbonate (PC) and acetonitrile (AN). AN was often preferred, as its ESR is a factor of three lower than that of PC. However, there is continuing controversy over the safety of AN, in particular in vehicles, due to its high toxicity and flammability. Research aimed at developing a non-toxic electrolyte with low resistivity has proven unsuccessful (Hall et al., 2010; Burke A., 2007). Most organic electrolyte based ECs available commercially utilize PC as the solvent. Tetraalkylammonium salts of anions PF_6^- and BF_4^- are preferred for the production of organic electrolytes due to their high solubility, higher dielectric conductivity, and good conductivity (relative to other organic solutions) (Hall et al., 2010).

5.3 Room temperature ionic liquid electrolytes

Ionic liquids are room-temperature liquid solvent-free electrolytes; their voltage window stability is thus only driven by the electrochemical stability of the ions. A careful choice of both the anion and the cation allows the design of high-voltage supercapacitors, and 3 V, 1,000 F commercial devices are already available (Simon & Gogotsi, 2008; Tsuda & Hussey, 2007). However, the ionic conductivity of these liquids at room temperature is just a few milliSiemens per centimetre, so they are mainly used at higher temperatures. For applications in the temperature range –30 °C to +60 °C, where batteries and supercapacitors are mainly used, and ionic liquids still fail to satisfy the requirements because of their low ionic conductivity. However, the research on ionic liquids for ECs is expected to have an important role in the improvement of capacitor performance in the coming years (Simon & Gogotsi, 2008).

5.4 Solid electrolytes

Replacing liquid electrolytes with solid electrolytes will enhance the safety of ECs, as it can solve problems associated with electrolyte leakage and corrosion. Gel polymer electrolytes (GPE) (Lee et al., 2009; Nagatomo et al., 1987; Peramunage et al., 1995) are capable of giving sufficient rigidity for enhancing device safety and provide much higher ionic conductivity than the polymeric solid electrolytes. There has recently been increasing interest in applying GPE to Ecs (Wada et al., 2004; Wada et al., 2006; Choudhury et al., 2006). More recently, a gel-type electrolyte has been tentatively applied in MnO_2-based systems. The gel electrolyte will prevent leaking of the electrolyte, which is superior to a liquid electrolyte. Although the manufacturing cost of such gel-type device will undoubtedly be high, it will open up a new direction to develop reliable MnO_2-based devices (Lee, et al., 2008).

6. The current status and development of MnO_2-based supercapacitors

Metal oxides present promising potential because of high specific capacitance at low resistance, possibly making it easier to construct high energy, high power supercapacitors. Many metal oxides, such as NiO, $Ni(OH)_2$, MnO_2, Co_3O_4, IrO_2, Fe_3O_4, TiO_2, SnO_2, V_2O_5 and MoO_3 have been studied as SC electrode materials. However, none of these oxides are used in commercial production and they are still in lab-scale research (Jaualakshmi & Balasubramanian, 2008). Among of them, manganese oxide has been paid more attention.

Manganese can be present in three different valence states and its oxides are highly complex. The theoretical capacitance of manganese oxides reaches to ~1380 F/g but the electrochemical reversibility of redox transition of manganese dioxide is usually too low to be applicable and the pure manganese dioxide possess poor capacitive response due to its high resistance of bulk manganese oxide (Xu et al., 2010). In spite of this, manganese oxides are seen to be potential useful materials for pseudocapacitors not only due to their low cost but also to their environmental friendliness (Jaualakshmi & Balasubramanian, 2008).

As reported in literatures, the specific capacitance could reach 600 F/g for thin MnO_2 films (Chin et al., 2002; Pang et al., 2000). It is because thin MnO_2 film significantly reduces the diffusion distances for the solid-state transport of insertion cations and overcomes the limitations of the poor electronic conductivity of MnO_2. However, because of extremely low loading mass, thin films seem to be limited to microsystems for energy storage such as integrated devices or to be used to analyze the charge storage mechanism (Jaualakshmi &

Balasubramanian, 2008). The specific capacitance only reaches 150-300 F/g for MnO_2 powders (Xu et al., 2010). Prasad and Miura reported a capacitance value between 400 and 621 F/g for amorphous electrolytic manganese dioxide and MnO_2-based mixed oxides (Prasad et al., 2004; Prasad & Miura, 2004). But in fact, higher capacitance is expected to obtain. To further improve the performance of MnO_2-based supercapacitor, it is necessary to design MnO_2 materials into nanoarchitectures or composites with other materials, such as porous carbon or polymer.

In order to improve the electrical conductivity, chemical stability, mechanical stability, and flexibility of MnO_2 electrodes, poly (o-phenylenediamine), polyaniline, polypyrrole, and polythiophene and their derivatives have been introduced. The first MnO_2/PANI composite electrode was prepared through a two-step electrochemical route: nanostructured MnO_2 was potentiodynamically deposited on a polyaniline (PANI) matrix synthesized through an electrochemical method. The specific capacitance of the obtained MnO_2/PANI composite electrode reach to 715 F/g, and its energy density is about 200 Wh/ kg at a charge-discharge current density of 5 mA /cm^2 (Prasad & Miura, 2004). Electrochemical co-deposition process and chemical polymerization method were also applied to prepare MnO_2/PANI composite (Sun & Liu, 2008); Zhou et al., 2005). Through addition PANI, the electrochemical performance of MnO_2 as supercapacitor electrode material was enhanced significantly. MnO_2/Ppy nanocomposite electrode materials were also prepared by using different process, and the experimental results indicated that the improvement of composite electrodes was attributed to good conductivity of Ppy (Sharma et al., 2010; Sharma, et al., 2008; Zhang et al., 2009). Polythiophene and its derivatives generally possess excellent electronic conductivity, high chemical stability, and reasonable mechanical flexibility, but they provide low electrochemical energy density (Liu & Lee, 08). The combination of MnO_2 with its high energy storage capacity and highly conductive and flexible. MnO_2/ PThs have advantages over just the sum of the individual components (Sharma & Zhai, 2009). Hence, the conducting polymer and MnO_2 nanoarchitecture are very promising materials with potential applications as ECs, and more efforts have been dedicated to incorporate polyaniline, polypyrrole, and polythiophene conductive polymers to generate MnO2–polymer composite electrodes with desirable morphologies and electrochemical performance (Wei et al., 2011).

It is noted that the nanostructured carbons produced under different synthesizing conditions exhibit a variety of physicochemical features and electrochemical properties (Wei et al., 2011). Nanostructured carbons such as CNTs, nanographites, carbon nanofoams, and ordered mesoporous carbons, especially, grapheme, are widely used as high surface area and excellent electron conducting architectures for MnO_2-based composite electrodes (Dong et al., 2006; Zhang et al., 2009; Lei et al., 2008). The combination of MnO_2 with different carbon is another method to improve the performance of MnO_2 based supercapacitors.

Hybrid systems offer an attractive alternative to conventional pseudocapacitors or EDLCs by combining a battery-like electrode (energy source) with a capacitor-like electrode (power source) in the same cell. An appropriate electrode combination can even increase the cell voltage, further contributing to improvement in energy and power densities (Simon & Gogotsi, 2008). In 2002, Hong et al. initially considered activated carbon (AC) as the negative electrode material for MnO_2-based supercapacitor (Be´guin et al., 2006). Except for the effort to find new negative electrodes, attempts have been made to use the new forms of MnO_2 as new positive electrode material. These capacitors showed promising performance. Another challenge for this system is to use organic electrolytes to reach higher cell voltage, thus improving the energy density.

7. Opportunities and challenges in future

ECs are being used across a vast swath of commercial and industrial equipment. Future generations of ECs are expected to come close to current Li-ion batteries in energy density, maintaining their high power density. This may be achieved by using ionic liquids with a voltage window of more than 4 V, by discovering new materials that combine double-layer capacitance and pseudo-capacitance, and by developing hybrid devices. ECs will have a key role in energy storage and harvesting, decreasing the total energy consumption and minimizing the use of hydrocarbon fuels. In some instances they will replace batteries, but in many cases they will either complement batteries, increasing their efficiency and lifetime, or serve as energy solutions where an extremely large number of cycles, long lifetime and fast power delivery are required.

Concerning the materials issues, MnO_2 is one of the most studied materials as a low-cost alternative to RuO_2. It has a very high theoretical capacitance of ~1380 F/g but suffer from poor conductivity, only 30% or even lower of theoretical value can be obtained. To further improve the performance of MnO_2-based supercapacitor, it is necessary to design MnO_2 materials into nanoarchitectures with desirable physicochemical features or composites with other materials, such as porous carbon or conductive polymer.

8. Acknowledgment

We would like to thank the Natural Science Foundation of China under Grant Nos. 20963011 and 21063014 for the financial support. We also appreciate the financial support from Science and Technology Key Project of Chinese Ministry of Education (210247).

9. References

Simon, P. & Gogotsi, Y. (2008). Materials for electrochemical capacitors. *Nature Materials,*Vol. 7, pp. 845-848, ISSN 1476-1122

Xu, C. J.; Kang, F. Y.; Li, B. H. & Du, H.D. (2010). Recent process on manganese dioxide based supercapacitors. *Journal of Materials Research,* Vol. 25, pp. 1421-1425, ISSN 0884-2914

Conway, B.E. (1999). Electrochemical Supercapacitor: Scientific Fundamentals and Technological Applications (Kluwer Academic/Plenum Publishers, *New York*). ISBN-10: 0306457369

Burke, A. (2000). Ultracapacitors: Why, how, and where is the technology. *Journal of Power Sources,* Vol.91, pp. 37-45, ISSN 0378-775

Yoda, S. & Lshihara, K. (1997). The advent of battery-based societies and the global environment in the 21st century. *Journal of Power Sources,* Vol. 68, pp. 3-7, ISSN 0378-7753

Becker, H. (1957). An electrochemical capacitor. U. S. *Patent* No. 2800616

Yoshino, A. Tsubata, T.; Shimoyamada, M.; Satake, H.; Okano, Y.; Mori S. & Yata, S. (2004). Development of a lithium-type advanced energy storage device. *Journal of The Electrochemical Society* ,Vol.151, pp.2180-2184, ISSN 0013-4651

Whittingham, M. S.; Savinelli, R. F. & Zawodzinski, T. (2004). Batteries and Fuel Cells.*Chemical Reviews*, Vol.104, pp.4243–4886, ISSN:0009-2665

Toupin M.; Brousse, T. & Bélanger, D. (2002). Influence of Microstucture on the Charge Storage Properties of Chemically Synthesized Manganese Dioxide. *Chemistry of Materials*, Vol.14, pp.3946-3952, ISSN 0897-4756

Wang, D.W.; Li F.; Liu M.; Lu, G.Q. & Cheng, H.M. (2008).3D Aperiodic Hierarchical Porous Graphitic Carbon for High Rate Electrochemical Capacitive Energy Storage. *Angew. Chemie. Int. Ed.* Vol.47, pp.373-376, ISSN 0570-0833

Reddy, A.L.M.; Shaijumon,M.M.; Gowda, S.R.& Ajayan, P.M. (2009). Coaxial MnO_2/Carbon Nanotube Array Electrodes for High-Performance Lithium Batteries.*Nano Lett.*, Vol.9 (3), February 2, 2009, pp.1002–1006 , ISSN 1530-6984

Conway, B.E.; Birss,V. & Wojtowicz, J. (1997). The role and utilization of pseudocapacitance for energy storage by supercapacitors. *Journal of Power Sources*, 66,1-14, ISSN 0378-7753

Lin, C.; Ritter, J.A. & Popov, B.N.(1998). Characterization of Sol-Gel-Derived Cobalt Oxide Xerogels as Electrochemical Capacitors. *Journal of The Electrochemical Society*, Vol.145, pp.4097-4103, ISSN 0013-4651

Xing, W.; Li, F.; Yan, Z.F.; & Lu, G.Q. (2004). Synthesis and electrochemical properties of mesoporous nickel oxide. *Journal of Power Sources*, Vol. 134, pp.324-329, ISSN 0378-7753

Jeong, Y.U. & Manthiram, A. (2002).Nanocrystalline Manganese Oxides for Electrochemical Capacitors with Neutral Electrolytes. *Journal of The Electrochemical Society*, Vol.149, pp.1419-1422, ISSN 0013-4651

Reddy, R.N. & Reddy, R.G. (2003). Sol–gel MnO_2 as an electrode material for electrochemical capacitors. *Journal of Power Sources*,Vol.124, pp.330-335, ISSN 0378-7753

Reddy, R.N. & Reddy, R.G. (2004). Synthesis and electrochemical characterization of amorphous MnO_2 electrochemical capacitor electrode material. *Journal of Power Sources*, Vol.132, pp. 315-320, ISSN 0378-7753

Long, J.W.; Young, A.L. & Rolison, D.R. (2003). Spectroelectrochemical Characterization of Nanostructured,Mesoporous Manganese Oxide in Aqueous Electrolytes. *Journal of The Electrochemical Society* , Vol.150, pp.1161-1165, ISSN 0013-4651

Qi, F.; Hirofumi, K. & Kenta, O. (1999). Manganese oxide porous crystals. *J. Mater. Chem.*, Vol.9, pp.319-333, ISSN 0897-4756

Shen, Y. F.; Zerger, R. P.; DeGuzman, R. N.; Suib S. L.; McCurdy L.; Potter, D. I. & O'Young C.L.(1993). Manganese Oxide Octahedral Molecular Sieves:Preparation, Characterization, and Applications. *Science*, Vol.23, pp.511-515., ISSN 0036-8075

Cao H. & Suib, S. L. (1994). Highly efficient heterogeneous photooxidation of 2-propanol to acetone with amorphous manganese oxide catalysts. *Journal of the American Chemical Society*, Vol.116, pp.5334-5342, ISSN 0002-7863

Tsuj, M. & Abe, M. (1985).Manganese(IV) Oxide and some Chromatographic Applications. *Bull. Chem. Soc. Jpn.* Vol.58, pp.1109-1114., ISSN 0009-2673

Wang X. & Li Y.D.(2002). Selected-Control Hydrothermal Synthesis of α- and β-MnO_2 Single Crystal Nanowires, *Journal of the American Chemical Society*, Vol. 124, pp.2880–2881, ISSN 0002-7863

Johnson, C.S.; Mansuetto, M. F.; Thackeray, M. M.; Yang, S.H. & Hackney, S. A.(1997). Stabilized Alpha-MnO_2 Electrodes for Rechargeable 3 V Lithium Batteries. *Journal of The Electrochemical Society* , Vol. 144, pp.2279-2283, ISSN 0013-4651

Rossouw, M. H.; Liles, D. C.; Thackeray, M. M.; I. W.; David, F & Hull, S. (1992). Alpha manganese dioxide for lithium batteries: A structural and electrochemical study. *Mater. Res. Bull.*, Vol.27, pp.221-230, ISSN 0025-5408

Brock, S.L; Duan, N.G; Tian, Z.R; Giraldo, O.; Zhou, H & Suib, S. L.(1998). A review of porous manganese oxide materials. *Chemistry of Materials*, Vol. 10, pp.2619-2628, ISSN 0897-4756

Devaraj, S. & Munichandraiah, N. (2008). Effect of Crystallographic Structure of MnO_2 on Its Electrochemical Capacitance Properties. *The Journal of Physical Chemistry C*, Vol.112, pp.4406-4417, ISSN 1932-7447

Portehault, D.; Cassaignon, S.; Baudrin, E. & Jolivet, J.P. (2009).Twinning Driven Growth of Manganese Oxide Hollow Cones through Self-Assembly of Nanorods in Water.*Cryst Growth Des*, Vol.9, pp.2562–2565, ISSN 1528-7483

Hill, L. I. & Verbaere, A. (2004) . On the structural defects in synthetic γ-MnO_2. *Journal of Solid State Chemistry*, Vol.177, pp.4706–4723, ISSN 0022- 4596

Wolff, D. & P. M. (1959) . Interpretation of some γ-MnO2 diffraction patterns, *Acta crystallogr*. Vol.12, pp.341−345, ISSN 1600-5724

Ma,R; Bando, Y; Zhang, L & Sasaki, T. (2004) . Layered MnO_2 Nanobelts: Hydrothermal Synthesis and the Electrochemical Measurements.*Adv. Mater.*, Vol. 16,pp.918–922, ISSN 0935-9648

Zhou, M; Zhang, X; Wei, J.M; Zhao, S. L.; Wang, L.&Feng, X. (2011).Morphology-Controlled Synthesis and Novel Microwave Absorption Properties of Hollow Urchinlike alpha-MnO_2 Nanostructures. *The Journal of Physical Chemistry C*, Vol.115, pp.1398–1402, ISSN 1932-7447

Zhang, J. T.; Chu, W.; Jiang, J. W. & Zhao, X. S. (2011). Synthesis, characterization and capacitive performance of hydrous manganese dioxide nanostructures. *Nanotechnology*, Vol.22, pp.125703- ISSN 0957-4484

Jiang, H; Zhao, T; Ma, J; Yan C. Y. & Li C. Z. (2011).Ultrafine manganese dioxide nanowire network for high-performance supercapacitors. *Chemical Communication*, Vol.47, pp.1264-1266, ISSN 1359-7345

Tang, X.H.; Li, H.J.; Liu, Z.H.; Yang, Z. P. & Wang, Z. L. (2011). Preparation and capacitive property of manganese oxide nanobelt bundles with birnessite-type structure. *Journal of Power Sources*, Vol. 196, pp.855-859, ISSN 0378-7753

Song, M. S; Lee, K. M; Lee, Y.R, Kim, I.Y.; Kim, T.W. Gunjakar, J.L. & Hwang S.J.(2010). Porously Assembled 2D Nanosheets of Alkali Metal Manganese Oxides with Highly Reversible Pseudocapacitance Behaviors, *The Journal of Physical Chemistry C*, Vol.114, pp.22134–22140, ISSN 1932-7447

Liu, Y; Zhang, M; Zhang, J.H; Qian Y.T. (2006).A simple method of fabricating large-area α-MnO_2 nanowires and nanorods .*Journal of Solid State Chemistry*, Vol.179, pp.1757-1761, ISSN 0022- 4596

Yang, M.Y.; Ni, P; Li, Y.; He, X. X. & Liu Z.H. (2010). Synthesis and electrochemical performance of beta-MnO_2 with semi-tubular morphology, *Mater. Chem. Phys.* , Vol.124, pp.155-158, ISSN 0254-0584

Xu, M.W.; Kong, L.B.; Zhou, W.J. & Li, H.L.(2007). Hydrothermal Synthesis and Pseudocapacitance Properties of Hollow Spheres and Hollow Urchins a-MnO_2. *The Journal of Physical Chemistry C,* Vol.111, pp.19141-19147, ISSN 1932-7447

Wei W.F.; Cui, X.W; Chen,W.X. & Ivey D. G. (2011), Manganese oxide-based materials as electrochemical supercapacitor electrodes. Chemical Society Reviews, Vol.40,pp.1697-1721, ISSN 0306-0012

Subramanian,V., Zhu H. W. & Wei B. Q.(2006) Electrochemiluminescence studies of the cyclometalated iridium(III) complexes with substituted 2-phenylbenzothiazole ligands. *Electrochem. Commun.,* Vol. 8, pp.827–832, ISSN 1388-2481

Yan, J.; Fan, Z. J.; Wei, T.; Cheng, J.; Shao, B.; Wang, K.; Song, L. P. & Zhang, M. L.(2009).Carbon nanotube/MnO_2 composites synthesized by microwave-assisted method for supercapacitors with high power and energy densities. *Journal of Power Sources,* Vol.194, pp.1202–1207, ISSN 0378-7753

Ma, S. B.; Nam, K. W.; Yoon, W. S.; Yang, X. Q.; Ahn, K. Y.; Oh, K. H. & Kim, K. B. (2008). Electrochemical properties of manganese oxide coated onto carbon nanotubes for energy-storage applications. *Journal of Power Sources,* Vol.178, pp.483–489. ISSN 0378-7753

Jiang, R. R; Huang, T.; Tang, Y.; Liu, J.; Xue, L. G.; Zhuang J. H. & Yu, A. H. (2009).Factors influencing MnO_2/multi-walled carbon nanotubes composite's electrochemical performance as supercapacitor electrode.*Electrochim. Acta,* Vol.54, pp.7173–7179, ISSN 0013-4686

Dong, X. P.; Shen, W. H.; Gu, J. L.; Xiong, L. M.; Zhu,Y. F.; Li Z. & Shi, J. L.(2006).MnO_2-Embedded-in-Mesoporous-Carbon-Wall Structure for Use as Electrochemical Capacitors. *The Journal of Physical Chemistry B,* Vol.110, pp.6015–6019, ISSN 1520-6106

Zhang, L. L.; Wei,T. X.; Wang W. J. & Zhao, X. S. (2009).Manganese oxide /carbon composite as supercapacitor electrode materials. *Microporous Mesoporous Mater.,* Vol. 123, pp.260–267, ISSN 1387-1811

Zhu, S.; Zhou, H.; Hibino, M. Honma, I. & Ichihara,M. (2005). Synthesis of MnO_2 nanoparticles confined in ordered mesoporous carbon using a sonochemical method. *Adv. Funct. Mater.* Vol.15, pp.381-386, ISSN 1616-3028

Lei, Y.; Fournier, C.; Pascal, J. L.; Favier, F.(2008). Mesoporous carbon-manganese oxide composite as negative electrode material for supercapacitors. *Microporous and Mesoporous Materials,* Vol. 110, pp.167-176, ISSN 1387-1811

Xu, M.W; Jia W.;Bao, S.J; Su, Z.& Dong, B.(2010). Novel mesoporous MnO_2 for high-rate electrochemical capacitive energy storage. *Electrochimica Acta,* Vol.55, pp.5117–5122, ISSN 0013-4686

Wang, Y.Q; Yin, L.X.; Palchik, O; Hacohen, Y.R; Koitypin, Y. & Gedanken, A. (2001). Rapid Synthesis of Mesoporous Yttrium−Zirconium Oxides with Ultrasound Irradiation. *Langmuir,* Vol.17, pp.4131–4133, ISSN 0743-7463

Lee, B.; Lu, D.; Kondo, J.N. & Domen, K. (2002). Three-Dimensionally Ordered Mesoporous Niobium Oxide. *Journal of the American Chemical Society,*124,pp.11256-11257, ISSN 0002-7863

Xu, M.W.; Zhao, D.D.; Bao, S. J. & Li, H. L.(2007). Mesoporous amorphous MnO_2 as electrode material for supercapacitor. *J Solid State Electrochem,* Vol. 11, pp.1101–1107, ISSN 0022-4596

Zhang, D.Z.; Luo, L.A.; Liao, Q.; Wang, H.; Fu H.B & Yao, J.N. (2011).Polypyrrole/ZnS Core/Shell Coaxial Nanowires Prepared by Anodic Aluminum Oxide Template Methods. *The Journal of Physical Chemistry,* Vol. 115 .pp.2360–2365, ISSN 1932-7447

Zhao, QA; Wen, G.H; Liu, Z.G, Fan, Y.B, Zou, G.T., Li L, Zheng, Rk, Ringer S.P & Mao ,H.K. (2011).Synthesis of dense, single-crystalline CrO_2 nanowire arrays using AAO template-assisted chemical vapor deposition. *Nanotechnology,* Vol.22,pp.125603- ISSN 0957-4484

Pan, H; Gao, H; Lim, S.H, Feng,Y.P & Lin, J. (2004). Highly ordered carbon nanotubes based on porous aluminum oxide.*Journal of Nanoscience and Nanotechnology,* Vol.4, pp.1014-1018, ISSN 1550-7033

Sui, Y. C., Acosta, D. R.; González-León,J. A.; Bermúdez, A. ; Feuchtwanger, J.; Cui, B. Z.; Flores, J. O. & Saniger J. M. (2001). Structure, thermal stability, and deformation of multibranched carbon nanotubes synthesized by CVD in the AAO template. The Journal of Physical Chemistry B , Vol.105, pp. 1523-1527, ISSN 1520-6106

West, W.C.; Myung,N.V.; Whitacre, J. F. & Ratnakumar, B.V. (2004). Electrodeposited amorphous manganese oxide nanowire arrays for high energy and power density electrodes. *Journal of Power Sources,* 126, pp.203–206. ISSN 0378-7753

Xu, C.L.; Bao; S.J.; Kong, L.B.; Li, H. & Li, H.L.(2006). Highly ordered MnO_2 nanowire array thin films on Ti/Si substrate as an electrode for electrochemical capacitor.*Journal of Solid State Chemistry,* Vol.179, pp.1351–1355, ISSN 0022-4596

Elliott, J. M.; Attard, G. S.; Bartlett, P. N.; Coleman, N. R. B.; Merckel, . D. A. S.; Owen, J. R. (1999). "Nanostructured platinum (H-I-ePt) films: Effects of electrodeposition conditions on film properties".*Chemical Materials.,* Vol.11, pp.3602-3609, ISSN 0897-4756

Yamauchi, Y.; Takai,A.; Nagaura,T.; Inoue, S. & Kurod,K. (2008) Pt Fibers with Stacked Donut-Like Mesospace by Assembling Pt Nanoparticles: Guided Deposition in Physically Confined Self-Assembly of Surfactants. *Journal of the American Chemical Society,* Vol.130, pp.5426–5427, ISSN 0002-7863

Xu, C.L.; Zhao, Y.Q.; Yang, G.W.; Li, F.S & Li, H.L. (2009). Mesoporous nanowire array architecture of manganese dioxide for electrochemical capacitor applications.Chemical Communications, Vol.48, pp.7575-7577, ISSN 1359-7345

Choy, J.H.; Kim, D.H.; Kwon, C.W.; Hwang, S.J. & Kim, Y.I. (1999) Physical and electrochemical characterization of nanocrystalline $LiMn_2O_4$ prepared by a modified citrate route. *Journal of Power Sources*, Vol.77, pp.1-11, ISSN 0378-7753

Wu, X.M.; Li, X.H.; Xiao, Z.B.; Liu, J.B.; Yan, W.B. & Ma, M.Y. (2004). Synthesis and characterization of $LiMn_2O_4$ powders by the combustion-assisted sol–gel technique.*Materials Chemistry and Physics,*Vol.84, pp.182-186, ISSN 0254-0584

Pang, S. C.; Anderson, M. A. & Chapman T. W., (2000). Novel Electrode Materials for Thin-Film Ultracapacitors: Comparison of Electrochemical Properties of Sol-Gel-Derived and Electrodeposited Manganese Dioxide. *Journal of The Electrochemical Society,* Vol.147, pp.444–450, ISSN 0013-4651

Lide, D. R.; Handbook of P. Ragupathy, Park, D. H.; Campet, G.; Vasan, H. N. ; Hwang,S. J.; Choy, J. H. & Munichandraiah, N. (2009). Remarkable Capacity Retention of Nanostructured Manganese Oxide upon Cycling as an Electrode Material for Supercapacitor. *The Journal of Physical Chemistry C,* Vol.113, pp. 6303–6309, ISSN 1932-7447

Chin,S. F.; Pang, S. C. & Anderson, M. A. (2002).Material and Electrochemical Characterization of Tetrapropylammonium Manganese Oxide Thin Films as Novel Electrode Materials for Electrochemical Capacitors. *Journal of The Electrochemical Society,* Vol. 149, pp.A379–A384, ISSN 0013-4651

Long, J. W.; Young, A. L. & Rolison, D. R.(2003). Spectroelectrochemical Characterization of Nanostructured,Mesoporous Manganese Oxide in Aqueous Electrolytes. *Journal of The Electrochemical Society,* Vol.150, A1161–A1165, ISSN 0013-4651

Lin, C. K.; Chuang,K. H.; Lin,C. Y.; Tsay, C. Y. & Chen, C. Y.(2007). Manganese oxide films prepared by sol–gel process for supercapacitor application .*Surface and Coatings Technology,* Vol.202, pp.1272–1276, ISSN 0257-8972.

Lin, C. C. & Chen, H. W. (2009). Coating manganese oxide onto graphite electrodes by immersion for electrochemical capacitors. *Electrochim. Acta,* Vol.54,pp.3073–3077, ISSN 0013-4686

Lin, C. C. & Chen, H. W. (2009). Electrochemical characteristics of manganese oxide electrodes prepared by an immersion technique, *Journal of Applied Electrochemistry,* Vol.39, pp.1877–1881, ISSN 0021-891X

Be´langer, D.; Brousse,T. & Jeffrey,W.L.(2008). Manganese dioxides: Battery materials make the leap to electrochemical capacitors. *The Electrochemical Society Interface • Spring,* pp.49-52, ISSN 1064-8208

Trasatti, S. (1991). Physical electrochemistry of ceramic oxides. *Electrochim. Acta* Vol.36, pp.225-241, ISSN 0013-4686

Wu, X. M.; Li, X.H.; Xiao, Z.B.; Liu, J.B.; Yan, W.B. & Ma, M.Y. (2004). Synthesis and characterization of LiMn2O4 powders by the combustion-assisted sol–gel technique . *Materials Chemistry and Physics,* Vol. 84, pp.182-186, ISSN 0897-4756

Armand, M.; Dalard, F.; Reroo, D. & Mouliom,C. (1985). Modelling the voltammetric study of intercalation in a host structure: application to lithium intercalation in RuO_2. *Solid State Ionics,* Vol.15, pp.205-210, ISSN 0167-2738

Bao, S.J.; Liang, Y.Y. & Li,H.L. (2005).Synthesis and electrochemical properties of $LiMn_2O_4$ by microwave-assisted sol–gel method.*Materials Letters,* Vol. 59,pp.3761 - 3765, ISSN: 0167-577X

Kadoma,Y.; Oshitari, S.; Ui,K. & Kumagai,N.(2007).Synthesis of hollandite-type $LixMnO_2$ by Lit ion-exchange in molten salt and lithium insertion characteristics. *Electrochim. Acta,* Vol.53, pp.1697-1702, ISSN 0013-4686

Johnson, C.S. (2007). Development and utility of manganese oxides as cathodes in lithium batteries. *Journal of Power Sources,* Vol. 165, pp.559-565. ISSN 0378-7753

Wu, M.Q.; Snook, G.A.; Chen, G.Z. & Fray, D.J. (2004). Redox deposition of manganese oxide on graphite for supercapacitors. *Electrochemistry Communications,* Vol.6, pp.499-504, ISSN 1388-2481

Hu, C.C. & Tsou, T.W.(2002).Ideal capacitive behavior of hydrous manganese oxide prepared by anodic deposition. *Electrochemistry Communications,* Volume 4, Issue 2, February 2002, pp.105-109 ISSN 1388-2481

Bao, S.J.; He, B.L.; Liang, Y.Y.; Zhou,W.J. & Li, H.L. (2005). Synthesis and electrochemical characterization ofamorphous MnO_2 for electrochemical capacitor.*Materials Science and Engineering A,* Vol.397, pp.305–309, ISSN: 0921-5093.

Xu, C.; Li, B.; Du,H.; Kang, F. & Zeng, Y.(2008).Supercapacitive studies on amorphous MnO_2 in mild solutions. *Journal of Power Sources,* Vol. 184, pp.691-694, ISSN 0378-7753

Wen, S.; Lee, J.W.; Yeo, I.H.; Park, J. & Mho, S.I. (2004). The role of cations of the electrolyte for the pseudocapacitive behavior of metal oxide electrodes, MnO_2 and RuO_2. *Electrochimica Acta*, Vol.50, pp.849-855, ISSN 0013-4686

Toupin, M.; Brousse, T. & Be´langer, D.(2004).Charge storage mechanism of MnO_2 electrode used in aqueous electrochemical capacitor. *Chemistry of Materials*, 16, pp.3184–3190, ISSN 0897-4756

Kuo, S.L. & Wu, N.L. (2006).Investigation of pseudocapacitive charge-storage reaction of $MnO_2 \bullet nH_2O$ supercapacitors in aqueous electrolytes. *Journal of The Electrochemical Society*, Vol.153, pp.1317 -1324, ISSN 0013-4651

Athouel, L.; Moser,F.; Dugas, R.; Crosnier, O.; Belanger, D. & Brousse, T.(2008).Variation of the MnO_2 birnessite structure upon charge/discharge in an electrochemical supercapacitor electrode in aqueous Na_2SO_4 electrolyte. *The Journal of Physical Chemistry C*, Vol.112, pp.7270–7277, ISSN 1932-7447

Cheng, F.Y.; Su,Y.; Liang, J.; Tao Z.L. & Chen J. (2010). MnO_2-Based Nanostructures as Catalysts for Electrochemical Oxygen Reduction in Alkaline Media. *Chemistry of Materials*,Vol. 22 (3), pp 898–905, ISSN 0897-4756

Brousse, T.; Toupin, M.; Dugas, R.; Athoue¨l, L.; Crosnier, O. & Be´langer, D. (2006). Crystalline MnO_2 as Possible Alternatives to Amorphous Compounds in Electrochemical Supercapacitors, *Journal of The Electrochemical Society*, Vol.153, pp.2171-2180, ISSN 0013-4651

Devaraj, S. & Munichandraiah, N.(2008).Effect of crystallographic structure of MnO_2 on its electrochemical capacitance properties. *The Journal of Physical Chemistry C*, Vol.112, pp.4406–4417, ISSN 1932-7447

Ghodbane, O. Pascal, J.L. & Favier, F. (2009).Microstructural effects on charge-storage properties in MnO_2-Based electrochemical supercapacitors. *Applied Materials & Interfaces*, Vol.1, pp.1130–1139, ISSN: 1944-8244

Chang, J.K.; Lee, M.T.; Tsai, W.T.; Deng, M.J.& Sun I.W. (2009).X-ray photoelectron spectroscopy and in situ x-ray absorption spectroscopy studies on reversible insertion-desertion of dicyanamide anions into-from manganese oxide in ionic liquid. *Chemistry of Materials*, Vol.21,pp. 2688-2696., ISSN 0897-4756

Lee, M.T.; Tsai, W.T.; Deng, M.J.; Cheng, H.F.; Sun, I.W. & Chang J.K. (2010). Pseudocapacitance of MnO_2 originates from reversible insertion-desertion of thiocyanate anions studied using in situ x-ray absorption spectroscopy in ionic liquid electrolyte. *Journal of Power Sources*, Vol.195, pp.919-922, ISSN 0378-7753

Xu, C.; Li, B.; Du,H.; Kang, F. & Zeng Y.(2009). The reversible insertion properties of zinc ion into manganese dioxide and its application for energy storage. *Electrochemical and Solid-State Letters*, Vol.12, pp.A61-A65, ISSN 1099-0062

Xu, C.J.; Du, H.D; Li, B.H.; Kang, F. & Zeng Y. (2009). Asymmetric activated carbon-manganese dioxide capacitors in mild aqueous electrolytes containing alkaline-earth cations. *Journal of The Electrochemical Society* , Vol.156, pp. A435-A441, ISSN 0013-4651

Hall, P.J.; Mirzaeian, M.; Isobel Fletcher, S.; Sillars, F.B. Anthony; Rennie, J. R. Shitta-Bey, G. O.; Wilson, G.; Cruden, A & Carter, R. (2010). Energy storage in electrochemical capacitors: designing functional materials to improve performance. *Energy & Environmental Science*, Vol.3, pp.1238–1251, ISSN:1754-5692

K€otz, R. & Carlen, M. (2000). Principles and applications of electrochemical capacitors. *Electrochim. Acta,* Vol. 45, pp.2483–2498, ISSN 0013-4686

Burke, A. (2007). R & Dconsiderations for the performance and application of electrochemical capacitors. *Electrochim. Acta,* Vol.53, pp.1083–1091, ISSN 0013-4686

Tsuda, T. & Hussey, C. L. (2007).Electrochemical applications of room-temperature ionic liquids. *The Electrochemical Society interface,* Vol.16, pp.42–49, ISSN 1064-8208

Lee, K.T.; Lee, J.F. & Wu N.L.(2009). Electrochemical characterizations on MnO_2 supercapacitors with potassium polyacrylate and potassium polyacrylate-co-polyacrylamide gel polymer electrolytes. *Electrochimica Acta* Vol. 54, pp.6148-6153, ISSN 0013-4686

Nagatomo, T.; Ichikawa & Omato O.(1987). All-Plastic Batteries with Polyacetylene Electrodes. *Journal of The Electrochemical Society,*Vol.134 pp.305.-308, ISSN 0013-4651

Peramunage, D.; Pasquariello, D.M. & Abraham, K.M. (1995). Polyacrylonitrile-Based Electrolytes with Ternary Solvent Mixtures as Plasticizers. *Journal of The Electrochemical Society* , Vol.142, pp.1789-1798, ISSN 0013-4651

Wada, H.; Nohara, S.; Furukawa, N.; Inoue, H.; Sugoh, N.; Iwasaki, H.; Morita, M.; C. Iwakura, (2004). Electrochemical characteristics of electric double layer capacitor using sulfonated polypropylene separator impregnated with polymer hydrogel electrolyte. *Electrochimica Acta,* Vol. 49, pp.4871-4875, ISSN 0013-4686

Wada, H.; Yoshikawa, K.; Nohara, S.; Furukawa, N.; Inoue, H.; Sugoh, N.; Iwasaki, H. & Iwakura, C. (2006). Electrochemical characteristics of new electric double layer capacitor with acidic polymer hydrogel electrolyte. *Journal of Power Sources,* Vol. 159, pp.1464-1467, ISSN 0378-7753

Choudhury, N.A.; Sukula, A.K.; Sampath, S. & Pitchumani, S. (2006).Cross-Linked Polymer Hydrogel Electrolytes for Electrochemical Capacitors. *Journal of The Electrochemical Society,*Vol.153, pp.614-616,ISSN 0013-4651

Lee, K.T. & Wu, N.L.(2008).Manganese oxide electrochemical capacitor with potassium poly(acrylate) hydrogel electrolyte. *Journal of Power Sources,*Vol.179 pp.430–434. ISSN 0378-7753

Jaualakshmi, M; & Balasubramanian, K. (2008). Simple capacitors tosupercapacitors-An Overview. *International Journal of Electrochemical Science,* Vol.3,1196-1217, ISSN 1452-3981

Prasad, K. R. & Miura N.,(2004). Potentiodynamically deposited nanostructured manganese dioxide as electrode material for electrochemical redox supercapacitors. *Journal of Power Sources,* Vol. 135, pp.354-360, ISSN 0378-7753

Prasad, K. R. & Miura, N. (2004). Electrochemically synthesized MnO_2-based mixed oxides for high performance redox supercapacitors. *Electrochemistry Communications,* Vol.6, pp.1004-1008, ISSN 1388-2481

Prasad K. R. & Miura, N. (2004).Polyaniline-MnO_2 Composite Electrode for High Energy Density Electrochemical Capacitor. *Electrochemical and Solid-State Letters,* Vol. 7,pp. 425–428, ISSN 1099-0062

Sun,L. J. & Liu, X. X.(2008). *European Polymer Journal,* 44, pp.219–224. *ISSN: 0014-3057*

Zhou,Z. H.; Cai,N. C. & Zhou, Y. H. (2005).*Mater. Chem. Phys.,* Vol. 94, pp.371–375, ISSN 0254-0584

Sharma, R. K.; Karakoti, A.; Seal S. & Zhai, L.(2010). Multiwall carbon nanotube-poly(4-styrenesulfonic acid) supported polypyrrole/manganese oxide nano-composites for high performance electrochemical electrodes. *Journal of Power Sources,* Vol. 195, pp.1256–1262, ISSN 0378-7753

Sharma, R. K.; Rastogi, A. C. & Desu S. B.(2008).Manganese oxide embedded polypyrrole nanocomposites for electrochemical supercapacitor. *Electrochimica Acta,* Vol.53, pp.7690–7695, ISSN 0013-4686

Zhang, X.; Yang W. S. & Ma, Y. W. (2009).Synthesis of Polypyrrole-Intercalated Layered Manganese Oxide Nanocomposite by a Delamination/Reassembling Method and Its Electrochemical Capacitance Performance.*Electrochem. Solid-State Lett.,* Vol.12, pp.95–98, ISSN 1099-0062.

Liu, R. & Lee,S. B (2008). MnO_2/Poly(3,4-ethylenedioxythiophene) Coaxial Nanowires by One-Step Coelectrodeposition for Electrochemical Energy Storage, *Journal of the American Chemical Society,* Vol. 130, pp.2942–2943, ISSN 0002-7863.

Sharma, R. K. & Zhai, L. (2009). Multiwall carbon nanotube supported poly(3,4-ethylenedioxythiophene)/manganese oxide nano-composite electrode for super-capacitors, *Electrochimica Acta,*Vol. 24, pp.7148–7155, ISSN 0013-4686

Dong, X. P.; Shen, W. H.; Gu, J. L.; Xiong,L. M.; Zhu, Y. F.; Li, Z. & Shi, J. L.(2006).MnO_2-Embedded-in-Mesoporous-Carbon-Wall Structure for Use as Electrochemical Capacitors. *The Journal of Physical Chemistry B,* Vol. 110, pp.6015–6019, ISSN 1520-6106

Zhang, L. L.; Wei, T. X.; Wang, W. J. & Zhao,X. S. (2009). Manganese oxide–carbon composite as supercapacitor electrode materials. Microporous Mesoporous Mater., Vol.123, pp.260–267, ISSN: 1387-1811.

Lei, Y.; Fournier, C.; Pascal J. L. & Favier,F. (2008). Mesoporous carbon–manganese oxide composite as negative electrode material for supercapacitors. *Microporous Mesoporous Mater.* Vol.110, pp.167–176, ISSN: 1387-1811.

Be´guin, F.; Khomenko, V. & Raymundo-Pinero E.(2006).Optimization of an asymmetric manganese oxide/activated carbon capacitor working at 2V in aqueous medium. *Journal of Power Sources,* Vol.153, pp.183-190, ISSN 0378-7753

13

High Temperature PEM Fuel Cells Based on Nafion®/SiO_2 Composite Membrane

XiaoJin Li, ChangChun Ke, ShuGuo Qu,
Jin Li, ZhiGang Shao and BaoLian Yi
*Laboratory of Fuel Cell System & Engineering, Dalian Institute of Chemical Physics,
Chinese Academy of Sciences,
China*

1. Introduction

1.1 High temperature PEMFC

Polymer electrolyte membrane fuel cell (PEMFC) is considered to be one of the most promising alternative energy conversion devices for motor vehicles and other stationary applications, due to its quick start, high energy efficiency, and environmentally friendly qualities(Marban and Vales-Solis 2007).

At present, most PEMFCs are operated at <80°C, due to the dependence of aboard used per-fluorinated sulfonic acid membrane (such as Nafion® series) on water. Even so, operating PEMFC at a high temperature (>100°C) has many benefits (Yang, Costamagna et al. 2001; Li, He et al. 2003). Firstly, it avoids the existence of two phase flow in the flow field, thus enhances the stability & reliability of PEMFCs system. Then, operating PEMFC at a high temperature reduces the power loss caused by the electrochemical polarization of cathode. In addition, high temperature operation is also beneficial to make use of the exhaust heat of PEMFC system effectively and enhance the CO endurance of anode(Yang, Costamagna et al. 2001), etc.

The key point of the HT-PEMFC is to develop a type of proton exchange membrane that can be endurable to the high temperature and still maintain high proton conduction. Because the most widely used commercial Nafion membrane is not competent for operating at high temperature due to dehydration. To solve this problem, many kinds of solutions have been proposed. Generally, these solutions can be divided into three catalogs as followings: a) to incorporate hydrophilic or proton conductive inorganic nano particles into the Nafion matrix to prepare so-called inorganic-organic composite membrane(Deng, Moore et al. 1998; Adjemian, Lee et al. 2002; Adjemian, Srinivasan et al. 2002; Costamagna, Yang et al. 2002; Shao, Joghee et al. 2004; Xu, Lu et al. 2005; Yonghao Liu, Baolian Yi et al. 2005; Adjemian, Dominey et al. 2006; Shao, Xu et al. 2006; Alberti, Casciola et al. 2007; Lin, Yen et al. 2007; Casciola, Capitani et al. 2008; Jian-Hua, Peng-Fei et al. 2008; Jin, Qiao et al. 2008; Jung, Weng et al. 2008; Rodgers, Shi et al. 2008; Wang, Yi et al. 2008; Yuan, Zhou et al. 2008; Santiago, Isidoro et al. 2009; Yan, Mei et al. 2009); b) to substitute the water in Nafion with non-volatile or low-volatile polar solvent; c) to prepare new material that can conduct proton independent of water(Deng, Moore et al. 1998; Adjemian, Lee et al. 2002; Yonghao Liu, Baolian Yi et al. 2005; Lin, Yen et al. 2007; Tang, Wan et al. 2007; Yen, Lee et al. 2007; Rodgers, Shi et al. 2008).

1.2 Nafion inorganic composite membrane

During the above solutions for high-temperature operation of PEMFC, the first solution that could be also called "Nafion-inorganic composite" (Costamagna, Yang et al. 2002; Mauritz, Mountz et al. 2004; Shao, Joghee et al. 2004; Adjemian, Dominey et al. 2006; Shao, Xu et al. 2006; Jung, Weng et al. 2008; Wang, McDermid et al. 2008; Wang, Zhao et al. 2008; Yan, Mei et al. 2009) is most widely investigated at present. Many composite membranes of this type were reported in literatures, such as Nafion composite membranes with SiO_2 (Mauritz, Stefanithis et al. 1991; Shao, Joghee et al. 2004; Yonghao Liu, Baolian Yi et al. 2005; Shao, Xu et al. 2006; Tang, Wan et al. 2007; Yen, Lee et al. 2007; Jin, Qiao et al. 2008; Jung, Weng et al. 2008; Rodgers, Shi et al. 2008; Wang, McDermid et al. 2008; Yuan, Zhou et al. 2008; Jin, Qiao et al. 2009), sulfonated-SiO_2(Wang, Zhao et al. 2008), TiO_2 (Jian-Hua, Peng-Fei et al. 2008; Santiago, Isidoro et al. 2009), and ZrP (Alberti, Casciola et al. 2007), and etc(Alberti and Casciola 2003; Jones and Rozière 2003). Among these composite membranes, Nafion/SiO_2 composite membrane was most extensively evaluated and promising, for its relative higher performance and less incidental problems. Several preparation methods have been reported, such as solution-recast route (Shao, Joghee et al. 2004), sol-gel method, self-assembling method (Tang, Wan et al. 2007), in-situ sol-gel method (Adjemian, Lee et al. 2002; Yonghao Liu, Baolian Yi et al. 2005), and so on. Among these methods, in-situ sol-gel method is a promising process route, because it could obtain composite membrane with higher uniformity, smaller SiO_2 particles and easy to carry out. K.A. Mauritz and co-workers (Mauritz, Stefanithis et al. 1991; Mauritz, Stefanithis et al. 1995) first proposed this method, then K.T. Adjemian (Adjemian, Srinivasan et al. 2001; Adjemian, Lee et al. 2002; Adjemian, Srinivasan et al. 2002) and many other investigators(Yonghao Liu, Baolian Yi et al. 2005; Yu, Pan et al. 2007) applied Nafion/SiO_2 composite membrane prepared by this or improved method to PEMFC and DMFC.

2. Nafion®/SiO_2 composite membrane fabrication

2.1 Membrane fabrication

Nafion/SiO_2 composite membranes were prepared via an in-situ sol-gel reaction of tetraethyl-orthosilicate (TEOS), as that of Ref (Mauritz, Stefanithis et al. 1991). In our previous work (Ke, Li et al. 2010; Ke, Li et al. 2011), a technique in controlling silicate nanoparticle diameter in preparation of Nafion/SiO_2 composite membranes was put forward. The detailed process is shown as follows. Firstly, the Nafion (NRE212) membrane (DuPont, USA) was dried in the vacuum drying oven at 80°C for 12h. Then, the membrane was dipped into the CH_3OH/H_2O solution (30°C) and kept in the CH_3OH/H_2O solution for 1 h. After that, the sample was taken out and the remnant liquid on the surface of the membrane was rubbed out with filter paper. The sample was then immerged into CH_3OH/TEOS solution (30°C) to carry out the in-situ sol-gel reaction for 3 minutes. The diameter of SiO_2 nanoparticle incorporated in Nafion was controlled by changing the in-situ sol-gel reaction reactant concentrations. After reaction, the sample was kept in the vacuum drying oven at 80°C for 48 h, and then Nafion/SiO_2 composite membrane was obtained. In this work, the content of the SiO_2 incorporated into Nafion is 8.5wt.% with a statistical diameter of 10 nm.

Before the subsequent measurements, all the composite membranes were pretreated with the process as follows. Firstly, membranes were kept in H_2O_2 (5wt%, 80°C) for 1 h, followed by rinsing them with de-ionized water (80°C) for two times. Then membranes were soaked in H_2SO_4 (0.5M, 80°C) for 1 h. Finally, membranes were rinsed with de-ionized water (80°C) repeatedly until the PH of the washing water was around 7.

For the ex-situ research of the mechanism of sulfonation, the powder of SiO_2 nano-particles was also sulfonated by concentrated H_2SO_4 (98%, 80 °C), just as the sulfonation process of Nafion/SiO_2 composite membrane. The sulfonation time was 10 h, for the powder SiO_2.

2.2 Membrane sulfonation

The above obtained Nafion/SiO_2 composite membranes were dried in vacuum drying chamber for 24 h. Then the samples were soaked in concentrated H_2SO_4 (98%, 80 °C) for certain time (4, 10, and 16 h) to sulfonate the composite membrane. The concentrated H_2SO_4 treated membranes were then soaked in de-ionized water (80 °C) for 12 h, and rinsed with de-ionized water until the washing water exhibited neutral.

Nafion/SiO_2 sulfonated for 4, 10 and 16 h are signed as Nafion/S-SiO_2-4h, Nafion/S-SiO_2-10h and Nafion/S-SiO_2-16h, respectively. For the ex-situ research of the mechanism of sulfonation, the powder of SiO_2 nano-particles was also sulfonated by concentrated H_2SO_4 (98%, 80 °C), just as the sulfonation process of Nafion/SiO_2 composite membrane. The sulfonation time was 10 h, for the powder SiO_2.

3. Spectroscopic studies of Nafion/SiO_2 composite membranes

3.1 FT-IR spectra

Spectroscopic methods are powerful tools for analysis of the chemical structure of the sample. In order to understand the mechanism of the sulfonation of the Nafion/SiO_2 composite membrane, several spectroscopic methods are used synthetically.

The FT-IR spectra of the membranes were collected from 400 cm^{-1} to 1200 cm^{-1}, using a Bruker Vector22 (Bruker Optics, Germany) FT-IR spectrometer at a resolution of 4 cm^{-1} in absorption mode. FT-IR of the powder SiO_2 and sulfonated SiO_2 (S-SiO_2) nano-particles were also performed, in order to avoid the background signal of the Nafion matrix. The IR spectra of the untreated SiO_2 and the sulfonated S-SiO_2 were recorded with potassium bromide tabletting in transmission mode.

Figure 1 (a) exhibits the FT-IR spectra of the Nafion, Nafion/SiO_2, Nafion/S-SiO_2-4h, Nafion/S-SiO_2-10h, and Nafion/S-SiO_2-16h. In fact, as we can see, it is a little difficult to distinguish the structure change of the Nafion/SiO_2 composite membrane after sulfonation from Figure 1 (a). This maybe caused by -SO_3H groups in the side chains of Nafion, which is much adverse for the identification of the -SO_3H that maybe attached to the surface of SiO_2 nano particles. Therefore, in this experiment, an ex-situ characterization method was adopted. Powder SiO_2 nano-particles were prepared outside the Nafion membrane via a sol-gel reaction of TEOS. Then the obtained SiO_2 was sulfonated (signed as S-SiO_2) by concentrated H_2SO_4 for 10 h at 80 °C as mentioned in section 2.2.

Figure 1 (b) illustrates the spectra of the obtained SiO_2 and sulfonated S-SiO_2. It is shown that the peak at 3400 cm^{-1}, which is the characteristic signal of -OH, is widened after sulfonation. It is generally acknowledged that the widening of the hydroxyl peak is caused by the hydrogen bond. It suggests that there are strong hydrogen bonds in the sulfonated S-SiO_2. This should be caused by the interaction between surface -OH groups of SiO_2 nano-particles and H_2SO_4 molecules. However, characteristic peaks of the -SO_3H group are located in 1000-1100 cm^{-1}, which coincides with the strong absorption band of symmetric and anti-symmetric vibration of the Si-O-Si. For this reason, the structure signal of -SO_3H is not clear by FT-IR.

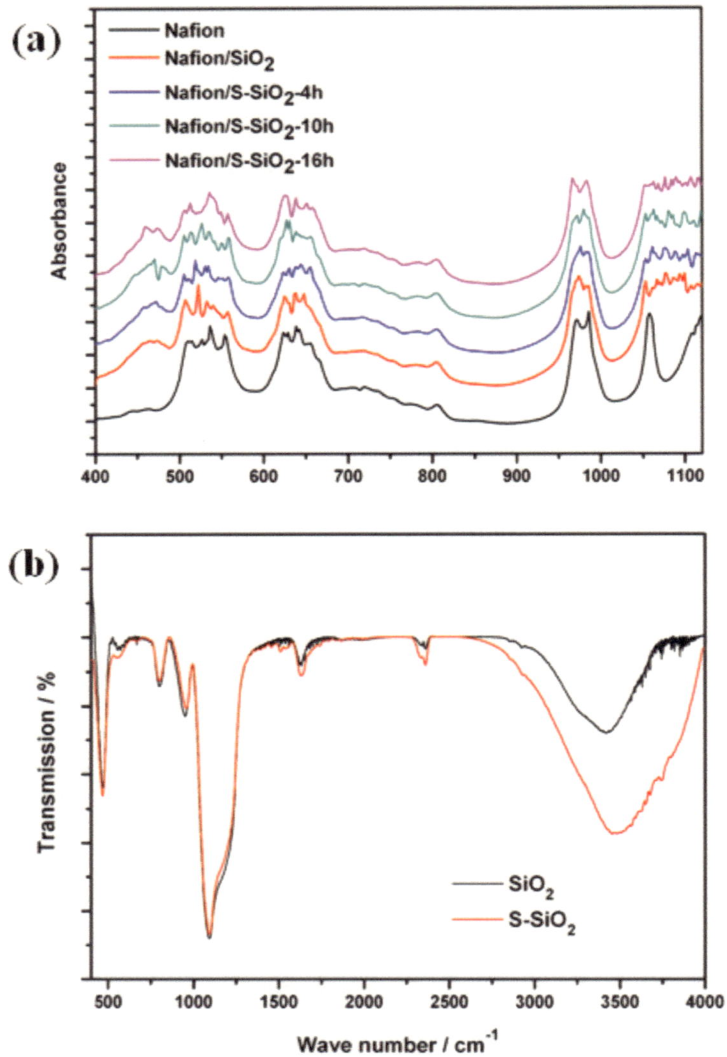

Fig. 1. (a) FT-IR spectra of Nafion, Nafion/SiO_2 and sulfonated Nafion/S-SiO_2 composite membranes; (b) FT-IR spectra of sol-gel derived SiO_2 and the sulfonated S-SiO_2. (Ke, Li, et al., 2011)

3.2 UV Raman spectra

UV resonance Raman spectra were recorded at room temperature using a home-made UV resonance Raman spectrograph of State Key Laboratory of Catalysis (Dalian institute of Chemical Physics) at a resolution of 2 cm^{-1}. The laser line at 325 nm of a He-Cd laser was used as an exciting source with an output of 25 mW.

From FT-IR, it suggests the existence of strong hydrogen bond in the S-SiO_2. However, whether the -SO_3H groups are chemically attached to the surface of the SiO_2 nano-particles, or how the -SO_3H groups interact with SiO_2 nano-particles remains unclear. UV resonance

Raman spectroscopy (UVRRS) is a powerful tool for surface species detection & identification. Therefore, in this experiment the UVRRS spectra of SiO_2 and S-SiO_2 were recorded, using a home-made UV Raman Resonance Spectrograph. The UVRRS spectra of SiO_2 and S-SiO_2 are shown as Figure 2. It can be seen that the scattering pattern of SiO_2 is much different from that of the sulfonated S-SiO_2. For the scattering pattern of the sulfonated S-SiO_2, the peaks at 873 cm^{-1} and ~620 cm^{-1} disappear, and a new peak at 710 cm^{-1} emerges. Besides, the band around 400 cm^{-1} is broadened after sulfonation.

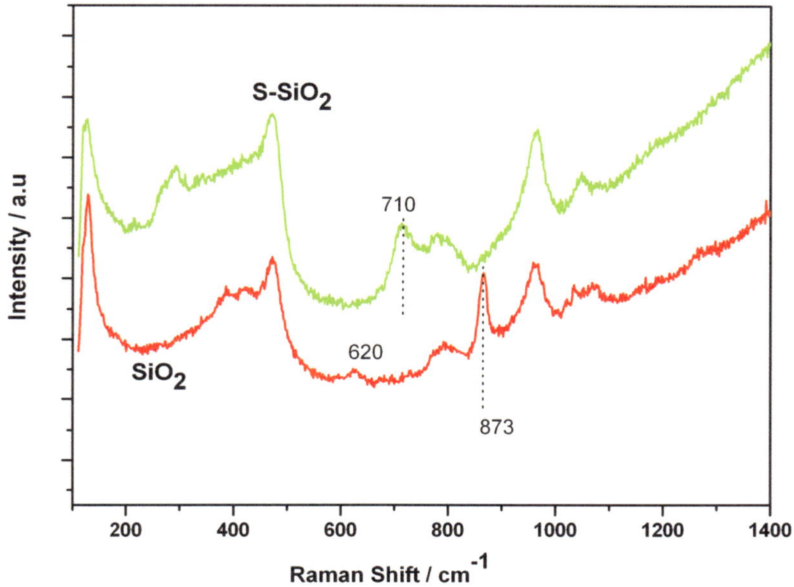

Fig. 2. UV Raman spectra of sol-gel derived SiO_2 and the sulfonated S- SiO_2. (Ke, Li, et al., 2011)

It is well known, the scattering peaks at ~620 cm^{-1} and 873 cm^{-1} could be attributed to the so called D2 defect structure (strained three-membered rings) and strained 2-fold rings (edge-sharing tetrahedra) respectively(Brinker, Kirkpatrick et al. 1988), which form primarily on the silica surface by condensation reactions involving isolated adjacent silanol groups(Brinker, Tallant et al. 1986). After sulfonation, the ~620 cm^{-1} and 873 cm^{-1} peaks disappear, suggesting that a reverse reaction of surface condensation occurs during sulfonation. And the new peak at 710 cm^{-1} should be due to the new surface specie (Si-O-SO_3H) forming.

From the UVRRS results, it is clearly seen that sulfonation affects the surface structure of the SiO_2 nano-particles to a large degree. The chemical bond assembled between sulfonic group and SiO_2 nano-particles is responsible for covering the proton conductivity loss of the Nafion/SiO_2 composite membrane.

4. Physic-chemical properties of Nafion/SiO_2 composite membranes

4.1 Water uptake

For the water-uptake evaluation, to remove the residual water, the membrane was firstly dried in vacuum drying oven for 12 h at 60°C. Then, the membrane was quickly taken out from the oven and weighed precisely. The weight of the dry membrane was signed as W_{dry}. After that, the membrane was soaked in de-ionized water at certain temperature (40°C, 60°C, or 80°C) for 24h. The weight of the wet membrane was signed as W_{wet}. The water-uptake (Wu) can be calculated by the following equation:

$$Wu = \frac{W_{wet} - W_{dry}}{W_{dry}} \times 100\% \tag{1}$$

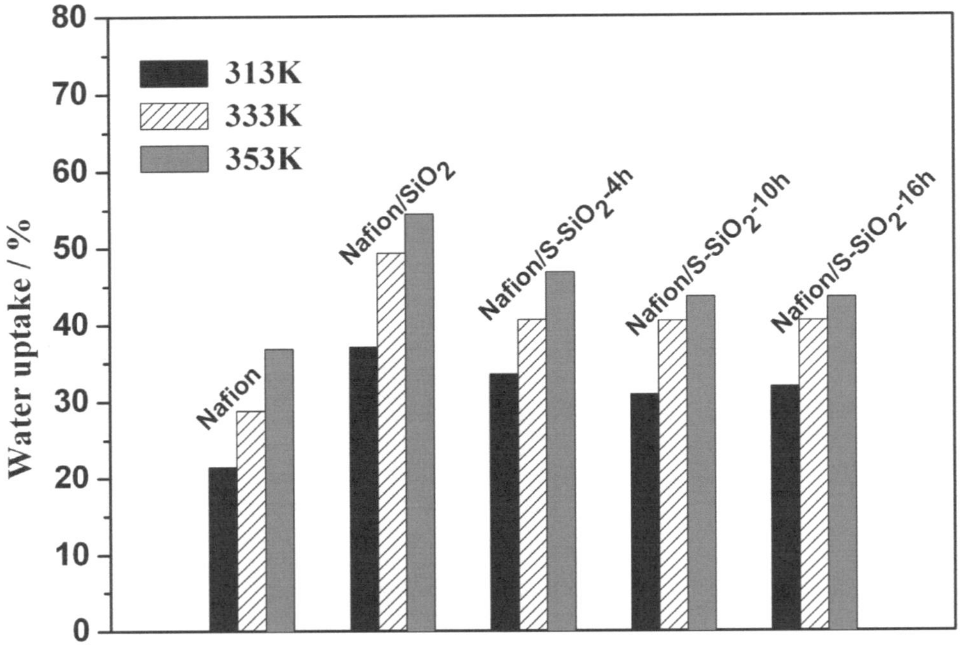

Fig. 3. Water uptakes of Nafion, Nafion/SiO_2 and Nafion/S-SiO_2 composite membranes sulfonated for 4h, 10h, and 16h. (Ke, Li, et al., 2011)

The water uptakes of Nafion, Nafion/SiO_2, Nafion/S-SiO_2-4h, Nafion/S-SiO_2-10h, and Nafion/S-SiO_2-16h at 40, 60, and 80 °C are shown in Figure 3. It shows that the Nafion/SiO_2 has a higher water uptake than the unmodified Nafion at the considering temperature region. At the temperature of 80 °C, the water uptake of the Nafion is 35%, while Nafion/SiO_2 has water up take of 56%, much higher than that of the unmodified Nafion. That should attribute to the hydrophily of the SiO_2 nano particles, of which the surface -OH groups have excellent capability of water maintenance. However, after treatment with concentrated sulfuric acid, the water uptake of the composite membrane decreases. And, it descends synchronously as the sulfonation time ascends.

4.2 Proton conductivity

Electrochemical Impedance Spectroscopy (EIS) was carried out to measure the proton conductivities of the membranes, using a Solartron impedance/Gain-phase Analyzer (model SI 1260) combined with a Solartron Electrochemical Interface (model SI 1287). Z-plot and Z-view software were used to control the measurement process. The amplitude of the AC signal was 20 mV, and the frequency was set from 100 Hz to 1 MHz. All samples were soaked in de-ionized water at 40 °C for 24 h and then sealed between two carbon-paper electrodes with an area of 0.332 cm^2 for measurement. The conductivity could be calculated using the following equation:

$$\sigma = \frac{L}{R \times A} \tag{2}$$

Where, σ is the proton conductivity of the membrane, R is the resistance of the membrane, and the sign L as well as A stands for the thickness of the membrane and the area of the electrode, respectively.

The EIS plots of the membranes are shown as Figure 4. It could be seen that the membrane resistance of Nafion/SiO_2 is higher than that of the unmodified Nafion. It suggests that SiO_2 nano-particles incorporated into the Nafion matrix block the conduction of protons. Therefore, in this chapter we sulfonated the Nafion/SiO_2 composite membrane on the purpose to cover the conductivity loss caused by the SiO_2. In Figure 4, it could be seen that the membrane resistances of the Nafion/S-SiO_2 (4h, 10h, and 16h) composite membranes are lower than that of the un-sulfonated Nafion/SiO_2 composite membrane. It proves that the sulfonation is an effective method to cover the shortcoming of Nafion/SiO_2 composite membrane on proton conductivity.

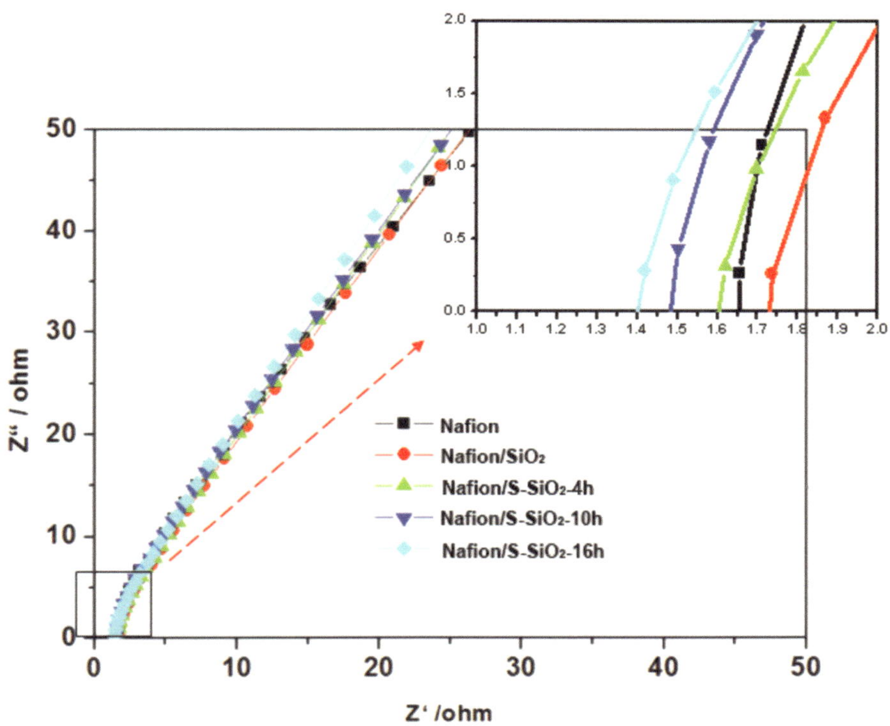

Fig. 4. Proton conductivities of Nafion, Nafion/SiO_2 and Nafion/S-SiO_2 composite membranes sulfonated for 4h, 10h, and 16h. (Ke, Li, et al., 2011)

4.3 Diffusion coefficient of water

Water transport in the Nafion membrane influences the proton transport in the membrane(Sumner, Creager et al. 1998; Inzelt, Pineri et al. 2000), so it is important to understand the transport process in the fuel cell. Water balance in the membrane depends on the interaction among the three kind of water transmission mechanisms---electroosmosis, back diffusion and hydraulic permeation. Membrane water diffusion coefficient as the PEMFC water transmission process's key parameter has attracted many attentions (Nguyen and White 1993; Zawodzinski, Jr. et al. 1993; Motupally and Becker 2000). However, most of the studies were based on the Nafion membrane. Researches on the water diffusion coefficient in the modified Nafion membrane are few. Nicotera (Nicotera, Zhang et al. 2007; Nicotera, Khalfan et al. 2008) et al used pulse gradient field spin echo nuclear magnetic resonance technology (PFGSE-NMR) obtain the water diffusion coefficient of the recast Nafion/SiO_2 composite membrane, but they did not connect the composite membrane waters' diffusion coefficient with the SiO_2 content and the temperature. In this work, the diffusion coefficient of water across the Nafion212/SiO_2 composite membranes with different SiO_2 content were measured through steady state permeation, and an empirical correlation of the water apparent diffusion coefficient of Nafion212/SiO_2 composite membrane as a function of SiO_2 content and temperature was presented.

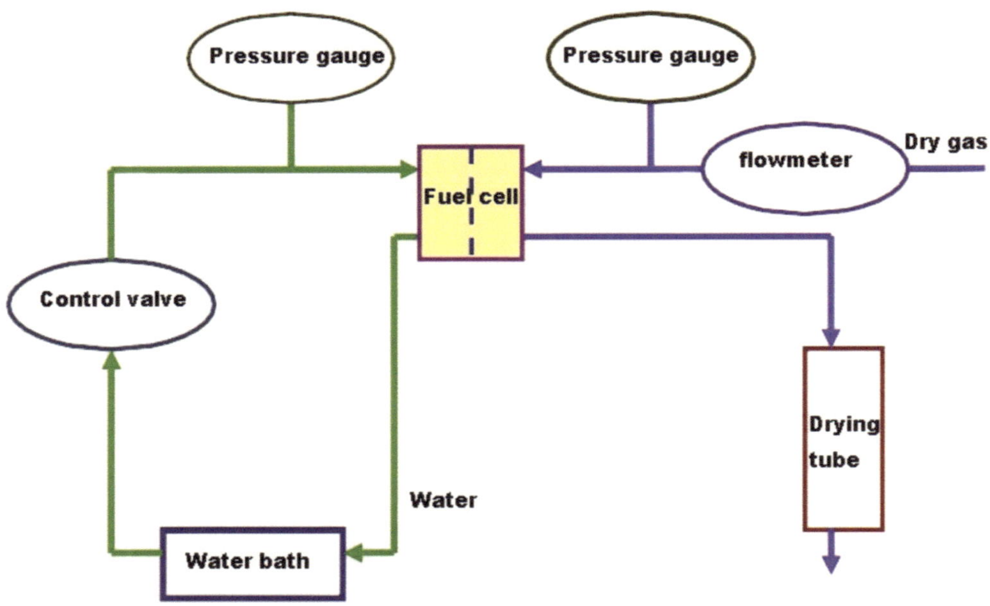

Fig. 5. Scheme of the experimental setup

A pseudo fuel cell, consisting of the measured membrane sandwiched between two GDLs, was used to measure the water flux through the membrane. Figure 5 is the scheme of the experimental setup. In the process, preheated water was circulated through one side of the fuel cell and dry air passed through the other side. As the gas traversed the length of the cell, its humidity increased due to the diffusion of water across the membrane. To measure the amount of water diffusion across the composite membrane, the air exiting the cell was passed through a tube filled with silica gel. Gravimetric analysis was conducted by precision analytical balance. The temperatures of intake air, the preheated water and the fuel cell were kept consistent. Water flux through the composite membrane under different operating conditions was measured by adjusting the air flow rate and temperature. In the experiment, water diffusion coefficients of Nafion/SiO_2 composite membranes with weight content of silica from 7.8% to 16.6% were measured and the area of the pseudo fuel cell is 5 cm^2.

Figure 6 displayed the water flux through the Nafion212/SiO_2 composite membrane with different content of SiO_2 as a function of the inlet gas flow rate of air at different operating temperatures. From this figure, the water flux through the composite membrane increases with the increasing of the temperature. At the same temperature, the water flux through the composite membrane increases with the increasing of the SiO_2 content, which may be caused by the water apparent diffusion coefficient.

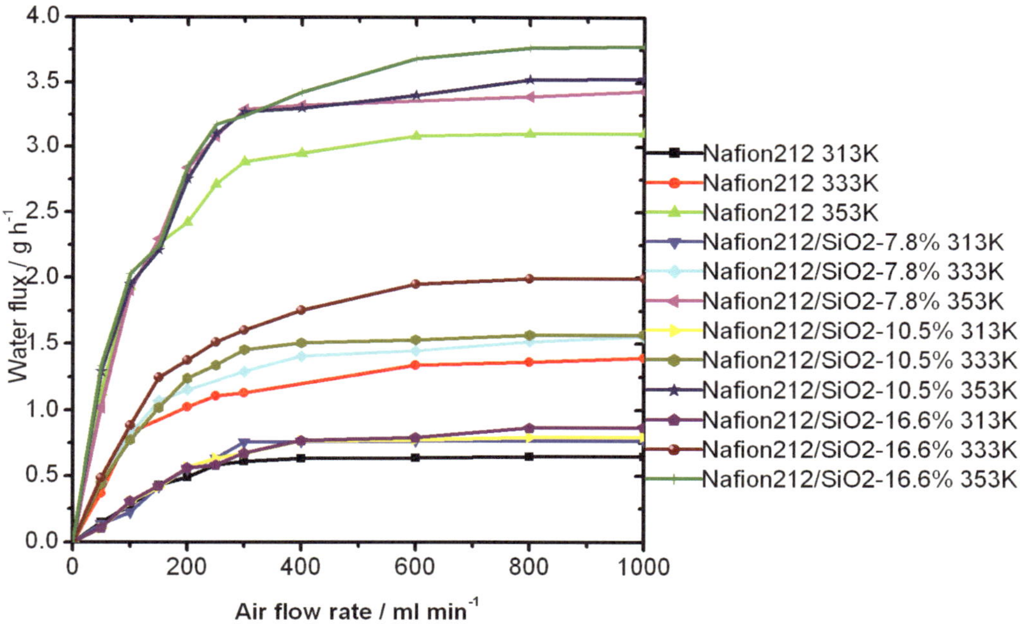

Fig. 6. Water flux of Nafion212/SiO_2 composite membrane with different content of SiO_2 as a function of the inlet gas flow rate of air at different operating temperatures. (Qu, Li, et al. , 2011)

Assumed that the water transfer across the membrane was the control step, the water concentration gradient across the membrane was the driving force for the water diffusion. According to the Fick's first law, the water diffusion flux was

$$J = D_w \frac{c_{w,a} - c_{w,g}}{\delta_m} \tag{3}$$

where J is water flux, D_w water apparent diffusion coefficient, δ_m membrane thickness, $c_{w,a}$ pure water concentration, 55.6 mol/l, $c_{w,g}$ water concentration in the air.
The water flux across the composite membrane was measured at the air flow rate of 800ml/min under temperature from 303 to 353K. The water apparent diffusion coefficient calculated from Eq. 3 was showed in Figure 7. It was clearly seen that the water apparent diffusion coefficient of Nafion212 membrane was lower than that of Nafion/SiO_2 composite membrane, and the diffusion coefficient of composite membrane increased with the increase of SiO_2 content.

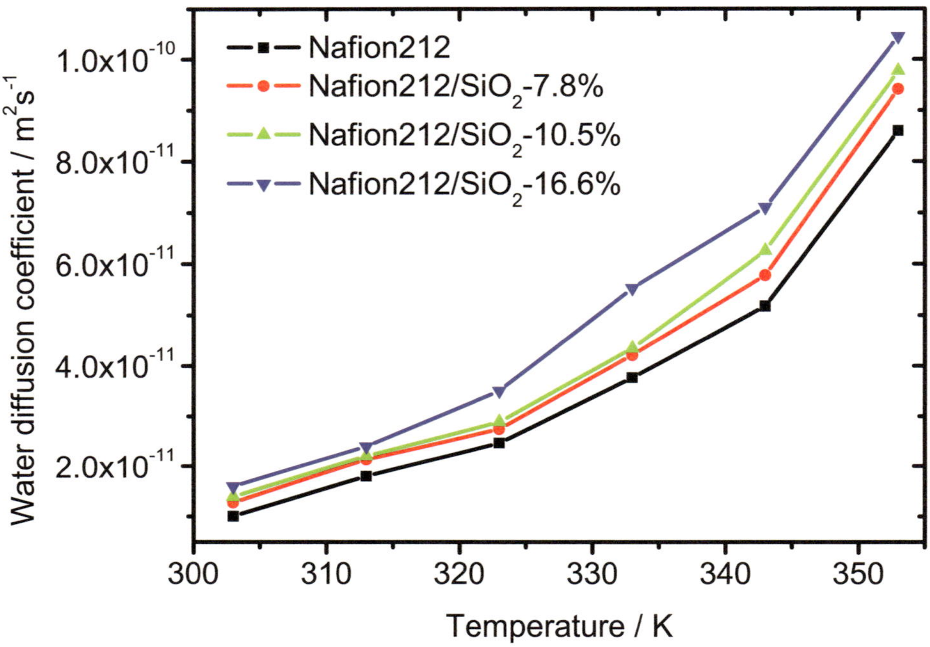

Fig. 7. The water apparent diffusion coefficient of Nafion212/SiO_2 composite membrane with different content of SiO_2 as a function of temperature. (Qu, Li, et al. , 2011)

Assumed that the water apparent diffusion coefficient of the Nafion212/SiO_2 composite membrane and temperature were consistent with Arrhenius equation, that was

$$lnD_w = lnD_0 - \frac{Ea}{RT} \tag{4}$$

Equation (4) was carried on the linear regression to Nafion212/SiO_2 composite membrane's lnD_w and 1000/T, as shown in Figure 8. The results showed that the lnD_w was linear correlation with 1000/T in the temperature range from 303 to 353K. This demonstrated that the assumption was right. The apparent diffusion activation energy (Ea) and pre-exponential factor of the composite membrane was obtained through linear fit in Figure 8, listed in Table 1. The apparent diffusion activation energy of the Nafion212 membrane was higher than that of the Nafion212/SiO_2 composite membrane, indicated that water diffusion was easier in the Nafion212/SiO_2 composite membrane.

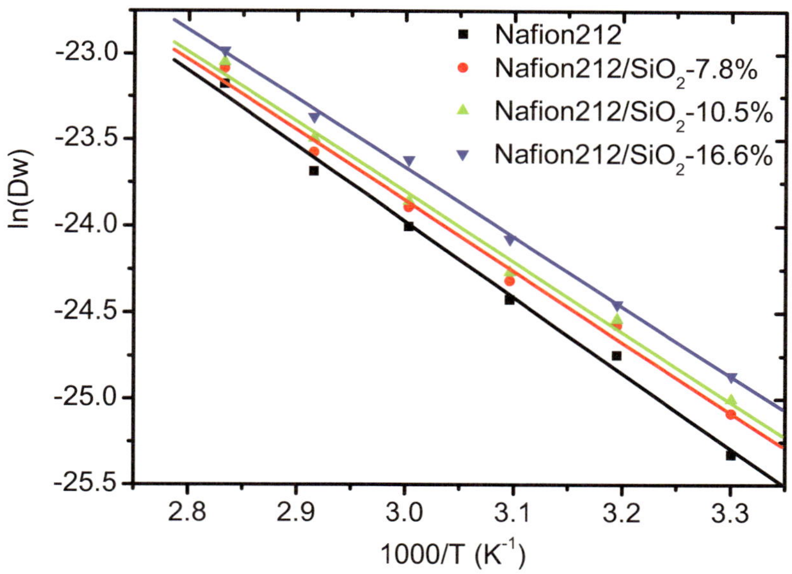

Fig. 8. Arrhenius plot of water apparent diffusion coefficient of Nafion212/SiO_2 composite membrane with different content of SiO_2. (Qu, Li, et al. , 2011)

	E_a / KJ mol^{-1}	D_0 / m^2 s^{-1}
Nafion212	36.43	1.977×10^{-5}
Nafion212/SiO_2-7.8%	34.03	9.398×10^{-6}
Nafion212/SiO_2-10.5%	33.77	8.999×10^{-6}
Nafion212/SiO_2-16.6%	33.36	8.939×10^{-6}

Table 1. Apparent activation energy and pre-exponential factor of composite membrane

Associate the apparent activation energy and the pre-exponential factor of composite membrane with the SiO_2 content, getting the following relationship

$$Ea = 33.16882 + 3.25989\exp\left(-\frac{W_{SiO_2}}{0.05989}\right) \tag{5}$$

$$D_0 = 8.89161\times10^{-6} + 1.08831\times10^{-5}\exp\left(-\frac{W_{SiO_2}}{0.02504}\right) \tag{6}$$

Then the empirical correlation of apparent diffusion coefficient of the Nafion212/SiO_2 composite membrane was

$$D_w = \left(8.89161\times10^{-6} + 1.08831\times10^{-5}\exp\left(-\frac{W_{SiO_2}}{0.02504}\right)\right) \exp\left(-1000\left(33.16882 + 3.25989\exp\left(-\frac{W_{SiO_2}}{0.05989}\right)\right)\Big/RT\right)\left(m^2/s\right) \tag{7}$$

The comparison of water apparent diffusion coefficient of the Nafion212/SiO_2 composite membrane calculated using this empirical correlation and the experimental data was shown in Figure 9, with a difference in less than 10%. Therefore this empirical correlation can be used to represent the relationship between the water apparent diffusion coefficient of the Nafion212/SiO_2 composite membrane and SiO_2 content and temperature.

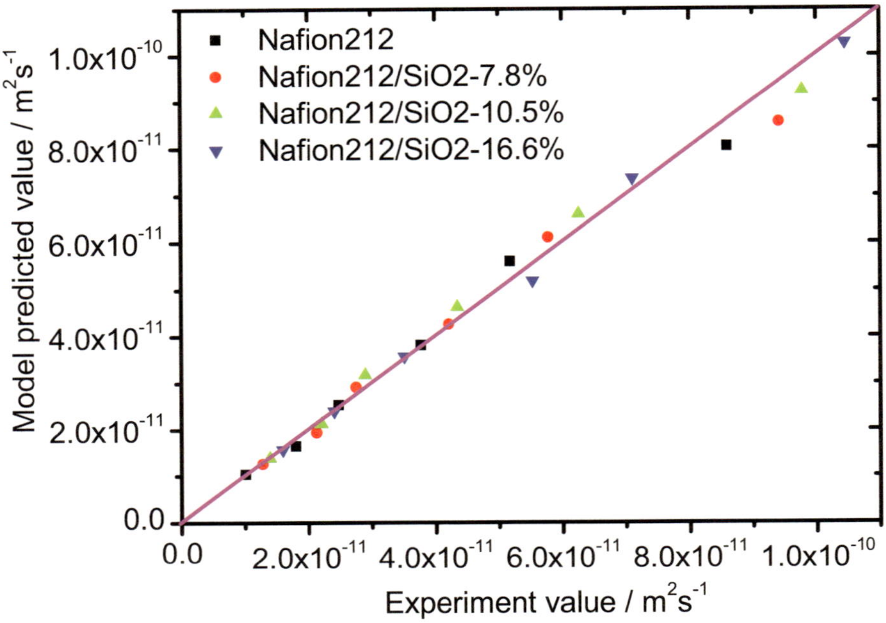

Fig. 9. Comparison of the empirical correlation predicted and experimental water apparent diffusion coefficient of composite membrane. (Qu, Li, et al. , 2011)

5. High temperature PEM fuel cell technologies

5.1 Fuel cell performance

All membrane electrode assemblies (MEAs) were prepared via a hot-pressing process. The Gas Diffused Electrode (GDE) was prepared first. In preparation of GDE, carbon paper from Toray, 20wt.% Pt/C from E-TEK, PTFE suspension and Nafion solution (DuPont, USA) were used. Carbon powder and PTFE suspension (DuPont, USA) were well mixed with ethanol under ultrasonic for 30 min, and spray-coated on side of the previous PTFE-treated carbon paper (Toray, Japan). The carbon paper was desiccated at 245 °C for 30min to eliminate the surfactant of PTFE suspension, and then 350 °C for 1 h. Catalyst inks were prepared by mixing 20wt.% Pt/C (E-TEK), 5 wt.% Nafion solution, and isopropanol. The mixture was sonicated for 30min and spray-deposited onto the carbon powder layer of the above treated carbon paper. Then the carbon paper was desiccated at 245 °C for 30min to eliminate the residual solvent, and then 350 °C for 1 h. Finally, Nafion solution (~2% with ethanol) was spray-deposited on to the catalyst layer. After dried at 100 °C for 2 h, the electrode is obtained. The contents of Nafion and PTFE in the catalyst layer are both 23.6 wt. %.
Both the loadings of Pt/C catalyst on the anode and cathode were 0.4 mgPt.cm^{-2}. Then, two pieces of gas diffused electrode with effective area of 5 cm^2 were hot-pressed onto one piece

of membrane to fabricate an MEA. The MEA was sandwiched into a single cell with stainless steel end plates and graphite groove flow fields as current collectors. The performance of the fuel cell was evaluated by polarization curve measurement at the temperatures of 60°C and 110°C, respectively. The H_2 and O_2 were fed into the fuel cell in co-flow mode. When the cell was operated at the temperature of 60°C with fully humidified H_2/O_2 gases, the flow rates of inlet gases were adjusted with current density to maintain the utilization of H_2 at 70% and O_2 at 40% for various current densities. When the cell was operated at the temperature of 110°C with H_2/O_2 gases and the relative humidity (RH) of 59%, the inlet gases were controlled at fixed flow rates of 35 and 100ml.min^{-1} respectively. All MEAs were evaluated under an absolute pressure of 0.3 MPa.

Figure 10 gives the I-V curves of these membranes: Nafion NRE212, Nafion/SiO_2, Nafion/S-SiO_2-4h, Nafion/S-SiO_2-10h, and Nafion/S-SiO_2-16h. It can be seen from the Figure 10 shows that the PEMFC single cell performance of each of the three sulfonated Nafion/S-SiO_2 membranes is enhanced compared to that of the un-treated Nafion/SiO_2 during almost the whole considering current density range, at 110 °C and 59% RH. Particularly, in the high current density region above 800 mA cm^{-2}, Nafion/S-SiO_2 outperforms the un-treated Nafion/SiO_2 significantly. At the current density of 1000 mA cm-2, the output voltage is measured as 0.598, 0.690, and 0.667V when the Nafion/SiO_2 membrane is sulfonated for 4, 10, and 16 h, respectively. While for the untreated Nafion/SiO_2, the output voltage is 0.587 V at the same current density. It demonstrates_that sulfonation is obviously beneficial to the elevated temperature & low humidity performance of the membranes.

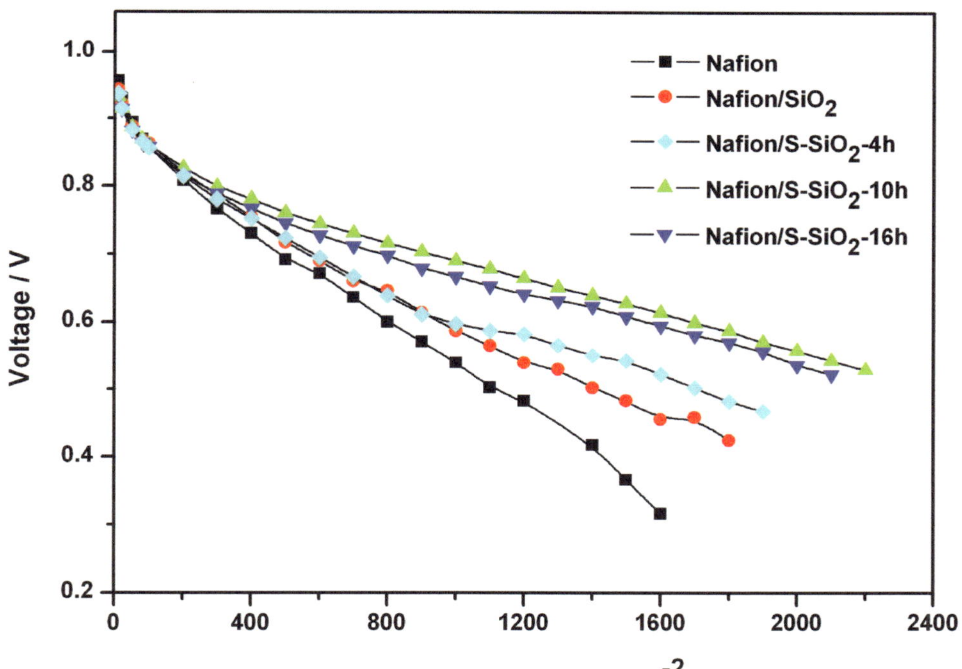

Fig. 10. I-V curves of Nafion, Nafion/SiO_2 and sulfonated Nafion/S-SiO_2 composite membranes at 110 °C, 59%RH.

5.2 High temperature PEMFC system

The fuel cell system consists of gas supply subsystem, water and heat management subsystem, fuel cell stack, automatic control subsystem. A ten-cell stack based on sulfonation Nafion/S-SiO_2-10h composite membrane was used in the fuel cell system, with single cell surface active area of 276 cm^2. 90%(vol) glycol water solution was used as coolant in the heat management system, which has a boiling point of 140.6 °C.

Figure 11 gives assembled high temperature PEMFCs system.

Fig. 11. Assembled high temperature PEMFCs system

Operate condition	content
Oxygen inlet	No humidification, 0.3 MPa (gauge pressure)
Hydrogen inlet	No humidification, 0.3 MPa (gauge pressure)
Cooling medium	90 % glycol water solution
Operate temperature	110 °C
Usage rate of hydrogen	≥98 %
Usage rate of oxygen	≥98 %

Table 2. The operate parameter of the high temperature PEMFCs system

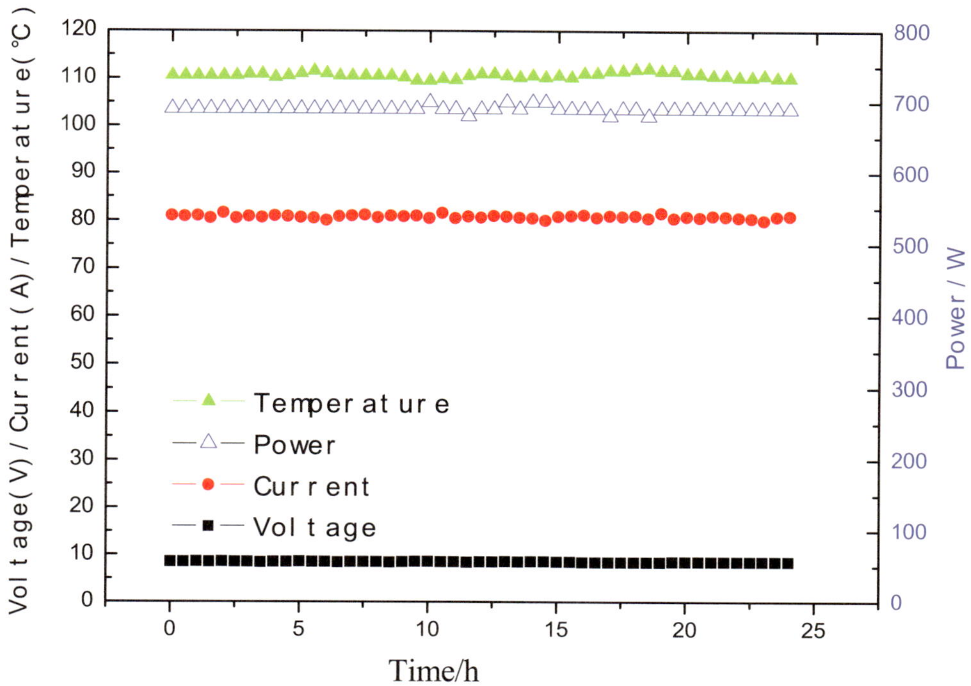

Fig. 12. Stabilization testing of the high temperature PEMFCs (H_2/O_2)

Table 2 gives the operate parameter of the high temperature PEMFCs system operating with H_2/O_2. Figure 12 gives stabilization testing of the high temperature PEMFCs operating. From Figure 12, the performance of fuel cell stack based on Nafion/S-SiO_2-10h composite membrane showed good stability under high operating temperature.
The fuel cell system is also running with hydrogen and air. Table 3 gives the operate parameters of the system with H_2/Air.

Operate condition	content
Air inlet	0.25 MPa, saturation humidification at 90 °C
Hydrogen inlet	0.27 MPa, Saturation humidification at 35 °C
Cooling medium	90 % glycol water solution
Operate temperature	110 °C
Usage rate of hydrogen	≥98 %
Usage rate of air	≥50 %

Table 3. The operate parameter of the high temperature PEMFCs system

Figure 13 gives stabilization testing of the high temperature PEMFCs operating at 110 °C. From Figure 13, the fluctuation of temperature is 110 °C±3 °C, the fluctuation of voltage is 5.75±0.25 V, 600±20 W for the power, and 108±0.5 A for the current. The fluctuation of the fuel cell stack's parameter is bigger than the system operating with hydrogen and oxygen, mainly due to the unsteady humidification. However, the performance of the fuel cell stack shows no significant degradation.

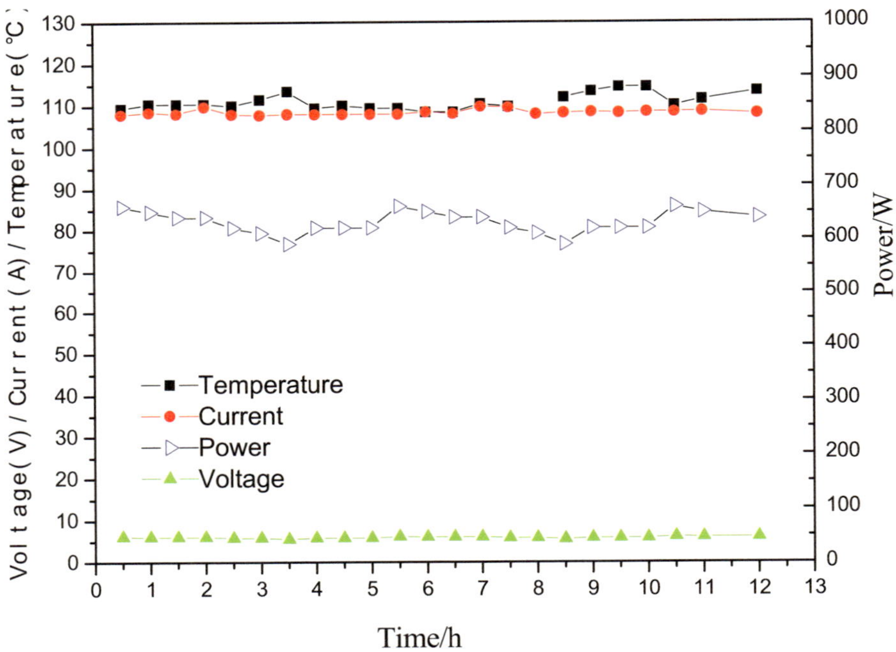

Fig. 13. Stabilization testing of the high temperature PEMFCs (H_2/air)

From the high temperature PEMFCs system testing in this work, no significant degradation of fuel cell performance was observed. However, further life-testing is required for investigating the mechanism of composite membrane degradation under high operating temperature.

6. Conclusion

The sulfonated Nafion/SiO_2 composite membranes were prepared by sulfonation of Nafion/SiO_2 composite membranes with concentrated sulfuric acid. Compared to untreated Nafion/SiO_2, the sulfonated Nafion/SiO_2 composite membranes have lower water uptake and higher proton conductivities. The lower water uptake of sulfonation Nafion/SiO_2 composite membrane is due to the partly elimination of the surface -OH groups of the SiO_2 nano-particles. And, the high proton conductivity of Nafion/S-SiO_2 composite membranes can be well explained by the formation of the chemical bonds between -SO_3H and the surface of SiO_2 nano-particles

The diffusion fluxes of water across the Nafion212/SiO_2 composite membranes with different SiO_2 content were measured through steady state permeation. An empirical correlation of the water apparent diffusion coefficient of Nafion212/SiO_2 composite membrane as a function of SiO_2 content and temperature was obtained.

Single cells based on Nafion/SiO_2 composite membranes have been tested at temperature of up to 110°C and compared with the experimental results based on sulfnoation Nafion/SiO_2 composite membranes. Single cell tests show that the performance of Nafion/S-SiO_2 composite membranes substantially exceeds Nafion/SiO_2 at 110 °C and 59%RH, and in the initial testing stage no performance reduction is observed.

Related fuel cell system technologies have been further developed, especially on water and heat management and an exper-imental high temperature PEMFC system has been integrated based on a 10-cell stack with a single cell active area of 276 cm^2. From the test results, the system show good reliability and maneuverability, and fuel cells show good stability and uniformity.

7. Acknowledgement

The authors thank for the joint support by the National High Technology Research and Development Program of China (863 Program No. 2007AA05Z131) and the National Natural Science Foundation of China (No. 20206030).

8. References

Adjemian, K. T., Dominey, R., et al. (2006). Function and characterization of metal oxide-naflon composite membranes for elevated-temperature H-2/O-2 PEM fuel cells. *Chemistry of Materials*, Vol.18,No.9,pp. 2238-2248

Adjemian, K. T., Lee, S. J., et al. (2002). Silicon oxide Nafion composite membranes for proton-exchange membrane fuel cell operation at 80-140 degrees C. *Journal of the Electrochemical Society*, Vol.149,No.3,pp. A256-A261

Adjemian, K. T., Srinivasan, S., et al. (2002). Investigation of PEMFC operation above 100 degrees C employing perfluorosulfonic acid silicon oxide composite membranes. *Journal of Power Sources*, Vol.109,No.2,pp. 356-364

Adjemian, K. T., Srinivasan, S., et al. (2001). Investigation of proton-exchange membrane fuel cells operating above 100 degrees C. *Abstracts of Papers of the American Chemical Society*, Vol.221,No.U382-U382

Alberti, G. and Casciola, M. (2003). Composite membranes for medium-temperature PEM fuel cells. *Annual Review of Materials Research*, Vol.33,No.129-154

Alberti, G., Casciola, M., et al. (2007). Novel Nafion-zirconium phosphate nanocomposite membranes with enhanced stability of proton conductivity at medium temperature and high relative humidity. *Electrochimica Acta*, Vol.52,No.28,pp. 8125-8132

Brinker, C. J., Kirkpatrick, R. J., et al. (1988). Nmr Confirmation of Strained Defects in Amorphous Silica. *Journal of Non-Crystalline Solids*, Vol.99,No.2-3,pp. 418-428

Brinker, C. J., Tallant, D. R., et al. (1986). Sol-Gel Transition in Simple Silicates .3. Structural Studies during Densification. *Journal of Non-Crystalline Solids*, Vol.82,No.1-3,pp. 117-126

Casciola, M., Capitani, D., et al. (2008). Nafion-zirconium phosphate nanocomposite membranes with high filler loadings: Conductivity and mechanical properties. *Fuel Cells*, Vol.8,No.3-4,pp. 217-224

Costamagna, P., Yang, C., et al. (2002). Nafion (R) 115/zirconium phosphate composite membranes for operation of PEMFCs above 100 degrees C. *Electrochimica Acta*, Vol.47,No.7,pp. 1023-1033

Deng, Q., Moore, R. B., et al. (1998). Nafion (R) (SiO2, ORMOSIL, and dimethylsiloxane) hybrids via in situ sol-gel reactions: Characterization of fundamental properties. *Journal of Applied Polymer Science*, Vol.68,No.5,pp. 747-763

Inzelt, G., Pineri, M., et al. (2000). Electron and proton conducting polymers: recent developments and prospects. *Electrochimica Acta*, Vol.45,No.15-16,pp. 2403-2421

Jian-Hua, T., Peng-Fei, G., et al. (2008). Preparation and performance evaluation of a Nafion-TiO2 composite membrane for PEMFCs. *International Journal of Hydrogen Energy*, Vol.33,No.20,pp. 5686-5690

Jin, Y. G., Qiao, S. Z., et al. (2009). Porous Silica Nanospheres Functionalized with Phosphonic Acid as Intermediate-Temperature Proton Conductors. *Journal of Physical Chemistry C*, Vol.113,No.8,pp. 3157-3163

Jin, Y. G., Qiao, S. Z., et al. (2008). Novel Nafion composite membranes with mesoporous silica nanospheres as inorganic fillers. *Journal of Power Sources*, Vol.185,No.2,pp. 664-669

Jones, D. J. and Rozière, J. (2003). In Handbook of Fuel Cells - Fundamental, Technology and Applications. *Vielstick, W. Gasteiger, H.A. Lamm, A. Eds. John Wiley & Sons: New York*, Vol.3,No.447-455

Jung, G. B., Weng, F. B., et al. (2008). Nafion/PTFE/silicate membranes for high-temperature proton exchange membrane fuel cells. *International Journal of Hydrogen Energy*, Vol.33,No.9,pp. 2413-2417

Ke, C.C., Li, X.J., et al. (2010). Preparation and properties of Nafion/SiO2 composite membrane derived via in situ sol–gel reaction: size controlling and size effects of SiO2 nano-particles. *Polymers for Advanced Technologies.* doi:10.1002/pat.1828

Ke, C.C., Li, X.J., et al. (2011). Investigation on sulfuric acid sulfonation of in-situ sol-gel derived Nafion/SiO2 composite membrane. *International Journal of Hydrogen Energy*, Vol.36,No.5,pp. 3606-3613

Li, Q. F., He, R. H., et al. (2003). Approaches and recent development of polymer electrolyte membranes for fuel cells operating above 100 degrees C. *Chemistry of Materials*, Vol.15,No.26,pp. 4896-4915

Lin, Y. F., Yen, C. Y., et al. (2007). High proton-conducting Nafion (R)/-SO3H functionalized mesoporous silica composite membranes. *Journal of Power Sources*, Vol.171,No.2,pp. 388-395

Marban, G. and Vales-Solis, T. (2007). Towards the hydrogen economy? *International Journal of Hydrogen Energy*, Vol.32,No.12,pp. 1625-1637

Mauritz, K. A., Mountz, D. A., et al. (2004). Self-assembled organic/inorganic hybrids as membrane materials. *Electrochimica Acta*, Vol.50,No.2-3,pp. 565-569

Mauritz, K. A., Stefanithis, I. D., et al. (1995). Microstructural Evolution of a Silicon-Oxide Phase in a Perfluorosulfonic Acid Ionomer by an in-Situ Sol-Gel Reaction. *Journal of Applied Polymer Science*, Vol.55,No.1,pp. 181-190

Mauritz, K. A., Stefanithis, I. D., et al. (1991). Microstructural Evolution of a Silicon-Oxide Phase in Nafion Membranes by an Insitu Sol-Gel Reaction. *Abstracts of Papers of the American Chemical Society*, Vol.202,No.285-POLY

Motupally, S. and Becker, A. J. (2000). Diffusion of water in Nafion 115 membranes. *Journal of the Electrochemical Society*, Vol.147,No.9,pp. 3171-3177

Nguyen, T. V. and White, R. E. (1993). A Water and Heat Management Model for Proton-Exchange-Membrane Fuel Cells. *Journal of the Electrochemical Society*, Vol.140,No.2178-2186

Nicotera, I., Khalfan, A., et al. (2008). NMR investigation of water and methanol mobility in nanocomposite fuel cell membranes. *Ionics*, Vol.14,No.3,pp. 243-253

Nicotera, I., Zhang, T., et al. (2007). NMR characterization of composite polymer membranes for low-humidity PEM fuel cells. *Journal of the Electrochemical Society*, Vol.154,No.B466-B473

Qu, S.G., Li, X. J., et al. (2011). Investigation on water diffusion coefficient of Nafion/SiO2 composite membrane. *Chinese Journal of Power Sources*, (in press, in Chinese)

Rodgers, M. P., Shi, Z. Q., et al. (2008). Transport properties of composite membranes containing silicon dioxide and Nafion (R). *Journal of Membrane Science*, Vol.325,No.1,pp. 346-356

Santiago, E. J., Isidoro, R. A., et al. (2009). Nafion-TiO2 hybrid electrolytes for stable operation of PEM fuel cells at high temperature. *Electrochimica Acta*, Vol.54,No.16,pp. 4111-4117

Shao, Z. G., Joghee, P., et al. (2004). Preparation and characterization of hybrid Nafion-silica membrane doped with phosphotungstic acid for high temperature operation of proton exchange membrane fuel cells. *Journal of Membrane Science*, Vol.229,No.1-2,pp. 43-51

Shao, Z. G., Xu, H. F., et al. (2006). Hybrid Nafion-inorganic oxides membrane doped with heteropolyacids for high temperature operation of proton exchange membrane fuel cell. *Solid State Ionics*, Vol.177,No.7-8,pp. 779-785

Sumner, J. J., Creager, S. E., et al. (1998). Proton conductivity in Nafion 117 and in a novel Bis[(perfluoroalkyl)sulfonyl]imide ionomer membrane. *Journal of the Electrochemical Society*, Vol.145,No.107-110

Tang, H., Wan, Z., et al. (2007). Self-assembled Nafion-silica nanoparticles for elevated-high temperature polymer electrolyte membrane fuel cells. *Electrochemistry Communications*, Vol.9,No.8,pp. 2003-2008

Wang, K. P., McDermid, S., et al. (2008). Preparation and performance of nano silica/Nafion composite membrane for proton exchange membrane fuel cells. *Journal of Power Sources*, Vol.184,No.1,pp. 99-103

Wang, L., Yi, B. L., et al. (2008). Pt/SiO2 as addition to multilayer SPSU/PTFE composite membrane for fuel cells. *Polymers for Advanced Technologies*, Vol.19,No.12,pp. 1809-1815

Wang, L., Zhao, D., et al. (2008). Water-retention effect of composite membranes with different types of nanometer silicon dioxide. *Electrochemical and Solid State Letters*, Vol.11,No.11,pp. B201-B204

Xu, W. L., Lu, T. H., et al. (2005). Low methanol permeable composite Nafion/silica/PWA membranes for low temperature direct methanol fuel cells. *Electrochimica Acta*, Vol.50,No.16-17,pp. 3280-3285

Yan, X. M., Mei, P., et al. (2009). Proton exchange membrane with hydrophilic capillaries for elevated temperature PEM fuel cells. *Electrochemistry Communications*, Vol.11,No.1,pp. 71-74

Yang, C., Costamagna, P., et al. (2001). Approaches and technical challenges to high temperature operation of proton exchange membrane fuel cells. *Journal of Power Sources*, Vol.103,No.1,pp. 1-9

Yen, C. Y., Lee, C. H., et al. (2007). Sol-gel derived sulfonated-silica/Nafion (R) composite membrane for direct methanol fuel cell. *Journal of Power Sources*, Vol.173,No.1,pp. 36-44

Yonghao Liu, Baolian Yi, et al. (2005). Study on silicon oxide Nafion composite membranes for proton exchange membrane fuel cell operation. *Chinese Journal of Power Sources*, Vol.29,No.2,pp. 92-94,112

Yu, J., Pan, M., et al. (2007). Nafion/silicon oxide composite membrane for high temperature proton exchange membrane fuel cell. *Journal of Wuhan University of Technology-Materials Science Edition*, Vol.22,No.3,pp. 478-481

Yuan, J. J., Zhou, G. B., et al. (2008). Preparation and properties of Nafion (R)/hollow silica spheres composite membranes. *Journal of Membrane Science*, Vol.325,No.2,pp. 742-748

Zawodzinski, T. A., Jr., C. D., et al. (1993). Water uptake by and transport through Nafion® 117 membranes. *Journal of the Electrochemical Society*, Vol.140,No.4,pp. 1041-1047

Part 3

Practical Applications of Energy Storage

14

Energy Storage for Balancing a Local Distribution Network Area

I. Grau Unda, P. Papadopoulos, S. Skarvelis-Kazakos, L. M. Cipcigan and N. Jenkins
Cardiff University, United Kingdom

1. Introduction

According to a study carried out for the formerly Department of Trade and Industry (DTI) and the British regulatory authority (OFGEM), installed micro-generation capacity in the UK could grow to as much as 8GW by 2015 (Energy Saving Trust, 2005; Ingram & Probert, 2003). This will require a new and highly decentralised approach to energy planning and policy. In addition to resolving the everyday causes of power disturbances and outages at local and regional levels, the electricity industry has to be able to restore service after major events (e.g. floods and storms). Ensuring reliability has and will continue to be a priority. The recent blackouts in UK (Blades, 2008; British Broadcasting Corporation [BBC], 2010) highlight the significant public and private interest in electricity reliability. During times of crisis caused by the rolling blackouts in California in 2001, it was found that customers are willing to pay very high prices for reliable power (Lawton et al., 2003). Backup generation ownership reveals that the owners place a high value on reliability, and therefore are willing to pay for the cost of operating backup generation to reduce the impact of an actual outage. Micro-generation can be used to sustain the distribution system during unavailability of the transmission in-feeds and brings opportunities for the improvement of security of supply to customers in major events, such as floods or storms (Kato et al., 2007; Kondo et al., 2008). Energy storage systems have been identified as a potential source to support micro-generation, to improve their carbon performance (Skarvelis-Kazakos et al., 2010) and to offset the intermittency of renewable energy micro-sources (Grau et al., 2010).
Integration of individual small scale distributed generators in the low voltage (LV) side of the grid could provide benefits for the customers, not only from an economic point of view but also as an electric supply guarantee. Supply continuity could be reached by associating a number of loads (customers) and micro-sources into different subsystems (micro-grids). Thus, a micro-grid will be able to work in parallel with the grid, with the capability to switch to off-grid mode (intentional islanded) in case of a grid emergency (Grau et al., 2009a).
Technical recommendations, such as G59/1 (Energy Network Association, 1991), G83/1 (Energy Network Association, 2008) and the standard 1547 from Institute of Electrical and Electronics Engineers (IEEE, 2003), specify that micro-generation sources must be disconnected in the case of loss of the grid. The fact that the micro-generation sources would not be controlled by the utility grid in an islanded mode, could lead to operation beyond the grid requirements. This could prove hazardous, not only for the utility equipment, but also

for the personnel's health (Chowdhury et al., 2008). Embracing a reliable control strategy for the islanded mode, could be defined as the key issue in the micro-grid development and expansion. Such a strategy can maximise the micro-grid benefits, not only from the customer's point of view, but also from the utility system, since planned islanding operation may be part of the utility planning and operation strategies.

This chapter is organised in the following way: Section 2 is concerned with the technical challenges arising from intentional islanding of micro-grids that include micro-generation sources. An overview of these challenges is provided together with the possible solutions identified from the literature. A combination of an Energy Storage System (ESS) and a backup generator is proposed as a solution for intentional islanding. A micro-grid model is defined and used for steady state voltage studies using IPSA+ and PSCAD/EMTDC power systems simulation software.

Section 3 analyses the ESS requirements to balance a local area, defined as a micro-grid in this study. A methodology drawn from the literature (Abu-Sharkh et al., 2005) is used for calculating the ESS requirements. Case studies using a micro-grid model are defined.

Section 4 evaluates the combined use of a backup generator and an ESS for balancing a local area. A Java-based software tool performing sequential power flows was developed to examine the ESS and the backup generator requirements under different micro-grid load/generation conditions.

Section 5 uses the results from the previous sections to evaluate the use of ESS for electricity market participation. The MATLAB Optimisation Toolbox is used to obtain the optimal behavior of a pre-defined rated Energy Storage System, based on the requirements of a given micro-grid.

The chapter concludes with the main results and summarises the conclusions of each section.

2. Micro-grid intentional islanding

An overview of the main technical challenges regarding the grid-connected and islanded mode of micro-grids is provided. Appropriate solutions drawn from the literature are discussed. The use of energy storage and a backup generator is analysed. Part of a typical LV power distribution network is used for steady state voltage studies. Case studies are described and simulation results are analysed.

2.1 Technical challenges

When a micro-grid is to be operated at both grid-connected and islanded mode, frequency, steady state voltage, protection and earthing issues arise. These issues are discussed for each mode.

2.1.1 Frequency

2.1.1.1 Grid connected

Large centralised synchronous machines are equipped with speed governors, which are responsible for ensuring the balance of the system they belong to and hence the network frequency stability. Some micro-generation sources, are designed to operate with constant power without contributing to frequency control, therefore large penetration of such sources may lead to a less stiff system, determining the utility frequency stability (Lopes et al., 2006).

2.1.1.2 Intentional islanding

During islanded mode the micro-generation sources connected to the micro-grid may not be able to provide frequency control. The low inertia of the synchronous micro-sources together with the constant power output ones may be not sufficient for the micro-grid frequency stability without the utility support. The smaller size of the resulted micro-grid after islanding compared to the utility grids, give rise to a micro-system more sensitive to power variations, where small unbalances may be translated in big and fast frequency variations (Abu-Sharkh et al., 2006). The output of some micro-sources, such as PV and wind turbines, depends on the intermittency of their renewable resources; therefore changes in the power balance will not be only dependent on the load variation but also on the available micro-source power output.

2.1.2 Steady state voltage

2.1.2.1 Grid connected

The operation of a micro-source within the LV side of the network is associated with a rise of voltage at the point of connection (Conti et al., 2003). This can be seen as an opportunity for micro-source penetration, since a higher margin against under voltages is achieved. On the other hand, a high level of micro-source penetration could imply a violation of the upper voltage statutory limits (+ 10% in UK [Ingram & Probert, 2003]) at the point of connection (Jenkins et al., 2000). When networks are lightly loaded, voltage is more likely to violate statutory limits. That is due to the tap settings of MV/LV transformers being traditionally set to keep the voltage at the most remote customer just below the maximum limits (Jenkins et al., 2000).

2.1.2.2 Intentional islanding

In general, micro-sources operate in slave mode (i.e. they set the grid voltage as reference for their power electronics interfaces) when grid-connected. If due to the conditions prior to the intentional islanding, the micro-sources have to be disconnected, a voltage source is required to re-energise the micro-grid. Alternatively, the power electronics interface of one micro-source has to operate in master mode. This may prove a complex task when more micro-sources are added to the network, as the planning procedures of the Distribution Network Operators (DNOs) should be enhanced. Moreover the micro-grid's voltage will be dependent on the micro-sources power output, which may prove inadequate in the case of intermittent renewable micro-sources.
In high voltage electricity systems, reactive power compensation is used for the voltage control. In the low voltage side of the network, though, active power flow control will be critical to keep the voltage between statutory limits, due to the low X/R ratio. Therefore, balance inside the micro-grid may not be achieved in islanded mode, since the maximum active power flow along feeders will be limited (Zhou et al., 2007). The number of micro-sources, the penetration level and their location along the micro-grid will determine the voltage profile.

2.1.3 Protection

2.1.3.1 Grid connected

The electrical protection equipment of an electricity network is rated and operated according to the fault levels and fault clearance times of faulty currents inserted from the upstream network (Boutsika et al., 2005). When micro-sources are embedded in the

distribution network, an increase in the fault current levels is anticipated (Boutsika et al., 2005). Therefore, the rating and characteristics of electrical protection equipment may no longer be adequate to cope with the new fault current levels.
Traditional network protection schemes are based on unidirectional fault current flow. Embedded micro-sources in LV networks may reverse power flows especially when generation occurs at lightly loaded periods (Chowdhury et al., 2008).

2.1.3.2 Intentional islanding

An essential condition for the operation of micro-grids in an islanded mode is their compliance to the same safety requirements as those traditional centralised-generation operated networks (Jayawarna et al., 2005). In the case of a fault, the traditional grid rotational generators are injecting large fault currents, thus protection devices in the distribution network are mainly over-current sensing. The fault current emitted through the power electronics interfaced micro-sources inside the micro-grid, will be below the levels of traditional generators fault current (Jayawarna et al., 2005). Thus, possible faulty currents from micro-sources may not be detected by existing over-current relays.

2.1.4 Unearthed neutral

Current practices allow micro-sources to operate with their neutral earthed or not, while being synchronised to the utility system. The common practice is to earth only the neutral of the low voltage side of the MV/LV transformer (Dexters et al., 2007). The main reason for this is the degree of complexity in earth fault currents control, added by the neutral earth connection of the micro-sources (Dexters et al., 2007). When the micro-grid is intentionally islanded, an earth reference point should be provided. The lack of an earthed reference could lead to over voltages and safety problems for the personnel in case of a fault.

2.2 Solutions for intentional islanding

2.2.1 Frequency and voltage, micro-source control strategy

It is anticipated that future micro-grids will comprise a Micro-Grid Central Controller (MGCC) and dispersed micro-controllers for each micro-source and controllable load (Lopes et al., 2006). The micro-source controllers will control the power electronic interfaces (inverters) of the micro-sources. Two main strategies are currently used in inverter based control schemes: PQ inverter control, where specified P and Q values are delivered by the inverter, and Voltage Source Inverter (VSI), where voltage and frequency through P/f and Q/V droops are controlled under predetermined limits. The VSI strategy could be considered as more appropriate for islanded mode operation, since its behavior is similar to that of synchronous machines. Nevertheless, both types of inverter control strategies can coexist in an efficient way (Lopes et al., 2006). Load controllers will be mainly responsible for load shedding when the power generated inside the system cannot match the demand. Energy storage will play a key role in order to keep the frequency at the desired levels due to its bi-directional power flow capability (Ito et al., 2007).

2.2.2 Protection

In (Wu et al., 2008) a method for adjusting the settings of relays is proposed. This method can be used to modify the protection scheme during the transition from grid-connected to islanded mode. Energy storage controllers could be programmed to introduce higher fault currents to be detected by conventional protection devices. In (Jayawarna et al., 2005) a flywheel is presented as a solution for the desirable fault currents while operating in islanded mode.

2.2.3 Unearthed neutral

The earth reference of the micro-grids is usually located at the low voltage side of the MV/LV transformer. The control of a micro-source neutral connection for switching on during intentional islanding is proposed by (Dexters et al., 2007)

2.3 Energy storage and backup generator for intentional islanding operation

The connection of a backup generator and an energy storage device at the point of common coupling is proposed to replace the grid after disconnection (Fig.1) and will be similar with the spinning reserve of large generators in the conventional grid. The role of the energy storage will be: (i) to absorb any excess of energy supplied by the micro-grid; (ii) to cope with fast balance changes, and (iii) to ride-through the gap between the failure of grid power and the start-up of the generator (Grau et al., 2009b). The generator will be responsible for injecting power to the micro-grid for balancing purposes. Portable generator for backup measures offers a feasible solution when permanent deployment is not possible. The combination of energy storage, backup generator and micro-generators is anticipated to manage the micro-grid demand requirements.

The loads may be domestic loads, commercial loads or electric vehicles.

Ideal intentional islanding control strategies should achieve a smooth transition without use of load shedding. However, depending on the conditions prior to the intentional islanding, the micro-sources may need to be disconnected. In such case, the micro-grid will need to integrate black start capability in order to re-energise the system and re-start the micro-sources.

2.3.1 Micro-grid model for voltage studies

A LV micro-grid model, based on the UK generic network presented in (Ingram & Probert, 2003) is used. The model consists of a LV feeder modeled in detail, which supplies 96 residential customers, uniformly distributed among the 3-phases. Details of the whole network can be found in (Ingram & Probert, 2003; Papadopoulos et al., 2010).

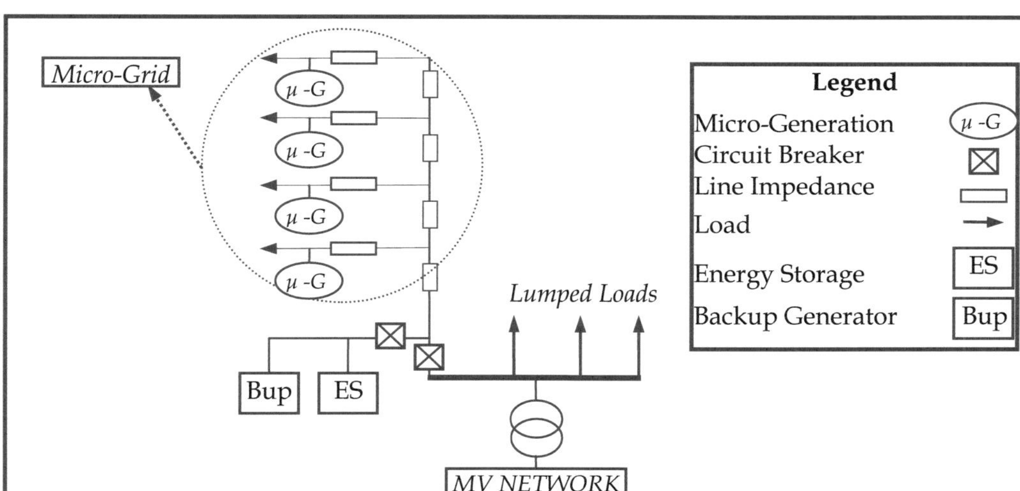

Fig. 1. Case study micro-grid

2.3.2 Case study

During islanded operation the micro-grid's demand is assumed to be covered by the power generated from the micro-sources, the backup generator and the energy storage. Simplified generation models were used to emulate the behavior of micro-generators. The domestic loads were modeled as purely resistive with typical minimum and maximum values acquired from the Electricity Association (Ingram & Probert, 2003). These are 0.16kW for a summer minimum and 1.3kW for a winter maximum residential load. An After Diversity Maximum Demand (ADMD) factor was applied per 100 customers.

Different levels of micro-generation penetration with a base scenario of 1.1kW per customer were studied. 100% penetration is equivalent to each customer having installed a micro-generator of 1.1 kW. The simulations were run for minimum and maximum loading conditions and the steady state voltage measurements recorded. The voltage set at the point of common coupling with the micro-grid was kept fixed at 1 p.u. IPSA+ and PSCAD/EMTDC micro-grid system models were used to cross-check the simulation results.

2.3.3 Simulation results

The simulation results are presented in Fig. 2 and Fig. 3, in which the measurement of each LV Segment is abbreviated with Seg, while Seg4 is the most remote segment in the micro-grid.

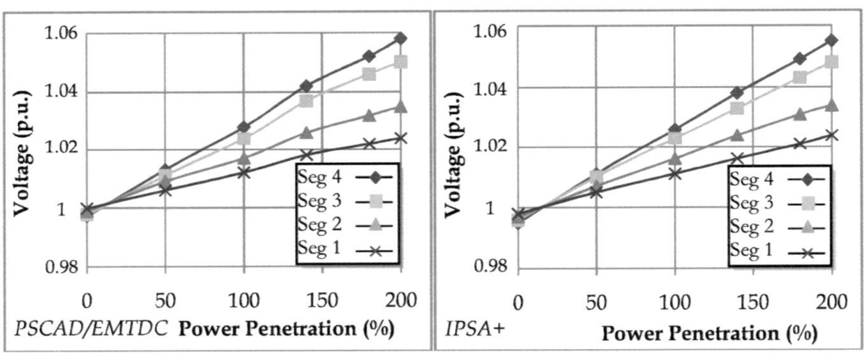

Fig. 2. Steady state voltage measurements at each LV segment during minimum load conditions

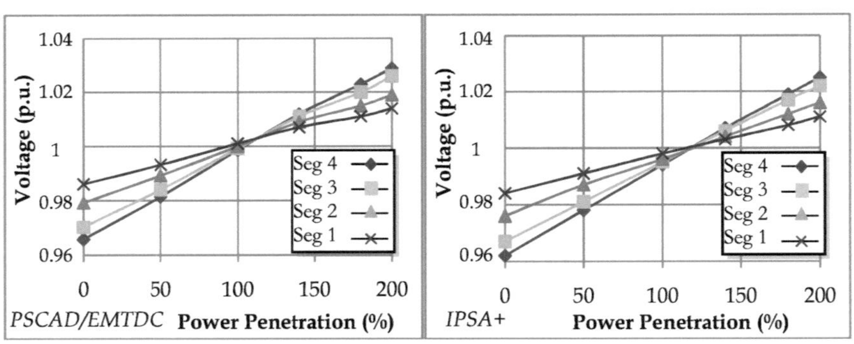

Fig. 3. Steady state voltage measurements at each LV segment during maximum load conditions

2.3.4 Discussion of the results

The results presented in Fig. 2 and Fig. 3 showed no violation of the statutory limits (1.1p.u. - 0.94p.u. in UK) for different levels of penetration. This could be a consequence of the following simulated conditions:

- Small size of the micro-grid;
- 1 p.u. constant voltage reference at the point of common coupling;
- Proximity of the reference voltage to the remotest part of the segment;
- Backup generator and energy storage capability to absorb and to inject the required power.

For maximum load scenario, it is found that until approximately 100% penetration, the power is flowing from the energy storage and backup generator to the loads. After 100% penetration, the power is flowing from the micro-sources towards the energy storage (Fig. 3).

3. Balancing a local area with energy storage

The energy storage power and capacity requirements to achieve a power balanced area are analysed. A balanced area is a part of an electricity network, self-sufficient in terms of electricity demand. To achieve self-sufficiency, the generation should always meet the demand. The methodology used to obtain the micro-generation penetration levels and the energy storage characteristics in the power balanced area is based on the work presented in (Abu-Sharkh et al., 2005). In this approach, the required combination of micro-generation sources and energy storage within a single dwelling are calculated. These requirements are then extrapolated to the number of households forming the micro-grid.

3.1 Calculation of self-sufficiency requirements

The micro-generation capacity required to satisfy the dwelling daily demand was calculated. The daily demand of the dwelling was assumed to be supplied by the micro-generators connected to it.

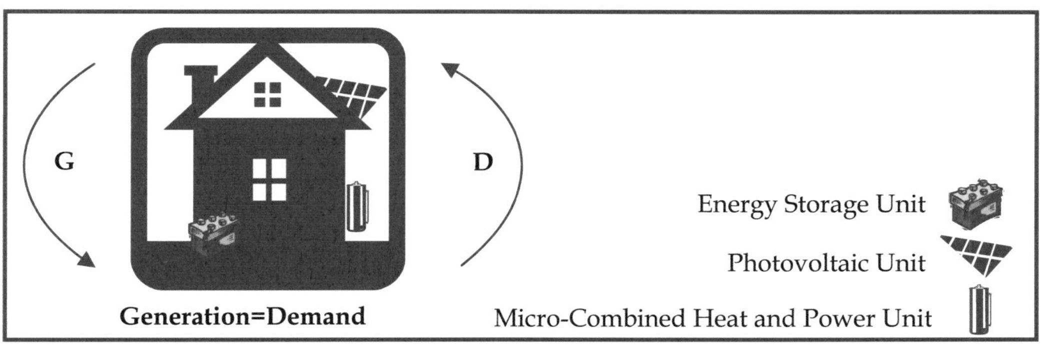

Fig. 4. Self-sufficient dwelling

Average generation and demand profiles for winter and summer seasons were used. Generation was assumed to come from PhotoVoltaics (PV) and micro-Combined Heat and Power (μ -CHP) units. The energy generated should satisfy the demand of the dwelling for both seasons (equations 1 and 2). By solving both inequalities, the penetration levels of the studied micro-sources were derived.

$$nE_{CHPwin} + mE_{PVwin} \geq E_{dwin} \quad (1)$$

$$nE_{CHPsum} + mE_{PVsum} \geq E_{dsum} \quad (2)$$

Where

E_{CHP} is the electric energy generated by the μ-CHP,
E_{PV} is the energy generated by the PV,
E_d is the energy demand,
n is the penetration levels for μ-CHP and
m the penetration levels for PV.
The sub indexs win and sum stand for winter and summer respectively.

Once the generation values were input, the characteristics of the required energy storage (maximum power and capacity requirements) were acquired. The value for a single household was then extrapolated to the number of customers forming the micro-grid.

3.2 Case study input data

Half-hourly residential load profiles where drawn from (UK Energy Research Center, 1997), and scaled to values for the specific model (from 0.16kVA to 1.3kVA per customer) provided in section 2. These profiles are deemed to be representative for the UK residential loads. Generation profiles for the PV and μ-CHP were used from (Mott McDonald, 2004) for winter and summer average days. The μ -CHP profiles were scaled to a maximum electrical power output of 1.5kWe. The data are shown in Fig. 5 and Fig. 6.

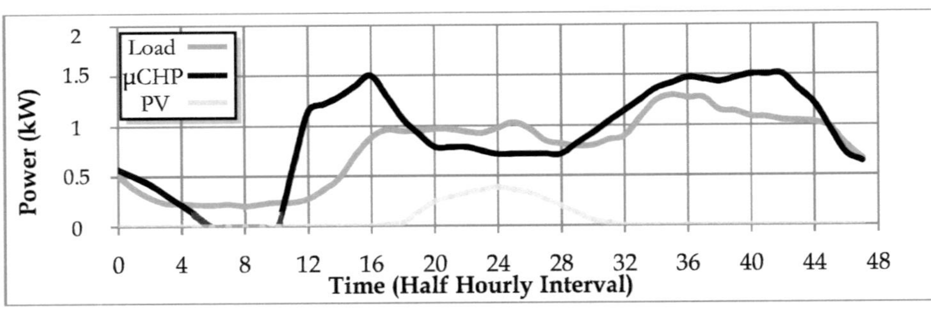

Fig. 5. Winter generation and demand profiles

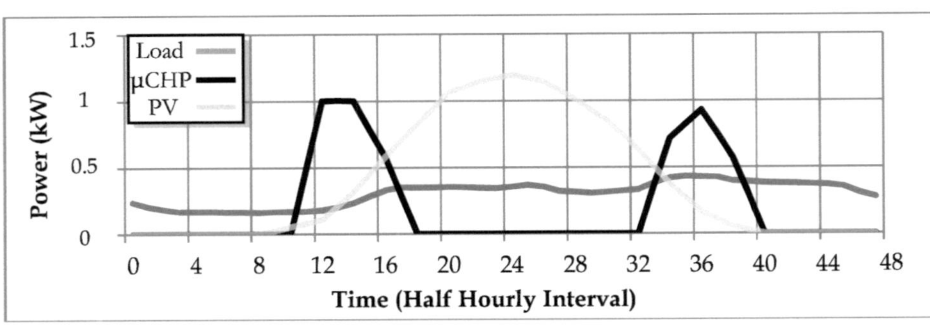

Fig. 6. Summer generation and demand profiles

3.3 Case study results

The methodology described in section 3.1 was applied and the results were extrapolated to a micro-grid consisting of 96 customers and one energy storage unit. Fig. 7 shows the graphical solutions of equation 1 and equation 2, using the data from section 3.2, which gives the micro-generation penetration levels required in each season to achieve the micro-grid balance. The intersection point indicates the optimal penetration levels required in order to achieve the balance in both seasons (Abu-Sharkh et al., 2005).

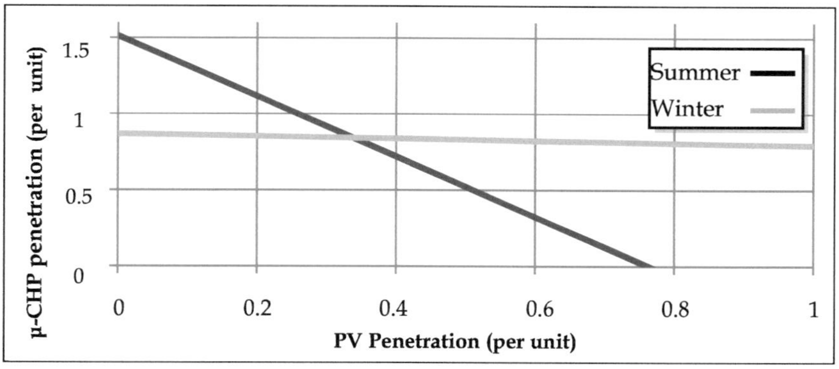

Fig. 7. Micro-generation penetration required to achieve the balance in the micro-grid

The intersection co-ordinates are m=0.34 and n=0.84, where m is the required PV penetration and n is the required µ-CHP penetration level. These penetration levels were applied to the generation profiles. By comparing the generation against the demand the energy imported from the grid and to the grid at each time interval was obtained.

The maximum energy storage capacity for each season was determined by plotting the evolution of the energy storage State of Charge (SoC), as described below. The power was assumed to be constant at each half-hourly interval; therefore the periods where the maximum energy was absorbed or injected to the grid determined the maximum power requirements.

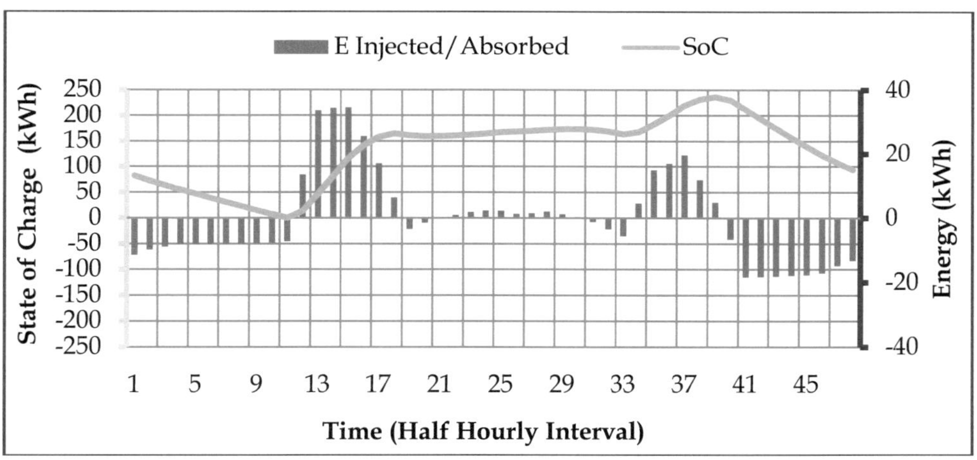

Fig. 8. Evolution of the energy storage during summer load conditions

Fig. 8 and Fig. 9 show the utilisation of the energy storage during typical summer and winter days for the optimal micro-generation penetration levels (Fig. 7). The black columns correspond to the energy injected or absorbed by the energy storage unit at each time step. The negative values represent the energy injected by the energy storage unit; conversely the positive values represent the energy absorbed by the energy storage unit. The energy injected/absorbed is presented on the figures' right axis (black). The grey line represents the evolution of the energy storage State of Charge (SoC) during the day. The SoC values are presented on the figures' left axis (grey). The maximum value of the grey line determines the capacity required by the energy storage in each season. The sizing of the energy storage in order to achieve self-sufficiency was found to be 236.27 kWh, which is the maximum value of the two seasons.

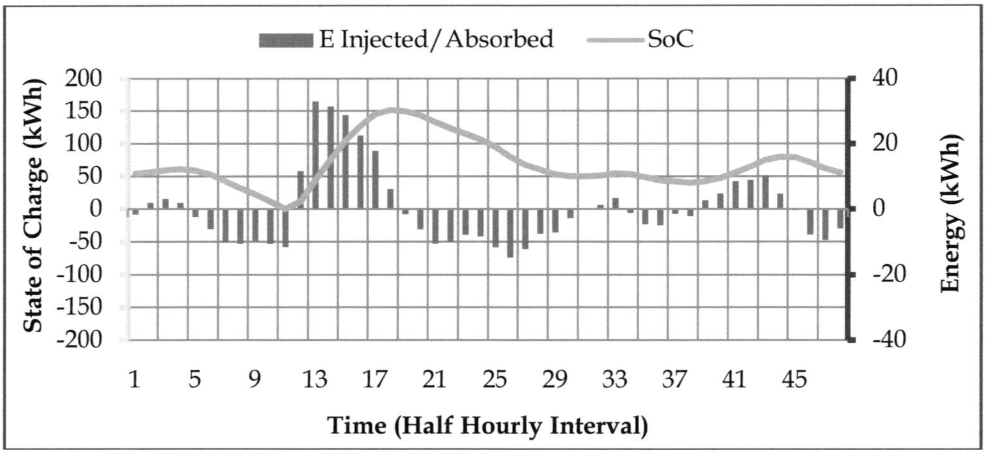

Fig. 9. Evolution of the energy storage during winter load conditions

The power and energy requirements of the energy storage system are summarised in Table 1.

Season	Maximum Power injected (kW)	Maximum Power absorbed (kW)	Capacity required (kWh)	Total Energy injected (kWh)	Total Energy absorbed (kWh)
Summer	-36.80	68.88	236.27	-251.68	251.68
Winter	-29.47	65.96	151.17	-202.55	202.55

Table 1. Energy storage requirements for summer and winter scenarios for the micro-grid

4. Balancing a local area with energy storage and backup generator

This section investigates the power rating and energy capacity requirements for the energy storage and backup generator to achieve a balanced area. Different penetration levels of PV and μ-CHP were used. The micro-grid model shown in Fig.1 was considered.
The steady state voltage changes and the system efficiency (power line losses) are evaluated for different penetration levels of micro-generation. A study case is presented, where the use of backup generator is minimised.

4.1 Methodology

The penetration levels considered for each micro-source range from 0% to 100% in steps of 10%. For each combination of micro-sources penetration level, sequential power flows were

performed at every time interval. The procedure was repeated for both summer and winter system conditions. The power line losses and the voltage steady state measurements were recorded for each time interval. The policy examined was to minimise the use of the backup generator. Once the losses were recorded, the need of a backup generator was estimated. The parameters considered were: (i) the system energy generation and demand, the (ii) energy storage efficiencies and (iii) the line losses. A round-trip efficiency of 72% was assumed for the ESS (Oudalov et al., 2007).
A backup generator was required in the case that:

$$\sum_{t=1}^{48} Egen_t - \sum_{t=1}^{48} Edem_t < 0 \tag{3}$$

Where:

$$Egen_t = \begin{cases} [(Gen_t - Dem_t) - Losses_t] * \eta ch * \eta dch, & Gen_t - Dem_t > 0 \\ 0, & Gen_t - Dem_t < 0 \end{cases}$$

$$Edem_t = \begin{cases} 0, & Gen_t - Dem_t > 0 \\ [(Dem_t - Gen_t) + Losses_t], & Gen_t - Dem_t < 0 \end{cases}$$

Where for each time step t:

Egen is the usable energy stored in the energy storage system,
Edem is the micro-grid required energy,
Gen is the energy generated by the micro-sources,
Dem is the domestic demand,
ηch is the energy storage charging efficiency and
ηdch is the energy storage discharging efficiency.

The flow chart in Fig. 10 depicts the developed algorithm. This chart shows the case where the overall system demand is higher than the generation. The calculations consider two consecutive days (i.e. 96 half hour time steps). The procedure is done for each combination of micro-sources for summer and winter season.
In time steps where the generation is higher than the demand, the energy stored in the ESS, is:

$$Ein = [(Gen_t - Dem_t) - losses_t] * \eta ch \tag{4}$$

If the demand is higher than the generation, the algorithm checks if there is energy available in the energy storage system ($SoC_t > 0$). In this case the energy supplied by the ESS is:

$$Eout = [(Dem_t - Gen_t) + Losses_t] / \eta dch \tag{5}$$

If additional energy is required than the available energy in the ESS ($SoC_{t+1} < 0$), the ESS will supply its available energy and the rest is supplied by the backup generator.
If the demand is higher than the generation and there is no energy available in the ESS, the energy is supplied by the backup generator:

$$EBup = (Dem_t - Gen_t) + Losses_t \tag{6}$$

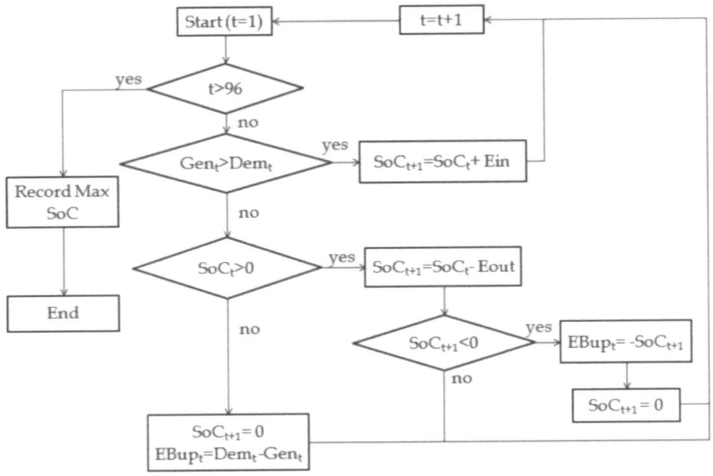

Fig. 10. Algorithm flow chart

The evolution of the energy storage system SoC was recorded; the maximum value of all the time steps determined the required capacity of the ESS. The maximum power requirements were determined by the amount of energy injected or absorbed by the ESS. The power rating of the backup generator was determined by the time step where maximum energy was generated.

4.2 Sensitivity studies

The sensitivity of the variables described in section 4.1 was examined, with respect to the energy losses and the voltage violation within the micro-grid. The capacity requirements of the ESS and the energy requirements of the backup generator were studied for various cases. The micro-generation levels were varied from 0-100% in 10% steps. The results are presented below.

4.2.1 System losses

The power line losses for all micro-generation penetration levels were recorded. Fig. 11 shows the results for both summer and winter season loading cases.

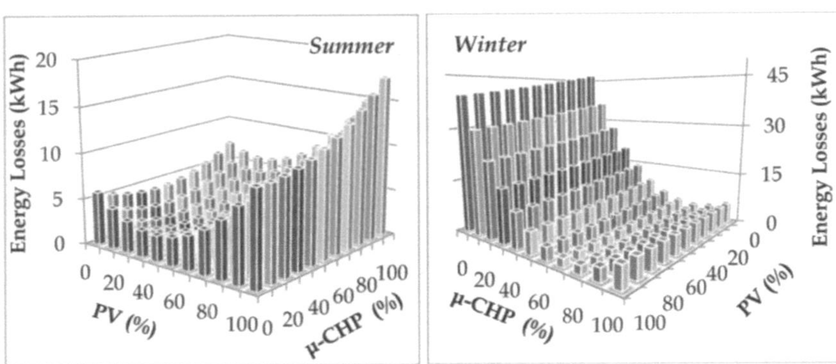

Fig. 11. Energy losses in the micro-grid

4.2.2 Steady state voltage limits violations

The voltage measurements of the remote feeder were recorded. Voltage violations were found only during summer loading conditions. Fig. 12 shows the cases when voltage violated statutory limits. Combinations of PV and μ-CHP penetration levels, marked with an "X" in Fig. 12, exceeded the voltage statutory limits.

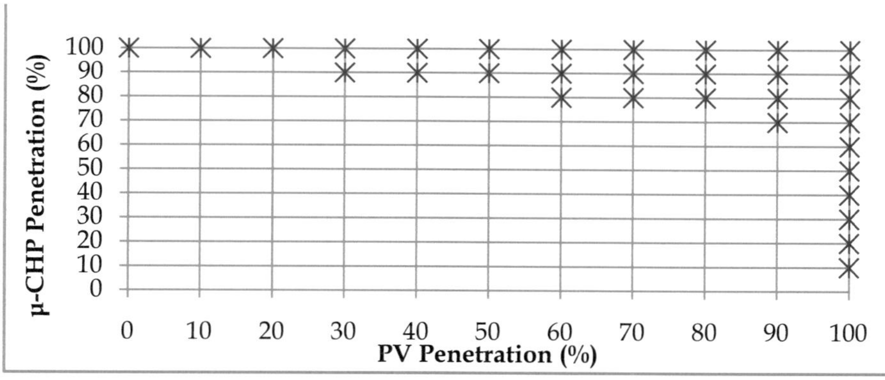

Fig. 12. Voltage statutory limits violation for summer season system loading

4.2.3 Backup generator energy requirements

Fig. 13 shows the backup generator energy requirements to achieve an energy balanced micro-grid for all studied micro-generation penetration levels. It can be observed that during the winter season, the backup generator requirements are determined by the μ-CHP penetration. In combinations where no backup generator is required, the energy balance is achieved only with the ESS.

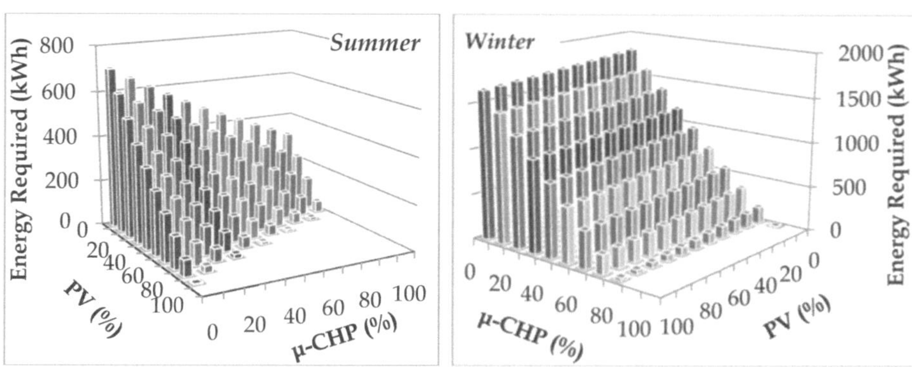

Fig. 13. Backup generator energy requirements for different penetration levels of micro-generation

4.2.4 Energy storage system capacity requirements

Fig. 14 shows the ESS capacity requirements to achieve an energy balanced micro-grid for all studied micro-generation penetration levels. It can be observed that for low micro-generation penetration levels, the energy storage capacity requirements are limited due to the reduced energy exported from the micro-generators. During summer loading conditions, for high micro-generation penetration levels, the ESS requirements are determined by the time period when no energy is generated from the micro-generators (time step 40 to time step 9 in Fig.6).

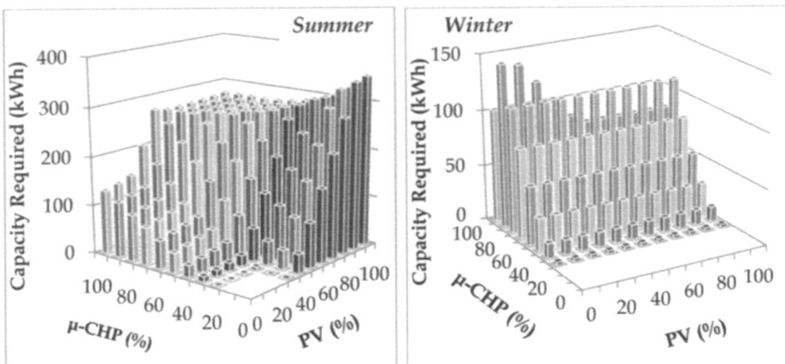

Fig. 14. ESS capacity requirements for different penetration levels of micro-generation

4.2.5 Minimum energy imported from the backup generator

The sensitivity studies of the above subsections showed that the minimum energy required from the backup generator, within voltage statutory limits, would be achieved with a μ-CHP penetration of 80% and a PV penetration of 50%. In this case, the total amount generated during summer was found to be higher than the demand. The technical ESS and backup generator requirements for this case are shown in Table 2.

	Energy (kWh)		Power (kW)	
Back up Generator	Summer	Winter	Summer	Winter
	0.00	108.48	0.00	23.27
	Capacity (kWh)		Power (kW)	
Energy Storage System	Summer	Winter	Summer	Winter
	279.64	112.74	68.59	60.35

Table 2. Energy storage system and backup generator's requirements

5. Energy storage market participation

The use of the energy storage device to participate in the electricity markets is presented. The minimum SoC required at every time interval to ensure energy balance in the micro-grid is obtained from Section 4. The MATLAB Optimisation Toolbox (MathWorks™, 2011) is used to obtain the optimal behavior of a pre-defined rated ESS, based on the requirements of the studied micro-grid.

5.1 Methodology

In this study, the "fmincon" tool from the MATLAB Optimisation Toolbox (MathWorks™, 2011) was used. Fmincon uses a gradient based Non-Linear Programming method. The algorithm that was used was the Interior Point algorithm. The algorithm probes the objective function variables to draw the gradient of the variables. In this way, the direction of the optimal point is estimated.

The objective function was:

$$L = \sum_{t=1}^{48} (Esell_t * Psell_t - Ebuy_t * Pbuy_t) \quad (7)$$

Where for each time step t:

Esell is the energy sold to the market,
Psell is the electricity price of the energy sold,
Ebuy is the energy bought from the market and
Pbuy is the electricity price of the energy bought.

The evolution of the energy storage SoC is constrained to the energy traded at each time period (equation 8):

$$SoC_{t+1} = SoC_t - Esell_t + Ebuy_t \quad (8)$$

The maximum energy traded is constrained by the minimum SoC requirements at every time step (equation 9). After each trading period the SoC available in the energy storage has to be at least the minimum value calculated in order to achieve an energy balance.

$$MinSoC_{t+1} \leq SoC_t - Esell_t + Ebuy_t \quad (9)$$

Where

MinSoC is the minimum State of Charge required at the specific time step

The maximum energy to be traded will be also constrained to the energy storage unit maximum capacity (equation 10) and its maximum power rating (equation 11 and 12). Since the power is assumed to be constant during each time step, the maximum power will be determined by the energy traded divided by the duration of the half an hour time step.

$$MaxSoC \geq SoC_t - Esell_t + Ebuy_t \quad (10)$$

$$Pmax \geq Esell_t / 0.5 \quad (11)$$

$$Pmax \geq Ebuy_t / 0.5 \quad (12)$$

Where

MaxSoC is the energy storage maximum capacity and
Pmax is the maximum power rating of the energy storage.

The final constraint is presented in equation 13; the aim of this constraint is to avoid selling and buying electricity at the same time step.

$$Min(Esell_t, Ebuy_t) = 0 \quad (13)$$

5.2 Case study

5.2.1 Assumptions

The micro-grid under study in this section, together with the generation and demand profiles are those presented in Section 2 and 3. The energy storage SoC requirements are the values obtained in Section 4.2.5, where an 80% µ-CHP and 50% PV penetration were considered. The energy storage unit considered has a capacity of 300kWh and a maximum power output of 100kW and is considered to be grid-connected. Average electricity prices were drawn from the literature (New Electricity Trading Arrangements, 2011), the electricity buy price (from the market) is assumed to be 10% higher than the electricity sell price (to the market) (Dimeas, 2006). The parameters considered in the optimisation algorithm are the electricity prices and the energy storage SoC requirements presented in section 4.2.5. Neither operational and maintenance costs, nor the ageing effects in the energy storage unit have been considered. It has been assumed that the energy traded is not constrained to a minimum energy amount (i.e. any amount of energy can be traded). The backup generator has not been considered in the optimisation.

5.2.2 Results

The amount of energy to be traded at different time steps, which would maximise the benefits of the energy storage unit under the given conditions, is shown in Fig. 15 for winter and in Fig. 16 for summer scenario. The grey line presents the SoC evolution of the energy storage, the black line shows the minimum SoC requirements of the energy storage at every time step. The black bars represent the amount of energy sold; the grey bars represent the amount of energy bought and the dotted black line shows the market prices (presented in the secondary axis).

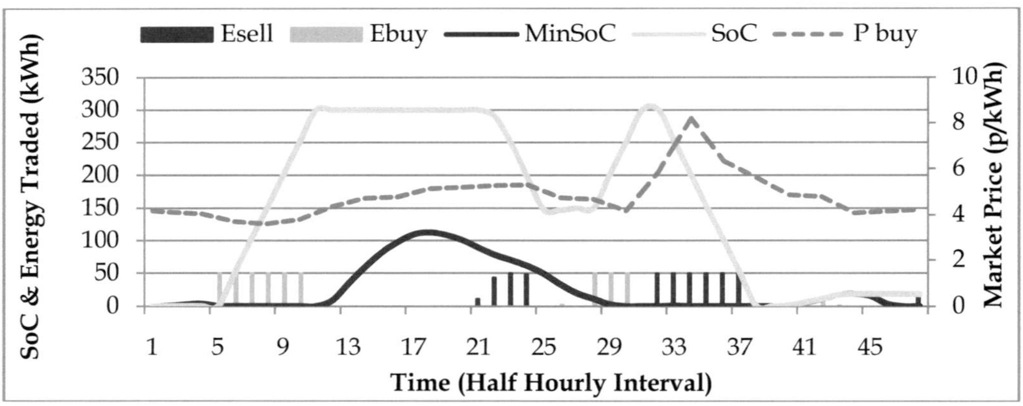

Fig. 15. Optimisation results for winter scenario

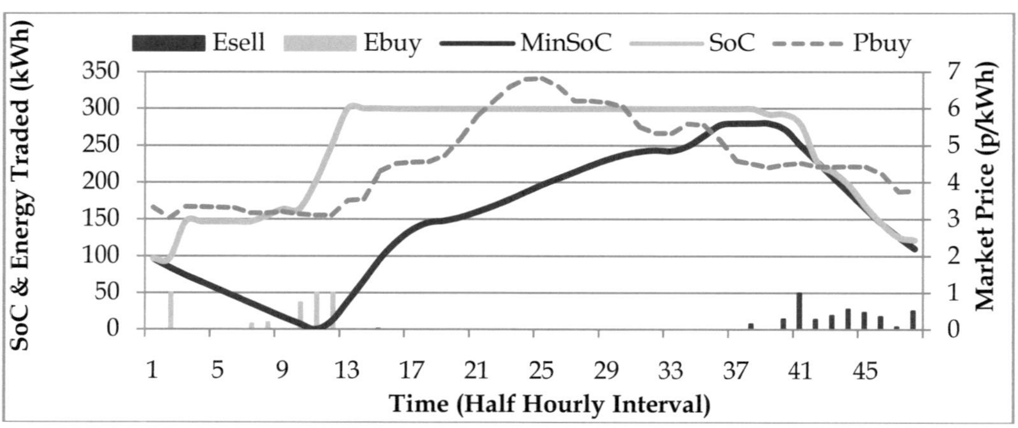

Fig. 16. Optimisation results for summer scenario

Fig. 15 and Fig. 16 show how the maximum energy that can be traded at each time step is constrained by the maximum power output of the energy storage; having a maximum power output of 100 kW, the maximum energy that can be imported or exported per time step is 50 kWh. It can be seen during summer loading conditions, that the market participation is constrained by the minimum SoC requirements. For winter loading conditions, the higher differences between the maximum energy storage capacity and the SoC requirements enables higher market participation.

6. Conclusions

Energy systems are anticipated to face changes with the increase of power generation devices embedded in the LV networks. Micro-grid is a concept developed in the last decade. It describes LV electricity networks that may deliver electricity to the customers supplied in it, either connected to the main grid or islanded in autonomous mode. The islanded mode would be feasible with the use of power generated by dispersed generation embedded in LV networks, usually at the customer premises. To ensure reliability and security of supply, the concept of intentional islanding was introduced. Intentional islanding is the forced transition to islanded mode due to a major event in the network.

The main technical issues associated with intentional islanding were addressed in Section 2. The impact of micro-generation sources on intentional islanding was analysed. Frequency, voltage, earthing and protection issues were discussed. Literature based solutions were analysed for each of the discussed impacts. A UK micro-grid electrical model was defined. The combination of an Energy Storage device with a Backup Generator for supporting islanded mode operation was proposed. This combination would assist the smooth transition between grid connected and islanded modes. Steady state voltage studies were conducted for different micro-generation penetration levels and micro-grid loading conditions. The results showed that up to 1.1 kW per customer micro-generation penetration can be safely integrated with regards to the voltage statutory limits.

To achieve energy autonomy in a micro-grid, the energy generated by micro-sources should cover the load demand. A methodology was presented in Section 3, to obtain the optimal ESS power and energy capacity within a micro-grid. The ESS would have to bring energy balance to the micro-grid, for given micro-generation penetration levels. A case study was defined based on the micro-grid model of Section 2. The combined use of a backup generator and the ESS to achieve the energy balance was investigated in Section 4. Optimal combinations were calculated for different micro-generation penetration levels and micro-grid loading conditions.

To maximise the benefits of an ESS, its participation in the electricity market was investigated in Section 5. This feature would entail the dual use of ESS for both local support and energy trading. To achieve this, a minimum State of Charge should always be available in the ESS. This minimum SoC for each time interval was obtained from Section 4. Assuming that the rest of the ESS capacity is available for market participation, an optimisation tool was used to schedule the energy storage charging and discharging.

7. Acknowledgements

This research was supported by the UK's Engineering and Physical Science Research Council (EPSRC), through a Science and Innovation award.

8. References

Abu-Sharkh, S.; Li, R.; Markvart, T.; Ross, N.; Wilson, P.; Yao, R.; Steemers, K.; Kohler, J. & Arnold R. (2005). Microgrids: Distributed on-site Generation, Tyndall Centre for Climate Change Research, Technical Report 22.

Abu-Sharkh, S.; Arnold, R.J.; Kohler, J.; Li, R.; Markvart, T.; Ross, J.N.; Steemers, K.; Wilson, P.; & Yao, R. (2006). Can Microgrids Make a Major Contribution to UK Energy Supply?, *Renewable and Sustainable Energy Reviews*, , Pages 78-127, ISSN 1364-0321, DOI: 10.1016/j.rser.2004.09.013. April 2006

Blades, H. (2008). Power cut in South London after Fire at Sub-Station. *The Times [online]*. Available from: http://www.timesonline.co.uk/tol/news/uk/article4090867.ece

Boutsika, Th.; Papathanassiou, S. & Drossos, N. (2005). Calculation of the Fault Level Contribution of Distributed Generation According to IEC Standard 60909, *Proceedings of CIGRE Symposium Power Systems with Dispersed Generation*, Athens, April 2005.

British Broadcasting Corporation [BBC] (2010). Fire cuts power to thousands of Portsmouth homes. *British Broadcasting Corporation [BBC]*. Available from: http://www.bbc.co.uk/news/10423764

Chowdhury, S.P.; Chowdhury, S.; Chui Fen Ten & Crossley, P.A. (2008). Islanding Protection of Distribution Systems with Distributed Generators – A Comprehensive Survey Report, *IEEE Power and Energy Society General Meeting* - Conversion and Delivery of Electrical Energy in the 21st Century, pp.1-8, July 2008

Conti, S.; Raiti, S. & Tina, G. (2003). Small-Scale Embedded Generation Effect on Voltage Profile: an Analytical Method, *IEE Proceedings Generation, Transmission and Distribution*, pp. 78-86, January 2003.

Dexters, A.; Loix, T.; Driesen, J. & Belmans, R. (2007). A Comparison of Grounding Techniques for Distributed Generators Implemented in Four-Wire Distribution Grids, UPS Systems and Microgrids. *Proceedings of CIRED 2007, 19th International* Conference and Exhibition on Electricity Distribution. Austria, May 2007

Dimeas, A. L. (2006), Contribution to the Distributed Control of Power Systems with Distributed Energy Resources in Low Voltage. *Doctorate thesis*, National Technical University of Athens - NTUA, Athens, 2006

Energy Saving Trust. (2005). Potential for Microgeneration Study and Analysis, Final Report, Econnect, ElementEnergy

Energy Network Association (1991) G59/1, Recommendation for the Connection of Embedded Generating Plant to Public Distribution Systems above 20 kV or with Outputs over 5MW

Energy Network Association (2008) G83/1, Recommendation for the Connection of Small Scale Embedded Generators (up to 16A per phase) in parallel with public low-voltage distribution networks

Grau, I.; Cipcigan, L.M. Jenkins, N. & Papadopoulos, P. (2009a). Microgrid Intentional Islanding for Network Emergencies, *44th International Universities Power Engineering Conference, (UPEC 2009)*, Glasgow, September 2009.

Grau, I.; Skarvelis-Kazakos, S; Papadopoulos, P.; Cipcigan, L.M. & Jenkins, N. (2009b). Electric vehicles (EV) Support for Intentional Islanding. A Prediction for 2030, *North American Power Symposium*, Mississippi, October, 2009.

Grau, I.; Papadopoulos, P.; Skarvelis-Kazakos, S; Cipcigan, L.M. & Jenkins, N. (2010). Virtual Power Plants with Electric Vehicles, *2nd European Conference SmartGrids and E-Mobility*, Brussels, October 2010

Ingram, S. & Probert, S. (2003). The Impact of Small Scale Embedded Generation on the Operating Parameters of Distribution Networks, P B Power, Department of Trade and Industry

Institute of Electrical and Electronics Engineers (2003). IEEE 1547 Standard for Interconnecting Distributed Resources with the Electric Power System

Ito, T.; Tanabe, T.; & Sasaki, N. (2007). Development of Microgrid Control System, Meiden Review, International Edition, 2007. 140(2): p. 4-7.

Jayawarna, N.; Jenkins, N.; Barnes, M.; Lorentzou, M.; Papthanassiou, S. & Hatziagyriou, N. (2005). Safety Analysis of a Microgrid, *International Conference on Future Power Systems*, pp.-7, November 2005.

Jenkins, N.; Allan, R.; Crossley, P.; Kirschen, D. & Strbac, G. (2000). Embedded Generation, IEE Power and Energy Series 31, ISBN: 0-85296-774-8 & 978-0-85296-774-4, London, UK.

Kato, T.; Suzuoki, Y.; Funabashi, T.; Cipcigan, L. & Taylor, P. (2008). Microgrid Configuration for Major Network Events, SmartGrids for Distribution, *IET-CIRED Seminar*, pp.1-4, Frankfurt, June 2008

Kondo, Y.; Kato, T.; Suzuoki, Y. & Funabashi, T. (2007). A Study on Composition of Emergency Microgrid Consisting of Residential PV Systems, *proceedings of the 23rd Conf. on Energy, Economy, and, Environment*, pp.61-64, Tokyo, January 2007.

Lawton, L.; Sullivan, M.; Van Liere, K. & Katz, A. (2003). A Framework and Review of Customer Outage Costs: Integration and Analysis of Electric Utility Outage Cost Surveys', U.S. Department of Energy, Contract No. DE-AC03-76SF00098

Lopes, J.A.P.; Moreira, C.L. & Madureira, A.G. Defining Control Strategies for MicroGrids Islanded Operation. (2006). *Power Systems, IEEE Transactions*, pp. 916- 924, May 2006 DOI: 10.1109/TPWRS.2006.873018

MathWorksTM. (2011) MATLAB® Optimization Toolbox, Available from http://www.mathworks.com/products/optimization, (visited 01.04.2011)

Mott McDonald. (2004). System Integration of Additional Micro-generation (SIAM), Department of Trade and Industry, September 2004, available from the DGCG website www.distributedgeneration.gov.uk.

New Electricity Trading Arrangements (2011). Available from http://www.bmreports.com/

Oudalov, A.; Cherkaoui, R. & Beguin, A. (2007). Sizing and Optimal Operation of Battery Energy Storage System for Peak Shaving Application, *IEEE Power Tech*, , vol., no., pp.621-625, 1-5 July 2007 DOI: 10.1109/PCT.2007.4538388, Lausanne, July 2007

Papadopoulos, P.; Skarvelis-Kazakos, S.; Grau, I.; Cipcigan, L. M. & Jenkins, N. (2010). Predicting Electric Vehicle Impacts on Residential Distribution Networks with

Distributed Generation, *IEE Vehicle Power and Propulsion Conference (VPPC)*, Lille, September, 2010.

Skarvelis-Kazakos, S.; Papadopoulos, P.; Grau, I.; Gerber, A.; Cipcigan, L. M.; Jenkins, N. & Carradore, L. (2010) Carbon Optimized Virtual Power Plant with Electric Vehicles, *45th International Universities Power Engineering Conference (UPEC),* Cardiff, September, 2010

UK Energy Research Centre. (1997). Electricity user load profiles by profile class, Available from http://data.ukedc.rl.ac.uk/browse/edc/Electricity/LoadProfile/data

Wu, C.X.; Wen, F.S. & Lou, Y.L. (2008). The Existed Problems and Possible Solutions of Micro-Grid Based on Distributed Generation, *Conference on Electric Utility Deregulation and Restructuring and Power Technologies (DRPT)* 2008., pp.2763-2768, Nanjing, April 2008

Zhou, Y.; Ferreira, J.A. & Bauer, P. (2007). Grid-Connected and Islanded Operation of a Hybrid Power System, *IEEE Power Engineering Society Conference and Exposition in Africa, 2007. PowerAfrica*, pp.1-6, Johannesburg, July 2007

15

Sizing and Management of Energy Storage for a 100% Renewable Supply in Large Electric Systems

Oscar Alonso, Santiago Galbete and Miriam Sotés
Acciona, Spain

1. Introduction

Many developed countries are moving towards a low carbon economy and are therefore demanding higher levels of renewable energy sources. These energy sources include wind, solar, biomass, etc. to supply the energy demand. Nevertheless, there are still some aspects that warrant further technical and economical feasibility studies for those renewable energy sources to be considered sustainable alternatives.

The random nature of renewable energy sources, mainly solar and wind is the major limiting factor in achieving significant penetration in any electric system. This limiting factor has different consequences depending on the ratio between the amount of renewable generation and the demand level. This has been studied by many authors from different perspectives and in many cases the key element was energy storage. For example, energy storage can be used to reduce the production fluctuations of large scale wind farms, to move a certain production amount to better remunerated periods, to reduce prediction errors to minimize penalties, to increase the power predictability, to participate in secondary power markets and to achieve fully controllable energy production through any renewable primary source. In such way, any renewable generator would offer guaranteed production and may participate in electric markets on equal terms with non-renewable generators.

Most analyses of isolated large electric systems with renewable supply and storage are performed based on energy balance results over several years (Bremen et al., 2009) (Alonso et al., 2009, 2010). This methodology has been extended in order to include real data measurements from various renewable technologies and also storages with different dynamic and rates. In particular, this method allows the resolution of multiple scenarios of renewable penetration levels, profiles, technologies, etc in order to obtain the minimum storage service that will reduce the conventional production or even satisfy a total supply of the demand through only renewable producers. This methodology has been firstly used to analyse a suitable large system: the Spanish electric system. Although Spain is electrically connected with other countries, the low rates in exchanged energy allow the simplification to consider it as an isolated system. As it will be shown, Spain offers excellent opportunities to produce large amounts of renewable production. However, the integration will require some storage to guarantee the electrical supply meets the demand, especially in future scenarios where 100% is proposed only using renewable sources. Nevertheless, the existence of current hydro systems with huge storages strongly reduce the need of additional units.

The transition from current generating mix (renewable and non-renewable) to a likely future generation mix (with only renewable) has been also analyzed. As it will be explained, the prompt introduction of these storage systems, better if it is arranged in a very disperse way, will ease the replacement of conventional generation, starting with those units especially pollutant. Finally, some proposals of future scenarios are also presented and discussed regarding technical feasibility.

2. Random nature of renewable energy sources

Renewable energy from wind and solar sources are rapidly increasing their influence in the electric grid system worldwide. Both energy sources are uncontrollable by nature. Because of that when their contributions become important, locally and at greater scales, several and well reported grid integration problems may arise. Although energy demand and renewable production are random in nature, demand usually maintains a clear tendency that allows its reliable forecasting, especially in developed countries. For example, figure 1 shows the typical demand profile during a working and a non working day for Spain during 2010. Demand follows a similar profile experiencing variations during the year; season dependent, and also from one year to another, figure 2. These variations normally depend on the economical situation and the development level of the analyzed system. For example, 2006 Spanish electric demand versus 2005 verified an increase around 2.9%, while 2006 versus 2007 was about 3.8% (REE, 2009).

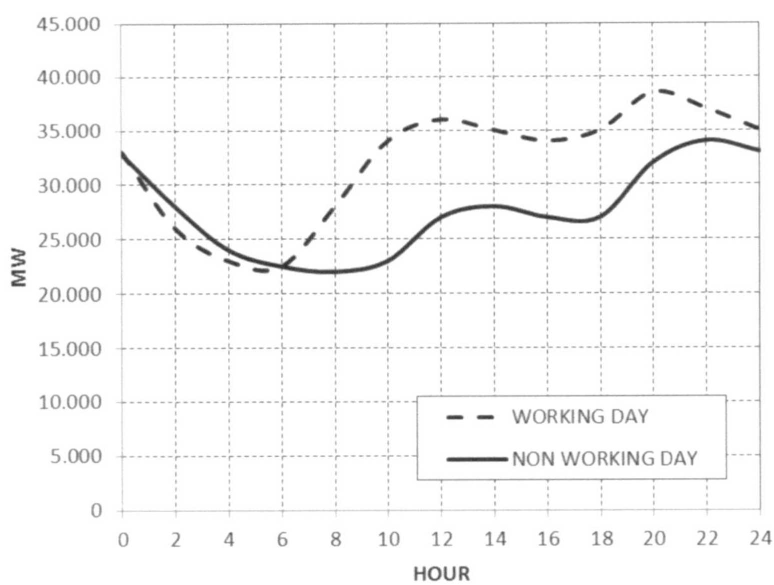

Fig. 1. Example of daily electric demand in Spain (Working and non working day)

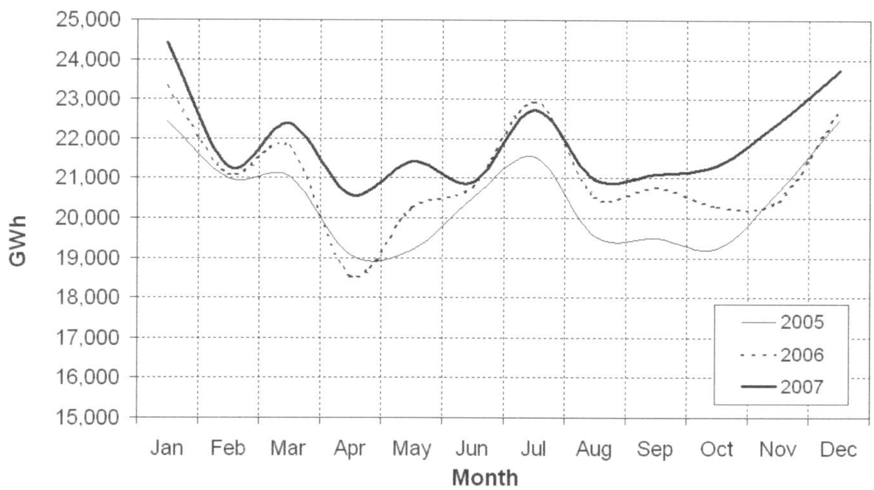

Fig. 2. Monthly demand in Spain (2005 - 2007)

In a preliminary stage the following basic balance is valid for any isolated electric system without storage.

$$\begin{gathered} \textit{Non Renewable Power} + \textit{Potential Renewable Power} \\ = \\ \textit{Power Demand} + \textit{System Power Losses} + \textit{Renewable Power Losses} \end{gathered} \tag{1}$$

In this equation, *Non Renewable Power* represents all power contributions coming from conventional energy sources such as coal, gas, nuclear, co-generation, etc -based power plants. *Potential Renewable Power* is the power that could be produced at any moment taking into account all operative power plants of wind, solar, hydro, etc. This amount of power depends on the primary source availability and the technology efficiency. *Power demand* is the electric power demanded by users, *System Power Losses* represents power losses in lines, transformers, etc, and *Renewable Power Losses* represents the renewable power that was not transformed into the electric system. Several and important aspects are involved with this concept. The integration of all *Renewable Power Losses* during a year from now on will be called the *Renewable Energy Losses*.

A few years ago the Renewable Energy Losses were insignificant because grid operation rules in most countries were establishing priority for the new renewable sources in detriment of conventional ones. Normally local overload on some lines has been wielded to temporally stop some generators. However, nowadays several countries with high wind energy penetration have been changing the rules to allow grid operators to stop wind generation under the excuse of low quality of service or low reliability of the grid operation. Despite of the ethical debate about all these particular aspects, these new losses are becoming important and will increase during the next years unless newer grid stability solutions are provided.

The above mentioned energy losses are hard to be either calculated or estimated. Nevertheless, apart from those losses, as non-controllable renewable energy penetration increases, energy opportunity production losses due to demand limitation shall be added. It may happen that non-controllable renewable production exceeds the electric demand and as a consequence renewable generation must be limited. Thus, *Renewable Energy Losses* are mainly due to two different causes, grid operation and over production stops (locally or globally).

Figure 3 shows normalized hourly potential productions coming from non-controllable renewable energy sources versus maximum hourly demand value. To illustrate this two different cases have been considered: Spain and Navarra. Navarra is a Spanish region with 620,000 inhabitants with a high renewable penetration level (in 2010 more than 80% of its electric demand was supplied by renewable generators). In Spain´s case it is appreciated that the potential production is always well below the demand curve (also normalized). Potential Renewable production never reaches the electric demand, which means that in principle *Renewable Energy Losses* should be zero. However, analyzing the case of Navarra; potential renewable production is higher than electric demand in a large number of hourly intervals. Assuming Navarra was an isolated electric system; non-controllable renewable plants should be stopped due to the considerably increasing *Renewable Energy Losses*.

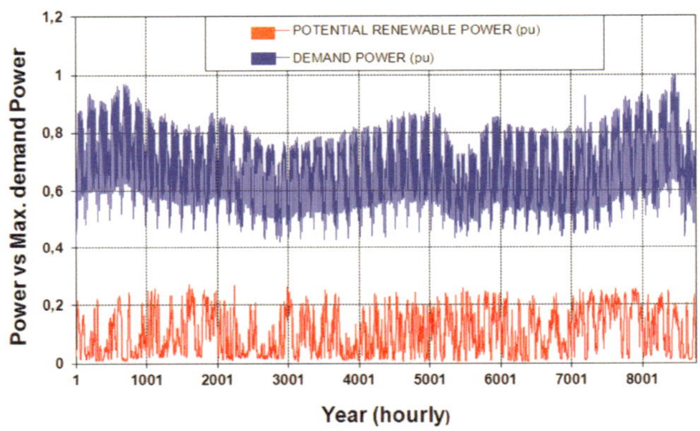

(a) Spain

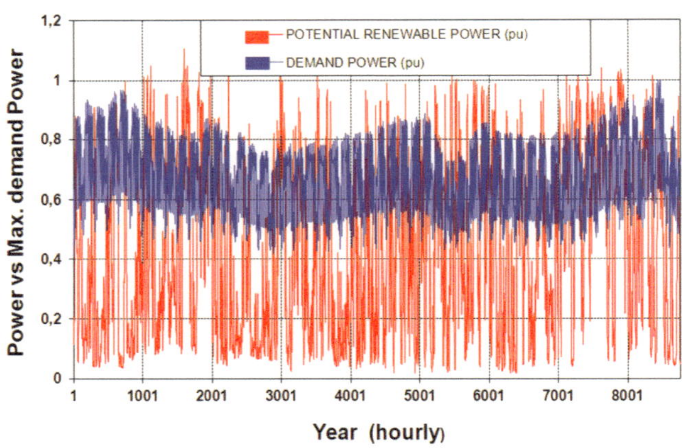

(b) Navarra

Fig. 3. Normalized Renewable Production and demand

Figure 4 shows the yearly evaluation of the *Renewable Energy Delivered to Grid* (a) and the *Renewable Energy Losses* (b) in function of the *Potential Renewable Production* versus electric demand. From now on, this last relation will be named as *RPPR, Renewable Potential Production Ratio*. In this analysis, no energy storage system of any kind was considered.

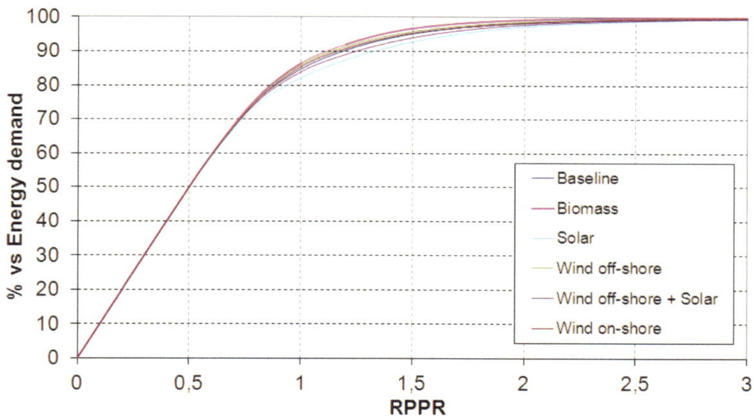

(a) Yearly Renewable production delivered to grid

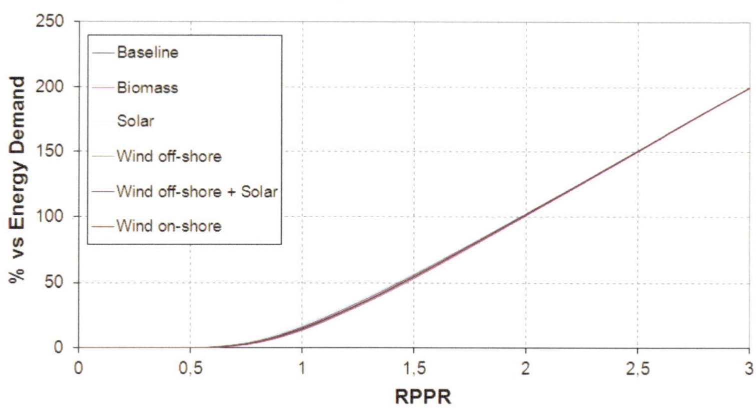

(b) Renewable Energy Losses

Fig. 4. Normalized index at different *RPPR*

Therefore, this graph is showing a hypothetical substitution of non-renewable energy with renewable in an isolated electric system. Different combinations of likely on-shore wind, off-shore wind, solar, biomass and hydraulic power plants were included in these renewable mixes (large hydraulic power plants were not considered). Every combination is built up using a differently scaled data series of real production of each technology. Trought this way it is possible to prepare combinations where one or two technologies get highlighted; according to different and likely future perspectives that seem to consider more feasibility on some technologies than others. Detailed contribution of each renewable technology for any combination will be defined later in section 4.2.3. For instance, renewable mix named "Baseline" includes all technologies according to current levels of each one in Spain. For higher *RPPR* levels this set of series is scaled according to available official development plans (Spanish Ministry of Industry, 2010) and some assumptions. However, the mix named "Solar" was prepared to follow a different tendency. For low *RPPR* the combination of technologies is the same as in the Baseline case, but as the *RPPR* is increased the high solar power contribution is highlighted with respect to the baseline case becoming the more relevant renewable influence. Same concepts apply to the rest of combinations already prepared. Although results depend on the specific production mixes and demand profiles used, no big differences have been found as it can be seen in figure 4. It is remarkable that only very big ratios of *RPPR* achieve complete demand fulfillment without storage systems, which then involves extremely huge *Renewable Energy Losses*, besides of an unacceptable cost effective energy. This aspect can also be seen in figure 5 where the minimum *RPPR* to get a 100% renewable-based supply for every combination of renewable technologies has been presented. The main reason for those big *Renewable Energy Losses* is the necessity of stopping a lot of renewable production plants as demand is lower than available production. Even though current scenarios of renewable production differ considerably with correspondents shown on figure 4, clearly it is appreciated that not only renewable production plants will be required, but additional elements to optimize the global energy management such as energy storages.

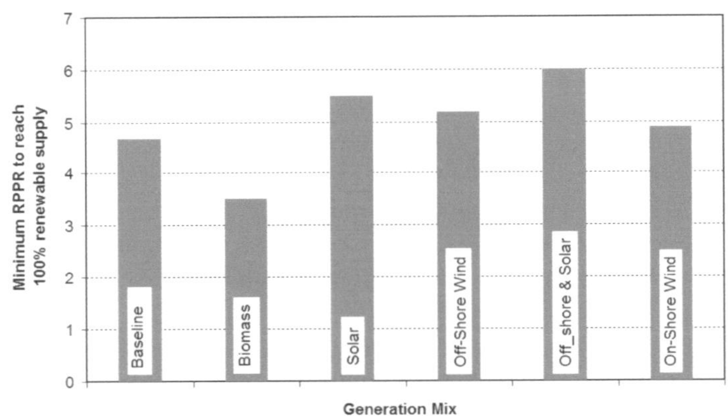

Fig. 5. Minimum *RPPR* to get a 100% renewable-based supply for different combinations of renewable technologies.

According to the results exposed in figure 4 and 5 the following conclusions can be obtained:

- Renewable real production does not fulfill the electric demand within a reasonable renewable overproduction. Analyzing extrapolations of real electric systems with high penetration of renewable producers (without any storage) to guarantee the demand excessively large and unviable *RPPR* levels (large renewable system) are required. Higher levels of off-shore generators seem to diminish the variability (Tipping & Sinclair, 2009) thus, improving the demand tracking capabilities with lower *RPPR*.
- *Renewable Energy Losses*, become considerable with *RPPR* ratios higher than 0.5. Those losses are only consequences of a potential renewable production that is higher than the electric demand. Actually, *Renewable Energy Losses* may be higher than those shown on the graph, since grid operators may reduce renewable generation arguing low

reliability of the operation. For the Spanish electric system, figure 3, *Renewable Energy Losses* are negligible since current *RPPR* ratio is still very small. However, for Navarra with a *RPPR* higher than 0.6 in 2009, in case this region was electrically isolated a 5% of renewable losses should be expected.

- The distance between the *Real Renewable Production* (Inst. Demand supply line) and the 100% line is the percentage of the required non-renewable energy contribution. For a certain future free of non-renewable generators this energy should be based on controllable renewable energy (biomass, hydro, H_2 plants, etc.). However, in most countries it will not be feasible to get this amount of controllable renewable production. Therefore to get the 100% renewable target for most *RPPR* a great amount of non-controllable production would be required, which again would produce excessive *Renewable Energy Losses*.

Renewable Energy Losses and the necessary controllable renewable energy may also be strongly reduced for any *RPPR* when energy storages are included in the electrical equation. As it will be demonstrated such energy storages are not necessarily huge. Besides, renewable based systems are usually presented on a highly dispersed basis across the territory and a system of smaller and strategically distributed storages may definitely increase the global operation, performance and reliability. For example, the combination of a wind power plant and a relatively close to site storage under a common control system makes possible the operation of the whole system like a controllable power plant. Thus, combined solutions may offer a similar performance to actual conventional plants, opening new technical features for a better integration and also new possibilities on the electric market. The analysis about how different storage systems may help with the integration of more renewable production is the main objective of the following study. This analysis has been extended to the ideal situation of a total demand supply only with renewable producers.

3. Analysis methodology

A mathematical model properly defined offers the possibility to explore opportunities of higher integration levels of renewable sources in large and isolated electric systems using energy storages. Therefore, equation (1) must be completed with the storage contributions as it is established on equation (2). Here, *Storage Power* responds to control system needs and can be positive or negative, producing the corresponding decrement or increment of the storage energy level. This power must also satisfy the condition of a long term integral tendency to a constant value (steady state operation). Of course, the storage model also must include limits in power and levels to represent real systems.

$$\begin{gathered} \textit{Non Renewable Power} + \textit{Potential Renewable Power} \\ = \\ \textit{Power Demand} + \textit{System Power Losses} + \textit{Renewable Power Losses} + \boldsymbol{\textit{Storage Power}} \end{gathered} \tag{2}$$

The mathematical modeling of these systems requires power sources (renewable and non renewable), power sinks (demand and losses), storages and a set of control rules. Renewable power sources can be modeled using scalable power profiles depending on the specific technology. This approach will increase the reliability of the whole modeling, especially if said power profiles are based on real production measurements (Acciona Remote Control

Centre, 2010) recorded over the years on a number of existing power plants. Power demand profiles, series, etc, are usually available from grid operators (REE, 2010), while storage dynamics are usually simple involving a few global parameters. Control or management rules may also be quite complex, in particular when forecasting and territory distribution are considered. The complete model should be appropriate to describe current scenarios but, more importantly, to advance as much as realistically possible future opportunities. Of course, this approach does not consider likely improvements in future renewable technologies thus offering lower performance solutions. This chapter introduces some basics about the mathematical model and real data used in these studies. In any case, both the modeling process and analysis has been carried out considering the following and important directives:

- To guarantee the electrical demand at any moment.
- To minimize the non-renewable production.
- To minimize *Renewable Energy Losses*.
- To offer reliable solutions based on stationary multi-annual study.

The simulation platform may be prepared to work with two storages: hydro-based and a generic solution of reversible storage. The process to get in and out in this reversible storage is also performed by generic electric power drives. In this way it is avoided to mention any specific technology: water, batteries, heat, compress-air, etc. The future seems to be opened to most of these technologies, although size and grid penetration of every one will finally depend on aspects such as technical development, economical ratios, environmental or territorial limitations, etc (Price, 2010). Water-based storage size is fixed according to real plants already in operation in Spain and only power drivers (pumps and turbines) are available to be changed for further analysis. However, reversible storages have parameters for storage size, power drives, performances, etc. Both storages can operate together coordinately. The program includes a set of rules to determine when and how much a specific storage shall work. The decisions for that are taken considering current levels of energy on each storage, the required turbine power (or drive out power), etc. In any case, the main objective of such rules is to offer globally as much available storage as possible and also help to increase as much renewable energy as possible. The system also offers parameters to freely setup levels of different renewable producers: solar, wind, biomass, etc. Therefore, the whole system shows programmed results suitable for exploring current and future renewable-based electric systems with any grid penetration level.

3.1 Renewable power profiles

Renewable power sources have been modeled using production measurements recorded from a number of real power plants (Acciona Remote Control Centre, 2010). These facilities have been in operation in Spain for years and include solar, on-shore wind, biomass and hydro sources. For every different plant there was available several years of hourly-based series of power production. All data were arranged and compiled technology by technology in order to prepare independent time-series during a *normal year*. Thus, this equivalent or normal year was determined to represent a set of years in terms of average and standard deviation production. The final result was a set of hourly-series of renewable power production by technology also capable to be scaled to any power. This set of models has been very useful to determine the sensitivity of storage systems depending on the specific composition of renewable power mixes.

Off-shore hourly-series were calculated using ocean wind speed measurements and power curves of most promising multi-megawatt off-shore wind turbines. However, for other important renewable technologies, like tidal waves, geothermal, etc, it was not available neither time series nor reliable technical information to produce useful data. Therefore, due to the huge potential of these technologies (García & Linares, 2005) better future perspectives are expected than already obtained.

All available hourly-series of real measurements were obtained from databases property of Acciona Energy. This company is also owner of large renewable systems covering a total of 8,500 MW. Most facilities have been running for years providing invaluable data of renewable production, losses and other technical aspects of great importance.

3.2 Simulation basis

The simulation program has been setup to simulate the operation of a large electric system over several years until steady-state is reached. This program includes several preparatory stages prior to the running of the multi-annual simulation:

1. Setup. Definition of the global system size by means of some parameters:
   a. Renewable Potential Production Ratio (RPPR).
   b. Reversible storage size and power.
   c. Renewable mix. Weighting factors are available for every renewable technology. Through them a specific profile of the global mix can be established. For example, it is possible to prepare a renewable mix to strengthen the influence of solar generators, or off-shore influence, etc. Therefore, a set of different mixes with distinct characters can be specifically prepared which are useful for sensitivity analysis.
2. Initialization. Setting up of initial values on all state variables, focused to get as close as possible to the steady state.
3. Simulation of one year. The electrical system is simulated hour by hour using the algorithm in figure 6. This complex decision-making model is based on equation (2) where physical system limits (pump or turbine rates, storage size and current level, yield factors, etc) and control rules are also included. These rules are based on the set of directives introduced before which basically try to maximize renewable production with minimum storage.
4. Steady state analysis. This one year simulation will be continuously run until steady conditions are reached. The program also includes options to simulate demand and production variations over the years according to statistical information. In order to get representative results at least three years with different data should be executed (Acciona Remote Control Centre, 2010).

The algorithm in figure 6 clearly distinguishes two situations depending on the potential renewable production with respect to the demand. For potential over production it is necessary to check whether the remaining energy can be stored. In this sense, limiting factors are a full storage or insufficient pump power. On the other hand, energy deficit can be compensated using stored energy, when available, or conventional production. The program establishes priority over the stored energy to minimize non renewable contributions. Bearing in mind that also turbines have a limit on their power rate. Besides basic magnitudes such as energy storage level, other important magnitudes are also evaluated on every simulation step, in particular the *Non Renewable Contribution* and the *Renewable Energy Losses*.

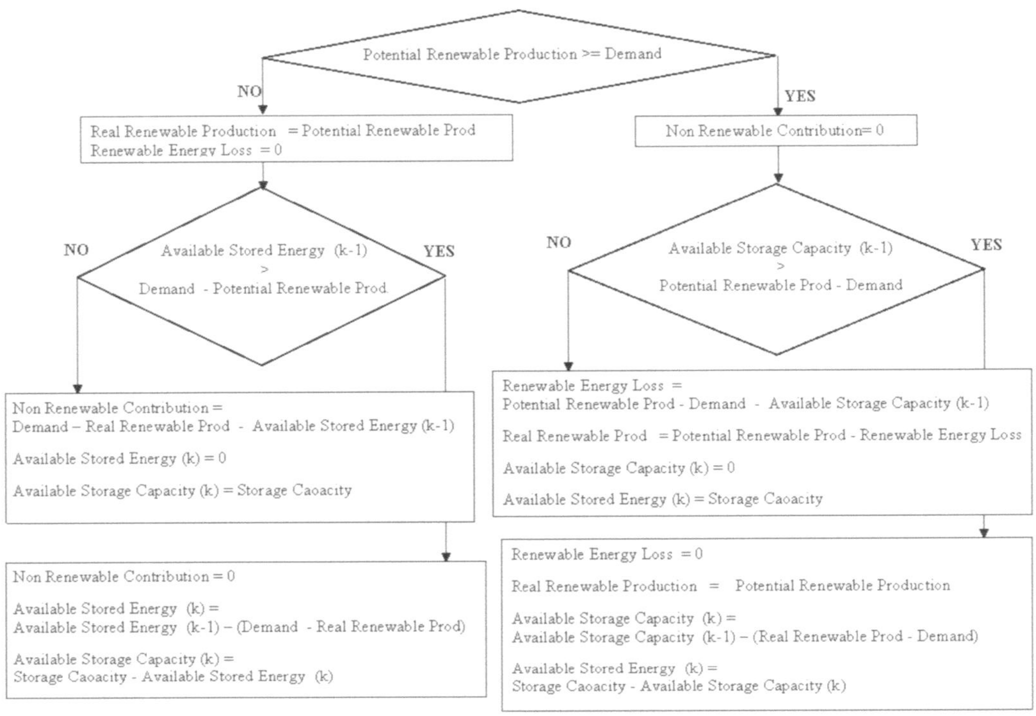

Fig. 6. Hour-by-hour energy balance algorithm.

Figure 7 shows examples of a same baseline case (defined later) simulated at different *RPPR*. In all of the time-graphs the results in the last year of simulation are presented (steady conditions). As it can be observed, normalized potential renewable and demand power with respect to the maximum power demand are shown. These graphs also show the storage level normalized with respect to the maximum storage (obtained for case with *RPPR* =1.0). As it can be observed, for low *RPPR* (figure 7.a) during most part of the year the renewable production gets below demand and on a limited number of occasions the storage is used. Moreover, the lack of energy must be covered by conventional generators. However, for higher *RPPR*, figures 7.b and 7.c, the influence of the storage becomes relevant opening the possibility of demand fulfillment without conventional contribution (100% renewable supply). Figure 7.b corresponds with a critical point where the yearly potential renewable production exactly generates the yearly demand. In this situation to assure the demand power a huge storage with a high powered pump or turbine is continuously required. Nevertheless, as the *RPPR* increases (for example, in figure 7.c for *RPPR* = 1.2) smaller sized storages will be necessary to fulfill demand. In this figure, the minimum storage to satisfy demand is just 30% of the required in critical situations. It is verified that the higher the *RPPR* the lower the optimum storage but also the higher the *Renewable Energy Losses*. (balance renewable overproduction, energy storage size).

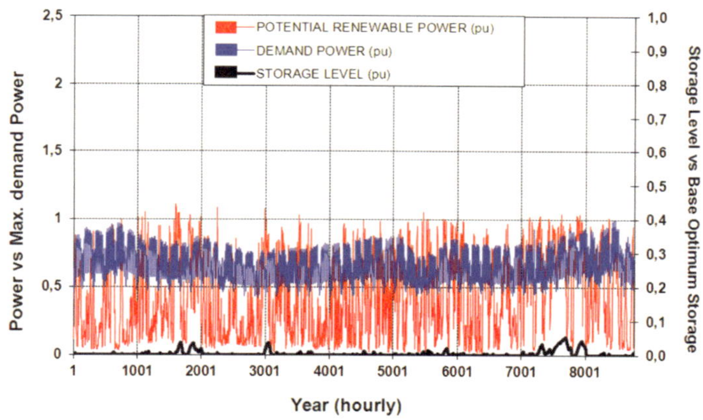

(7.a) *RPPR* = 0.64

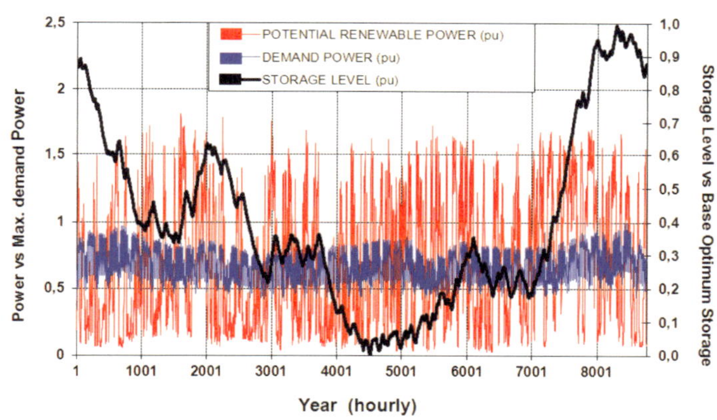

(7.b) *RPPR* = 1.0

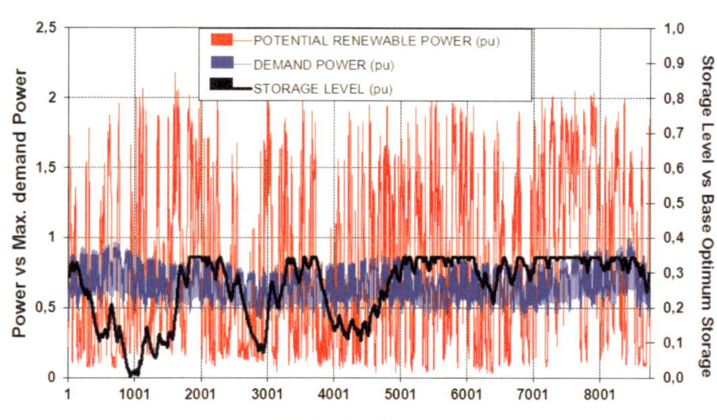

(7.c) *RPPR* = 1.2

Fig. 7. Normalized renewable production, demand and energy stored at different *RPPR*.

3.3 Critical Storage Curve

The minimum storage required to optimize the renewable production (reduction of potential losses) while minimizing the conventional contribution changes depending on the *RPPR* (figure 7). It is possible to determine the optimal storage for any *RPPR* following the procedure explained before. Figure 8 shows an example of this optimal storage for a generic electric system. This curve will be named from now on as *Critical Storage Curve*, and as it can be observed there are two different ranges:

- *RPPR* ≤ 1: Within this range the renewable potential production does not reach the demand energy. However, there are no *Renewable Energy Losses* and the non renewable contribution gets minimized (figure 8, point 1).
- *RPPR* ≥ 1: Within this other range it is feasible to fulfill the electric demand just with renewable energy. Non renewable contribution is not necessary. However, *Renewable Energy Losses* are produced for *RPPR* higher than 1 (figure 8, point 4).

For *RPPR* =1 it was verified that the non renewable contribution and the *Renewable Energy Losses* were both zero although a huge storage seemed to be required. For *RPPR* higher than 1, above the *Critical Storage Curve* (figure 8, point 5) storage is higher than the minimum required. However, below the critical curve (figure 8, points 2 and 3) the storage capacity is not enough to assure minimal values of non renewable contribution and *Renewable Energy Losses*. Point 0 corresponds to very low *RPPR* values where no storage is required to optimize the potential renewable production. This is a consequence of very limited production sequences with no overproduction at any moment during the year. A similar situation was shown on figure 3 a. Another remarkable point is 6, which correspond with very high *RPPR* values. In this unfeasible situation, no storage is required because the minimal renewable production is always over the demand at any moment. However, almost all potential production will become *Renewable Energy Losses*. A similar situation was pointed out in figure 4 for very high *RPPR* levels. Moreover, different renewable combinations, like those shown in figure 5, will produce different *Critical Storage Curves*. Therefore, this curve can be used to compare the tendency of different future scenarios with different renewable mixes. More specifically, this tool reveals important information about likely storage needs and their dependency on the *RPPR* and the specific renewable mix. Thus, offering a particular vision today, about the best politics or incentives in renewable investments for the future.

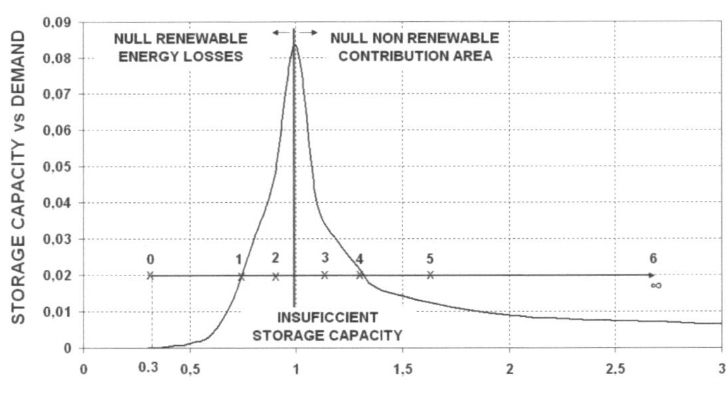

Fig. 8. Normalized *Critical Storage Curve* at different *RPPR*

3.3.1 Sampling time influence

In order to verify the reliability of the results obtained, several analyses have been performed. The first one attempts to determine the minimal measurement interval advisable (monthly, daily, hourly, etc). Figure 9 shows the *Critical Storage Curve* calculated with different measurement intervals. As it were foreseen, monthly measurement intervals lead to significantly lower storage than the obtained with shorter intervals (daily, hourly). This is a consequence of a higher percentage of renewable production directly satisfying the demand with respect to analyses performed with lower sampling times. *Critical Storage Curves* calculated in hourly and daily measurement intervals basis differ minimally. Even when using ten-minute sampling data the *Critical Storage Curve* is nearly the same as those calculated based on daily or hourly data. As a consequence of this analysis, it was decided to perform all analyses always based on hourly series as it was mentioned before.

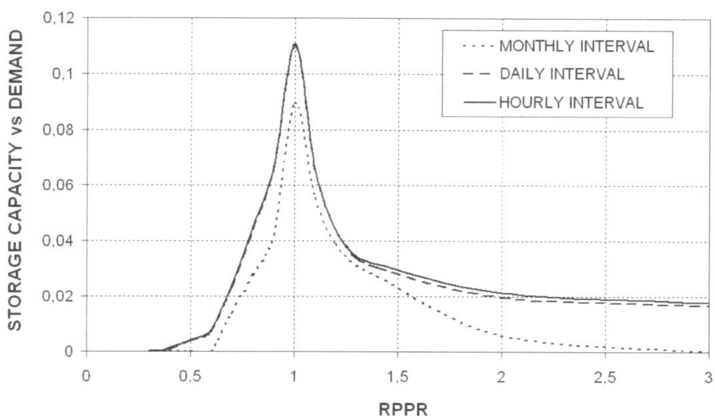

Fig. 9. *Critical Storage Curve* for different sampling times

3.3.2 Wind profile influence

The optimal energy storage sizing depends on both demand and renewable production profiles. However, demand profiles present a certain hourly and yearly regularity as it was shown previously, while renewable production profiles may be very different according to locations and scales (the lower the scale the higher the difference). In order to show such dependence, *Critical Storage Curves* for a typical demand profile and four real wind production profiles (Acciona Remote Control Centre, 2010) have been calculated, figure 10. These wind productions correspond with four wind farms placed in different Spanish regions. In this figure it is clearly appreciated that such difference may be very relevant. When comparing wind farm "S" and "T" at *RPPR* = 1, the required storage capacity for wind farm "S" is almost twice the required for wind farm "T". Also the analysis has been performed for a case which considers an average production profile of the four wind farms mentioned previously and which represents the expected evolution in wider regions. This profile's results are much smoother when it is compared with the independent wind farms, which in turn also leads to smaller storage requirements ("VRST" curve).

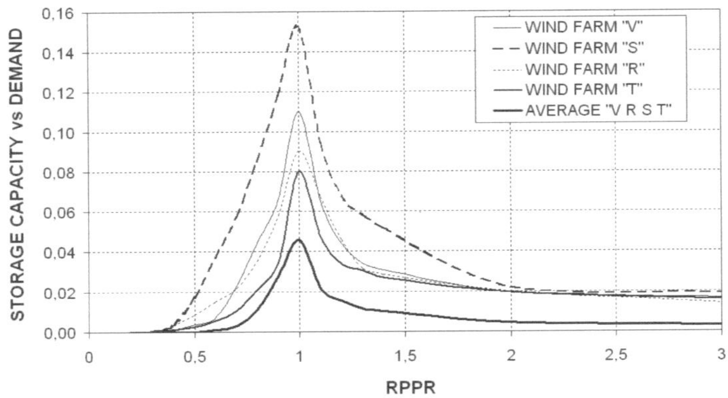

Fig. 10. Influence of different wind generation profiles.

3.3.3 Influence of storage power drives

Any energy storage system is necessarily equipped with drives to in and out power. These drives have limited powers that play an important role for the whole system's efficiency. Moreover, the efficient use of the storage will depend on the drive's power. Drives rated very low will not be able to pump or turbine, in the case of water, the necessary power; which finally will imply increasing system losses (renewable generators that must be stopped) or increasing the contribution of non-renewable generators (not enough turbines to meet demand). Figure 11 shows the dependence of the power drives on the *Critical Storage Curve*. As it can be observed, as the drive power diminishes the curves move towards the right-hand side. This means that more renewable contribution will be needed to compensate losses. But more importantly, the right part of the curves does not mean anymore zero conventional contribution. In fact, the *Critical Storage Curve* just represents the best storage that corresponds with the rated drive power. Besides, the intrinsic efficiency of these drives is also important as some real technologies do really have low rates; introducing a new element to worsen the situation. Therefore, the drive powers limits and efficiency must be clearly taken into consideration in further analyses due to notable influent results.

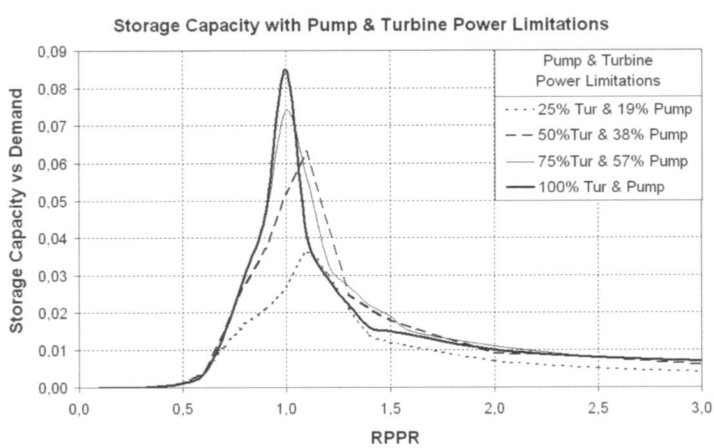

Fig. 11. Influence of storage drives limit power.

4. Energy storage analysis in large isolated electric systems

4.1 Introduction of the study case. Spain

Spain had in 2010 a population of 47 million inhabitants in an area of 504,790 km^2, and a total electrical demand of 251.43 GWh (REE, 2009). In this country, renewable sources (including large hydro) suppled 31% of the total electrical energy demand in 2009. Thus, the *RPPR* for Spain in 2009 was around 0.31; however, this number does not correlate with real renewable production due to the different curtailments introduced in various nodes as a consequence of power limitations.

Spanish electric system is connected to the European electric grid through France and Andorra, to the African grid through Morocco, and finally with Portugal. The total energy flow along those connections is just 3.2% of the Spanish electric demand (REE, 2009). Therefore, in order to analyze how energy storage may optimise the total renewable production, Spain has been considered an isolated electric country.

Figure 12 shows the geographical and climatic map. Spain has a variety of regions: mountainous, plains, long shore perimeter, etc., being the general climate continental, with some regions being mediterranean, oceanic, etc

(a) geography (b) climate

Fig. 12. Maps of Spain.

Spain offers excellent features for the basis of this study for the following reasons:

- Spain presents a high development level in all economic sectors. Its electrical demand profile can be considered representative of a modern, industrial and diverse society.
- Spain currently has a diversified renewable energy production comprising high levels of wind, solar, hydro, mini-hydro, biomass, etc.
- Due to its condition of being one of the leader countries in renewable generation, several years of complete hourly production data from many renewable generators in various technologies are available. For example, wind power hourly series are available since 1992.
- There are two different energy storage systems already available: classic hydro and reversible storage. The first type includes all classic hydro generators without pump systems only based on rain water. These systems include huge water storage able to supply rated power during several weeks. The total storage capacity of these systems is around 7.1% of the yearly Spanish electric demand while the current power installed is 16,657 MW (17.8% of the total power installed) [3]. The second type corresponds with

reversible storage systems (turbine plus pump) with a total capacity of around 74,201 MWh, which represents 0.03% of the yearly electric demand. Power drives are rated at 6% of the maximum demand power during a year (around 45 GW).

- Geography and climate of the country are very diverse among the different constituting areas. The South of the country is especially suitable for solar (photovoltaic and Thermo-solar plants) while the north of the country may concentrate most biomass plants. However, wind power is abundant and available across the whole country. The same applies for hydro although the north concentrates more facilities. Table 1 show how many times the potential renewable resource of each technology could satisfy the Spanish electric demand (according to the situation during 2009) (García & Linares, 2005).

Renewable Technology	Potential satisfaction of the electric demand (times)
Thermo-Solar	31.8
Photovoltaic	4.44
Wind on-shore	7.34
Wind off-shore	1.1
Waves	0.9
Hydraulic	0.1

Table 1. Theoretical potential renewable resources in Spain.

4.2 Baseline critical storage curve

A Baseline case has been defined basically according to the current mix of renewable producers, table 2. It includes a set of hourly series taken from real measurement productions (solar, wind and Biomass) and also calculated from meteorological information (wind off-shore). For future scenarios these data series have been scaled increasing the global potential production according to known official (Spanish, Ministry of Industry, 2010) plans of renewable developments for the future. As it can be observed, these plans concede good opportunities for solar developments although not very optimistic ones for biomass plants.

Renewable Technology	Current power (year 2010, MW)	Hypothetical Power for RPPR = 1.3
Solar (PV & Thermal)	3,700	60,000
Wind on-shore	19,000	40,000
Wind off-shore	-	10,000
Biomass	2,000	10,000

Table 2. Baseline definition

The *Critical Storage Curve* for the baseline case is presented in figure 13.a. This curve presents the strictly minimum reversible-type storage for every *RPPR*. Therefore, large existent hydro storages currently spread throughout the country were not included in this analysis because there are not reversible in power. Their influence is crucial as it will be shown but they need a new set of operation rules to efficiently work with reversible storages. Figure 13.b shows the minimum required drive power to be installed in accordance with the critical storage. Due to system power losses (power drives losses particularly), the minimum *RPPR* for 100% renewable supply is around 1.1 instead of 1.0 (critical). Nevertheless, around these numbers both storage size and driver power presents unfeasible values.

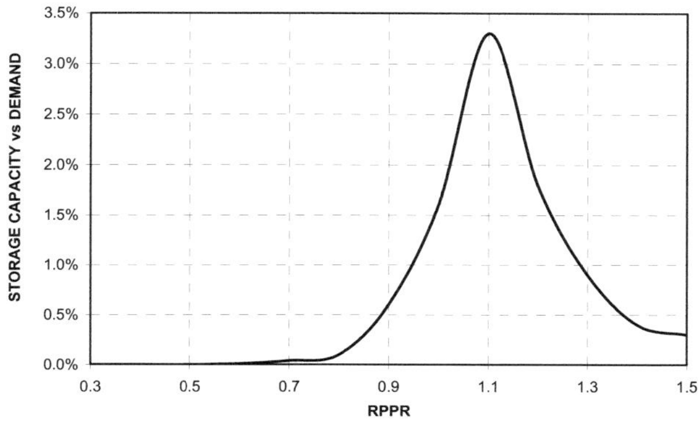

(a) Critical Storage Curve

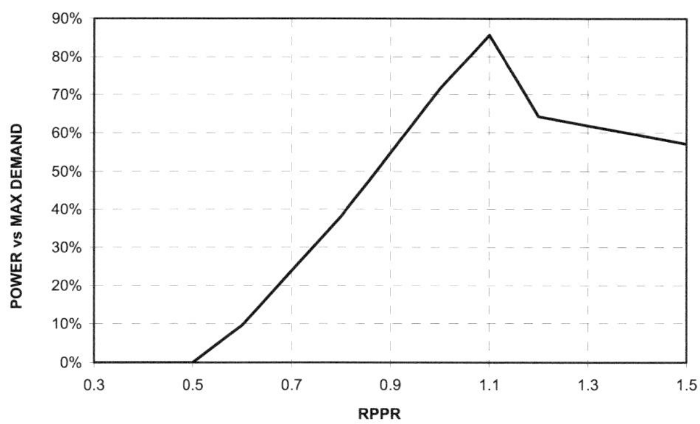

(b) Critical drives power curve

Fig. 13. Critical Storage and Power Curves for baseline case.

4.2.1 Influence of large hydro storages

Current large hydro plants are usually operated under market-based rules or on grid operator needs. However, the existing set of plants conform a huge energy storage that can be operated to improve the integration of the rest of renewable technologies. These strategies have been included on the simulation platform explained before. Therefore, now the system includes two different storages: hydro-based and reversible. The control rules for each one have been coordinated in order to make the most efficient use of both storages at any instant. Logically, behind such rules there is a general objective to maximise the renewable production, reducing losses and conventional contributions. Figure 14 shows the *Critical Storage Curves* when the large hydro storage is also considered. As it can be observed, the new reversible storage is now clearly smaller than the previous case, figure 13. This reduction also affects the required drives power as demonstrated on the curve in figure 15. Both reductions also imply fewer losses moving the whole curve towards the left-hand side.

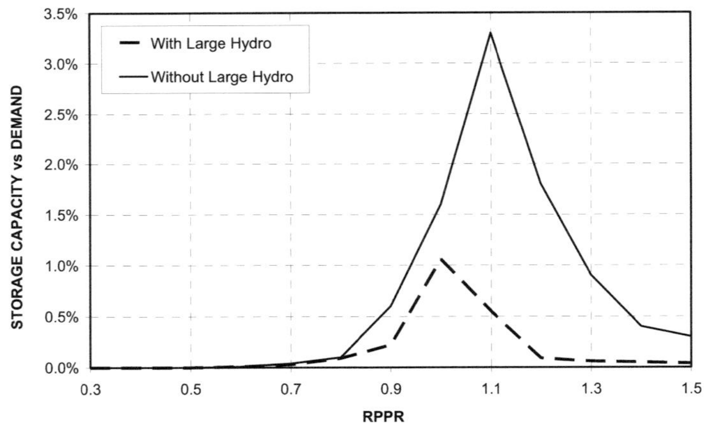

Fig. 14. *Critical Storage Curves* for baseline case.

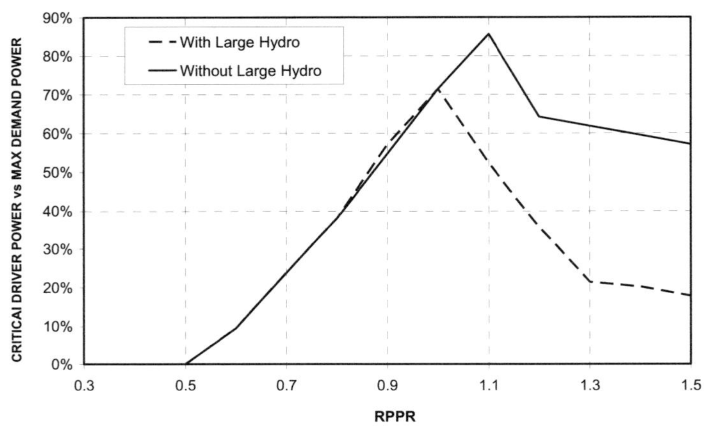

Fig. 15. Critical Power Drives for baseline case.

A 100% renewable scenario requires *RPPR* over 1.1 to be technical and economically feasible. A reasonable value according to all curves shown on figure 14 and 15 could be around *RPPR* = 1.3. Table 3 summarizes the difference between both scenarios (with and without large hydro). As it can be observed, the efficient and coordinated use of current large hydro systems strongly reduces the need for large reversible storages, both in size and power. Moreover, an over sizing like the one proposed in table 3 (*RPPR* = 1.3) should lead to a global reversible storage need which is only twice the existing one. Considering that such increment in renewable contribution from current rates to the correspondent for *RPPR* = 1.3 may take some decades, the proposed increment in storages could clearly be feasible.

The positive influence of current large hydro storages can be incremented if some more actions were taken. Particularly interesting is the possibility of increasing turbine ratings of the whole system. Current power is around 16.6 GW but some plants still admit additional powering to reach a maximum of around 20 GW. Figures 16 and 17 show the influence on the critical storage and power curves respectively due to an increase in uniquely large hydro power.

RPPR =**1.3**	**Reversible Storage Capacity vs Yearly Energy Demand**	**Drives Power vs Max Demand Power**
No large hydro storage	0.9%	62%
Large hydro storage (7,1%)	0.06%	21%

Table 3. Reversible storage and power requirements considering large hydro.

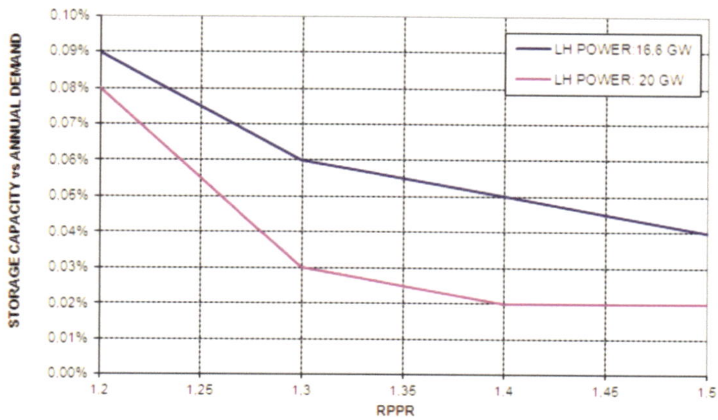

Fig. 16. *Critical Storages Curves* for baseline case depending on large hydro power.

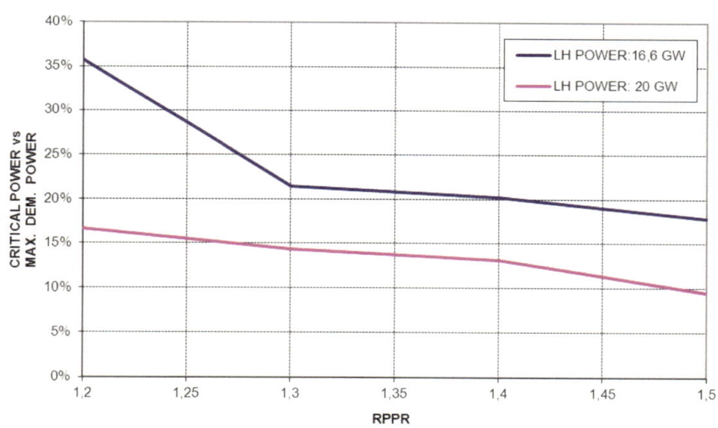

Fig. 17. Critical Power for baseline case depending on large hydro power.

Table 4 summarises the influence of higher turbine power ratings at *RPPR* = 1.3. Again a notable reduction of reversible storage needs is observed. Moreover, with the proposed turbine empowering, already available reversible systems should almost be enough. This hypothetical scenario with 20 GW large hydro turbine ratings makes it almost possible to achieve a 100% renewable supply with the existing hydro pump storage system. It should be enough to increase power in such systems from the current 4 GW to around 6 GW. Moreover, enormous technical improvements within R&D divisions have taken place over the last decade that offer commercial power plants of up to 50 MW the ability to deliver energy peaks during 60 minutes. A proper distribution of these units would fulfil the required storage and also improve the system efficiency (locally and globally) although some control complications may arise.

RPPR =1.3	Reversible Storage Capacity vs Yearly Energy Demand	Drives Power vs Max Demand Power
Large Hydro Power: 16.6 GW	0.06%	21%
Large Hydro Power: 20 GW	0.03%	14%

Table 4. Reversible storage and power requirements depending on large hydro power.

4.2.2 Influence of demand control

Demand control is sometimes used by grid operators in order to compensate energy unbalances under certain circumstances. These techniques require fast communications and commitment with some users which receive economical compensation for this service. This possibility has been explored in order to understand the influence on the proposed renewable-based system for any RPPR. The demand control algorithm establishes a daily schedule with the percentage of demand to be shared among different intervals throughout the day. The decisions are made using the forecasting of the renewable production (already available in simulation). In this way, on every hour the best relation between demand and available renewable production is established. This method offers the better perspectives that can be expected with demand controls as such because it uses perfect information. Thus, these results should be considered as a theoretical maximum positive influence. Figures 18 and 19 show the critical storage and power curves respectively depending on different ranges of demand control over the daily demand (0%, 15% and 30%). The first conclusion is the little influence that seems to offer demand control implementation. For *RPPR* = 1.3, the likely benefits maybe do not compensate the complication introduced in these demand control programs. However, further analysis with other demand control politics should be made to get definitive conclusions.

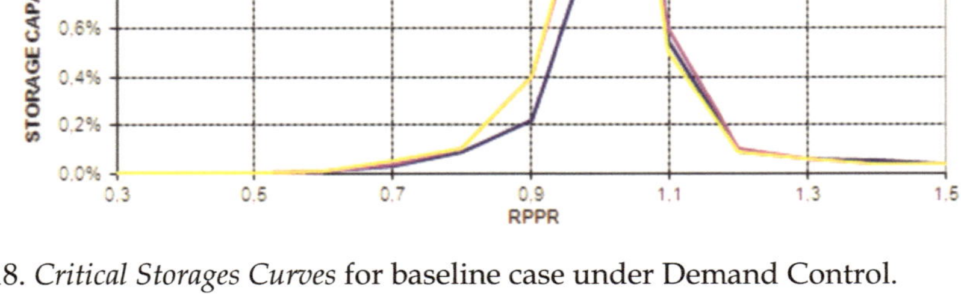

Fig. 18. *Critical Storages Curves* for baseline case under Demand Control.

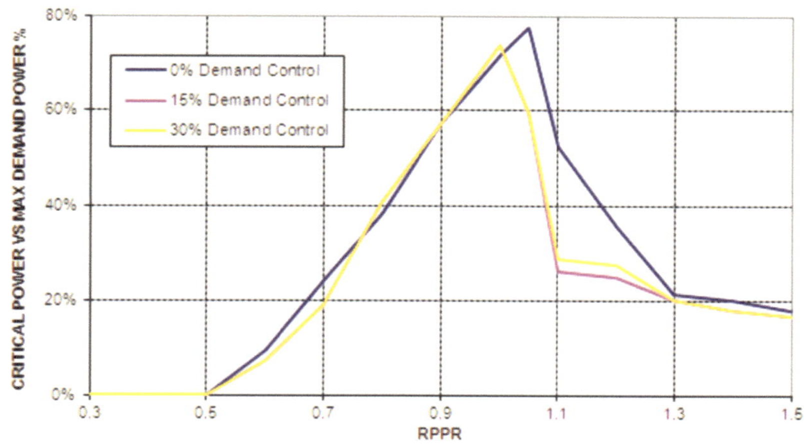

Fig. 19. Critical Power for baseline case under Demand Control.

4.2.3 Influence of the renewable mix generation

The analysis of potential renewable capabilities in Spain sorted by technology (table 1) evaluated by (García & Linares, 2005) opens multiple likely future scenarios of renewable mixes. The evolution in one and another direction will depend on many aspects such as political decisions, economical or technological incentives, etc. In that sense, one aspect that also should be considered for incentives could be the influence on future storage needs. As it will be demonstrated storage size or drive power is influenced by the mix characteristics for a same *RPPR*. Figure 20 shows the evolution of the total power installed by technology during several years, from 2000 until 2009. This graph also presents proposals of hypothetical extrapolations of likely future mixes according to known development plans (baseline) or other assumptions that concede higher development opportunities to some technologies.

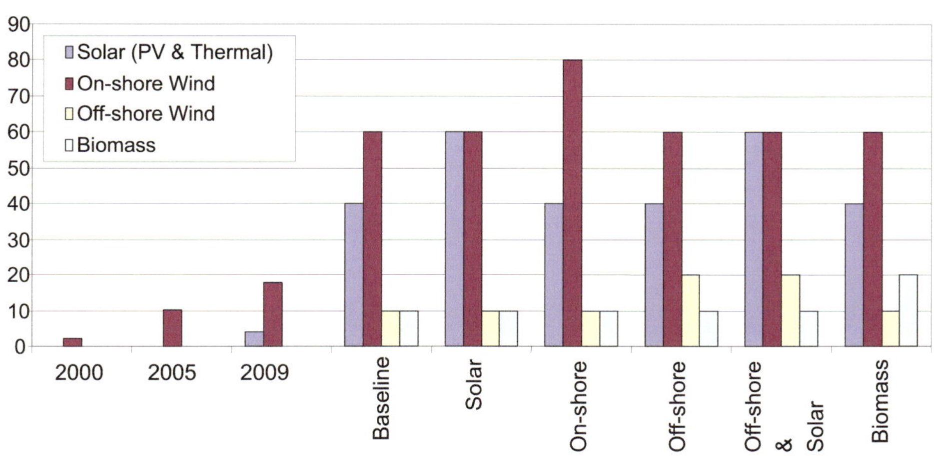

Fig. 20. Renewable mix generation scenarios.

The *Critical Storage Curve* of all renewable mixes has been calculated for two different large hydro powers: 16.6 GW (current) and 20 GW (hypothetical). Results are exposed in figure 21 where several aspects could be highlighted:

- All renewable mixes show a similar tendency offering good opportunities to reduce storage size as the *RPPR* or hydro power increases.
- Biomass case seems to offer the better opportunities to reduce future needs of storage. In fact, the higher the influence of controllable renewable producers the lower the storage needs. Consequently, more incentive of these technologies should be seriously considered.
- In general, greater hydro powers lead to lower storage needs. However, the differences are in some cases very little, being also important the characteristics of the renewable mix. In fact, around *RRPR* = 1.3 almost all mixes confluent on a narrow range of storage size, where only the Solar case seems to be out of range.
- Solar contribution must be increased for future systems. However, above certain levels the opportunity to reduce storage size or power diminishes considerably. This means other technologies become more relevant. Nevertheless, it is important to remember that in all mixes of figure 20, the expected contribution of this technology has been planned considerably high and above 40 GW, a rating that means around 1 kW per habitant to be installed.

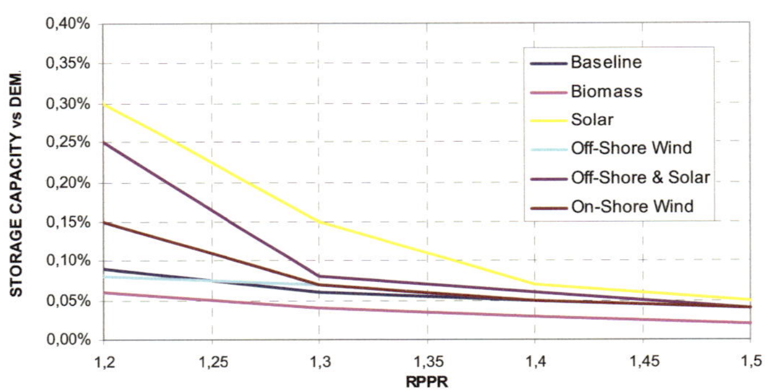

Large Hydro Power: 16,6 GW

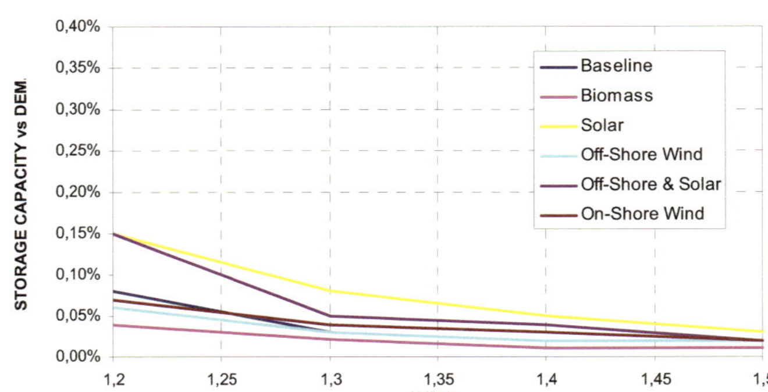

Large Hydro Power: 20 GW

Fig. 21. *Critical Storage Curves* for 6 renewable combinations and 2 large hydro powers.

5. Transition process

The transition towards a 100% renewable-based electric system starting from the current situation has critical implications and a lot of questions to be answered. Some of these questions could find a reasonable and acceptable answer through the analysis already made. The required renewable system, to guarantee the electric supply, seems to be clearly feasible in terms of natural resources. Moreover, the final system power to be installed in generators, storages, etc also seems to be feasible and surely justifiable from an economic perspective. However, there are also political and economic implications with pollutant technologies still pending to be cleared up, aspects that are beyond the scope of this analysis. Some official and non official plans (García & Linares, 2005) seem to consider a total substitution of current generation plants in around 40 years. Thus, the new renewable system must be planned today to reach the desired point in time. Therefore, it is important to have some

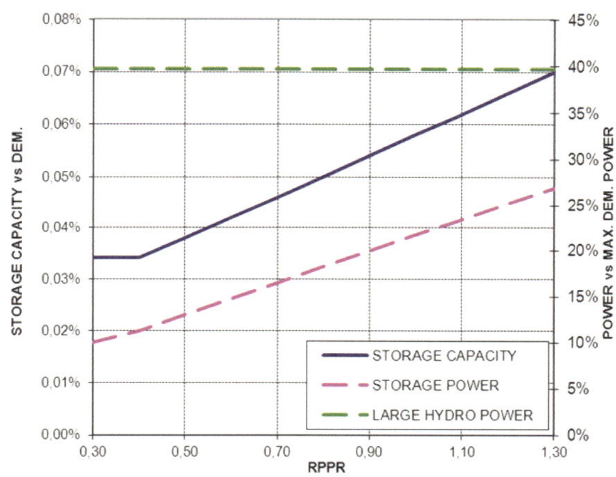

(a) No hydro empowering

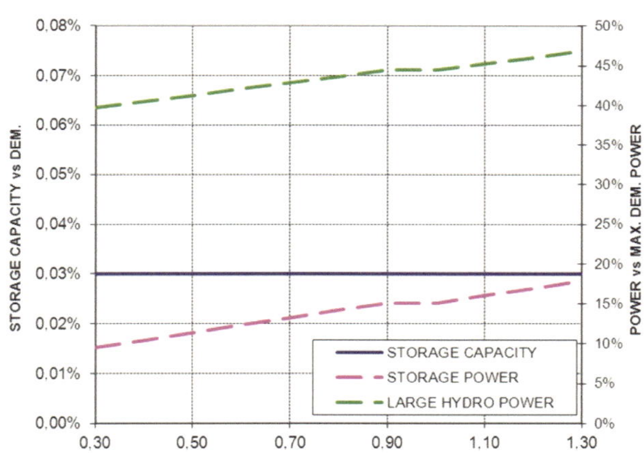

(b) Hydro empowered to 20 GW

Fig. 22. Proposals of reversible and hydro storages evolution possibilities.

clear numbers that help to define targets. During this study some numbers have arisen like a final RPPR around 1.3 and storages not really much more capable than current systems. The transition to such final system seems to require simply the implementation of more renewable plants coordinated with the decommissioning of existing ones, sensibly starting first with those more dangerous, pollutants and older. In any case, during this process the intermediate electric system must guarantee the supply which will have several implications as the following study will show. Figure 22 shows a proposal of increasing storages (both reversible and hydro) as the RPPR is going to be increased. The evolution of required storages power or size will be different if large hydro plants decide to increase the turbine power. If no additional power is to be installed, (figure 22.a) the reversible storage must be increased in size and power. On the contrary (figure 22.b), only the reversible power should be increased. Certainly, in this sense final decisions will also depend on likely economic advantages.

Figure 23 shows a hypothetical sequence of the different energy supplies during the transition process. Four different energy sources fulfill the electric demand for any *RPPR*:

1. Generation coming from conventional base plants (nuclear and coal mostly)
2. Generation coming from conventional controllable plants (gas, fuel, etc)
3. Renewable generation instantaneously delivered to the grid.
4. Renewable generation delivered to the grid through the storage.

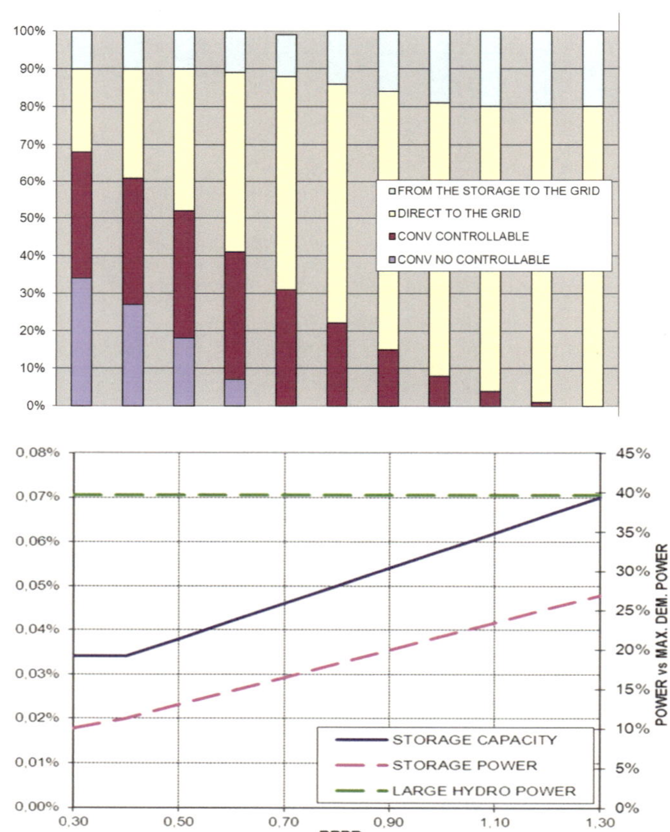

Fig. 23. Transition process with baseline case and 16,6 GW hydro power.

While the reduction of the conventional energy contribution maintains a gradual shape, unfortunately controllable conventional power plants must be maintained operative in order to assure peak demands. Figure 24 shows the sequence of conventional plants decommissioning correspondent with the process proposed in figure 23 for two different renewable mixes (wind and Biomass). Here it can be appreciated that only when *RPPR* values reach 0.9 a considerable controllable conventional power can be diminished, irrespective of the renewable mix. The energy produced for such plants is progressively decaying but must be operative because of power peaks needs. This would certainly require special economic incentives and politics to be feasible. Nevertheless, in this sense the situation in figure 24 corresponds with the two extreme cases richer in wind and biomass than the baseline case. Anyway, most controllable conventional plants can be slightly changed to use biofuel, biogas, etc. improving the total biomass service while fulfilling the required power to cover demand peaks.

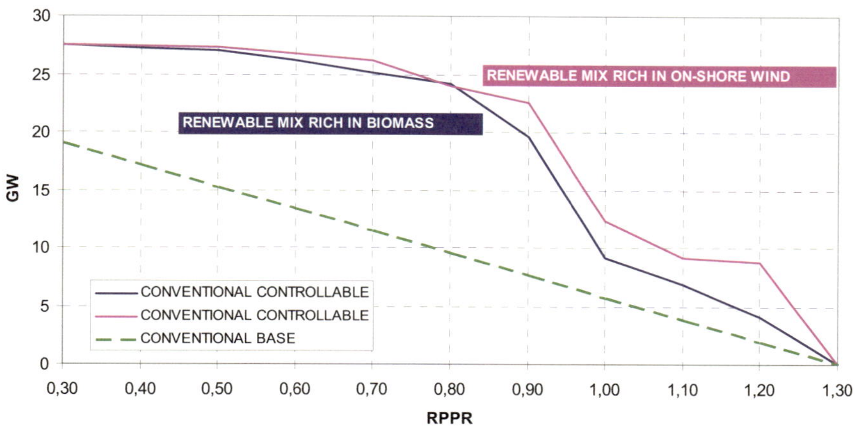

Fig. 24. Conventional power plants decommissioning sequence for two mix renewable generation scenarios.

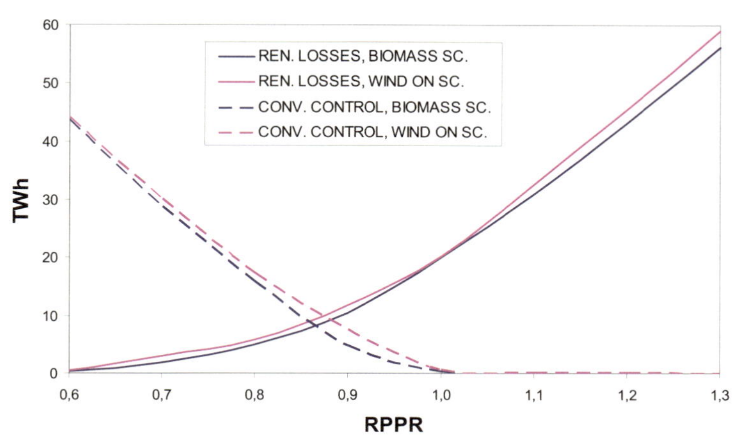

Fig. 25. Renewable energy losses and controllable conventional generation. Transition process.

During this transition certain inefficiencies must be admitted, such as increments in *Renewable Energy Losses*. Figure 25 shows extreme results (again for those combinations richer in wind and biomass than the baseline case) of these losses together with the evolution of the expected controllable conventional energy. The Biomass case requires from any *RPPR* value the minimum controllable conventional generation and creates the minimum renewable losses comparing with any other renewable mixes.

6. Conclusions

The use of storage systems is essential to allow future higher grid integration levels of renewable energy. Even a total substitution of current conventional generators to achieve a 100% renewable supply seems to be technically feasible. The Spanish electric system has been used as a base case for this study, due to the significant and diverse renewable energy technologies already installed and the high renewable resource available. However, the ratings per capita in Spain regarding electrical demand and renewable resources availability (hydro power, equivalent sun hours, wind on-shore and off-shore, biomass, etc) can be compared with a lot of countries with different levels of development. Therefore, most of the conclusions obtained from of these studies could also be extended for those in other countries and regions. The following ones are remarkable:

- Large hydro storage systems available in many countries may reduce significantly the reversible storage capacity needed to reach 100% renewable supply. However, in places without this possibility it will be advisable to plan a highly distributed reversible storage system. Two benefits would be expected: reduced transport losses and easier integration of these systems in both remote and civilised areas.
- The renewable mix clearly influences the size and drives power of reversible storages. Current political incentives should take care of these implications for future planning, especially regarding biomass systems because they offer the better opportunities to reduce future storage needs. Biomass also helps to smooth the transition from current to higher renewable systems.
- Increasing renewable rates will need the operation of storages accordingly for better grid integration. Both systems, hydro-based and reversible, must be adapted and coordinated to support and service such integration. In this moment the level of renewable production is relatively low without requiring storage service. However, countries like Spain with important levels of wind-based renewable production should already start to take care for future close needs.
- Above certain renewable integration the theoretical reversible storage needs are strongly reduced. Penetration levels around *RPPR* = 1.3 seems to offer good technical perspectives although final decisions will also be made considering the economic or environmental implications.
- The transition from current electrical systems, mostly based on non-renewable sources, towards 100% renewable should be carefully planned. As it has been pointed out, depending on the future renewable mix, available hydro, available renewable resources, etc, the storage system, the electrical transport and transmission system, the communications and other important aspects should be determined.

For years the never ending debate about the feasibility of a 100% renewable-based electric system has been taking place worldwide. This work has had the intention to clarify as much as possible in that sense, completing contributions of many other personalities and institutions. The study here devised offers good perspectives for future renewable energy developments. Moreover, from the point of view of technical feasibility some important aspects have been explored and analysed, seeking good opportunities and a path to be followed for a successful change.

7. Future research lines

Several research lines have been opened as a consequence of this work. Some of them related with technical aspects that still need more study or detail, such as new, more capable control techniques (especially forecasting), improvement of the simulation platform, influences of grid ancillary services, total and local effect of the dispersion of producers and storages, etc. However, other lines related with economic issues have also arisen becoming quite relevant. Technically, the proposal here exposed has developed a lot of points to defend it but, it is necessary to complement it with economic arguments. During the last years an important work to sustain the renewable option in regulated or free markets has been presented by many authors (Makarov et al. 2009). One basic analysis, introduced in (García & Linare,2006), compares the current electric system costs with respect to one only based on renewable energy, where important aspects are revealed that strongly support. However, one interesting line to be followed in future research will try to characterise the opportunities of a hypothetical electric company that makes use of the storage in order to maximise benefits. Market conditions, financial costs, etc must be determined for optimal economic exploitation.

8. References

Bremen, L., Greiner, M. & Knorr, K. (2009). A full renewable power supply scenario for Europe: The weather determines storage and transport, *Conf. Rec. International Workshop on Large-Scale Integration of Wind Power into Power System, 2009.* Bremen, Gemany, 2009.

Alonso, O. & Galbete, S. (2009). Sizing and Analysis of Big Scale and Isolated Electric Systems based on Renewable Sources with Energy Storage, *Conf. Rec. IEEE SAE* Valencia, Spain, *2009.*

Alonso, O., Galbete, S., & Sotes , M. (2010). 100% Renewable electric supply and transition analysis, *Proc.Rec IEE Large scale integration of wind power into power systems.* Amman,Jordan, 2010.

REE, Spanish Electric Grid Operator. (2009). Annual Report of Spanish electric system.

Tipping, J. & Sinclair, D. (2009). The benefits of marine technologies within a diversified renewable mix", *British Wind Energy Association.*

García, X. & Linares, P. (2005). *Report of Potential renewable energy resources in Spain,,* Greenpeace.

Spanish Ministry of Industry (2010). List of renewable projects awarded.

Acciona Remote Control Centre (2010). Hourly renewable production data register 2007-2010. Private base data, contact authors for availability.

Price, A. (2010). *The economics of electrical energy storage in the future European Network*. Advanced Energy Storage Forum. Marcus Evans. Amsterdam, Holland, 2010.

Makarov, V., Yang, B. & DeSteese, G. (2009). Wide Area Energy Storage and Management System to Balance Intermittent Resources in the Bonneville Power Administration and California, *Conf. Rec. IEEE SAE* Valencia, Spain, *2009.*

García, X. & Linares, P. (2006). *100% Renewable electric system for Spain and its economical feasibility,* Greenpeace.

16

Complementary Control of Intermittently Operating Renewable Sources with Short- and Long-Term Storage Plants

E. F. Fuchs and W. L. Fuchs
University of Colorado at Boulder, USA

1. Introduction

The deployment of renewable energy sources requires that neighborhood short- and long-term storage plants which complement the intermittent nature of their output, that is, produce electricity when the renewable plants are inactive and store electric energy when renewable plants generate more than the grid can accept. This chapter focuses on the interaction between conventional power plants with renewable sources and storage facilities. The various plants are modeled in the time domain and the resulting differential equation system is solved by available software. Results confirm that a complementary control permits the operation of renewable sources, paired with storage plants, within the frequency band of 59-61 Hz. The proper design of the time constants, choice of switching times/commands of storage plants, and the sufficient capability of the transmission lines (e.g., tie lines) are essential for steady-state and dynamic stability of a smart/micro grid as investigated in this chapter. Future research must concentrate on the measurement of the output powers of the various plants as a function of time and use these measurements for the timed switching commands.

1.1 Energy efficiency and reliability increases through interconnected power system

Prior to 1940 there were a limited number of interconnected power systems. Servicing loads was simple, as the systems were primarily radial circuits (Fig. 1a) and many power systems were operated in islanding mode. In more recent years interconnection of power systems has been favored (Fig. 1b), and within the US three power grids (Western, Eastern and Texan generation systems [ERCOT]) were established. Within these three power pools, many loop circuits with many load/generation buses and high levels of power exchange between neighboring companies exist. The latter point relates closely to interconnection advantages.

With no addition of actual generation capacity, it is possible to increase generation through interconnection. In the case of an outage of a generating unit, for example, power may be purchased from a neighboring company. The cost savings realized from lower installed capacity usually far outweigh the cost of the transmission circuits required to access neighboring companies. Fundamental and harmonic steady-state power flows are discussed in a recent book on power quality (E.F. Fuchs & Masoum, 2008a). Yet while energy efficiency of an interconnected system is increased by more fully loading existing generation plants, efficiency decreases by transporting energy via transmission lines over larger distances. The average transmission loss within interconnected systems is about 8%.

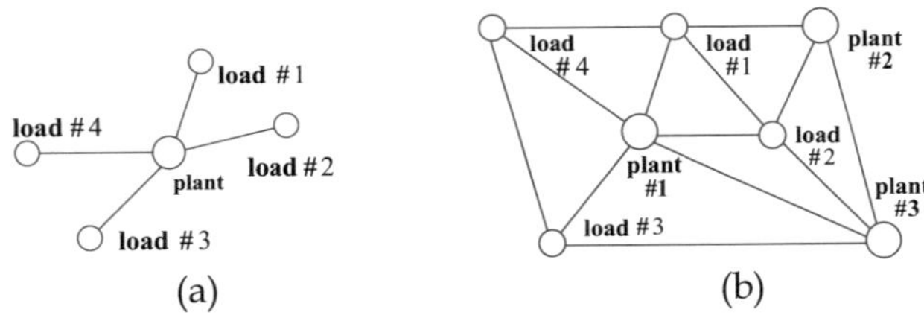

Figs. 1a,b. (a) Radial power system; (b) Interconnected power system

1.2 Future conventional and renewable energy sources

The generation mix (e.g., coal, natural gas, nuclear, hydro plants) of existing power systems will change in the future to mainly natural-gas fired plants, distributed renewable generation facilities (E. F. Fuchs & H. A. Fuchs, 2007) and storage plants. The ability to transition from the interconnected system to an islanding mode of operation must also be possible (E. F. Fuchs & F. S. Fuchs, 2008) to increase reliability. This means that any islanding system must have a frequency-leading plant ("frequency leader") in addition to renewable plants and storage plants. Renewable and storage plants cannot be frequency leaders because of their intermittent and limited output powers, respectively.

2. Review of current methods and issues of present-day frequency and voltage control

Present-day frequency/load and voltage control of interconnected systems takes place at the transmission level and is based on load-sharing and demand-side management. Load sharing relies on drooping characteristics (Wood, Wollenberg, 1984) (Fig. 2 and Figs. 4a, b, c) where natural gas, coal, nuclear and hydro plants or those with spinning reserves supply the additional load demand. If this additional load cannot be served by the interconnected plants, demand-side management (load shedding) will set in and some of the less important loads will be disconnected. This method of frequency/load control cannot

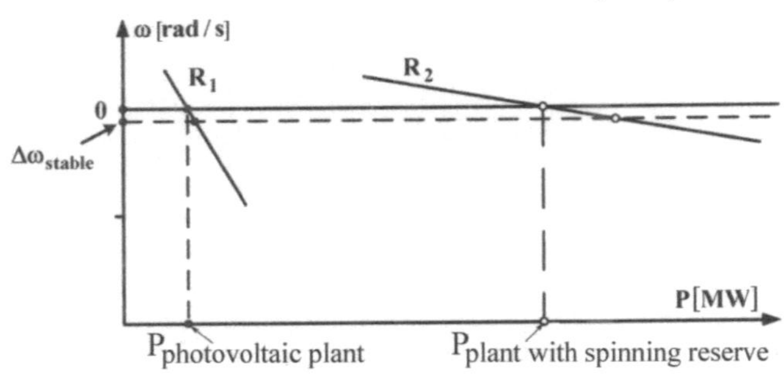

Fig. 2. Angular frequency versus output power: Stable frequency control relying on droop characteristics where a plant with spinning reserve operates continuously and a photovoltaic (PV) plant intermittently operates at its maximum output power due to peak-power tracking

be employed if renewable sources are operating at peak power in order to displace as much fuel as possible; it functions only as long as renewable energy represents a small fraction of overall generation capacity. As this fraction increases at either the distribution or transmission levels, frequency control problems will result. In today's interconnected systems, a frequency variation between f_{min}=59 Hz and f_{max}= 61 Hz, that is, $\Delta f = \pm$ 1.67% is acceptable (Fuller *et al.*, 1989). Voltage control is performed based on capacitor-bank switching, synchronous condensors and over/underexcitation of synchronous generators in power plants.

2.1 Isochronous control or frequency/load control of an isolated power plant with one generator only

Example #1: Fig. 3a illustrates the block diagram of governor, prime mover (steam turbine) and rotating mass (characterized by the momentum $M=\omega_o J$) and load of a turbo generator set [Wood & Wollenberg, 1984].

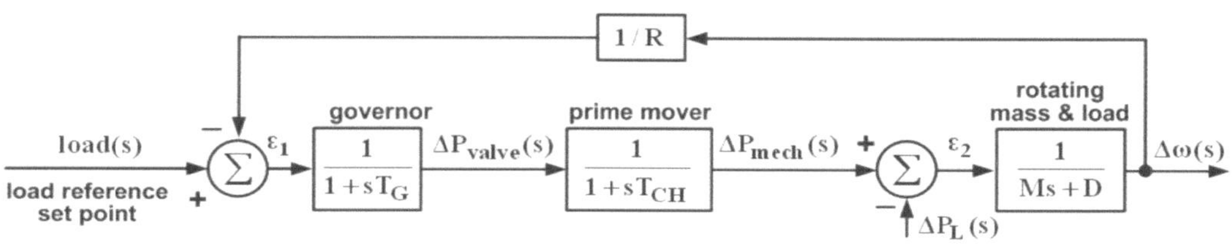

Fig. 3a. Block diagram of governor, prime mover and rotating mass & load at isochronous operation of power system, where D corresponds to the frequency-dependent load and $\Delta P_L(s)$ is the frequency-independent load

In Fig. 3a the parameters are as follows: Angular frequency change $\Delta\omega$ per change in generator output power ΔP, that is $R=\Delta\omega/\Delta P=0.01$ pu, the frequency-dependent load change ΔP_{L_freq} per angular frequency change $\Delta\omega$, that is $D=\Delta P_{L\_frequ}/\Delta\omega=0.8$ pu, step-load change $\Delta P_L(s)=\Delta P_L/s=0.2/s$ pu, angular momentum of steam turbine and generator set M=4.5, base apparent power S_{base} =500 MVA, governor time constant T_G=0.01 s, valve changing/charging time constant T_{CH}=1.0 s, and load reference set point load(s)=1.0 pu.

a. Derive for Fig. 3a $\Delta\omega_{steady\ state}$ by applying the final value theorem. You may assume load reference set point load(s) = 1.0 pu, and $\Delta P_L(s)=\Delta P_L/s=0.2/s$ pu. For the nominal frequency f*=60 Hz calculate the frequency f_{new} after the load change has taken place.
b. List the ordinary differential equations and the algebraic equations of the block diagram of Fig. 3a.
c. Use either Mathematica or Matlab to establish steady-state conditions by imposing a step function for load reference set point load(s) =1/s pu and run the program with a zero-step load change ΔP_L=0 (for 25 s) in order to establish the equilibrium condition without load step. After 25 s impose a positive step-load change of $\Delta P_L(s)=\Delta P_L/s=0.2/s$ pu to find the transient response of $\Delta\omega(t)$ for a total of 50 s, and at 50 s impose a negative step-load change of $\Delta P_L(s)=\Delta P_L/s=$ - 0.2/s pu to find the transient response of $\Delta\omega(t)$ for a total of 75 s.

The solution to this example is given in Application Example 12.7 of (E.F. Fuchs & Masoum, 2011) and illustrated in Fig. 3b.

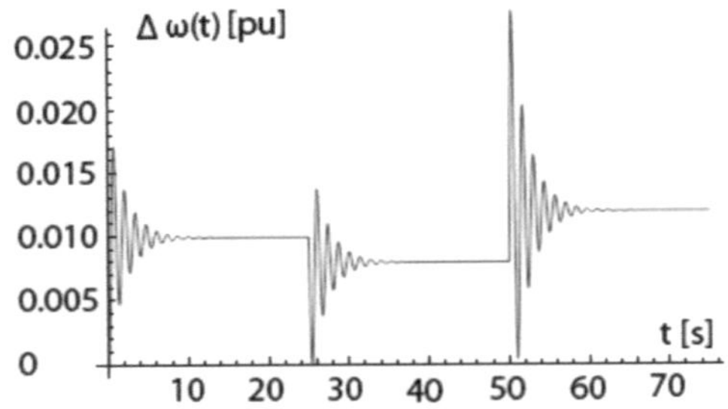

Fig. 3b. Angular frequency change $\Delta\omega(t)$ for a positive load-step at time t=25 s and a negative load step at time t=50 s

2.2 Load/frequency control with droop characteristics of an interconnected power system broken into two areas each having one generator

Example #2: Fig. 4a shows the block diagram of two generators interconnected by a transmission tie line (Wood & Wollenberg, 1984).

Data for generation set (steam turbine and generator) #1: Angular frequency change $\Delta\omega_1$ per change in generator output power ΔP_1, that is $R_1=\Delta\omega_1/\Delta P_1=0.01$ pu (e.g., coal-fired plant), the frequency-dependent load change ΔP_{L1_frequ} per angular frequency change $\Delta\omega_1$, that is $D_1=\Delta P_{L1\_frequ}/\Delta\omega_1=0.8$ pu, positive step-load change $\Delta P_{L1}(s)=\Delta P_{L1}/s=0.2/s$ pu, angular momentum of steam turbine and generator set $M_1=4.5$, base apparent power $S_{base}=500$ MVA, governor time constant $T_{G1}=0.01$ s, valve changing/charging time constant $T_{CH1}=0.5$ s, and load $ref_1(s)=0.8$ pu.

Data for generation set (steam turbine and generator) #2: Angular frequency change $\Delta\omega_2$ per change in generator output power ΔP_2, that is $R_2=\Delta\omega_2/\Delta P_2=0.02$ pu (e.g., coal-fired plant), the frequency-dependent load change ΔP_{L2_frequ} per angular frequency change $\Delta\omega_2$, that is $D_2=\Delta P_{L2\_frequ}/\Delta\omega_2=1.0$ pu, negative step-load change $\Delta P_{L2}(s)=\Delta P_{L2}/s=-0.2/s$ pu, angular momentum of steam turbine and generator set $M_2=6$, base apparent power $S_{base}=500$ MVA, governor time constant $T_{G2}=0.02$ s, valve changing/charging time constant $T_{CH2}=0.75$ s, and load $ref_2(s)=0.8$ pu.

Data for tie line: $T=377/X_{tie}$ with $X_{tie}=0.2$ pu.

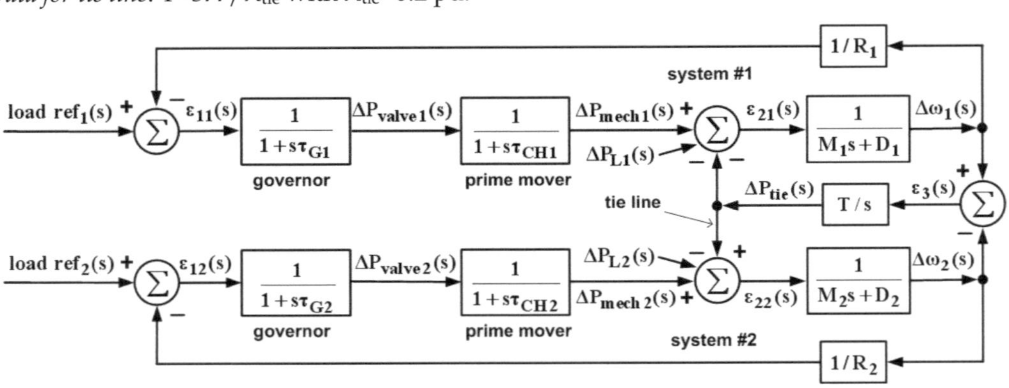

Fig. 4a. Block diagram of two interconnected generators through a tie (transmission) line, where D_1 and D_2 correspond to the frequency-dependent loads and $\Delta P_{L1}(s)$ and $\Delta P_{L2}(s)$ are the frequency-independent loads

a. List the ordinary differential equations and the algebraic equations of the block diagram of Fig. 4a.
b. Use either Mathematica or Matlab to establish steady-state conditions by imposing a step function for load $ref_1(s)=0.8/s$ pu, load $ref_2(s)=0.8/s$ pu and run the program with zero step-load changes $\Delta P_{L1}=0$, $\Delta P_{L2}=0$ (for 10 s) in order to establish the equilibrium condition. After 10 s impose positive step-load change $\Delta P_{L1}(s)=\Delta P_{L1}/s=0.2/s$ pu, and after 30 s impose negative step-load change $\Delta P_{L2}(s)=\Delta P_{L2}/s= -0.2/s$ pu to find the transient response $\Delta\omega_1(t)=\Delta\omega_2(t)=\Delta\omega(t)$ for a total of 50 s. Repeat part b) for $R_1=0.5$ pu, (e.g., wind-power plant), and $R_2= 0.01$ pu (e.g., coal-fired plant).

The solution to this example is given in Application Example 12.8 of (E.F. Fuchs & Masoum, 2011) and illustrated in Figs. 4b,c.

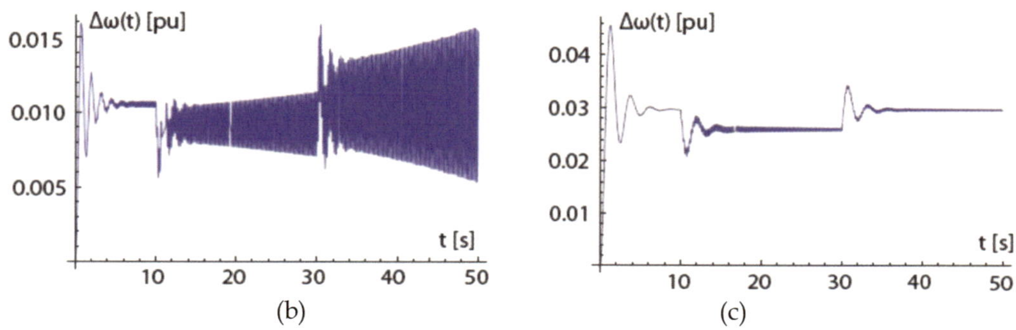

Figs. 4b,c. (b) Angular frequency change $\Delta\omega(t)$ for unstable operation due to the approximately same droop characteristics ($R_1=0.01$ pu) and ($R_2=0.02$ pu); (c) Angular frequency change $\Delta\omega(t)$ for stable operation due to the different droop characteristics ($R_1=0.5$ pu) and ($R_2=0.01$ pu)

3. Intermittently operating renewable plants

Portfolio standards will drive significant increases in renewable energy resources. Photovoltaic (PV) is expected to increase dramatically, with power penetration levels of 10-50% occurring in local areas within the next decade (Defree, 2009). Regions with large PV plants and relatively soft grids, such as the 8 MW SunEdison plant in Colorado's San Luis valley (on Xcel's Energy's grid) have experienced power quality issues primarily related to current harmonics at low PV output levels. Such PV systems can also experience rapidly fluctuating outputs. Fig. 5 (National Renewable Energy Laboratory [NREL], 2009) illustrates solar radiation levels on a horizontal plate during one recent 24-hour period. As can be seen from this time series (1-minute sampling rate) data, rapid transients occur within minutes. For the date shown, 67 transients occurred where radiation levels changed by more than 30% of the peak radiation level within one minute. These radiation transients directly affect PV output and place additional demands on spinning reserves. Control of renewable sources occurs at the distribution level and communication between transmission and distribution levels becomes important. A new frequency-control algorithm must therefore be designed to replace some of the large conventional plants (e.g., coal, nuclear, natural gas) by a great number of much smaller renewable and storage plants. While PV revenue growth slowed slightly in the last year, Gartner Group projects that PV implementation, measured on a power basis, will grow to 23.4 GW by 2013 (Defree, 2009) due both to consumer demand and portfolio standards requiring more renewable energy generation.

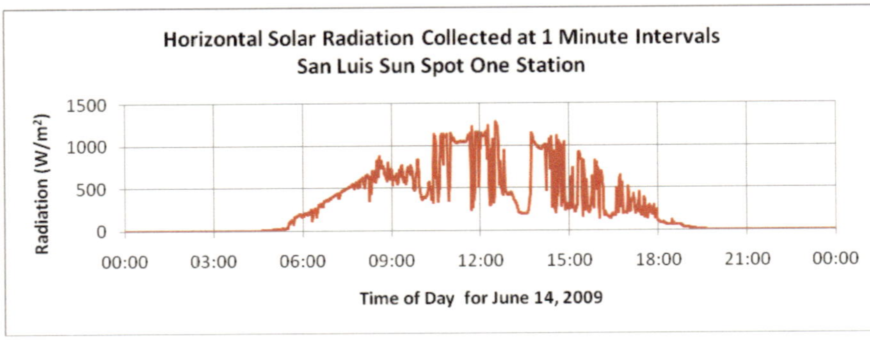

Fig. 5. Horizontal plate solar radiation in San Louis Valley, (NREL, 2009)

Similarly, high windpower (WP) penetrations have already occurred in Europe, and are likely to occur in regions of the US during the next decade. Areas relying on WP today are also experiencing power quality and control problems. PV systems experience different--typically faster--transients than wind systems because they are normally interconnected with inverters rather than rotating machines, which have virtually no inertia and low ride-through capacities. Residential-scale single-phase inverters are typically not designed to generate reactive power and are operated at unity power factor. A supply of reactive power to the grid would entail an increased DC voltage at the inverter input, as will be discussed in a later section. To date, most WP has been deployed in relatively large units from 1-5 MW dispersed turbines connected at the distribution level to hundreds of MW in wind farms connected at the transmission level. PV is frequently deployed in small units interspersed with residential or commercial loads in larger plants connected at the transmission level.

3.1 Design of PV plants

Fig. 6 illustrates the insolation within the contiguous United States in terms of kWh/m^2 per day. In Boulder, Colorado, one can expect (5.5-6.0) kWh/m^2 per day.

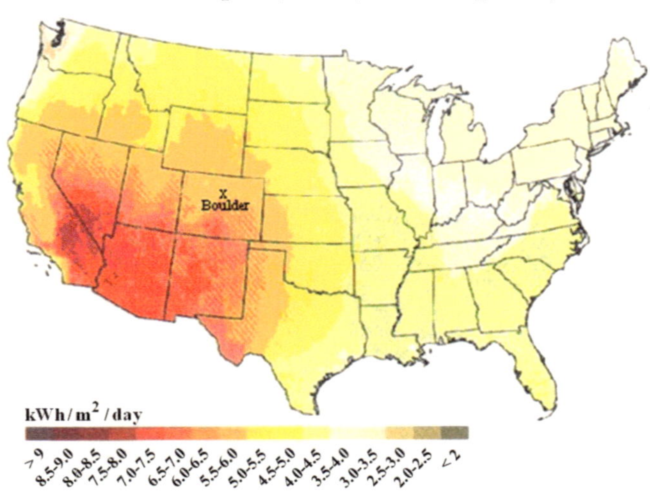

Fig. 6. Insolation or irradiance levels within the United States (created and prepared by NREL, 2008 for the U.S. Department of Energy)

The PV system of Fig. 7a installed and put on line in September 2007 generated during a recent 14-month period the data of Table 1 indicating the cumulative net meter reading $E_{net\ meter}$, the total generated AC energy $E_{generated}$, the AC energy supplied to the utility Xcel $E_{supplied\ Xcel}$, the AC energy consumed by the residence $E_{residence}$, and the CO_2 emissions avoided. The connection cost charged by the utility $C_{connection}$ is $8.55 per month. Fig. 7b illustrates the circuit components required for net metering.

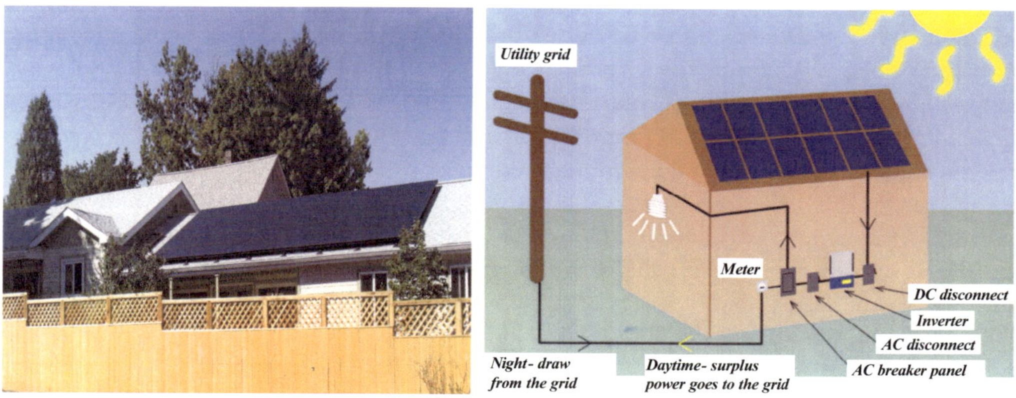

(a) (b)

Figs. 7a,b. (a) Residence with solar panels in Boulder, CO, 2007; (b) Net metering (Courtesy of Namasté Solar, 4571 North Broadway, Boulder, CO 80304), 2007

time period (readings recorded at end of each time period)	cumulative net meter reading $E_{net\ meter}$ [kWh]	cumulative total kWh generated $E_{generated}$ [kWh]	excess energy to Xcel $E_{supplied\ Xcel}$ [kWh]	energy consumed by residence $E_{residence}$ [kWh]	CO_2 emission avoided [lbs-force]
1/1/10-1/31/10	90,131	18,846	75	401	32,037
2/1/10-2/28/10	90,104	19,276	27	403	32,770
3/1/10-3/31/10	89,674	19,931	430	225	33,882
4/1/10-4/30/10	89,078	20,724	596	197	35,230
5/1/10-5/31/10	88,354	21,644	724	196	36,795
6/1/10-6/30/10	87,652	22,512	702	166	38,271
7/1/10-7/31/10	87,020	23,395	632	251	39,772
8/1/10-8/31/10	86,330	24,288	690	203	41,290
9/1/10-10/2/10	85,661	25,143	669	186	42,742
10/3/10-10/31/10	85,314	25,681	347	191	43,658
11/1/10-11/30/10	85,075	26,175	239	255	44,498
12/1/10-12/31/10	84,934	26,575	141	259	45,177
1/1/10-1/31/10	84,793	27,009	141	293	45,916
2/1/10-2/28/10	84,558	27,461	235	217	46,684

Table 1. Generated data of PV plant of Fig. 7a during a recent 14-month timeframe

Figs. 1.4, 1.5, and 1.6 of reference (E.F. Fuchs & Masoum, 2011) show the energy production of the 6.15 kW_{DC} plant of Fig. 7a during the entire year 2009, during October 2009, and during the 31st of October 2009, respectively. As can be seen from Fig. 1.5, since weather conditions make it likely there will be little energy production during periods of two to three days at a time, energy storage becomes important. The design data and the payback period information of the 6.15 kW_{DC} PV system of Fig. 7a are given in (E.F. Fuchs & Masoum, 2011).

Example #3: Design of a P_{AC}=5.61 kW PV power plant for a residence

A PV power plant consists of solar array, peak (maximum)-power tracker (Masoum *et al.*, 2002; Masoum *et al.*, 2004a), a step-up/step-down DC-to-DC converter, a deep-cycle battery for part f) only, a single-phase inverter, single-phase transformer, and residence which requires a maximum inverter AC output power of P_{inv}^{max} =5.61 kW as shown in Fig. 8. Note the maximum inverter output AC power has been specified because the entire power must pass through the inverter for all operating modes as explained below. In addition, inverters cannot be overloaded even for a short time due to the low heat capacity of the semiconductor switches. The modulation index of inverter is m=0.5 in order to guarantee a sinusoidal output current of the inverter neglecting pulse-width-modulated (PWM) switching harmonics.

Three operating modes will be investigated: In part f) the *operating mode #1* is a stand-alone configuration (700 kWh are consumed per month). In part g) the *operating mode #2* is a configuration where the entire energy (700 kWh) is consumed by the residence and the utility system is used as storage device only. In part h) the *operating mode #3* is a configuration where 300 kWh are consumed by the residence and 400 kWh are sold to the utility.

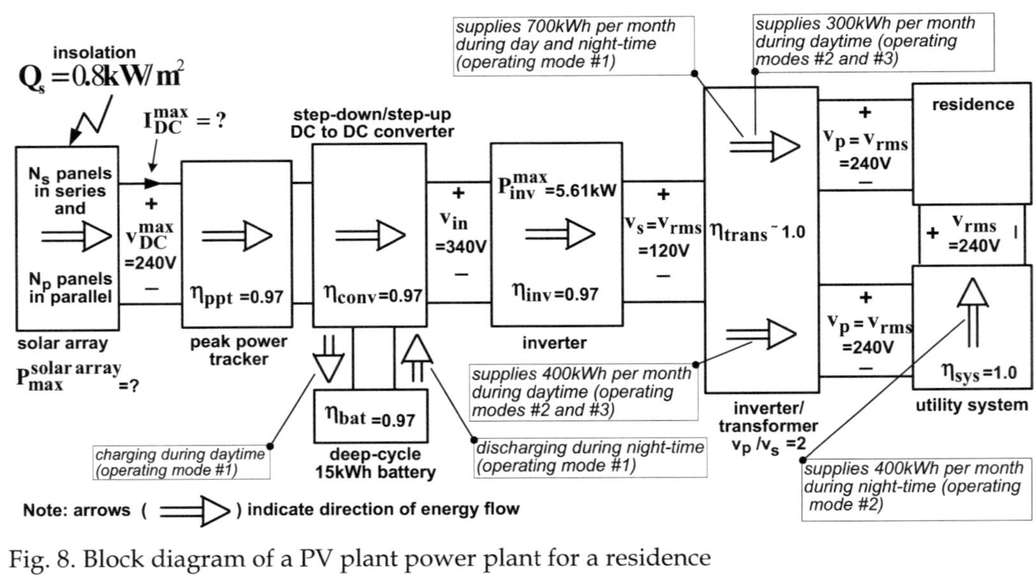

Fig. 8. Block diagram of a PV plant power plant for a residence

a. The power efficiencies of the maximum power tracker, the step-up/step-down DC-to-DC converter, the battery, and the inverter are 97 % each, while that of the transformer is about 1.00. What maximum power $P_{max}^{solar\ array}$ must be generated by the solar array, provided during daytime E_{month_day}=300 kWh will be delivered via the inverter to the residence (without storing this energy in battery), and sufficient energy will be stored in the battery so that the battery energy of E_{month_night}=400 kWh can be delivered during nighttime by the battery via the inverter to the residence: that is, a total of E_{month}= E_{month_day} + E_{month_night} =700 kWh can be delivered to the residence during one month?

b. For a commercially available solar panel the V-I characteristic of Fig. 9a was measured at an insolation of Q_s=0.8 kW/m². Plot the power curve of this solar panel: P_{panel}=f(I_{panel}).

c. At which point of the power curve P_{panel}=f(I_{panel}) would one operate assuming Q_s=0.8 kW/m² is constant? What values for power, voltage and current correspond to this point?

d. How many solar panels would one have to connect in series (N_s) in order to achieve a DC output voltage of $V_{DC}^{max} = 240$ V of the solar array? How many solar panels would one have to connect in parallel (N_p) in order to generate the inverter output power P_{inv}^{max} =5.61 kW?

e. How much would be the purchase price of this solar power plant, if 1 kW installed output capacity of the inverter (this includes the purchase and installation costs of solar cells + peak-power tracker + DC-to-DC converter + inverter) costs \$3,000 (after utility rebates and state/federal government tax-related subsidies)? Without tax rebates and subsidies the buyer would have to pay about \$5,000 per 1 kW installed output power capacity.

f. *Operating mode #1:* What is the payback period (in years, without taking into account interest payments) of this solar plant if the residence uses 700 kWh per month at an avoided cost of \$0.20/kWh (includes service fees and tax)? One may assume that this solar plant can generate every month 700 kWh and there is no need to buy electricity from the utility: 300 kWh per month will be used in the residence during daytime and during nighttime 400 kWh per month will be supplied via inverter from the battery to the residence. However, there is a need for the use of a 30 kWh deep-cycle battery as a storage element so that electricity will be available during hours after sunset. This battery must be replaced every four years at a cost of \$3,000.

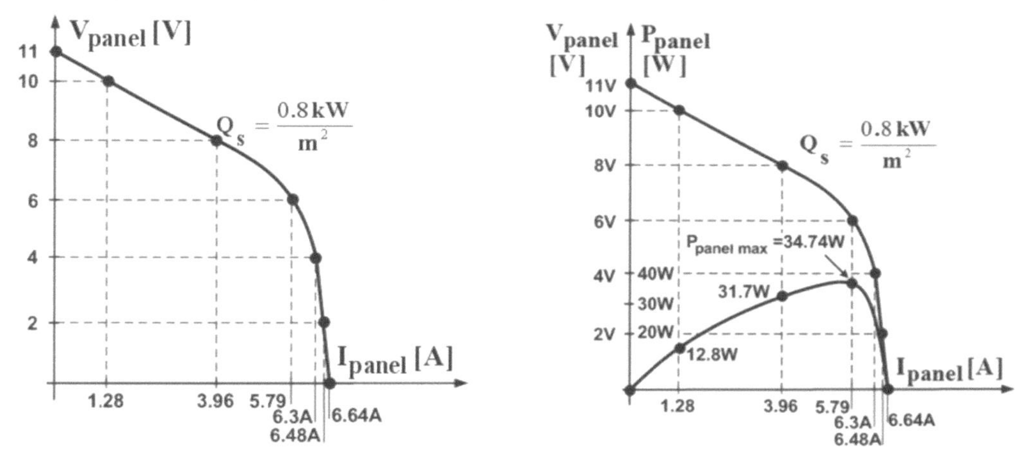

(a) (b)

Figs. 9a,b. (a) V-I characteristic of one solar panel; (b) P-I characteristic of one solar panel

g. *Operating mode #2:* What is the payback period (in years, without taking into account interest payments) of this solar plant if the residence uses 300 kWh per month at an avoided cost of \$0.20 per kWh? One may assume that this solar plant can generate 700 kWh per month and feeds 400 kWh into the power system of the utility company, which reimburses the plant owner \$0.20 per kWh (so-called "net metering"), in which case there is no need for a battery as a storage element because electricity can be supplied by the utility after sunset: 400 kWh at \$0.20 per kWh. There is a connection charge of \$8.55 per month.

h. *Operating mode #3:* What is the payback period (in years, without taking into account interest payments) of this solar plant if the residence uses 300 kWh per month at an avoided cost of \$0.20 per kWh? One may assume that this solar plant can generate every month 700 kWh of which every month the solar plant feeds 400 kWh into the power system of the utility company which reimburses the plant owner \$0.06 per kWh. There is a connection charge of \$8.55 per month.

i. Which power plant configuration (f, g or h) is more cost effective (e.g., has the shortest payback period)?

j. What is the total surface of the solar panels provided the efficiency of solar cells is 15% at Q_s=0.8 kW/m^2?

k. Instead of obtaining tax rebates and state/federal government subsidies the owner of a PV power plant obtains a higher price (feed-in tariff) for the electricity delivered to the utility: provided 700 kWh are fed into the utility grid at a reimbursement cost of \$0.75/kWh and the utility supplies 300 kWh to the residence at a cost of \$0.20/kWh, what is the payback period if the entire plant generating 5.61 kW_{AC} (there are no batteries required for storage) costs \$30,000? One may neglect interest payments, and there is a connection charge of \$8.55 per month.

l. Repeat part k) taking into account interest payments of 4.85%.

Solution:

a. The required maximum power output of the solar array is

$$P_{max}^{solar\ array} = \frac{\frac{300kWh}{700kWh} \cdot P_{inv}^{max}}{\eta_{ppt} \cdot \eta_{con} \cdot \eta_{inv}} + \frac{\frac{400kWh}{700kWh} \cdot P_{inv}^{max}}{\eta_{ppt} \cdot (\eta_{con})^2 \cdot (\eta_{bat})^2 \eta_{inv}}, \qquad (1)$$

whereby the first term is the power consumed by the residence during daytime, and the second term corresponds to the energy stored in the battery during daytime and consumed by the residence during nighttime (delivered by battery).

$$P_{max}^{solar\ array} = \frac{\frac{300kWh}{700kWh} \cdot 5.61}{(0.97)^3} + \frac{\frac{400kWh}{700kWh} \cdot 5.61}{(0.97)^6} = 2.634\ kW + 3.848\ kW = 6.483\ kW. \qquad (2)$$

b. Fig. 9b illustrates the P-I characteristic of one solar panel.

c. Operation at the knee of the V_{panel}-I_{panel} characteristic yields the voltage $V_{panel}^{max} = 6V$ and the current $I_{panel}^{max} = 5.79A$ resulting in the peak or maximum (Masoum *et al.*, 2002, and Masoum *et al.*, 2004a) power $P_{panel}^{max} = 34.74W$.

d. The number of in series connected panels is $N_s = V_{DC}^{max} / V_{panel}^{max} = 240 / 6 = 40$. The DC current delivered by the solar array is $I_{DC} = P_{max}^{solar\ array} / V_{DC}^{max} = 6483 / 240 = 27.01$ A. From this follows the number of panels in parallel $N_p = I_{DC} / I_{panel}^{max} = 27.01 / 5.79 = 4.66$. To be on the safe side one chooses $N_{p\_modified}=5$ panels in parallel. The choice of 5 panels in parallel increases the available maximum solar array output power to $P_{max\ modified}^{solar\ array} = (5 / 4.66) \cdot 6.483\text{kW} = 6.956\text{kW}$ and the inverter output power must be increased as well $P_{inv}^{max\ modified} = (5 / 4.66) \cdot 5.61\text{kW} = 6.02\text{kW}$.

e. Purchase cost of solar power plant (modified version, without battery) cost=($3000/kW) · 6.02kW =$18,058.

f. Payback period if battery is used as storage device: Avoided payments to utility per year are $700\text{kWh} \cdot 12 \cdot \$0.20 / \text{kWh} = \$1680$ resulting in the cost-benefit relation $1680(years)= $18058+($3000/4)(years) or $(years)_f$ =19.42.

g. Payback period if utility system is used a storage device and a connection charge to the utility ($8.55) is taken into account resulting in $1680(years) = $18058+$8.55 · 12(years) or $(years)_g$=11.45.

h. Payback period if 400 kWh per month are sold to utility: Avoided payments to utility are 300 kWh ·12·$0.20/kWh=$720 per year resulting in the cost-benefit relation $720 (years) + (400kWh·12·$0.06/kWh)(years)=$18,058+$8.55·12(years) or $(years)_h$=19.95.

i. Configuration with utility as storage device (case g) has the shortest payback period.

j. The required solar array area is area= $P_{max\ modified}^{solar\ array} / Q_s \cdot \eta_{cell} = \frac{6.956\text{kW}}{0.8 \cdot 0.15} = 58\text{m}^2$.

k. The income is $(700\text{kWh} \cdot \$0.75 / \text{kWh}) \cdot 12(\text{years}) = \$6300(\text{years})$ or payback period based on income=expenses is $6,300(years)=$30,000+$300 ·12 · 0.2 (years)+$8.55·12 (years) or $(years)_k$ =5.48.

l. Same as k) but with interest payments results in the cost-benefit relation (6,300-720-102.6)· (years)=30,000$(1.0485)^{(years)}$. No payback period exists.

3.2 Components of short-term and long-term storage and renewable energy plants

Electric storage components can store electricity in DC form only. For this reason AC-to-DC converters (rectifiers), DC–to-AC converters (inverters), and DC-to-DC converters (step-down and step-up) must be relied on. All types of converters are discussed in (E.F. Fuchs & Masoum, 2011).

3.2.1 The role and design of short-term and long-term storage plants

Short-term storage devices such as batteries, fuel cells, supercapacitors and flywheels can be put online within a few 60 Hz cycles, but cannot provide energy for more than about 10 minutes; flow batteries and variable-speed hydro plants, however, can change their load within a few 60 Hz cycles and are able to deliver power for days. Long-term storage plants such as constant-speed (pump)-hydro storage and compressed air plants require a start-up time of about 6-10 minutes, but can operate for several hours or even days. Two short-term storage plants will be analyzed in a later section.

3.2.2 Pulse-width-modulated (PWM) rectifier

Rectifiers (i.e., Fig. 10) are an integral part for the conversion of AC to DC of most intermittently operating renewable sources such as PV plants, WP plants and storage plants. Fig. 10 shows one type of rectifier which is used within a three-phase power system. This rectifier is analyzed with PSpice where the input and output voltage relations can be determined as a function of the duty cycle (E.F. Fuchs & Masoum, 2011).

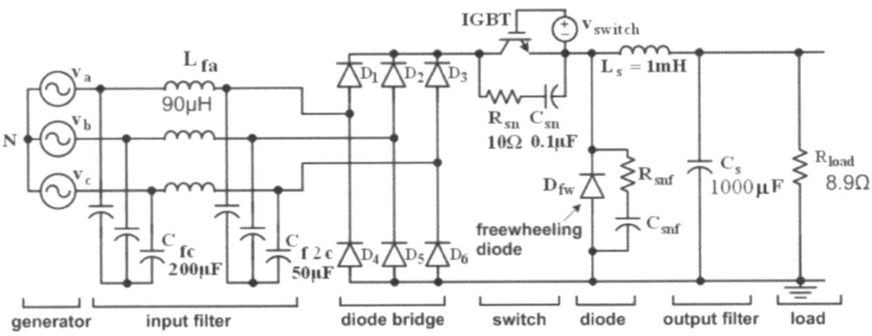

Fig. 10. Controlled three-phase rectifier with self-commutated switch

*Three-phase rectifier input voltages	*switch
Va 1 0 sin(0 465 60 0 0 0)	MOS 9 15 12 12 SMM
Vb 2 0 sin(0 465 60 0 0 -120)	*snubber resistors and capacitors
Vc 3 0 sin(0 465 60 0 0 -240)	Rsn 9 11 10
*switch gating signal of 3kHz	Csn 11 12 0.1u
vg 15 12 pulse(0 50 10u 0n 0n 166.6u	Rsnf 12 13 10
333u)	Csnf 13 10 0.1u
*diodes	*output filter and load resistor
D1 5 9 ideal	Ls 12 14 0.001
D2 10 7 ideal	Cs 14 10 1000u
D3 6 9 ideal	Rload 14 10 8.9
D4 10 5 ideal	*Model for MOSFET
D5 7 9 ideal	.model SMM NMOS(level=3 gamma=0
D6 10 6 ideal	kappa=0 tox=100n rs=0 kp=20.87u l=2u w=2.9
Dfw 10 12 ideal	+ delta=0 eta=0 theta=0 vmax=0 xj=0 uo=600
*input filter	phi=0.6 vto=0 rd=0 cbd=200n pb=0.8
Lfa 1 5 90u	+ mj=0.5 cgso=3.5n cgdo=100p rg=0 is=10f)
Lfb 2 6 90u	*diode model
Lfc 3 7 90u	.model ideal d(is=1p)
Cfa 1 4 200u	*options for improvement of convergence
Cfb 2 4 200u	.options abstol=10u chgtol=10p reltol=0.1
Cfc 3 4 200u	vntol=100m itl4=200 itl5=0
Rf1 4 10 10meg	.tran 0.5u 350m 300m 0.5m
Cf2a 5 8 50u	*plotting software
Cf2b 6 8 50u	.probe
Cf2c 7 8 50u	*fourier analysis
Rf2 8 10 10meg	.four 60 60 I(Rload)
	.end

Table 2. PSpice input program for PWM rectifier operation

The PSpice program is listed in Table 2 and the computed results are given in Table 3. Figs. 11a, b illustrate some of the results of Table 3. For a duty cycle of 0.05, a very high AC input voltage is required while at a duty cycle of 0.95 the AC input voltage is very low. In practice the duty cycle may vary between 25 and 75%. For very high power ranges thyristors or gate-turn-off (GTO) thyristors may be more suitable as discussed in Chapter 5 of (E.F. Fuchs & Masoum, 2011).

δ=0.05	$I_{DC\ load}$=45.05A	$V_{DC\ load}$=400.95 V	$V_{AN\ max}$=3000 V
δ=0.25	$I_{DC\ load}$=45.29A	$V_{DC\ load}$=403.08 V	$V_{AN\ max}$=890 V
δ=0.50	$I_{DC\ load}$=45.22A	$V_{DC\ load}$=402.46 V	$V_{AN\ max}$=465 V
δ=0.75	$I_{DC\ load}$=45.40A	$V_{DC\ load}$=404.06 V	$V_{AN\ max}$=320 V
δ=0.95	$I_{DC\ load}$=45.16A	$V_{DC\ load}$=401.92 V	$V_{AN\ max}$=256 V

Table 3. Dependency of the input voltage $V_{AN\ max}$ of a three-phase rectifier for given output voltages and currents, $V_{DC\ load}$, and $I_{DC\ load}$, respectively, at given duty cycles δ of the self-commutated switch (insulated gate bipolar transistor, IGBT).

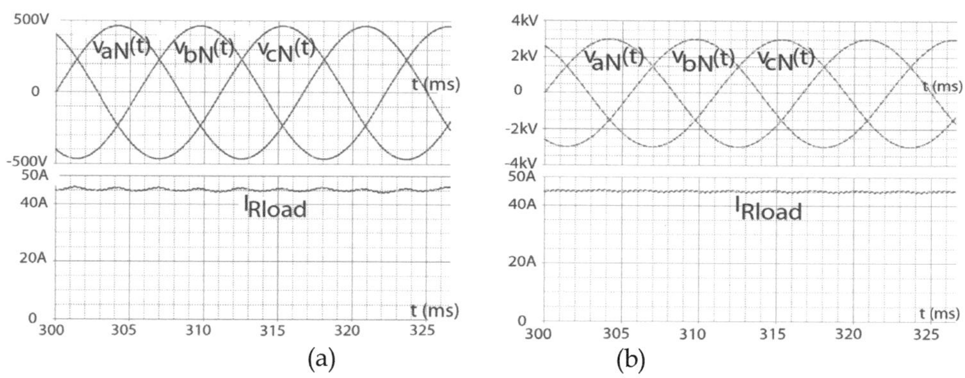

Figs. 11a,b. Rectifier input AC voltages and output DC current: (a) For a duty cycle of δ=50%; (b) For a duty cycle of δ=5% (see Table 3)

3.2.3 Current-controlled, PWM voltage–source inverter

Similar to rectifiers, inverters (see for example Figs. 12a, b) must be employed for some of the renewable energy sources and for storage plants to convert DC to AC.

Table 4 lists the PSpice program on which the results of Table 5 are based. This latter table illustrates the dependency of the input DC voltage V_{DC} of inverter as a function of the output power factor angle Φ, that is, the angle between phase current of inverter $I_{rms\ ph}$ and line-to-neutral voltage $V_{rms\ l\text{-}n}$ of power system as well as the modulation index m (E.F. Fuchs & Masoum, 2008a). Figs. 13a, b illustrate some of the results of Table 5. According to (IEEE Standard 519, 1992; IEC 61000-3-2, 2001-10; E.F. Fuchs & Masoum, 2008a) a total harmonic distortion of the inverter output current I_{THDi} of about 3% --ignoring the switching ripple which can be mitigated by an output filter as indicated in Fig. 12a-- is acceptable. By decreasing the modulation index m, say m=0.5, which requires an increased V_{DC}, and by increasing the wave shaping inductance L_w an almost ideal sinusoid for the output current can be achieved and reactive power can be supplied to the power system.

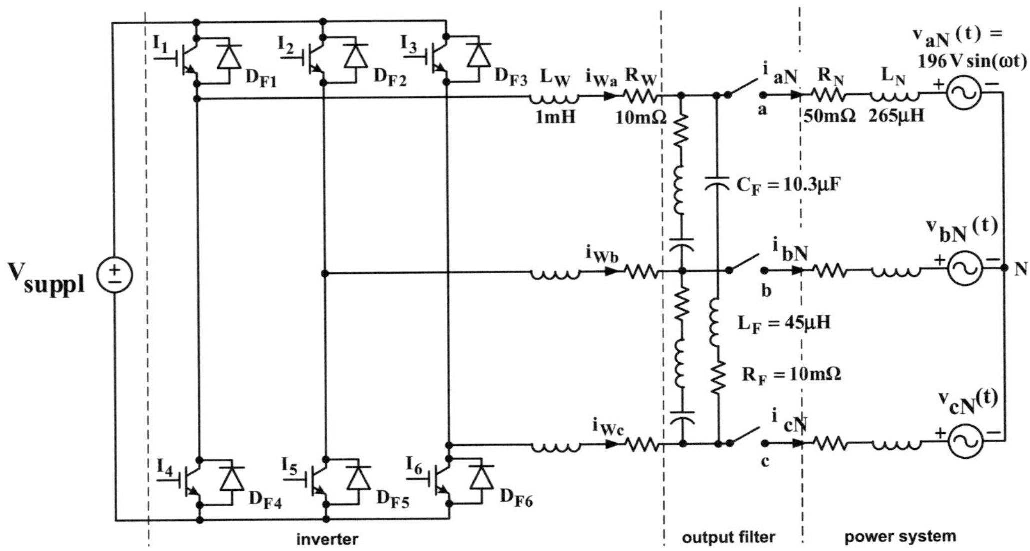

Fig. 12a. Current-controlled, PWM voltage-source inverter feeding power into utility system

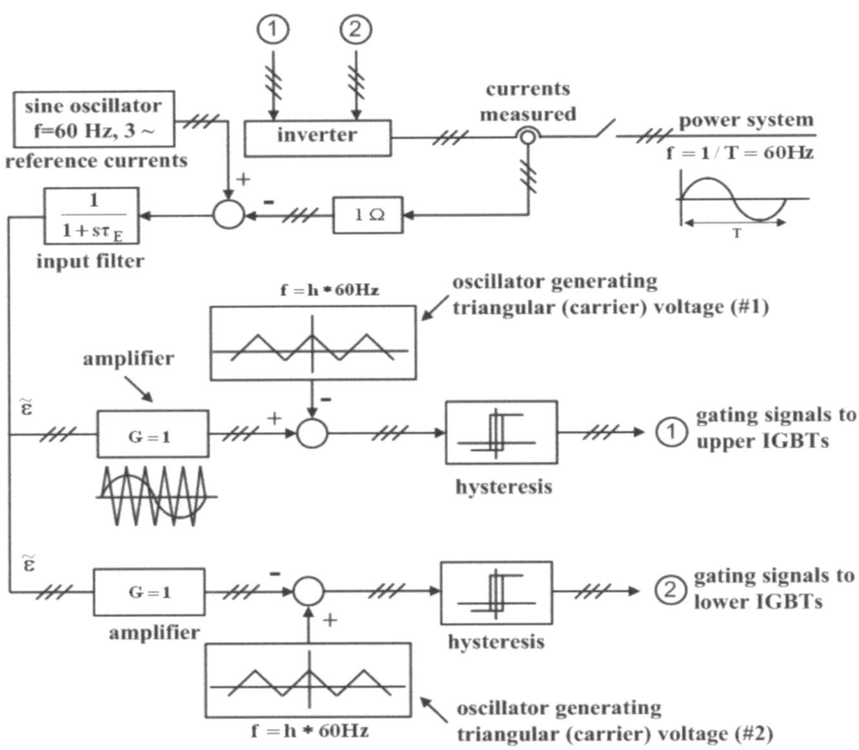

Fig. 12b. Block diagram of control circuit for current-controlled, PWM voltage-source inverter based on P-control

```
*Current-controlled pulse-width-
modulated (PWM) voltage-source
inverter
vsuppl 2 0 360
*switches
msw1 2 11 10 10 qfet
dsw1 10 2 diode
msw5 2 21 20 20 qfet
dsw5 20 2 diode
msw3 2 31 30 30 qfet
dsw3 30 2 diode
msw4 10 41 0 0 qfet
dsw4 0 10 diode
msw2 20 51 0 0 qfet
dsw2 0 20 diode
msw6 30 61 0 0 qfet
dsw6 0 30 diode
*waveshaping inductors
L_w1 10 15 1m
L_w2 20 25 1m
L_w3 30 35 1m
R_w1 15 16 10m
R_w2 25 26 10m
R_w3 35 36 10m
*current references in terms of voltages
vref1 12 0 sin(0 56.6 60 0 0 0)
vref2 22 0 sin(0 56.6 60 0 0 -120)
vref3 32 0 sin(0 56.6 60 0 0 -240)
eout1 13 0 15 16 100
eout2 23 0 25 26 100
eout3 33 0 35 36 100
*errorsignals produced as difference
between vr and eout
rdiff1 12 13a 1k
rdiff2 22 23a 1k
rdiff3 32 33a 1k
cdiff1 12 13a 1u
cdiff2 22 23a 1u
cdiff3 32 33a 1u
rdiff4 13a 13 1k
rdiff5 23a 23 1k
rdiff6 33a 33 1k
ecin1 14 0 12 13a 2
ecin2 24 0 22 23a 2
ecin3 34 0 32 33a 2
*Lfi1 16 15b 45u
*Lfi2 26 25b 45u
*Lfi3 36 35b 45u
*rfi1 15b 15c 0.01
*rfi2 25b 25c 0.01
*rfi3 35b 35c 0.01
*cfi1 15c 26 10.3u
*cfi2 25c 36 10.3u
*cfi3 35c 16 10.3u
*parameters of power system with a line-to-
line voltage of 340 V (amplitude)
RM1 16 18 50m
LM1 18 19 265u
Vout1 19 123 sin(0 196 60 0 0 -30)
RM2 26 28 50m
LM2 28 29 265u
Vout2 29 123 sin(0 196 60 0 0 -150)
RM3 36 38 50m
LM3 38 39 265u
Vout3 39 123 sin(0 196 60 0 0 -270)
*subcircuit for comparator
.subckt comp 1 2 9 10
rin 1 3 2.8k
r1 3 2 20meg
e2 4 2 3 2 50
r2 4 5 1k
d1 5 6 zenerdiode1
d2 2 6 zenerdiode2
e3 7 2 5 2 1
r3 7 8 10
c3 8 2 10n
r4 3 8 100k
e4 9 10 8 2 1
*models for zener diodes
.model zenerdiode1 D (Is=1p BV=0.1)
.model zenerdiode2 D (Is=1p BV=50)
.ends comp
*model for switch
.model qfet nmos(level=3 gamma=0
kappa=0 tox=100n rs=42.69m kp=20.87u
l=2u
+ w=2.9 delta=0 eta=0 theta=0 vmax=0 xj=0
uo=600 phi=0.6
+ vto=3.487 rd=0.19 cbd=200n pb=0.8
mj=0.5 cgso=3.5n cgdo=100p rg=1.2 is=10f)
```

Table 4. PSpice input program for PWM inverter operation

vtria1 5 0 pulse(-10 10 0 86.5u 86.5u 0.6u 173.6u) *gating signals for upper switches xgs1 14 5 11 10 comp xgs2 24 5 21 20 comp xgs3 34 5 31 30 comp *gating signals for lower switches egs4 41 0 poly(1) (11,10) 50 -1 egs5 51 0 poly(1) (21,20) 50 -1 egs6 61 0 poly(1) (31,30) 50 -1 *Filter is deleted because PSpice is limited to 64 nodes	*model for diodes .model diode d(is=1p) *options to aid convergence .options abstol=0.01m chgtol=0.01m reltol=50m vntol=1m itl5=0 itl4=200 *transient analysis .tran 5u 350m 300m 5u *plotting of traces .probe *Fourier analysis .four 60 12 I(L_w1) .end

Table 4. PSpice input program for PWM inverter operation (continuation)

Φ= 90 °	$V_{rms\ l\text{-}n}$=139 V	$I_{rms\ ph}$=54.90 A	V_{DC}=375 V	I_{THDi}=3.14%	m=1.04
Φ= 60 °	$V_{rms\ l\text{-}n}$=139 V	$I_{rms\ ph}$=48.57 A	V_{DC}=368 V	I_{THDi}=2.97%	m=1.07
Φ= 30 °	$V_{rms\ l\text{-}n}$=139 V	$I_{rms\ ph}$=44.80 A	V_{DC}=360 V	I_{THDi}=3.17%	m=1.09
Φ= 0 °	$V_{rms\ l\text{-}n}$=139 V	$I_{rms\ ph}$=45.08 A	V_{DC}=388 V	I_{THDi}=3.13%	m=1.01
Φ= -30 °	$V_{rms\ l\text{-}n}$=139 V	$I_{rms\ ph}$=48.59 A	V_{DC}=410 V	I_{THDi}=2.71%	m=0.96
Φ= -60 °	$V_{rms\ l\text{-}n}$=139 V	$I_{rms\ ph}$=53.58 A	V_{DC}=415 V	I_{THDi}=2.80%	m=0.95
Φ= -90 °	$V_{rms\ l\text{-}n}$=139 V	$I_{rms\ ph}$=59.09 A	V_{DC}=415 V	I_{THDi}=2.20%	m=0.95

Table 5. Dependency of the input DC voltage V_{DC} of inverter as a function of the output power factor angle Φ (generator notation, where a positive Φ corresponds to underexcited operation absorbing reactive power from the grid and a negative Φ corresponds to overexcited operation supplying reactive power to the grid) is the angle between phase current of inverter $I_{rms\ ph}$ and line-to-neutral voltage $V_{rms\ l\text{-}n}$ of power system, and m is the modulation index of inverter

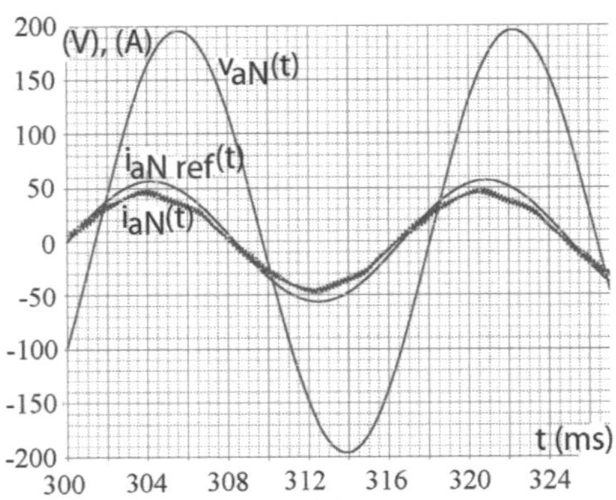

(a)

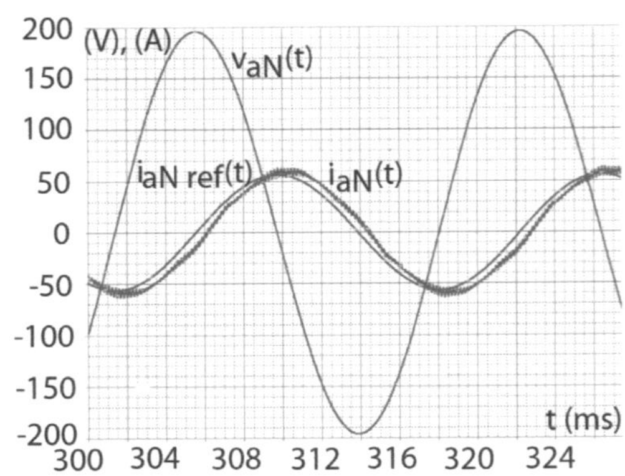

(b)

Figs. 13a,b. Inverter phase reference current, inverter phase output current, and power system phase voltage (about the same as inverter output voltage), see Fig.12a and Table 5; (a) Φ=30° (generator notation: leading power factor, underexcited, absorbing reactive power from grid); (b) Φ= - 90° (generator notation: lagging power factor, overexcited, delivering reactive power to grid)

3.2.4 Storage devices for short-term and long-term plants

Batteries, supercapacitors, ultracapacitors and flywheels serve as short-term storage devices while compressed air, flow batteries, (variable and constant-speed) hydro facilities and thermal storage are used for long-term storage facilities. Variable-speed WP and hydropower plants are based on the doubly-fed induction generator (DFIG) where energy can be provided under transient conditions by the rotating rotor slowing its angular velocity, and the DFIG may be used for short- and long-term storage. The required rotor excitation represents about 8% of the stator power (E.F. Fuchs & Masoum, 2011). Variable-speed drives (Yildirim *et al.*, 1998; E.F. Fuchs & Myat, 2010) are proposed for short- and long-term storage plants in order to eliminate mechanical gears.

4. Smartgrid and microgrids

Simultaneously, the implementation of smart/microgrids is accelerating. A key goal of SmartGrid City (Boulder, Colorado) and similar projects is to enable integration of significantly higher penetrations of renewable resources, both in concentrated utility-scale plants and dispersed residential-scale units. In addition to advanced utility controls such as smart meters, SmartGrid City combines a range of renewable resources. Specifically within Boulder, several hundred residential PV plants ranging from 2-10 kW are interconnected with the grid. Xcel's system includes an 8 MW PV plant in Colorado's San Luis Valley with an additional 17 MW PV plant (under construction) by Xcel Energy and SunPower.

Xcel's major Colorado WP resources include: Ponnequin/Weld County, 32 MW; Ridgecrest/Peetz, 30 MW; Colorado Green/Lamar, 162 MW; Spring Canyon/Peetz, 60 MW; Peetz Table/Peetz, 200 MW; Logan Wind/Peetz, 200 MW; Twin Buttes/Lamar, 75 MW; Cedar Creek/Grover, 300 MW; and Northern Colorado Wind/Peetz, 174 MW.

The Boulder municipal hydroelectric system includes 7 small hydropower plants generating more than 20,000 MWh per year (Cowdrey, 2004); these plants can operate intermittently depending upon the municipal treated water needs. Xcel also operates several pumped-storage (hydroelectric) plants, which can be relied upon for long-term storage. As a comparison Xcel's total installed power capacity in Colorado is 7 GW.

The smartgrid vision is the development of a smart (e.g., self-learning, self-healing, optimized, efficient, high power quality) power system, relying on existing proven distribution system technologies, adding new control and energy source/storage paradigms, and discarding approaches which interfere with intelligent control strategies, distributed generation and renewable resources. The smartgrid as envisioned by utilities must satisfy steady-state and transient operating conditions, avoiding the weak system effects (e.g., flicker, high-system impedance, voltage breakdown) encountered by the Danish grid, which has high WP plant penetration.

The rationale for the smart/microgrid lies in the integrative analysis of distributed energy sources--many of which will be intermittently operating--with the deployment of short-term and long-term storage plants. However, current load-sharing strategies will not work to integrate renewable sources and present-day storage plants because of the intermittent (peak-power) operation of renewable sources, and the online response time of different storage plants (E.F. Fuchs & Masoum, 2008a; E.F. Fuchs & Masoum, 2011). In addition, Fig. 4b illustrates that load-sharing utilizing standard droop strategies between two similarly-sized (e.g., PV) plants results in unstable frequency control due to similar response characteristics.

5. Design of short-term storage plants

Example # 4: Design a 10 MWh supercapacitor short-term storage plant

Relying solely on wind/solar energy is problematic because it may not be available when needed. A wind farm, for example, can lose as much as 60 MW within one minute. There are several scenarios of how the power change of 60 MW per minute can be mitigated through complementary, albeit more expensive, power sources. One is the combination of a (long-term) compressed-air storage (CAES) power plant with a (short-term) supercapacitor plant for bridging the time from when the WP plant output decreases (60 MW per minute) to when either a CAES plant (Mattick et al., 1975 and Vosburgh, 1978) or a pump-storage hydro plant (Glems, 1964; Raccoon Mountain, 1975; Cowdrey, 2004) can take over. A CAES plant requires a start-up time of about 6 minutes.

To bridge this 6-minute gap for a 100 MW compressed-air power plant, a supercapacitor plant can provide up to 100 MW during a 6-minute interval amounting to a required energy storage of 10 MWh. Inverters fed from a supercapacitor can deliver power within a few 60 Hz cycles to the power system, replacing the lost power of 60 MW per minute almost instantaneously. This combination of CAES plant and supercapacitor storage plant as a bridging energy source can be employed for peak-power operation as well as for improving power quality by preventing brownouts/blackouts. Fig. 14 depicts the block diagram of such a supercapacitor storage plant consisting of wind turbine, mechanical gear, synchronous generator, 3-phase transformer, 3-phase rectifier, supercapacitor bank, three-phase inverter, 3-phase transformer, and power system as discussed in Application Example 12.23 of (E.F. Fuchs & Masoum, 2011).

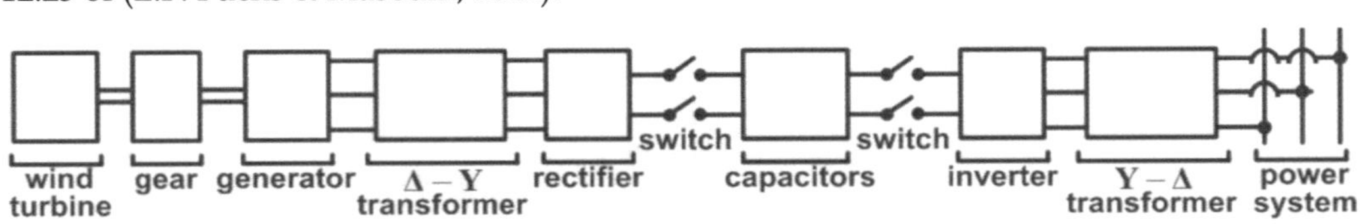

Fig. 14. Block diagram for charging and discharging supercapacitor bank

Example 5: Design of a 10MWh flywheel short-term storage power plant

Design a flywheel storage system which can provide for 6 minutes 100 MW, that is, energy of 10 MWh. The flywheel power plant consists (see Figs. 15a,b) of a flywheel, mechanical gear, synchronous machine, inverter-rectifier set and a step-up transformer (not shown in Fig. 15a). The individual components of this plant must be designed as follows: For the flywheel (made from steel) as shown in Fig. 15b (h=0.9 m, R_{1o}=1.5 m, R_{1i}=1.3 m, R_{2o}=0.50 m, R_{2i}=0.10 m, b=0.2 m), compute the energy stored as a function of angular velocity, see Application Example 12.26 of (E.F. Fuchs & Masoum, 2011).

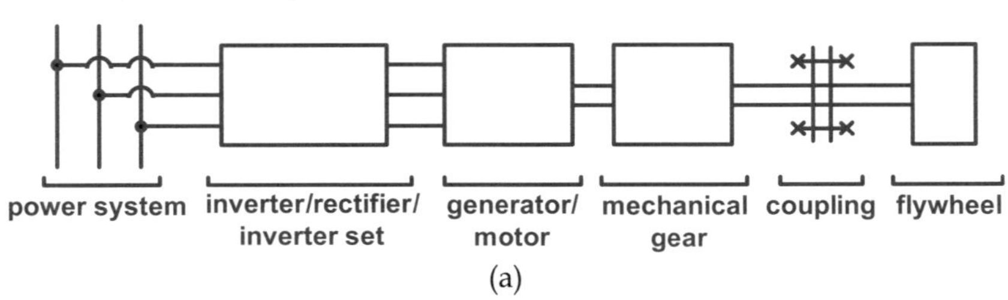

Fig. 15a. Flywheel power plant.

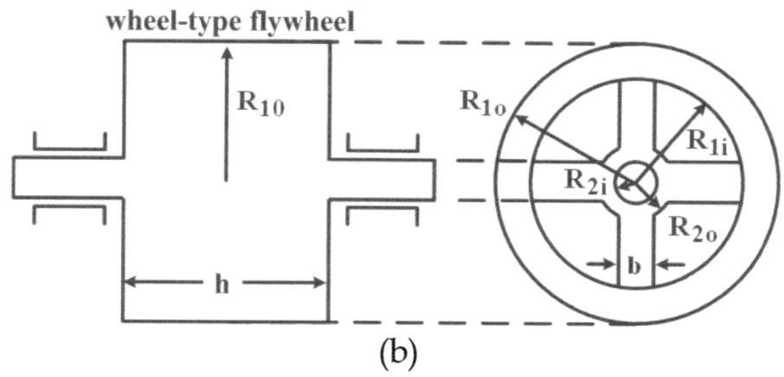

Fig. 15b. Flywheel power plant.

6. Complementary operation of renewable plants with short-term and long-term storage plants analyzed with either Mathematica or Matlab

The stability of a smart/microgrid consisting of a natural gas-fired power plant (the frequency leader), a long-term storage power plant, and two intermittently operating plants (i.e., PV and WP plants) with associated short-term storage plants is the objective of this section. In order to achieve stability for a given smart/micro- grid the following constraints must be satisfied:

1. instability and frequency variation is minimized through appropriate switching (in and out) of the short-term storage plants; in particular the time instant of switching is important.
2. transmission line (e.g., tie lines) parameters are optimized;
3. time constants of the governors and the valves must be within a feasible region;
4. droop characteristics of the individual plants must satisfy certain constraints.

Example 6: Operation of natural-gas fired and long-term storage plants with two renewable sources and two associated complementary short-term storage plants

Fig. 16 illustrates the sharing (increase) of the additional power among a short-term storage plant (e.g., R_2=0.01), a long-term-storage plant (e.g., R_3=10), and a PV plant (e.g., R_1=10) causing a frequency decrease. Stable frequency control is obtained when the short-term storage plant compensates the intermittent power output of the PV plant, and the plant with spinning reserve of Fig. 2 (natural-gas fired plant) is replaced by a long-term storage plant which is connected all the time to the power system and serves in this case as frequency leader. The PV plant and the short-term storage plant may operate intermittently but only in a complementary fashion.

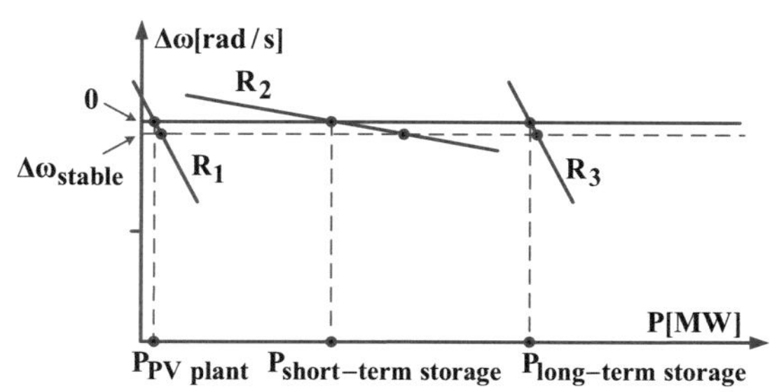

Fig. 16. Drooping characteristics of the short-term storage plant (e.g., R_2=0.01), the long-term storage plant (e.g., R_3=10), and intermittently operating PV (e.g., R_1=10) plant, accommodating additional demand/increase of power, where the PV plant is operated at peak power and cannot deliver additional power.

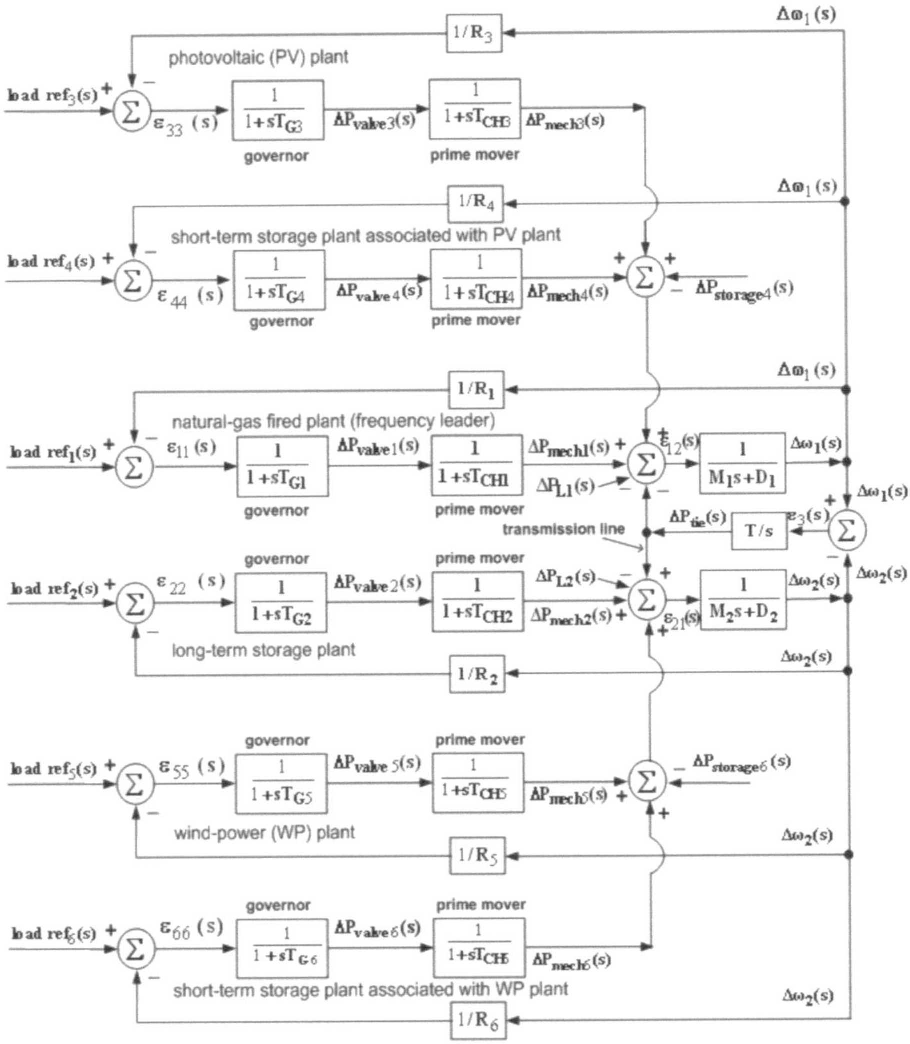

Fig. 17a. Block diagram of smart/micro grid resulting in stable frequency control.

In Fig. 2 the spinning reserve plant can be replaced by a natural-gas fired plant and a long-term storage plant (e.g., pump-hydro or compressed air facility), as shown in Figs. 17a, b, c. The intermittently operating PV and WP plants are complemented by short-term storage plants (e.g., battery-fed inverter). During times of high power demand, the storage plants supply power to maintain the power balance between generation and the served loads. At low power demand, the renewable sources will supply the storage plants. The generation increase of conventional peak-power plants (which can supply additional load through spinning reserve) are replaced by putting short-term (located next to renewable plants) and long-term storage plants (E.F. Fuchs & Masoum, 2008a, E.F. Fuchs & Masoum, 2011) online, reducing fossil-fuel consumption and contributing to renewable portfolio standards. Renewable sources are operated at their peak-power point e.g., the slopes of the droop characteristic R_3 and R_5 of Fig. 17b are large while those of R_1 and R_2 are relatively small to permit the increase of their output power upon demand. Thus the PV and WP plants cannot participate in frequency/load control. In Figu. 17c, all droop characteristics have a relatively small slope and all participating plants can output increased power and participate in frequency/load control. In the block diagram of the PV plant of Fig. 17a the governor and prime mover represent the solar array and inverter/rectifier while in the short-term storage plant associated with the PV plant, the governor and prime mover represent the storage device (e.g., battery, supercapacitor, flywheel) and inverter/rectifier. Similar considerations apply to the WP plant and its associated short-term storage plant.

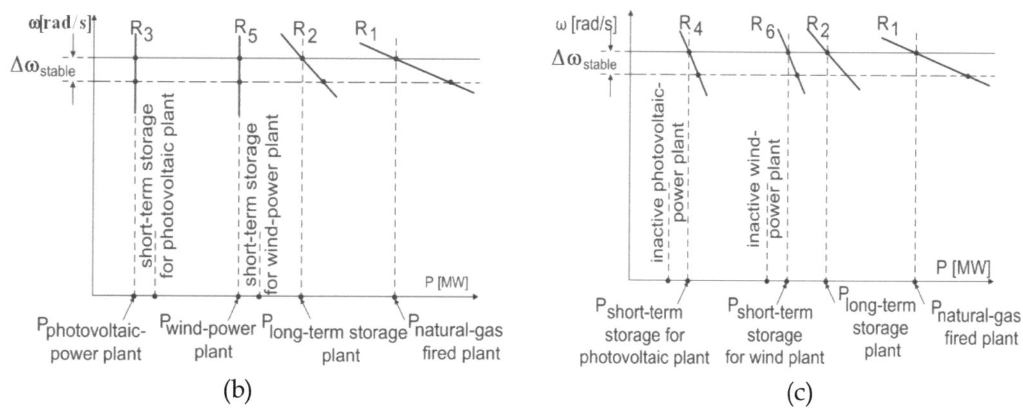

Figs. 17b,c. The natural-gas fired plant serves as frequency leader and the long-term storage plant serves as spinning reserve; (b) PV and WP plants operate at peak power and storage plants are disconnected; (c) Short-term storage plants compensate the zero power outputs of the PV and WP plants.

Data for natural-gas fired plant (system #1 of Fig.17a): Angular frequency per-unit (pu) change $\Delta\omega_1$ per change in generator output power ΔP_1, that is, $R_1=\Delta\omega_1/\Delta P_1=0.01$ pu, the frequency-dependent load change ΔP_{L1_frequ} per angular frequency change $\Delta\omega_1$, that is, $D_1=\Delta P_{L1\_frequ}/\Delta\omega_1=0.8$ pu, positive step-load change $\Delta P_{L1}/s=0.1/s$ pu, angular momentum of gas turbine and generator set $M_1=4.5$, base apparent power $S_{base}=500$ MVA, governor time constant $T_{G1}=0.3$ s, valve changing/charging time constant $T_{CH1}=0.9$ s, and load $ref_1(s)=0.8$ pu.
Data for long-term storage plant (system #2 of Fig.17a): Angular frequency per-unit (pu) change $\Delta\omega_2$ per change in generator output power ΔP_2, that is, $R_2=\Delta\omega_2/\Delta P_2=0.1$ pu (e.g., variable-speed pump hydropower plant), the frequency-dependent load change ΔP_{L2_frequ} per angular frequency change $\Delta\omega_2$, that is, $D_2=\Delta P_{L2\_frequ}/\Delta\omega_2=1.0$ pu, negative step-load change $\Delta P_{L2}(s)=\Delta P_{L2}/s= -0.2/s$ pu, angular momentum of hydro turbine and generator set $M_2=6$, base apparent power $S_{base}=500$ MVA, governor time constant $T_{G2}=0.2$ s, valve changing/ charging time constant $T_{CH2}=0.2$ s, and load $ref_2(s)=0.5$ pu.
Data for tie line: $T=\omega_o/X_{tie}$with $X_{tie}=0.2$ pu and $\omega_o=377$ rad/s for f=60Hz.
Data for PV plant (system #3 of Fig.17a): Angular frequency change $\Delta\omega_1$ per change in inverter output power ΔP_3, that is, $R_3=\Delta\omega_1/\Delta P_3=0.3$ pu, governor time constant $T_{G3}=0.1$ s, equivalent valve time constant $T_{CH3}=0.1$ s, and load $ref_3(s)=0.01$ pu.
Data for short-term storage plant associated with PV plant (system #4 of Fig.17a): Angular frequency change $\Delta\omega_1$ per change in generator output power ΔP_4, that is, $R_4=\Delta\omega_1/\Delta P_4=0.5$ pu, governor time constant $T_{G4}=0.2$ s, equivalent valve time constant $T_{CH4}=0.1$ s, and load $ref_4(s)=0.01$ pu.

Data for WP plant (system #5 of Fig.17a): Angular frequency change $\Delta\omega_2$ per change in generator output power ΔP_5, that is, $R_5=\Delta\omega_2/\Delta P_5=0.7$ pu, governor time constant $T_{G5}=0.1$ s, equivalent valve time constant $T_{CH5}=0.1$ s, and load $ref_5(s)=0.01$ pu.
Data for short-term storage plant associated with WP plant (system #6 of Fig.17a): Angular frequency change $\Delta\omega_2$ per change in generator output power ΔP_6, that is, $R_6=\Delta\omega_2/\Delta P_6=0.5$ pu, governor time constant $T_{G6}=0.2$ s, equivalent valve time constant $T_{CH6}=0.1$ s, and load $ref_6(s)=0.01$ pu.

a. List the ordinary differential equations and the algebraic equations of the block diagram of Fig. 17a. Derive the transfer function of the transmission (e.g., tie line) line.

b. Use either Mathematica or Matlab to establish steady-state conditions by imposing positive step functions for load $ref_1(s)=0.8/s$ pu, load $ref_2(s)=0.5/s$ pu, load $ref_3(s)=$ load $ref_4(s)$ = load $ref_5(s)$= load $ref_6(s)=0.01/s$ pu, and run the program with a zero step-load changes $\Delta P_{L1}=0$, $\Delta P_{L2}=0$ for 200 s. Save the steady-state values for all variables at 200 s. Plot the calculated angular frequency response $\Delta\omega(t)$ [pu]= $\Delta\omega_1(t)$ [pu]= $\Delta\omega_2(t)$ [pu].

c. Initialize the parameters with the steady-state values as obtained in Part b). After 300 s impose positive step-load change $\Delta P_{L1}(s)=\Delta P_{L1}/s=0.1/s$ pu, and after 400 s impose negative step-load change $\Delta P_{L2}(s)=\Delta P_{L2}/s=-0.1/s$ pu. Thereafter, for load $ref_3(s)$, load ref_4 (s), load $ref_3(s)$, load $ref_5(s)$, $DPstorage_4(s)$, $DPstorage_6(s)$ and load ref_4 (s):
Lr3[t_]:=If [t<600, 0, 0.06];
Lr4[t_]:=If [t<600.1, 0,- 0.6];
Lr3[t_]:=If [t<1200,If[t<1120,If[t<1000,If[t<940,If[t<910,0,0.03],0.09],0.05],0.03],0.0];
Lr5[t]:=If [t<1200,If[t<1120,If[t<1000,If[t<940,If[t<910,0,0.3],0.9],0.5],0.3],0.0];
DPstorage4[t_]:=If[t<1200.2,If[t<1120.2,If[t<1000.2,If[t<940.2,If[t<910.2,0,0.15],0.45],0.25], 0.15],0.0];
DPstorage6[t_]:=If[t<1200.2,If[t<1120.2,If[t<1000.2,If[t<940.2,If[t<910.2,0,0.15],0.45],0.25], 0.15],0.0];
Lr4[t_]:=If [t<700, 0,- 0.01].
Plot the given WP plant load reference Lr5[t] and calculated the transient response $\Delta\omega(t)$ for a total of 1,500 s.

d. Initialize the parameters with the steady-state values as obtained in Part b). After 300 s impose positive step-load change $\Delta P_{L1}(s)=\Delta P_{L1}/s=0.1/s$ pu, and after 400 s impose negative step-load change $\Delta P_{L2}(s)=\Delta P_{L2}/s=-0.1/s$ pu. Thereafter:
Lr3[t_]:=If [t<600, 0, 0.6];
Lr4[t_]:=If [t<600.1, 0,- 0.6];
Lr3[t_]:=If [t<1200,If[t<1120,If[t<1000,If[t<940,If[t<910,0,0.03],0.09],0.05],0.03],0.0];
Lr5[t_]:=If [t<1200,If[t<1120,If[t<1000,If[t<940,If[t<910,0,0.3],0.9],0.5],0.3],0.0];
DPstorage4[t_]:=If[t<1200.2,If[t<1120.2,If[t<1000.2,If[t<940.2,If[t<910.2,0,0.01],0.01],0.01], 0,0.01],0.0];
DPstorage6[t_]:=If[t<1200.2,If[t<1120.2,If[t<1000.2,If[t<940.2,If[t<910.2,0,0.30],0.90],0.50], 0.30],0.0];
Lr4[t_]:=If [t<700, 0,- 0.3].
Calculate and plot the transient response $\Delta\omega(t)$ for a total of 1,500 s.

e. Initialize the parameters with the steady-state values as obtained in Part b). After 300 s impose positive step-load change $\Delta P_{L1}(s)=\Delta P_{L1}/s=0.1/s$ pu, and after 400 s impose negative step-load change $\Delta P_{L2}(s)=\Delta P_{L2}/s=-0.1/s$ pu. Thereafter:

Lr3[t_]:=If [t<600, 0, 0.6];
Lr4[t_]:=If [t<600.1, 0,- 0.6];
Lr3[t_]:=If [t<1200,If[t<1120,If[t<1000,If[t<940,If[t<910,0,0.03],0.09],0.05],0.03],0.0];
Lr5[t_]:=If [t<1200,If[t<1120,If[t<1000,If[t<940,If[t<910,0,0.3],0.9],0.5],0.3],0.0];
DPstorage4[t_]:=If[t<1205.2,If[t<1125.2,If[t<1005.2,If[t<950.2,If[t<920.2,0,0.30],0.90],0.50], 0.30],0.0];
DPstorage6[t_]:=If[t<1200.2,If[t<1120.2,If[t<1000.2,If[t<940.2,If[t<910.2,0,0.01],0.01],0.01], 0,0.01],0.0];
Lr4[t_]:=If [t<700, 0,- 0.3].
Calculate and plot the transient response $\Delta\omega(t)$ for a total of 1,500 s.

f. Initialize the parameters with the steady-state values as obtained in Part b). After 300 s impose positive step-load change $\Delta P_{L1}(s)=\Delta P_{L1}/s=0.1/s$ pu, and after 400 s impose negative step-load change $\Delta P_{L2}(s)=\Delta P_{L2}/s = -0.1/s$ pu. Thereafter:
Lr3[t_]:=If [t<600, 0, 0.6];
Lr4[t_]:=If [t<600.1, 0,- 0.6];
Lr3[t_]:=If[t<1200,If[t<1120,If[t<1000,If[t<940,If[t<910,0,0.03],0.09],0.05],0.03],0.0];
Lr5[t_]:=If [t<1200,If[t<1120,If[t<1000,If[t<940,If[t<910,0,0.3],0.9],0.5],0.3],0.0];
DPstorage4[t_]:=If[t<1230,If[t<1150,If[t<1030,If[t<970,If[t<940,0,0.030],0.090],0.050],0.030],0.0];
DPstorage6[t_]:=If[t<1200.2,If[t<1120.2,If[t<1000.2,If[t<940.2,If[t<910.2,0,0.01],0.01],0.01], 0,0.01],0.0];
Lr4[t_]:=If [t<700, 0,- 0.3].
Calculate and plot the transient response $\Delta\omega(t)$ for a total of 1,500 s.

g. Investigate the influence of the power capability of the transmission line for the conditions as given in part c) by increasing X_{tie} from 0.2 pu to 0.237 pu.

Solution:

a. Differential and algebraic equations:
System #1: ε_{11}=load ref$_1$-$\Delta\omega_1/R_1$, $\Delta P_{valve\_1}+T_{G1}d(\Delta P_{valve\_1})/dt=\varepsilon_{11}$, $\Delta P_{mech\_1}+T_{CH1}d(\Delta P_{mech\_1})/dt=\Delta P_{valve\_1}$, $\varepsilon_{12}=\Delta P_{mech\_1}-\Delta P_{L1}-\Delta P_{tie}+\Delta P_{mech\_3}+\Delta P_{mech\_4}-\Delta P_{storage\_4}$, $\Delta\omega_1 D_1+M_1 d(\Delta\omega_1)/dt=\varepsilon_{12}$.
Coupling (tie, transmission) network: $(1/T)d(\Delta P_{tie})/dt=\varepsilon_3$, where $\varepsilon_3=\Delta\omega_1-\Delta\omega_2$.
System #2: ε_{22}=load ref$_2$-$\Delta\omega_2/R_2$, $\Delta P_{valve\_2}+T_{G2}d(\Delta P_{valve\_2})/dt=\varepsilon_{22}$, $\Delta P_{mech\_2}+T_{CH2}d(\Delta P_{mech\_2})/dt=\Delta P_{valve\_2}$, $\varepsilon_{21}=\Delta P_{mech\_2}-\Delta P_{L2}+\Delta P_{tie}+\Delta P_{mech\_5}+\Delta P_{mech\_6}-\Delta P_{storage\_6}$, $\Delta\omega_2 D_2+M_2 d(\Delta\omega_2)/dt=\varepsilon_{21}$.
System #3: ε_{33}=load ref$_3$-$\Delta\omega_1/R_3$, $\Delta P_{valve\_3}+T_{G3}d(\Delta P_{valve\_3})/dt=\varepsilon_{33}$, $\Delta P_{mech\_3}+T_{CH3}d(\Delta P_{mech\_3})/dt=\Delta P_{valve\_3}$.
System #4: ε_{44}=load ref$_4$-$\Delta\omega_1/R_4$, $\Delta P_{valve\_4}+T_{G4}d(\Delta P_{valve\_4})/dt=\varepsilon_{44}$, $\Delta P_{mech\_4}+T_{CH4}d(\Delta P_{mech\_4})/dt=\Delta P_{valve\_4}$.
System #5: ε_{55}=load ref$_5$-$\Delta\omega_2/R_5$, $\Delta P_{valve\_5}+T_{G5}d(\Delta P_{valve\_5})/dt=\varepsilon_{55}$, $\Delta P_{mech\_5}+T_{CH5}d(\Delta P_{mech\_5})/dt=\Delta P_{valve\_5}$.
System #6: ε_{66}=load ref$_6$-$\Delta\omega_2/R_6$, $\Delta P_{valve\_6}+T_{G6}d(\Delta P_{valve\_6})/dt=\varepsilon_{66}$, $\Delta P_{mech\_6}+T_{CH6}d(\Delta P_{mech\_6})/dt=\Delta P_{valve\_6}$.
Derivation of the tie-line transfer function:
The real power flow from bus # 1 with voltage $\tilde{V}_1$ to bus #2 with voltage $\tilde{V}_2$ is for the line reactance X_{line} neglecting the line resistance R_{line}: $P_{tie}=\frac{|\tilde{V}_1||\tilde{V}_2|}{X_{line}}\sin\theta$, where θ is the

angle between $\tilde{V}_1$ and $\tilde{V}_2$. For $|\tilde{V}_1| \approx |\tilde{V}_2|$ =constant their perturbations are $|\Delta\tilde{V}_1|$ = $|\Delta\tilde{V}_2|$ =0. Introducing perturbations for P and θ one obtains

$$P_{tieo} + \Delta P_{tie} = \frac{|\tilde{V}_{1o}||\tilde{V}_{2o}|}{X_{line}} \sin(\theta_o + \Delta\theta) = \frac{|\tilde{V}_{1o}||\tilde{V}_{2o}|}{X_{line}} \sin(\theta_o \cos\Delta\theta + \cos\theta_o \sin\Delta\theta) \quad (3)$$

$$P_{tieo} = \frac{|\tilde{V}_{1o}||\tilde{V}_{2o}|}{X_{line}} \sin\theta_o \text{ and } \Delta P_{tie} = \frac{|\tilde{V}_{1o}||\tilde{V}_{2o}|}{X_{line}} \cos(\theta_o)\ \Delta\theta = \frac{|\tilde{V}_{1o}||\tilde{V}_{2o}|}{X_{line} / \cos(\theta_o)}\ \Delta\theta = \frac{|\tilde{V}_{1o}||\tilde{V}_{2o}|}{X_{tie}}\ \Delta\theta$$

$$\Delta P_{tie} = \frac{|\tilde{V}_{1o}||\tilde{V}_{2o}|}{d(X_{tie})/dt} d(\Delta\theta)/dt, \text{ where } X_{tie} = X_{line} / \cos\theta o \quad (4)$$

or with the Laplace operator s=d(...)/dt

$$\Delta P_{tie} = \frac{|\tilde{V}_{1o}||\tilde{V}_{2o}|}{sX_{tie}} \Delta\omega(s) = \frac{|\tilde{V}_{1o}||\tilde{V}_{2o}|}{sX_{tie}} \{\Delta\omega_1(s) - \Delta\omega_2(s)\}. \quad (5)$$

In per unit with V_{base}, Z_{base} and ω_o=2πf=377 rad/s as references one obtains with $T=T_{tie}=\omega_o$[rad/s]/X_{tie} [pu]= 377/X_{tie} [pu] for f=60 Hz.

$$\Delta P_{tie}[pu] = (T/s)\{\Delta\omega_1(s)\,[pu] - \Delta\omega_2(s)\,[pu]\}. \quad (6)$$

The frequency variation of a power system should be within 59-61 Hz, that is, a frequency band of ±1.66 Hz is desirable (Fuller *et al.*, 1989). Table 6 lists the Mathematica program for Fig. 17a on which Figs. 18 and 19 are based.

b. Fig. 18a illustrates the establishment of steady-state conditions at time t=200 s for given systems parameters.
c. Fig. 18b illustrates the given WP plant load reference Lr5[t]. Fig.18c shows the transient response Δω(t) if storage plants 4 and 6 absorb the additional energy generated by WP plant with 0.2 s delay (see Table 6).

R1=0.01;	DPvalve4b=(DPvalve4[t]/.sol1[[1]]/.t->200);
d1=0.8;	DPmech5b=(DPmech5[t]/.sol1[[1]]/.t->200);
M1=4.5;	DPvalve5b=(DPvalve5[t]/.sol1[[1]]/.t->200);
Tg1=0.3;	DPmech6b=(DPmech6[t]/.sol1[[1]]/.t->200);
Tch1=0.9;	DPvalve6b=(DPvalve6[t]/.sol1[[1]]/.t->200);
Lr1=0.8;	ic16=Dw1[200]=Dw1b;
DPL1[t_]:=If [t<210, 0, 0.1];	ic17=DPmech1[200]=DPmech1b;
R2=0.1;	ic18=DPvalve1[200]=DPvalve1b;
d2=1.0;	ic19=DPtie[200]=DPtie1b;
M2=6;	ic20=Dw2[200]=Dw2b;
Tg2=0.2;	ic21=DPmech2[200]=DPmech2b;

Table 6. Mathematica input program for stability analysis of Figs. 17-19

Tch2=0.2;	ic22=DPvalve2[200]=DPvalve2b;
Lr2=0.5;	ic23=DPmech3[200]=DPmech3b;
DPL2[t_]:=If [t<400, 0, -0.1];	ic24=DPvalve3[200]=DPvalve3b;
Xtie=0.2;	ic25=DPmech4[200]=DPmech4b;
Ttie=377/Xtie;	ic26=DPvalve4[200]=DPvalve4b;
R3=0.3;	ic27=DPmech5[200]=DPmech5b;
Tg3=0.1;	ic28=DPvalve5[200]=DPvalve5b;
Tch3=0.1;	ic29=DPmech6[200]=DPmech6b;
Lr3[t_]:=If [t<600, 0, 0.06];	ic30=DPvalve6[200]=DPvalve6b;
R4=0.5;	R1=0.01;
Tg4=0.2;	d1=0.8;
Tch4=0.1;	M1=4.5;
Lr4[t_]:=If [t<600.1, 0,- 0.6];	Tg1=0.3;
R5=0.7;	Tch1=0.9;
Tg5=0.1;	Lr1=0.8;
Tch5=0.1;	DPL1[t_]:=If [t<300, 0, 0.1];
Lr5[t]:=If [t<600.2, 0, 0.3];	R2=0.1;
R6=0.5;	d2=1.0;
Tg6=0.2;	M2=6;
Tch6=0.1;	Tg2=0.2;
Lr6=0.01;	Tch2=0.2;
ic1=Dw1[0]= =0;	Lr2=0.5;
ic2=DPmech1[0]= =0;	DPL2[t_]:=If [t<400, 0, -0.1];
ic3=DPvalve1[0]= =0;	Xtie=0.2;
ic4=DPtie[0]= =0;	Tie=377/Xtie;
ic5=Dw2[0]= =0;	R3=0.3;
ic6=DPmech2[0]= =0;	Tg3=0.1;
ic7=DPvalve2[0]= =0;	Tch3=0.1;
ic8=DPmech3[0]= =0;	Lr3[t_]:=If[t<1200,If[t<1120,If[t<1000,If[t<940,
ic9=DPvalve3[0]= =0;	If[t<910,0,0.03],0.09],0.05],0.03],0.0];
ic10=DPmech4[0]= =0;	DPstorage4[t_]:=If[t<1200.2,If[t<1120.2,If[t<1000
ic11=DPvalve4[0]= =0;	.2,
ic12=DPmech5[0]= =0;	If[t<940.2,If[t<910.2,0,0.15],0.45],0.25],0.15],0.0];
ic13=DPvalve5[0]= =0;	DPstorage6[t_]:=If[t<1200.2,If[t<1120.2,If[t<1000
ic14=DPmech6[0]= =0;	.2,
ic15=DPvalve6[0]= =0;	If[t<940.2,If[t<910.2,0,0.15],0.45],0.25],0.15],0.0];
E11[t_]:=Lr1-Dw1[t]/R1;	R4=0.5;
E12[t_]:=DPmech1[t]-DPL1[t]-	Tg4=0.2;
DPtie[t]+ DPmech3[t]+DPmech4[t];	Tch4=0.1;
E3[t_]:=Dw1[t]-Dw2[t];	Lr4[t_]:=If [t<700, 0,- 0.01];
E22[t_]:=Lr2-Dw2[t]/R2;	R5=0.7;
E21[t_]:=DPmech2[t]-	Tg5=0.1;
DPL2[t]+DPtie[t] +	Tch5=0.1;
DPmech5[t]+DPmech6[t];	Lr5[t]:=If [t<1200,If[t<1120,If[t<1000,If[t<940,
E33[t_]:=Lr3[t]-Dw1[t]/R3;	If[t<910,0,0.3],0.9],0.5],0.3],0.0];

Table 6. Mathematica input program for stability analysis of Figs. 17-19 (continuation)

E44[t_]:=Lr4[t]-Dw1[t]/R4;
E55[t_]:=Lr5[t]-Dw2[t]/R5;
E66[t_]:=Lr6-Dw2[t]/R6;
eqn1=Dw1'[t]= =(1/M1)*(E12[t]-
d1*Dw1[t]);
eqn2=DPmech1'[t]=
=(1/Tch1)*(DPvalve1[t]-DPmech1[t]);
eqn3=DPvalve1'[t]= =(1/Tg1)*(E11[t]-
DPvalve1[t]);
eqn4=DPtie'[t]= =Tie*E3[t];
eqn5=Dw2'[t]= =(1/M2)*(E21[t]-
d2*Dw2[t]);
eqn6=DPmech2'[t]=
=(1/Tch2)*(DPvalve2[t]-DPmech2[t]);
eqn7=DPvalve2'[t]= =(1/Tg2)*(E22[t]-
DPvalve2[t]);
eqn8=DPmech3'[t]=
=(1/Tch3)*(DPvalve3 [t]-DPmech3[t]);
eqn9=DPvalve3'[t]= =(1/Tg3)*(E33[t]-
DPvalve3[t]);
eqn10=DPmech4'[t]=
=(1/Tch4)*(DPvalve4[t]-DPmech4[t]);
eqn11=DPvalve4'[t]=
=(1/Tg4)*(E44[t]-DPvalve4[t]);
eqn12=DPmech5'[t]=
=(1/Tch5)*(DPvalve5 [t]-DPmech5[t]);
eqn13=DPvalve5'[t]=
=(1/Tg5)*(E55[t]-DPvalve5[t]);
eqn14=DPmech6'[t]=
=(1/Tch6)*(DPvalve6[t]-DPmech6[t]);
eqn15=DPvalve6'[t]=
=(1/Tg6)*(E66[t]-DPvalve6[t]);
sol1=NDSolve[{eqn1,eqn2,eqn3,eqn4,eq
n5,eqn6,eqn7,eqn8,eqn9,eqn10,eqn11
,eqn12,eqn13,eqn14,eqn15,ic1,ic2,ic3,ic4
,ic5,ic6,ic7,ic8,ic9,ic10,ic11,ic12,ic13,ic14,
ic15},{Dw1[t],DPmech1[t],DPvalve1[t],
DPtie[t],Dw2[t],DPmech2[t],Dpval
ve2[t],DPmech3[t],DPvalve3[t],DPmech
4[t],DPvalve4[t],DPmech5[t],Dpvalv
e5[t],DPmech6[t],DPvalve6[t]},{t,0,200
},MaxSteps->100000000];
Plot[Dw1[t]/.sol1[[1]],{t,0,200},PlotRa
nge->All,AxesLabel-
>{"t[s]","Dw1[t][pu]"}]
	R6=0.5;
	Tg6=0.2;
	Tch6=0.1;
	Lr6=0.01;
	E11[t_]:=Lr1-Dw1[t]/R1;
	E12[t_]:=DPmech1[t]-DPL1[t]-DPtie[t]+
	DPmech3[t]+DPmech4[t]-DPstorage4[t];
	E3[t_]:=Dw1[t]-Dw2[t];
	E22[t_]:=Lr2-Dw2[t]/R2;
	E21[t_]:=DPmech2[t]-DPL2[t]+DPtie[t] +
	DPmech5[t]+DPmech6[t]-DPstorage6[t];
	E33[t_]:=Lr3[t]-Dw1[t]/R3;
	E44[t_]:=Lr4[t]-Dw1[t]/R4;
	E55[t_]:=Lr5[t]-Dw2[t]/R5;
	E66[t_]:=Lr6-Dw2[t]/R6;
	eqn1=Dw1'[t]= =(1/M1)*(E12[t]-d1*Dw1[t]);
	eqn2=DPmech1'[t]= =(1/Tch1)*(DPvalve1[t]-
	DPmech1[t]);
	eqn3=DPvalve1'[t]= =(1/Tg1)*(E11[t]-
	DPvalve1[t]);
	eqn4=DPtie'[t]= =Tie*E3[t];
	eqn5=Dw2'[t]= =(1/M2)*(E21[t]-d2*Dw2[t]);
	eqn6=DPmech2'[t]= =(1/Tch2)*(DPvalve2[t]-
	DPmech2[t]);
	eqn7=DPvalve2'[t]= =(1/Tg2)*(E22[t]-
	DPvalve2[t]);
	eqn8=DPmech3'[t]= =(1/Tch3)*(DPvalve3 [t]-
	DPmech3[t]);
	eqn9=DPvalve3'[t]= =(1/Tg3)*(E33[t]-
	DPvalve3[t]);
	eqn10=DPmech4'[t]= =(1/Tch4)*(DPvalve4[t]-
	DPmech4[t]);
	eqn11=DPvalve4'[t]= =(1/Tg4)*(E44[t]-
	DPvalve4[t]);
	eqn12=DPmech5'[t]= =(1/Tch5)*(DPvalve5 [t]-
	DPmech5[t]);
	eqn13=DPvalve5'[t]= =(1/Tg5)*(E55[t]-
	DPvalve5[t]);
	eqn14=DPmech6'[t]= =(1/Tch6)*(DPvalve6[t]-
	DPmech6[t]);
	eqn15=DPvalve6'[t]= =(1/Tg6)*(E66[t]-
	DPvalve6[t]);
	sol2=NDSolve[{eqn1,eqn2,eqn3,eqn4,eqn5,eqn6,
	eqn7,eqn8,
	eqn9,eqn10,eqn11,eqn12,eqn13,eqn14,eqn15,ic1,

Table 6. Mathematica input program for stability analysis of Figs. 17-19 (continuation)

Dw1b=(Dw1[t]/.sol1[[1]]/.t->200)	ic2,ic3,ic4,
DPmech1b=(DPmech1[t]/.sol1[[1]]/.t-	ic5,ic6,ic7,ic8,ic9,ic10,ic11,ic12,ic13,ic14,ic15},
>200);	{Dw1[t],
DPvalve1b=(DPvalve1[t]/.sol1[[1]]/.t-	DPmech1[t],DPvalve1[t],DPtie[t],Dw2[t],
>200);	DPmech2[t],
DPtie1b=(DPtie[t]/.sol1[[1]]/.t->200);	DPvalve2[t],DPmech3[t],DPvalve3[t],DPmech4[t
Dw2b=(Dw2[t]/.sol1[[1]]/.t->200)	],
DPmech2b=(DPmech2[t]/.sol1[[1]]/.t-	DPvalve4[t],
>200);	DPmech5[t],DPvalve5[t],DPmech6[t],DPvalve6[t
DPvalve2b=(DPvalve2[t]/.sol1[[1]]/.t-	]},
>200);	{t,200,1500},
DPmech3b=(DPmech3[t]/.sol1[[1]]/.t-	MaxSteps->100000000];
>200);	Plot[Evaluate[Lr5[t]],{t,200,1500},PlotRange->
DPvalve3b=(DPvalve3[t]/.sol1[[1]]/.t-	All,AxesLabel->
>200);	{"t (s)","Lr5[t] [pu]"}]
DPmech4b=(DPmech4[t]/.sol1[[1]]/.t-	Plot[Dw1[t]/.sol2[[1]],{t,200,1500},PlotRange->
>200);	All,AxesLabel->{"t[s]","Dw1[t][pu]"}]

Table 6. Mathematica input program for stability analysis of Figs. 17-19 (continuation)

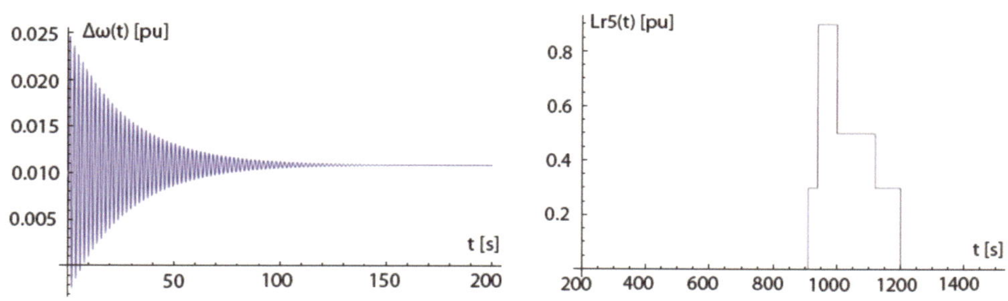

(a) (b)

Figs. 18a,b. (a) Establishment of steady-state conditions for Δω(t); (b) Plot of given WP plant load reference Lr5[t]

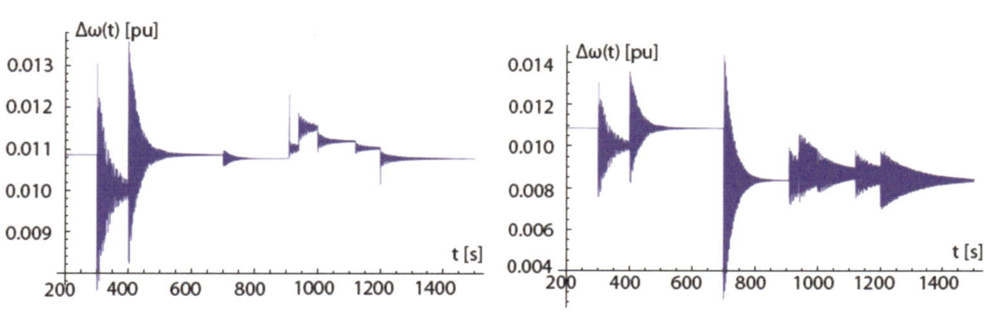

(c) (d)

Figs. 18c,d. (c) Transient response Δω(t) if storage plants 4 and 6 absorb the additional energy generated by WP plant with 0.2 s delay; (d) Transient response Δω(t) if only storage plant 6 absorbs the additional energy generated by WP plant with 0.2 s delay.

d. Fig. 18d depicts the transient response Δω(t) if only storage plant 6 absorbs the additional energy generated by WP plant with 0.2 s delay.
e. If the change of the WP plant is absorbed by the short-term storage plant 4 associated with the PV plant then Fig. 18e is obtained.
f. Fig. 18f shows the transient response Δω(t) if none of the storage plants absorbs a substantial amont of energy generated by WP plant.
g. Figs. 19a, b illustrate the influence of the reduction of the power capability of the transmission line increasing X_{tie} from 0.2 pu to 0.237 pu for the conditions of Fig.18c. Fig. 19a shows that steady-state stability is impaired, and after some load/demand changes the dynamic stability is lost (Fig. 19b) and the angular frequency/velocity diverges.

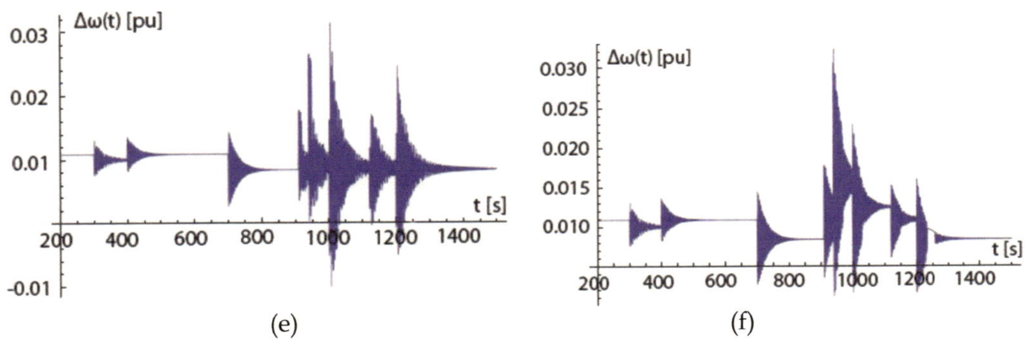

Figs. 18e,f. (e) Transient response Δω(t) if only storage plant 4 absorbs the additional energy generated by WP plant with delay; (f) Transient response Δω(t) if none of the storage plants absorbs a substantial amount of energy generated by WP plant

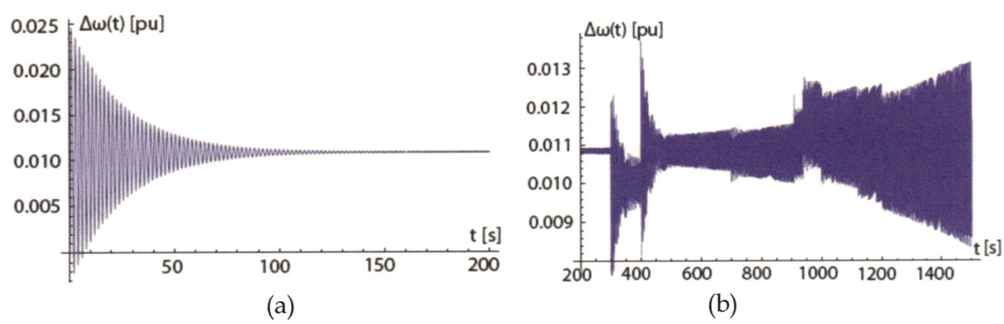

Figs. 19a,b. (a) Steady-state stability is impaired as can be noticed in Fig. 19a at 200s and in Fig. 19b at 200 s; (b) Dynamic stability is lost after some load/demand changes have occurred

7. Power quality issues (steady-state and transient phenomena)

In addition to power balance issues, distributed renewable sources lead to high system impedance at the distribution level because these plants cannot deliver additional transient currents when faults occur due to their operation at peak power (Masoum *et al.*, 2002; Masoum *et al.*, 2004a). Even with acceptable current harmonics, this high system impedance may result in unacceptably high voltage harmonics, single-time events (e.g., spikes due to network switching and synchronization) and non-periodic but repetitive (e.g., flicker) events, and contribute to power quality problems (E.F. Fuchs & Masoum, 2008a; E.F. Fuchs *et al.*, 2004; Masoum *et al.*, 2004b; E.F. Fuchs & Masoum, 2008b; Masoum *et al.*, 2004c; Masoum *et al.*, 2004d; E.F. Fuchs & H.A. Fuchs, 2008).

8. Conclusions

A review of existing frequency control practices based on droop characteristics and PSpice analyses of PWM rectifiers and inverters make clear that these components are ideal in a mix of short-term and long-term storage power plants. The complementary control algorithm as outlined in Figs. 17a,b,c and illustrated in Figs. 18, 19 permits intermittently-operating renewable sources to be matched with short-term and long-term storage plants so that the grid can be operated within a permissible frequency band of 59-61 Hz. However, if the intermittently operating plants are inactive for a longer time and the storage plants are depleted, demand-side management (e.g., load shedding) must be relied on.

In particular, the stability of a smart/microgrid consisting of natural gas-fired power plant (frequency leader), a long-term storage power plant and two intermittently operating plants (e.g., PV and WP plants) with associated short-term storage plants is assured if the following constraints are satisfied:

1. frequency variation is minimized through appropriate switching in and out of short-term storage plants;
2. transmission line parameters are optimized;
3. time constants of governors and valves are within feasible regions; and
4. droop characteristics of the individual plants must satisfy certain constraints.

9. References

Fuchs, E.F.; Masoum, M.A.S. (2008a). *Power Quality in Power Systems and Electrical Machines,* Elsevier/Academic Press, ISBN: 978-0-12-369536-9, Burlington, MA 01803, USA.

Fuchs, E. F.; Fuchs, H. A. (2007). Distributed Generation and Frequency/Load Control of Power Systems, *Proceedings of the 40th Annual Frontiers of Power Conference,* Oklahoma State University, Stillwater, Oklahoma, Oct. 29-30, 2007, pp. II-1 to II-9.

Fuchs E.F.; Fuchs, F.S. (2008). Intentional Islanding – Maintaining Power Systems Operation During Major Emergencies, *Proceedings of the 10th IASTED International Conference on Power and Energy Systems (PES 2008),* ISBN: 978-0-88986-738-3, Baltimore, Maryland, pp. 156-162.

Wood, A. J.; Wollenberg, B. F. (1984). *Power Generation Operation & Control,* John Wiley & Sons, Inc., ISBN 13: 9780471586999.

Fuller, J. F.; Fuchs, E. F.; Roesler, D. J. (1989). Influence of Harmonics on Power System Distribution Protection, *IEEE Trans. on Power Delivery,* April 1988, Vol. TPWRD-3, No. 2, pp. 546- 554.

Fuchs E. F.; Masoum, M. A. S. (2011). *Power Conversion of Renewable Energy Systems,* Springer-Verlag, New York, ISBN: 978-1-4419-7978-0.

Defree, S. (2009). http://www.edn.com/article/459288-Photovoltaic_market_to_see_17_growth_rate_Gartner_reports.php, 1 page.

NREL (2009). Data from NREL's "Sun Spot One" Data Collection Station in the Colorado San LuisValley, available from http://www.nrel.gov/midc/ss1/

Masoum, M. A. S.; Dehbonei, H.; Fuchs, E. F. (2002). Theoretical and Experimental Analyses of Photovoltaic Systems with Voltage-and Current-Based Maximum Power Point Tracking, *IEEE Trans. on Energy Conversion*, Vol. 17, No. 4, (Dec. 2002), pp. 514-522.

Masoum, M. A. S.; Mousavi Badejani, S. M.; Fuchs, E. F. (2004a). MicroProcessor-Controlled New Class of Optimal Battery Chargers for Photovoltaic Applications, *IEEE Trans. on Energy Conversion*, Vol. 19, No. 3, Sept. 2004, pp. 599-606.

IEEE Standard 519 (1992). IEEE Recommended Practices and Requirements for Harmonic Control in Electric Power Systems, IEEE-519, 1992.

IEC 61000-3-2 (2001-10). Consolidated Edition, Electromagnetic Compatibility (EMC) - Part 3-2: Limits for Harmonic Current Emissions.

Yildirim, D.; Fuchs, E. F.; Batan, T. (1998). Test Results of a 20 kW, Direct-Drive, Variable-Speed Wind Power Plant, *Proceedings of the International Conference on Electrical Machines, Istanbul*, Turkey, Sept. 2-4, 1998, pp. 2039-2044.

Fuchs, E. F.; Myat, M. H. (2010). Speed and Torque Range Increases of Electric Drives Through Compensation of Flux Weakening, *Proceedings of the 20th International Symposium on Power Electronics, Electrical Drives, Automation, and Motion,* Pisa, Italy, June 14 - 16, 2010,pp. 1569-1574.

Cowdrey, J. (2004). Boulder's Municipal Hydroelectric System, July 7, 2004, 16 pages.

Mattick, W.; Haddenhorst, H. G.; Weber, O.; Stys, Z.S. (1975). Huntorf- the World's First 290 MW Gas Turbine Air Storage Peaking Plant, *Proceedings of the American Power Conference,* Vol. 37, 1975, pp. 322-330.

Vosburgh, K. G. (1978). Compressed Air Energy Storage, *J. Energy*, Vol. 2, No. 2, March-April 1978, pp. 106-112.

Glems (1964). http://www.enbw.com/content/de/der_konzern/_media/pdf/imagebroschuere_standorte/Pumpspeicherwerk_Glems.pdf , 4 pages

Raccoon Mountain (1975). http://www.tva.gov/sites/raccoonmt.htm

Fuchs, E. F.; Roesler, D. J.; Masoum, M. A. S. (2004). Are Harmonic Recommendations According to IEEE and IEC Too Restrictive?, *IEEE Trans. on Power Delivery,* Vol. 19, No. 4, Oct. 2004, pp. 1775-1786.

Masoum, M. A. S.; Ladjevardi, M.; Jafarian, A.; Fuchs, E. F. (2004b). Optimal Placement, Replacement and Sizing of Capacitor Banks in Distorted Distribution Networks by Genetic Algorithms, *IEEE Trans. on Power Delivery,* Vol. 19, No. 4 , Oct. 2004, pp. 1794-1801.

Fuchs, E. F.; Masoum, M. A. S. (2008b). Torques in Induction Machines Due to Low-Frequency Voltage/Current Harmonics, *International Journal of Power and Energy Systems,* Vol. 28, Issue 2, 2008, pp. 212- 221.

Masoum, M. A. S.; Jafarian, A.; Ladjevardi, M.; Fuchs, E. F.; Grady, W. M. (2004c). Fuzzy Approach for Optimal Placement and Sizing of Capacitor Banks in the Presence of Harmonics, *IEEE Trans. on Power Delivery,*Vol. 19, No. 2, April 2004, pp. 822-829.

Masoum, M. A. S.; Ladjevardi, M.; Jafarian, A.; Fuchs, E. F. (2004d). Optimal Placement, Replacement and Sizing of Capacitor Banks in Distorted Distribution Networks by Genetic Algorithms, *IEEE Trans. on Power Delivery,* Vol. 19, No. 4, Oct. 2004, pp. 1794-1801.

Fuchs, E. F.; Fuchs, H. A. (2008). Power Quality of Electric Machines and Power Systems, *Proceedings of the 8thIASTED International Conference on Power and Energy Systems (EuroPES 2008),* June 16–18, 2008, Corfu, Greece, paper # 608-035, 9 pages.

17

Practical Application of Electrical Energy Storage System in Industry

Drabek, Streit and Blahnik
University of West Bohemia Plzen,
Faculty of Electrical Engineering,
Czech Republic

1. Introduction

1.1 Industry application of ESS

Storing of energy allows balancing of the supply and demand of energy. Today, the energy storage system (ESS) in commercial use can be broadly categorized as mechanical, electrical, chemical, biological and thermal. In further we will discuss electrical and partly mechanical system. Accumulation of electrical energy presents a big problem solved by a lot of specialists and scientists. Many ways of accumulation has been put into practice (for example: systems using flywheel, battery, supercapacitor etc.). Today commonly used vehicles of light traction (trams, trolleybuses) are not able to accumulate their kinetic energy. Saving of this energy is possible only by regenerative braking. This method is possible when another tram starts up on the same part of trolley line. Nowadays electrical energy storage system can be found in transport vehicles and in power engineering systems. In general the energy-storage devices charge during low power demands and discharge during high power demands, to provide energy boost when needed.

Carbon emissions, the depletion of natural resources by fast consumption, traffic congestion and the rising costs of fossil fuels are all issues pushing the world to search for alternative means of transportation. Mass-transit buses, fleet vehicles, long-haul trucks and other heavy-transportation vehicles such as trains, light rails, trams and subways, all benefit from the using of a hybrid power drives in relation to the energy storage system. The global number of vehicles around the World will triple in the next 50 years (Chan & Wong, 2004). Thus, methods of improving fuel economy have gained worldwide attention. The efficiency and all-electric range (AER) of hybrid electric vehicles (HEVs) depend on the capability of their energy-storage system (ESS), to store large amounts of energy and release it quickly according to load demands.

In transport ESS allows accumulation of the braking energy in conventional vehicles and special racing vehicles such as Formula 1. The kinetic energy is accumulated into the ESS during vehicle braking. This energy can be used to the vehicle acceleration again. It is important to save the energy in the vehicles accelerating very often such public transport vehicles (we can think about stabile or mobile version according to the local specific conditions).

Heavy transportation vehicles - such as trains, trams and subways - place particular demands on energy storage devices. Such devices must be very robust and reliable,

displaying both long operational lifetimes and low maintenance requirements. Further, the devices must operate efficiently under harsh conditions including handling peak currents, high duty cycles and frequent deep discharging.
ESS can absorb and store all kinetic energy from a braking system, depends on total energy designed in ESS. The emission-free stored electrical energy in ESS is then available to assist in acceleration, to reduce fuel consumption and accompanying emissions or energy drain, as well as drive the air conditioner, operate power steering or perform other electrical functions. As an added bonus, regenerative braking takes most of the mechanical brakes load off, reducing brake maintenance and replacement expenses.
Very important area of ESS is power engineering where the ESS covers small to large rated power range, including active PFC used to set the appropriate power factor and harmonic emission to the power grid.
The important characteristics of ESSs include energy density, power density, lifetime, cost, energy storage time and maintenance (see Fig. 3 and Fig. 4). Currently, batteries and ultracapacitors are the most common options for vehicular ESSs. Batteries usually have high energy densities and store the majority of onboard electric energy. On the other hand, ultracaps have high power densities and present a long life cycle with high efficiency and a fast response for charging/ discharging (Burke, 2007; Khaligh & Li, 2010; Zhang et al., 2008; Lu et al., 2007). A fuel cell (FC) is another clean energy source (Fig. 1 and Fig. 2); however, the long time constant of the FC limits its performance on vehicles. Very perspective ESS shows the flywheel system with low cost benefit and high energy density. However using flywheel in the transport is a bit disputable due to gyroscopic effect. Nevertheless for stationary systems the ESS with flywheels presents perspective topology with respect the low energy storage time (Fig. 4), this system cannot be used as back up power source as batteries and ultracaps.

Fig. 1. City bus powered by Fuel Cells (TriHyBus.cz, 2008).

Fig. 2. Detail on fuel cells, the ESS is composed of Li-ion battery and supecapacitors (TriHyBus.cz, 2008)

As design engineers have found, batteries have high-energy capability while the ultracapacitors have high power capability. In an optimal hybrid alternative drive system, both technologies could be combined in a way that maximizes the benefits of both. In general vehicles have batteries to provide energy back up for control systems, start engine etc.
The ESS of most of the commercially available HEVs is composed of only battery packs with a bidirectional converter connected to the high-voltage dc bus (the Toyota Prius, Honda Insight, and Ford Escape).
Start-stop technology enables the engine in conventional, electric or hybrid-electric delivery trucks and refuse vehicles to shut down when they come to a stop at a red light, picking up or dropping off passengers, or when sitting in traffic. ESS then provides a short burst of energy that restarts the motor. ESS can save millions of barrels of oil over conventional gasoline-only powered vehicles.

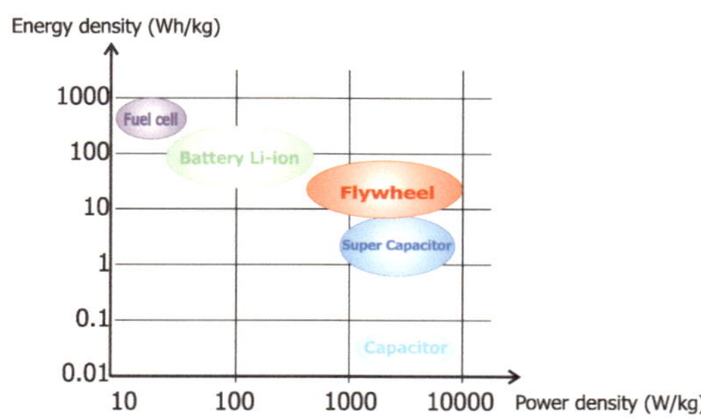

Fig. 3. Comparison of several ESS due to energy and power density

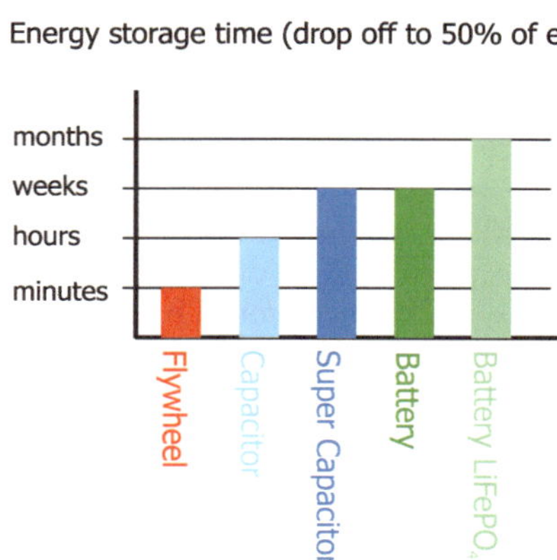

Fig. 4. Comparison of Energy storage time of several ESS

2. Battery

Batteries have widely been adopted in ground vehicles due to their characteristics in terms of high energy density, compact size, and reliability (Lukic et al, 2008).
The Batteries have a very wide field of applications in systems for the electrical ESS. They are used in stationary systems, as short time power sources (eg UPS).The other applications are in transport, used as the main source (eg, full-electric vehicle), or as the secondary sources (for covering of peak powers combinated with FC) - the main advantage is their

ability to accumulate electrical energy during braking. The batteries bonus proprieties are high energy density, excellent energy storage time and reliability. There must calculated with the fact that batteries are limited by operating temperature and exhibit the lower power density compared to the other electric ESS. Very important parameter of the battery is the price, that depends closely on the battery type. Among the other monitoring proprieties (batteries used in industry) there are charging and discharging currents, life time / cycles.

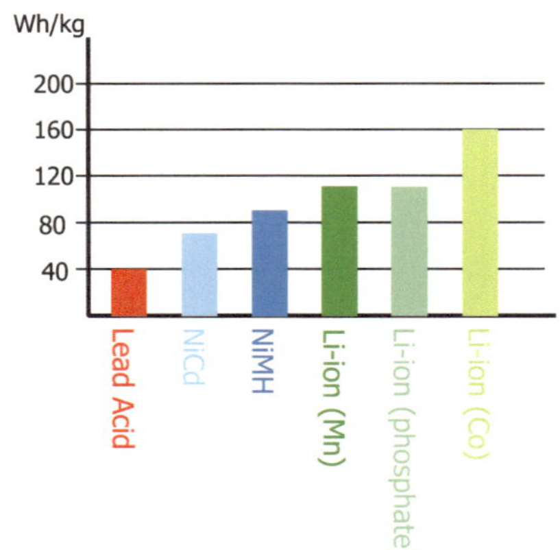

Fig. 5. The dependence of Energy density on battery weight

The basic types of batteries used in industry and their specifications are:

Lead-Acid Batteries

These types of batteries are composed of the spongy lead works as the negative active material, lead oxide is the positive active material, and diluted sulfuric acid is the electrolyte. During discharging, both materials (positive and negative) are transformed into lead sulfate (Williamson et al, 2005). The lead-acid battery presents several advantages for HEV applications. They are available in production volumes today, yielding a comparatively low-cost power source. In addition, lead-acid battery technology is a mature technique due to its wide use over the past 50 years. However, the lead-acid battery is not suitable for discharges over 20% of its rated capacity. When it operates at a deep rate of state of charge (SOC), the battery would have a limited life cycle.

These batteries are still one of the most popular for its low price. The disadvantage of this battery is using of toxic lead and very low charging currents for ESS.

Nickel-Metal Hydride (NiMH) Batteries

The NiMH batteries use an alkaline solution as the electrolyte. The battery is composed of nickel hydroxide on the positive electrode and the negative electrode consists of an alloy of vanadium, titanium, nickel, and other metals. The components of the battery are harmless to the environment; moreover, the batteries can be recycled. The advantages of these batteries are long lifetime, wide operating temperature ranges and the resistance to over charge and over discharge. The batteries should not be repeatedly discharged by high load currents, because the lifetime of battery is reduced to about 200 cycles. Another disadvantage is the battery memory effect.

Lithium-Ion Batteries

These types of batteries are characterized by high energy density and low memory effect. The Li-ion battery uses an oxidized cobalt material as a positive electrode, the negative electrode is made of a carbon material and lithium salt in an organic solvent is used as the electrolyte. The advantages of these batteries are high energy density of 120 Wh/kg, high power density of 300 W/kg, long battery life of 1000 cycles, low memory effect and environment harmless. The main disadvantage of this battery is still the high price and questionable availability in the future.

Nickel-Zinc (Ni-Zn) Batteries

Nickel-zinc batteries have high energy and power density, low-cost materials, and deep cycle capability and are environmentally friendly. The operation temperature of Ni-Zn batteries ranges from −10 ◦C to 50 ◦C, which means that they can be used under severe working circumstances. However, they suffer from poor life cycles due to the fast growth of dendrites, which prevents the development of Ni-Zn batteries in vehicular applications.

Nickel-Cadmium (Ni-Cd) Batteries

These types of batteries are characterized by long lifetime and can be fully discharged without damage. The specific energy of Ni-Cd batteries is around 55 Wh/kg. The disadvantage of these batteries is using of cadmium. These batteries can be recycled, but the cadmium is a kind of heavy metal that could cause environmental pollution if not properly disposed of.

A summary of the batteries properties used in industry can be seen in Table 1, or in detail (Buchmann, 2011).

	Lead Acid	NiCd	NiMH	Li-ion (Mn)	Li-ion (phosphate)	Li-ion (Co)
Energy density (Wh/kg)	30 - 50	45 - 80	60 - 120	100 - 135	90 - 120 *	150 - 190
Cycle life (80% of nominal capacity)	200 - 300	1500	300 - 500	300 - 500	1000 - 5000 *	300 - 500
Self discharge per month	5 %	20 %	30 %	low than 10 %	low than 3 % *	low than 10 %
Nominal voltage per cell	2 V	1.25 V	1.25 V	3.6 V	3.3 V *	3.6 V
Load current - peak -continuous	 5 C 0.2 C	 20 C 1 C	 5 C 0.5 C	 30 C 10 C	 20 C * 5 C *	 3 C 1 C
Operating temperature	-20°C to 60°C	-40°C to 60°C	-20°C to 60°C	-20°C to 60°C	-20°C to 60°C	-20°C to 60°C
Safety	Thermally stable	Thermally stable, fuse recommended	Thermally stable, fuse recommended	Stable to 250°C protection circuit recommended	Stable to 250°C protection circuit recommended	Stable to 150°C protection circuit mandatory
Toxicity	Toxic lead and acids, harmful to environment	Highly toxic, harmful to environment	Relatively, low toxicity, should be recycled	Low toxicity	Low toxicity	Low toxicity

* For the Li-ion (phosphate) battery it is necessary to know the type and used material

Table 1. Comparison of selected types of batteries usable for ESS

One of the main indicators of the battery quality is energy density (Wh/kg), Fig. 5 and Table 1 show that the best batteries are based on lithium-ion technology. In terms of safety and lifetime are today come to the fore a very advanced lithium-ion batteries with phosphate and that are widely using in electric traction. The leading manufacturers of these batteries are Valence Technology ($LiFeMgPO_4$) and Winston Battery Limited ($LiFeYPO_4$). In the data from the manufacturers can be found other important information about the batteries. Such as:

- Cycle life (80% of nominal capacity) depends on operating temperature Fig. 6.
- Discharge voltage behaviour (operable capacity) depends on temperature Fig. 7, Fig. 8.
- Discharge voltage behaviour (operable capacity) depends on load current Fig. 9.
- Battery self discharge behaviour per one year Fig. 10.

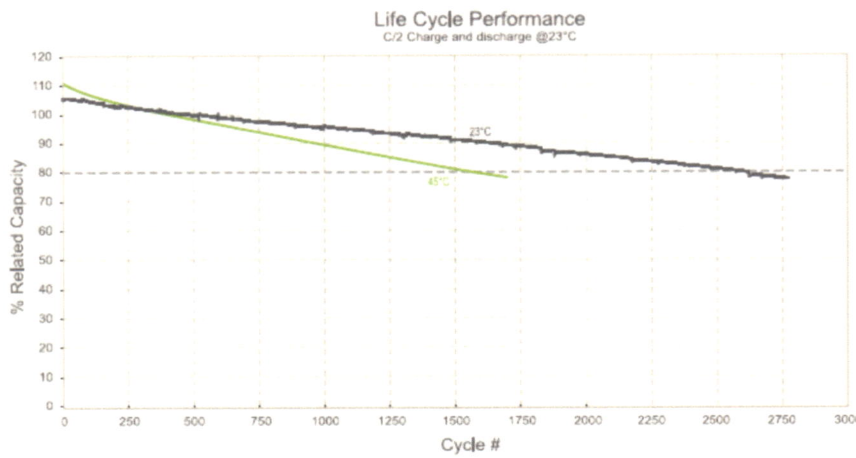

Fig. 6. Battery cycle life depends on operating temperature ($LiFeMgPO_4$) (Valence Technology, 2010)

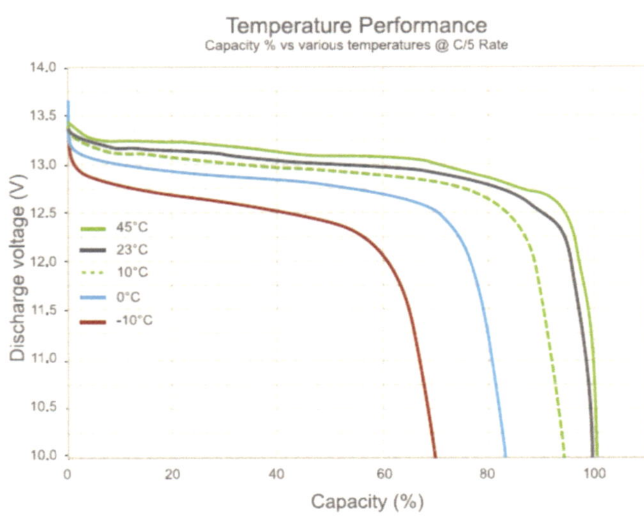

Fig. 7. Discharge Voltage depends on operating temperature ($LiFeMgPO_4$) (Valence Technology, 2010)

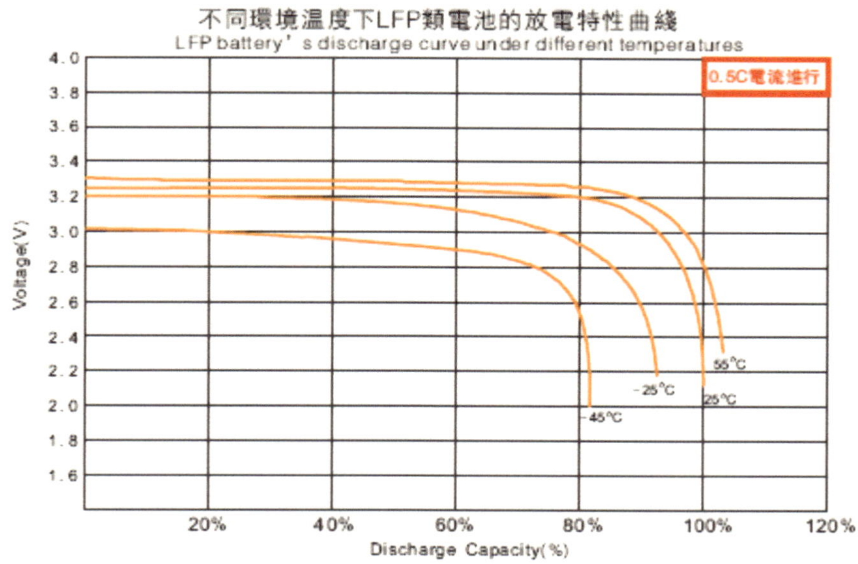

Fig. 8. Discharge Voltage depends on operating temperature (LiFeYPO4) (Winston Battery Limited, n.d.)

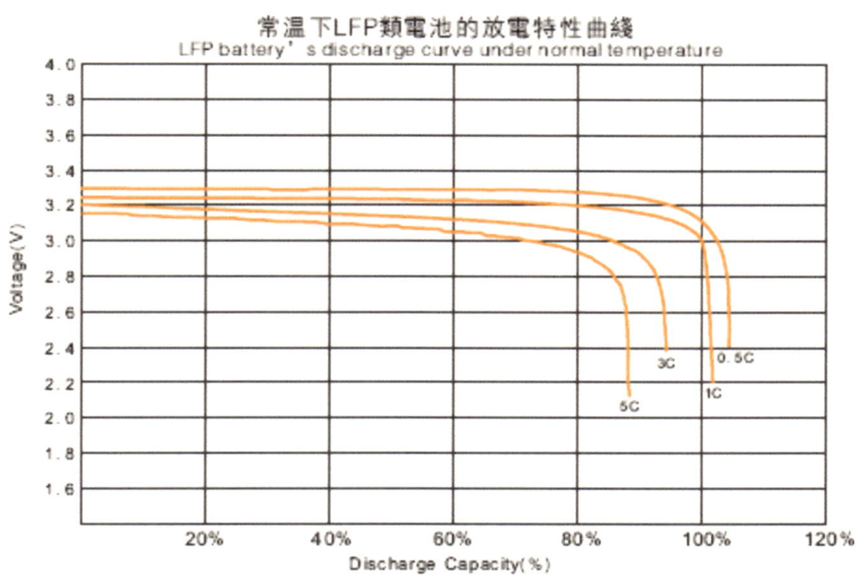

Fig. 9. Discharge Voltage depends on load current (LiFeYPO4) (Winston Battery Limited, n.d.)

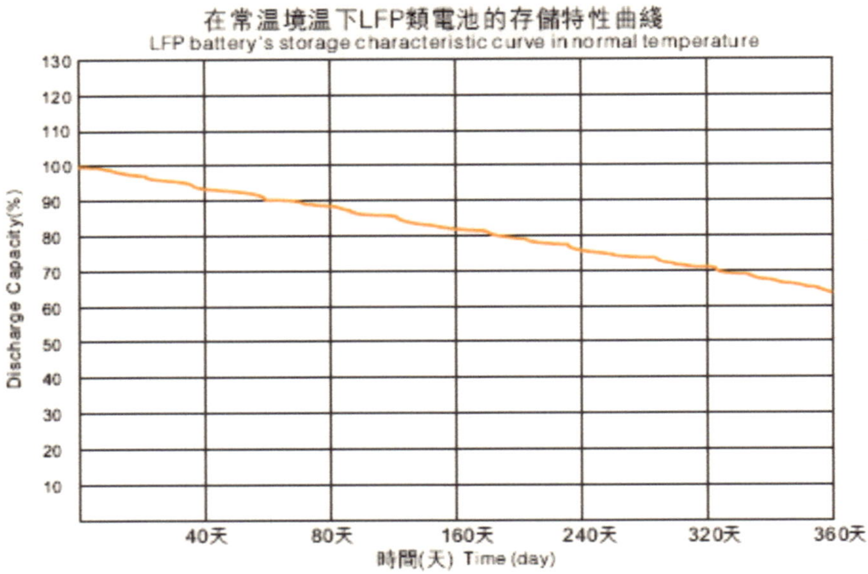

Fig. 10. Battery self discharge behaviour ($LiFeYPO_4$) (Winston Battery Limited, n.d.)

The electrical ESS based on battery must contain the special electric circuit, that ensures the battery charging and discharging at the specified limits. The basic topology is depicted in Fig. 11, battery cell is connected to load/source via buck/boost converter. The battery voltage is lower than DC-link voltage and for charging the battery is used the buck converter (T1, D2 elements). During discharging of battery cycle the voltage should be increased by boost converter (T2, D1 elements). This topology is simply and very often used at electric traction and at back-up system for example (Drabek & Streit, 2009).

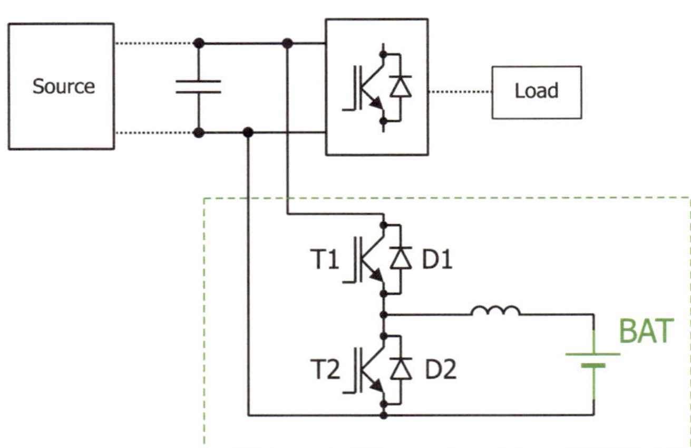

Fig. 11. The basic power circuit for ESS based on battery (standard buck/boost converter)

Sophisticated topology uses the N-phase buck/boost converter. The principle is similar as the previous one, only the current is divided into three parallel branches, that are controlled by shifted carriers in PWM.

The advantage of this topology is in reducing of the current ripple caused by shifted carriers PWM control. The example of three phase buck/boost converter is shown in Fig. 12.
This type of converter is practically used e.g. for photovoltaic systems (Lin et al., 2010), where is pressure to reduce converter losses. Another common use is at electric traction, where the reducing of the filtering inductor weight and size is welcome.

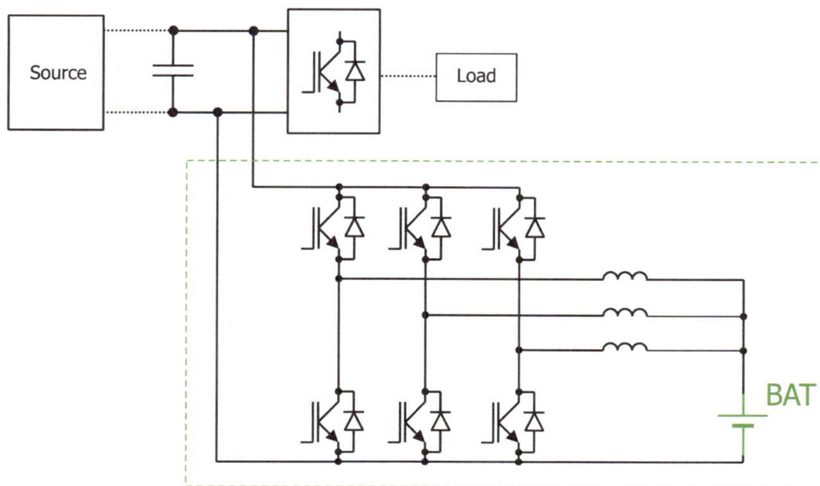

Fig. 12. The three-phase buck/boost converter for ESS based on battery

The modern power circuit topologies for electrical ESS based on battery is the bi-directional DC/DC converter with transformer binding for example Fig. 13. This type of converter is typically used for PV systems, because in relation to soft switching technology it has high efficiency (Shengyong et al., 2010).

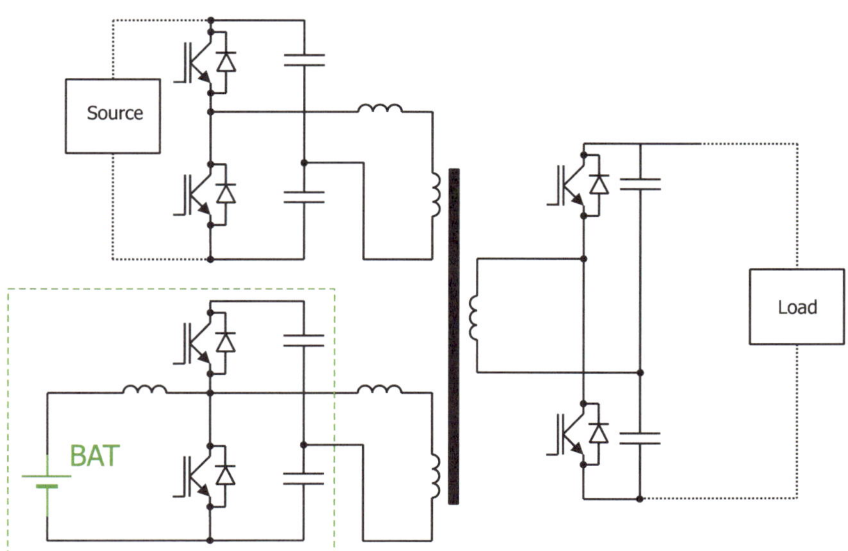

Fig. 13. Traction topology of ESS based on supercapacitors – high voltage version

The Fig. 14 - Fig. 16 show several transport vehicles powered by battery ESS.

Fig. 14. Battery scooter (Vectrix Corp, 2010) Battery: Li: LiFeP04, 30Ah capacity

Fig. 15. Battery car (CODA Automotive, 2010) Battery: LiFePO4, 728 cells (104s7p)

Fig. 16. Battery bus (Zhongtong Bus Holding Co, (n.d.) Battery: Li-ion, 540 Ah capacity

3. Supercapacitors (ultracapacitors)

Supercapacitors (SC) are electrochemical capacitors with an unusually high energy density compared to common capacitors, typically thousand times greater than a high-capacity electrolytic capacitor. They are based on a structure that contains an electrical double layer (anode (aluminium foil) - active carbon - separator - active carbon - cathode (aluminium foil) - it is shown in Fig. 17) and therefore they have the high capacity. The carbon nanotubes have very large surface area. One gram of this carbon presents surface area about 2000m2/g. The SC stores energy by physically separating positive and negative charges. The charges are stored on two parallel plates divided by an insulator. Since there are no chemical variations on the electrodes, therefore, UCs have a long cycle life but low energy density. The power density of the UC is considerably higher than that of the battery; this is due to the fact that the charges are physically stored on the electrodes. Low internal resistance gives UC high efficiency but can result in a large burst of output currents if the SC is charged at a very low SOC.

There are five UC technologies in development: carbon/metal fibre composites, foamed carbon, a carbon particulate with a binder, doped conducting polymer films on a carbon cloth, and mixed metal oxide coatings on a metal foil. Higher energy density can be achieved with a carbon composite electrode using an organic electrolyte rather than a carbon/metal fibre composite electrode with an aqueous electrolyte (Lu et al., 2007).

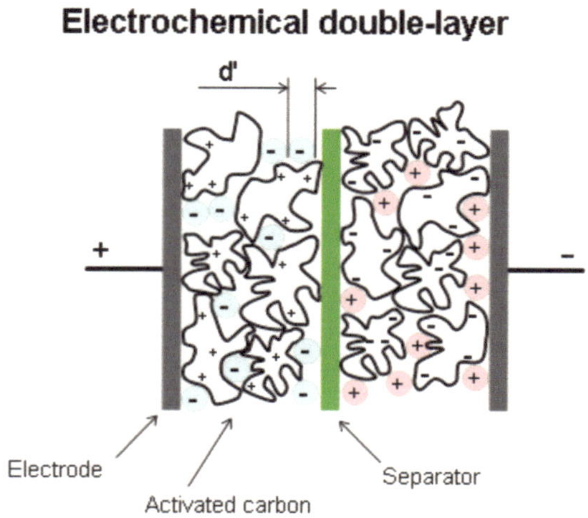

Fig. 17. Principle of supercapacitor (EPCOS AG, 2011)

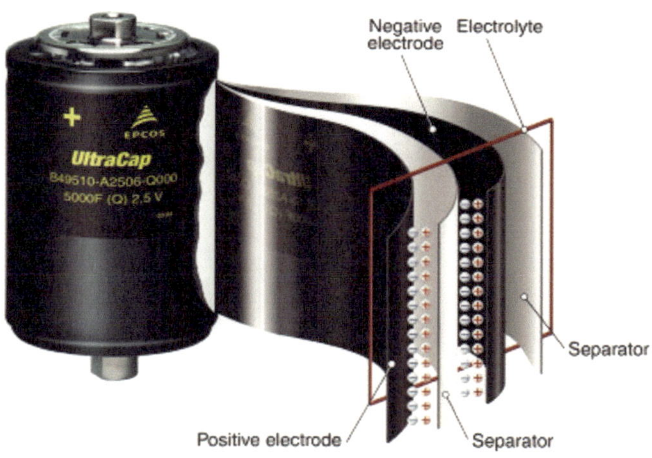

Fig. 18. Typical structure (layers) of supercapacitor (EPCOS AG, 2011)

Fig. 17 shows charge distribution in SC. Large surface area of carbon nanotube can soak up a lot of charge, therefore SC presents the high capacity. SC fills the gap between common capacitors and common batteries as you can see in the Fig. 3.

Ultracapacitors quickly capture energy from braking and then use that energy to provide a short burst of power during acceleration and to dramatically reduce the use of fuel in a conventional internal combustion engine or electrical energy drain in an all electric or hybrid system. Generally used ultracapacitors in transportation are compact, high-performance, have exceptionally long-life and fulfil many of the functions of batteries but with dramatically higher reliability and they are virtually impervious to any climate condition.

Fig. 19. The hybrid bus with supercapacitors (Maxwell Technologies, 2011)

Supercapacitors are made in different modules and packages. Because the voltage of one cell is only two volts, the cells must be connected to series connection for increase of the voltage. Large industry SC has a nominal voltage of 125V. This 125V transportation module is shown in Fig. 20.

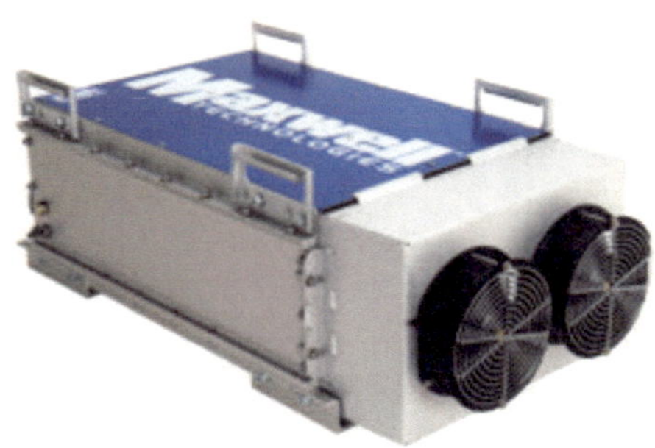

Fig. 20. 125V Transportation Module (Maxwell Technologies, 2011)

Fig. 21. SC on the roof of Scania bus (Green Car Congress - BioAge Group, 2010)

Features of 125V Transportation module:

- More than 1 million of charge - discharge cycles
- Operating temperature -40° C to +65° C
- Nominal capacitance 63F
- Internal resistance 18mΩ
- Constant current 150A
- Peak current 750A (1s 10% duty cycle)
- Energy about 100Wh (discharging to half voltage)
- Weight 59,5kg
- Size 619 x 425 x 265 mm

In principal the configuration of DC/DC converter should be following:

- Classical buck/boost converter (two-quadrant converter – one voltage polarity with current reversal) – Supercapacitor voltage cannot be higher than voltage in DC line due to diode (Fig. 22)
- Two buck/boost converters (two-quadrant converter according to C_{SC} – one voltage polarity with current reversal) – Supercapacitor voltage can be higher than voltage in DC line (Fig. 23).

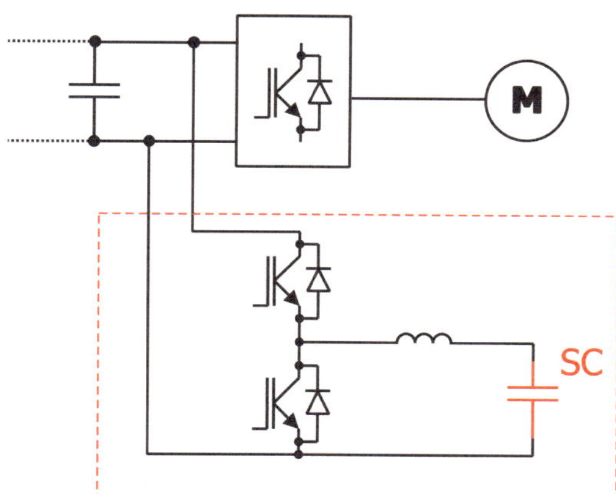

Fig. 22. Traction topology of ESS based on supercapacitors – low voltage version

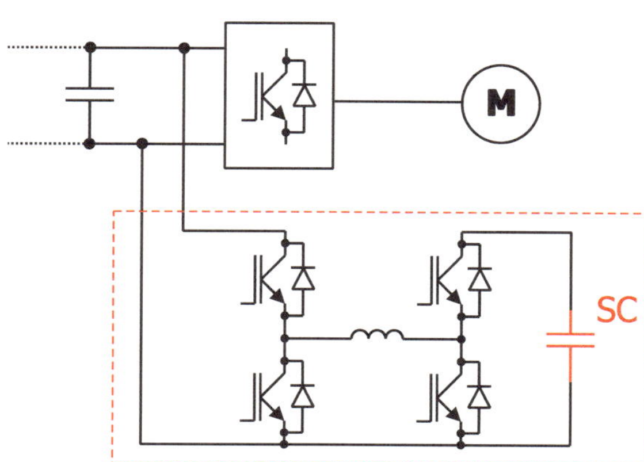

Fig. 23. Traction topology of ESS based on supercapacitors – high voltage version

Example of using SC in light rail

Fig. 24 shows principle schema of energy storage system with supercapacitors. This topic is described in the (Drabek & Streit, 2009).

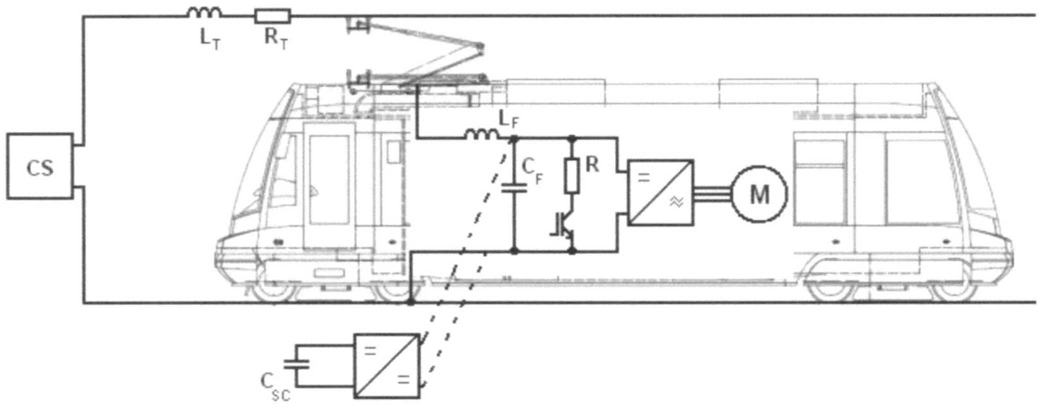

Fig. 24. Topology of energy storage system in light rail vehicle

The Energy Storage System is created with the DC/DC converter and with the block of supercapacitors. Charging of supercapacitors (braking of vehicle) and discharging (acceleration of vehicle) is controled by DC/DC pulse converter.

Voltage of supercapacitors corresponds with quantum of accumulated energy. It is necessary to connect SC by DC/DC converter to set voltage of SC independent of DC circuit of traction converter voltage. Controlling of DC/DC converter ensures using SC in right time. Energy accumulation during braking vehicle and using of the energy by first accelerating request appears as the best way of controlling idea.

Transferring of kinetic energy to the SC (means charging and discharging) can be done by appropriate control strategy of Buck/Boost converter. This accumulated energy can be used for acceleration of the vehicle. PI controller ensures current from ESS (Energy Storage System) as an equivalent with current of traction converter. It means that PI controller controls trolley line current to zero. This idea saves a lot of energy needed for repeated accelerations.

Simulations have been provided for a tram vehicle with rated power of 348 kW, weight 22t. Block of supercapacitors 125V, 63F by MAXWELL (Fig. 20) was chosen. We have used block of 4 supercapacitors - that means 500V and 15,75F (nominal voltage of trams is 600V and due to used converter in Fig. 22 it is necessary to use ESS with low voltage).

Basic settings for control of DC/DC convertor:

- U_{CS} has not exceed level of 500V and decrease under level of 250V (it depends on manual of SC MAXWELL, the common minimum voltage is 50% according to the lifetime cycles).
- ICS has not to exceed level of 750A (it depends on manual of SC MAXWELL according to the lifetime cycles).

It is possible to use few ways to control energy storage system. Firstly, to control voltage U_F to nominal value of trolley voltage (e.g. 600 V) – to discharge SC at lower voltage of UF and charge SC at lower voltage of U_F. The control of charging and discharging according to driving cycle of the vehicle (braking - charging, acceleration - discharging). The second method is used for described system.

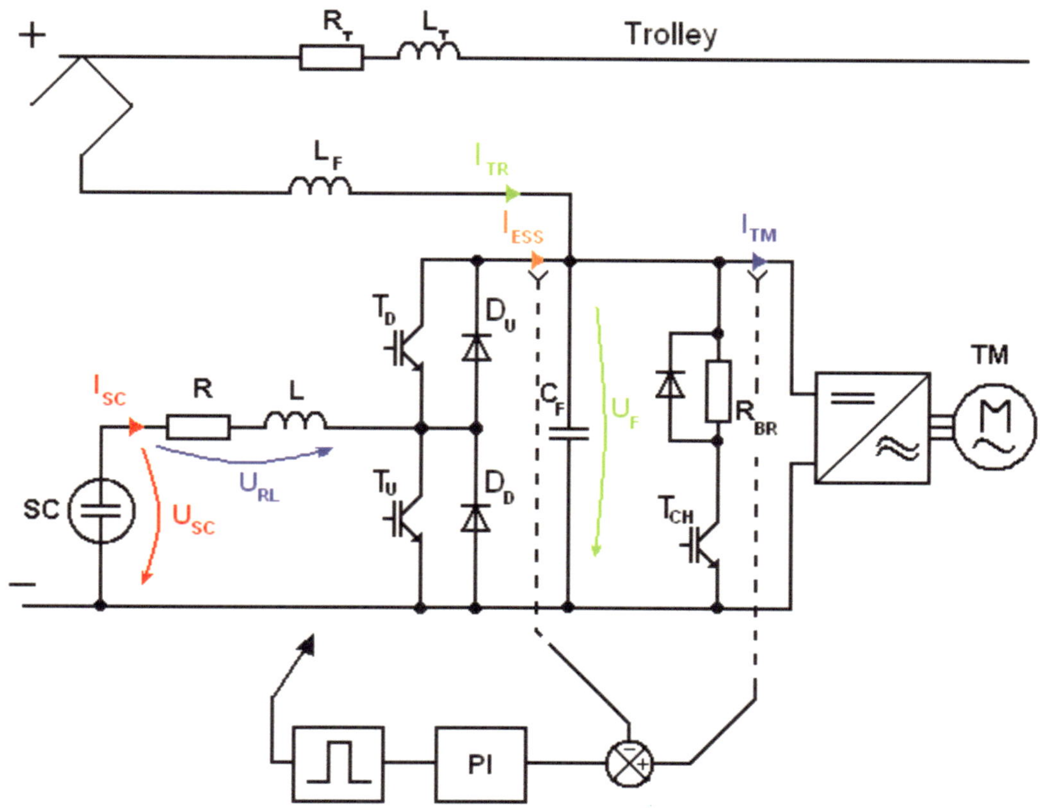

Fig. 25. Light rail topology - Buck (TD)/Boost (TU) convertor for SC

Proposed control strategy:

- Firstly to discharge SC by every request for acceleration and prepare SC for charging during braking.

Fig. 26 shows behaviour of energy storage system for chosen grading of track. Shown values correspond to schema in Fig. 25. Current is filtered to better view. In time period 1-2s current of traction motor is fully supplied by energy storage system SC as you can see in Fig. 26. In time 6s supercapacitor is fully discharged, therefore current for traction motor is fully supplied by trolley line. SC is charging by first braking in time 10-13s and it is discharging by next accelerating in time 14-17s. In time 23s SC is fully charge and redundant energy is taking out by braking resistors R_{BR} (if the recuperating to the trolley line is not possible due to over voltage).

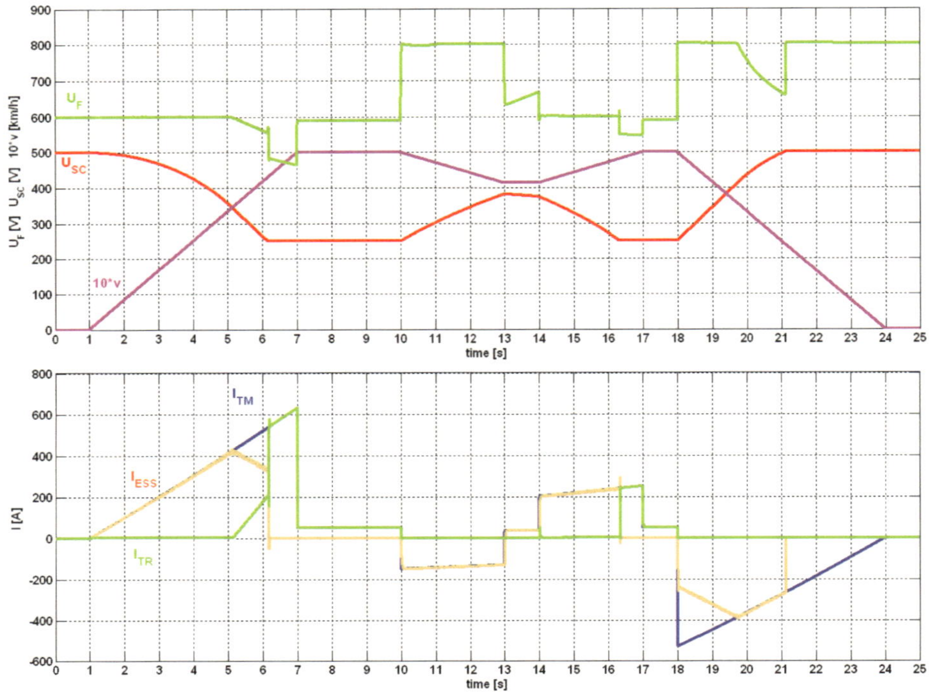

Fig. 26. Application of energy storage system for grading of track

4. EES with flywheel

Interesting energy storage system used in industry is flywheel. It is a mechanical ESS, which is again interesting for use in electric traction vehicle, as well as stationary ESS. The flywheel is a rotating device, this device accumulate kinetic energy as:

$$E = ½ \, . \, m.v^2 \tag{1}$$

Same as

$$E = ½ \, .J\omega^2 \tag{2}$$

E...energy [J]
m...weight [kg]
v...speed [$m.s^{-1}$]
J... moment of inertia [$kg.m^2$]
ω...angular velocity [s^{-1}]
The quantity of kinetic energy, that is accumulated by flywheel directly depends on the rotor moment of inertia, also it square depends on angular velocity. Therefore, it is advantageous to adjust the flywheel for high speed as possible. The aspect of the saving weight with maximizing of the flywheel energy is especially important in traction vehicles. Nowadays the well-known application is in the racing cars Formula 1, the system ESS is called KERS (Kinetic Energy Recovery System - Fig. 31).

As a perspective electric drive in ESS with flywheel is used the switching reluctance motor (SRM), that brings advantages in simple motor design, low cost, robust construction. On other hand it needs advanced control system (Talla & Stehlik, 2008). The energy storing and removing from flywheel can be done by mechanical means (Flybrid Systems, 2010), or by electrically Fig. 27.

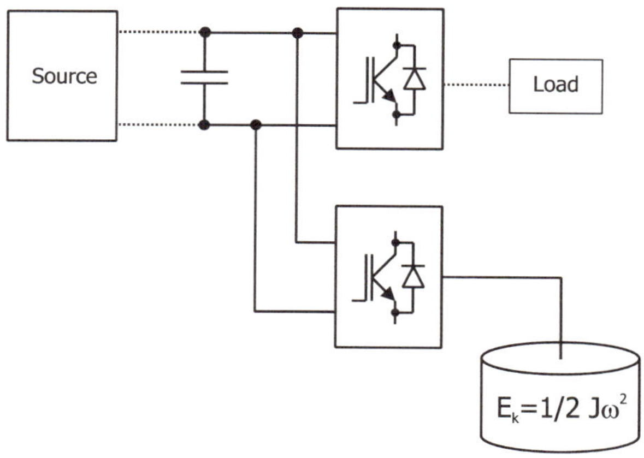

Fig. 27. Flywheel energy storage done electrically

The Fig. 28 shows a complete ESS device with flywheel and Fig. 29 shows the configuration of the flywheel as a presented by VYCON company (VYCON, 2011).

Fig. 28. Flywheel energy storage systems (VYCON, 2011)

A summary of the flywheel properties can be seen in Table 2, or in detail (AutoSpeed - Web Publications Pty Limited, 2011; Cibulka, 2009)

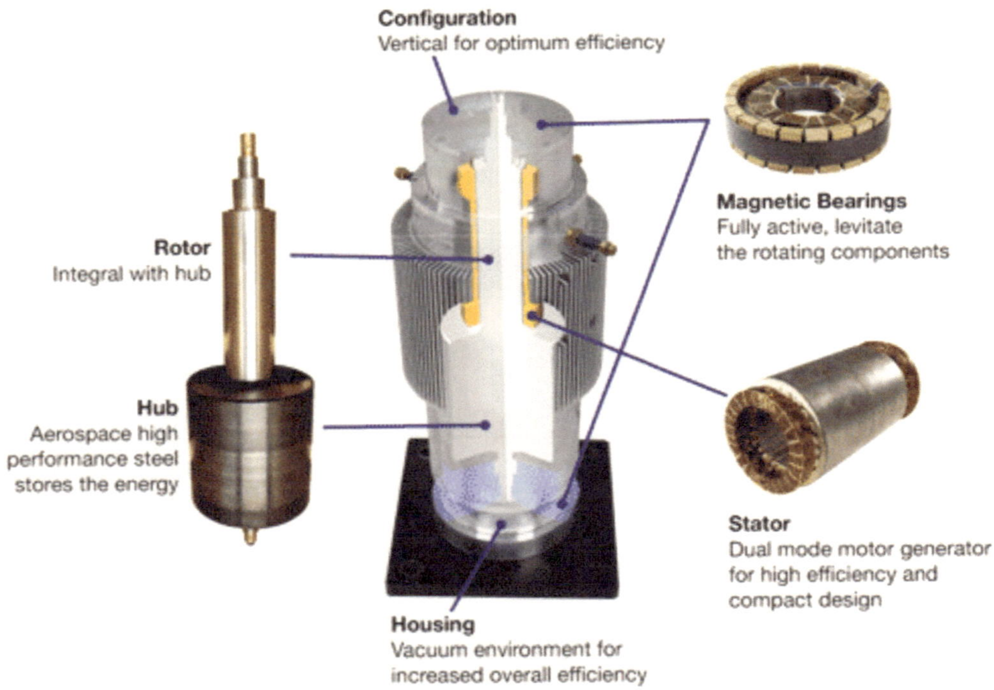

Fig. 29. The flywheel composition (VYCON, 2011)

	BATTERY $LiFePO_4$	SUPERCAPACITOR	FLYWHEEL
Energy density (Wh/kg)	90-120	1-4	40-60 metal 140-350 composite
Cycle life	500-1500 (3000)*	1 000 000	5 000 000
Self discharge	lower than 10 % (3%)* per month	50% of initial voltage per month	2-5% per minute
Operating temperature	-30°C +60°C (-45°C +85°C)*	-40°C +65°C	-40°C +150°C
Safety	+	+	-
Toxicity	-	+	+

* For the Li-ion (phosphate) battery is necessary to know the type and used ingredients

(donated)

Table 2. Comparison of ESS used in industry

Flywheel advantages and disadvantages

+	-
High reliability	Gyroscopic effect
Very low maintenence	High energy losses (2-5% per min)
High temperature range	Special (expensive) bearings
High power density	Dangerous if failure
No capacity decreasing in time	Vibration (producing and sensitive)
Enviromental friendly	Complicated control
Unneeded cooling	
Long lifetime	
Easy energy remaining information	

In Fig. 30 - Fig. 33 are shown different flywheel devices used at modern road vehicle.

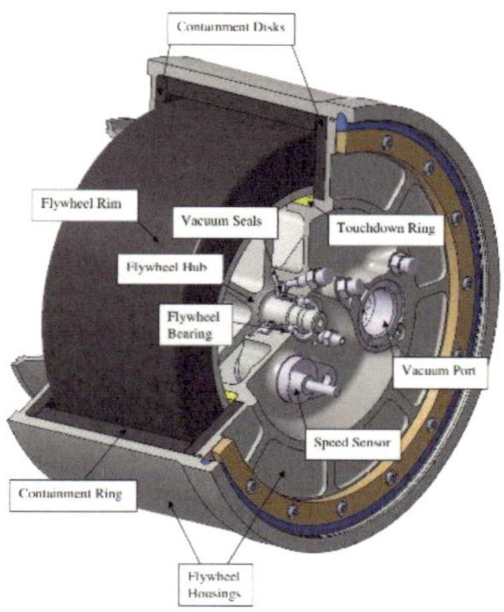

Fig. 30. Flywheel construction (Flybrid Systems, 2010)

Fig. 31. The original CVT based Formula One KERS (Flybrid Systems, 2010)

The next practical using of flywheel is at light traction vehicle as a tram. The fig. 34 presents hydrogen train using flywheel for energy storage.

Fig. 32. The Jaguar flywheel module with integrated vacuum and lubrication pumps (Flybrid Systems, 2010)

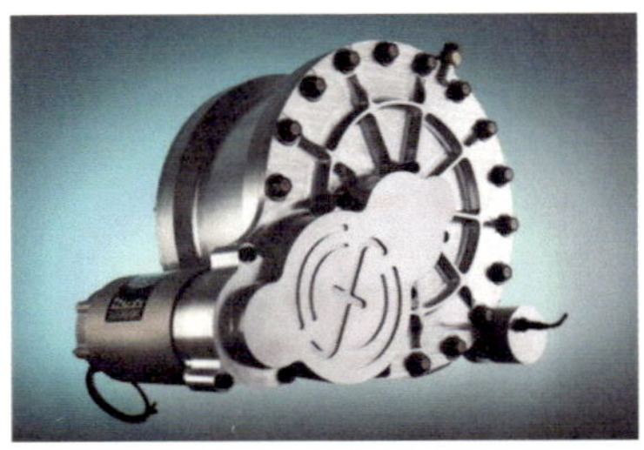

Fig. 33. The Flybrid Flywheel Capacitor (Flybrid Systems, 2010)

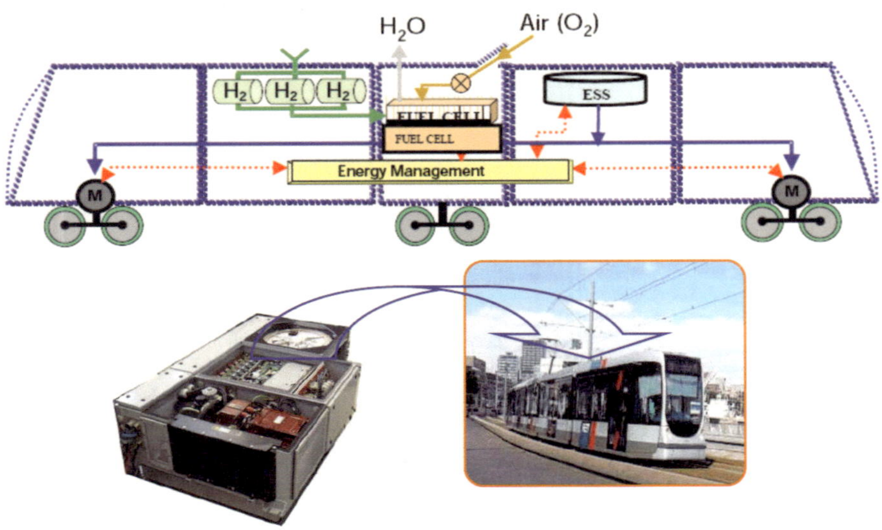

Fig. 34. Hydrogen train (ALSTOM, 2011)

5. Acknowledgment

This research work has been made within research project of Czech Science Foundation of the Czech Republic - No. GACR 102/09/1164.

6. References

ALSTOM. (2011). Power generation & Transmission, sustainable Rail Transport global leader - Alstom, In: *Alstom Transport,* January 2011, Available from: <http://www.alstom.com>

American Honda Motor Co., Inc. (2011). Honda Cars - New and Certified Used Cars from American Honda, In: *2011 Honda Insight Hybrid Overview - Official Honda Site,* January 2011, Available from: <http://automobiles.honda.com/>

AutoSpeed - Web Publications Pty Limited. (2011). AutoSpeed - Technology, Efficiency, Performance, In: *AutoSpeed - Using Flywheels for Energy Storage,* January 2011, Available from: <http://autospeed.com>

Buchmann, I. (2011). Battery information from Battery University, In: *What's the best battery?,* January 2011, Available from: <http://www.batteryuniversity.com>

Burke, A. F. (2007), Batteries and ultracapacitors for electric, hybrid, and fuel cell vehicles, *in Proc. IEEE,* Vol. 95, No. 4, (Apr. 2007), pp. 806–820, ISSN 0018-9219

Cibulka, J. (2009), Kinetic Energy Recovery System by Means of Flywheel Energy Storage, *International Journal Advanced Engineering.,* Vol. 3, No. 1, (2009), pp. 27-38, ISSN 1846-5900

CODA Automotive. (2010). Battery CODA Electric Vehicles, Electric Cars, Zero Emissions EV Car | CODA Automotive, In: *Battery system,* January 2011, Available from: <http://www.codaautomotive.com >

Drabek, P.; & Streit, L. (2009), The energy storage system for public transport vehicles, *Proceedings of 13th European Conference on Power Electronics and Applications, EPE '09,* ISBN 978-1-4244-4432-8, Barcelona, Sept 2009

EPCOS AG. (2011). EPCOS – Electronic Components, Modules and Systems, In: *Capacitors - Publications - EPCOS AG,* January 2011, Available from: <http://www.epcos.com>

Fisker. (2011). Fisker Automotive, In: *Technology – Fisker Automotive,* January 2011, Available from: <http://karma.fiskerautomotive.com/>

Flybrid Systems. (2010). Home - Flybrid Systems, In: *Original F1 System - Flybrid Systems,* January 2011, Available from: <http://www.flybridsystems.com>

Ford Motor Company. (2011). Ford – New Cars, Trucks, SUVs, Hybrids & Crossovers | Ford Vehicles, In: *2010 Ford Escape Hybrid SUV | A Fuel-Efficient Hybrid With SUV Capability | Ford.com,* January 2011, Available from: <http://www.ford.com/>

General Motors. (2010). 2011 Cars, SUVs, Trucks, Crossovers & Vans | Chevrolet, In: *2011 Chevy Volt | Electric Car | Chevrolet,* January 2011, Available from: <http://www.chevrolet.com>

Green Car Congress - BioAge Group. (2010). Green Car Congress, In: *Green Car Congress: Scania to Unveil Ethanol-Fueled Series Hybrid Bus Concept,* January 2011, Available from: < http://www.greencarcongress.com/>

Chan, C. C.; & Wong, Y. S. (2004), Electric vehicles charge forward, *IEEE Power Energy Mag.,* Vol. 2, No. 6, (Nov./Dec. 2004), pp. 24–33, ISSN 1540-7977

Khaligh, A.; & Li, Z. (2010), Battery, Ultracapacitor, Fuel Cell, and Hybrid Energy Storage Systems for Electric, Hybrid Electric, Fuel Cell, and Plug-In Hybrid Electric Vehicles: State of the Art, *IEEE Transactions on Vehicular Technology,* Vol. 59, No. 6, (July 2010), pp. 2806, ISSN 0018-9545

Lin, M.; Kai, S.; Teodorescu, P.; Guerrero, J.M. & Xinmin, J. (2010), An integrated multifunction DC/DC converter for PV generation systems, *Proceedings of IEEE International Symposium on Industrial Electronics ISIE,* ISBN 978-1-4244-6390-9, Bari, July 2010

Lu, Y.; Hess, H.; & Edwards, D. (2007), Adaptive control of an ultracapacitor energy storage system for hybrid electric vehicles, *Proceedings of Electric Machines & Drives Conference. IEMDC '07. IEEE International*, ISBN 1-4244-0742-7, Antalya, May 2007

Lukic, S.; Cao, J.; Bansal, R. C.; Rodriguez, F.; & Emadi, A. (2008), Energy storage systems for automotive applications, *IEEE Transactions on Industrial Electronics*, Vol. 55, No. 6, (June 2008), pp. 2258, ISSN 0278-0046

Maxwell Technologies. (2011). Maxwell Technologies - Ultracapacitors, Microelectronics, High Voltage, In: *Maxwell Technologies - Products - Ultracapacitor - 125 Volt Transportation Modules - Application Specific Modules,* January 2011, Available from: <http://maxwell.com>

Shengyong, L.; Xing, Z.; Haibin, G.; & Jun, X. (2010), Multiport DC/DC Converter for Stand-alone Photovoltaic Lighting System with Battery Storage, *Proceedings of International Conference on Electrical and Control Engineering ICECE*, ISBN 978-1-4244-6880-5, Wuhan, June 2010

Talla, J.; & Stehlik, J. (2008), FPGA based neural network position and speed estimator for switched reluctance motor drive, *Proceedings of International Conference on Circuits, Systems, Electronics, Control & Signal Processing WSEAS*, ISBN 978-960-474-035-2, Puerto de la Cruz, Dec 2008

TESLA. (2011). Tesla Motors | Premium Electric Vehicles, In: *Roadster Features and Specifications | Tesla Motors,* January 2011, Available from: <http://www.teslamotors.com/>

Toyota Motor Sales, U.S.A., Inc. (2011). Toyota Cars, Trucks, SUVs & Accessories, In: *Toyota Prius - 2011 Performance & Specifications,* January 2011, Available from: <http://www.toyota.com/>

TriHyBus.cz. (2008). TriHyBus: Oficiální web prvního vodíkového autobusu v ČR, In: *TriHyBus: Hybridní pohon,* January 2011, Available from: <http://trihybus.cz>

Valence Technology (2010). Lithium Phosphate Battery Suppliers | Valence Technology, In: *Battery Chemistry | Valence Technology,* January 2011, Available from: <http://www.valence.com >

Vectrix Corp. (2010). Vectrix EU :: :: Index, In: *Vectrix EU :: Products :: Vx-1,* January 2011, Available from: <http://www.vectrix.com>

VYCON. (2011). VYCON Flywheel Energy - Home, In: *VYCON - Energy Recycling Products,* January 2011, Available from: <http://www.vyconenergy.com>

Williamson, S.; Khaligh, A.; Oh, S.; & Emadi, A. (2005), Impact of energy storage device selection on the overall drive train efficiency and performance of heavy-duty hybrid vehicles, *Proceedings of Vehicle Power and Propulsion IEEE Conference VPPC*, ISBN 0-7803-9280-9, Chicago, Sept 2005

Winston Battery Limited (n.d.). Winston Battery Limited, In: *Winston Battery Limited,* January 2011, Available from: <http://www.thunder-sky.com >

Zhang, Y.; Wu, L.; Hu, X.; & Liang, H. (2008), Model and control for supercapacitor-based energy storage system for metro vehicles, *Proceedings of International Conference on Electrical Machines and Systems. ICEMS*, ISBN 978-1-4244-3826-6, Wuhan, Oct 2008

Zhongtong Bus Holding Co. (n.d.). Electric Tourism Coach, Express Bus in China, Zhongtong Bus, In: *Electric Tourism Coach,* January 2011, Available from: <http://www.zhongtongbuses.com>

18

Predictive Optimal Matrix Converter Control for a Dynamic Voltage Restorer with Flywheel Energy Storage

Paulo Gambôa[1,3,4], J. Fernando Silva[2,3],
S. Ferreira Pinto[2,3] and Elmano Margato[1,3,4]
[1]*Instituto Superior de Engenharia de Lisboa, DEEA,*
[2]*DEEC, Instituto Superior Técnico,*
[3]*Center for Innovation in Electrical and Energy Engineering,*
[4]*Centro de Electrotecnia e Electrónica Industrial,*
Portugal

1. Introduction

In recent years, Flywheel Energy Storage (FES) systems have been rediscovered by industry due to their advantages in comparison with other short term energy storage systems [1], [2], [3]. FES can be designed to protect critical loads against voltage sags, by using a Permanent Magnet Synchronous Motor (PMSM) and a matrix converter to control the power transfer process between the flywheel and the critical load. This arrangement eliminates the dc link capacitor needed in conventional back-to-back inverter rectifier systems.

The predictive discrete-time control of power converters has been outlined in [4], [5], [6], [7], and recently applied to matrix converters [8].

In this chapter, using FES and PMSM, a matrix converter based Dynamic Voltage Restorer (DVR) model is used to obtain the current references to control the matrix converter (section II). A detailed discrete-time dynamic model of the matrix converter is used to predict the expected values of the input and output currents for all the 27 possible output voltage vectors generated by the matrix converter (section III). A minimization procedure, based on a weighted cost functional, selects the optimal vector that minimizes both the output current errors and the input current errors.

The predictive optimal controllers here proposed can enhance the controllability of matrix converters by optimizing the vector choice to overcome the input-output coupling of matrix converters, due to their lack of enough stored energy.

A description and characterization of an experimental kinetic energy accumulator, based on FES is also presented (section IV). The matrix based DVR simulation results are shown in section V. Results show that FES with predictive optimal matrix converter control can be used as a DVR (Fig. 1) to excel in the mitigation of voltage sags and swells as well as voltage distortion at critical loads.

2. Dynamic voltage restorer

2.1 The concept of flywheel energy storage based DVRs

Power quality problems like voltage sags, swells and harmonics are a major concern of the industrial and commercial electrical consumers due to enormous loss in terms of time and money [10]. This is due to the advent of a large number of sophisticated electrical and electronic equipments, such as computers, programmable logic controllers, variable speed drives, and other accurate control systems.

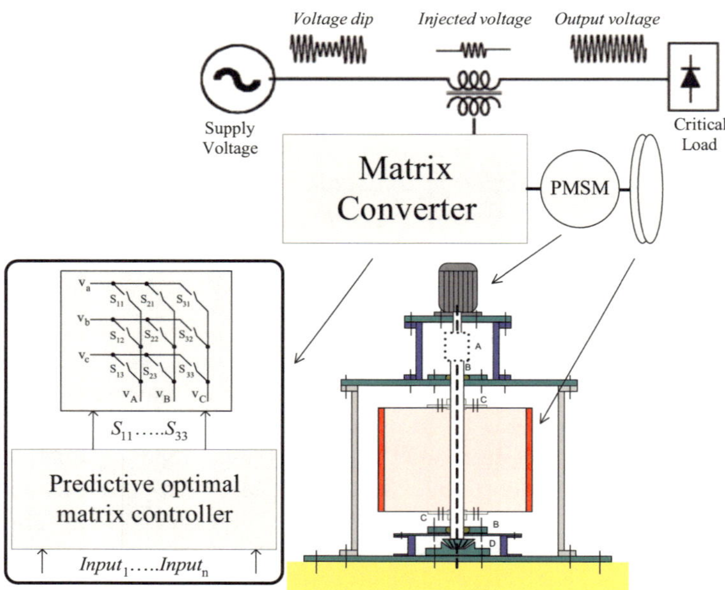

Fig. 1. Schematic diagram of a typical DVR.

The use of these equipments often requires very high quality power supplies. Some control equipments are highly sensitive to voltage disturbances, mainly voltage sags lasting several periods, which cause shut-downs and even failures. The adverse effects of voltage disturbances, such as sags and swells, dictated the need for effective mitigating devices. These devices include uninterruptible power supplies (UPS) and DVRs.

The DVR is one the most effective solutions for sags, since it only supplies the power difference between disturbed voltage and ideal voltages, not all the load power, as do UPSs. DVRs are series custom power devices, which should present excellent dynamic capabilities, to protect sensitive loads from voltage sags, swells and voltage harmonics, by inserting a compensating series voltage to restore the ideal network voltage waveform. Therefore, a DVR is basically a controlled voltage source installed in series between the supply and a sensitive load. It injects a voltage on the system in order to compensate any disturbance affecting the load voltage. Basic operating diagram of a DVR is as shown in Fig. 2, where the series voltage is inserted as the voltage on one winding of a transformer driven from the *RL* output filter of the matrix converter.

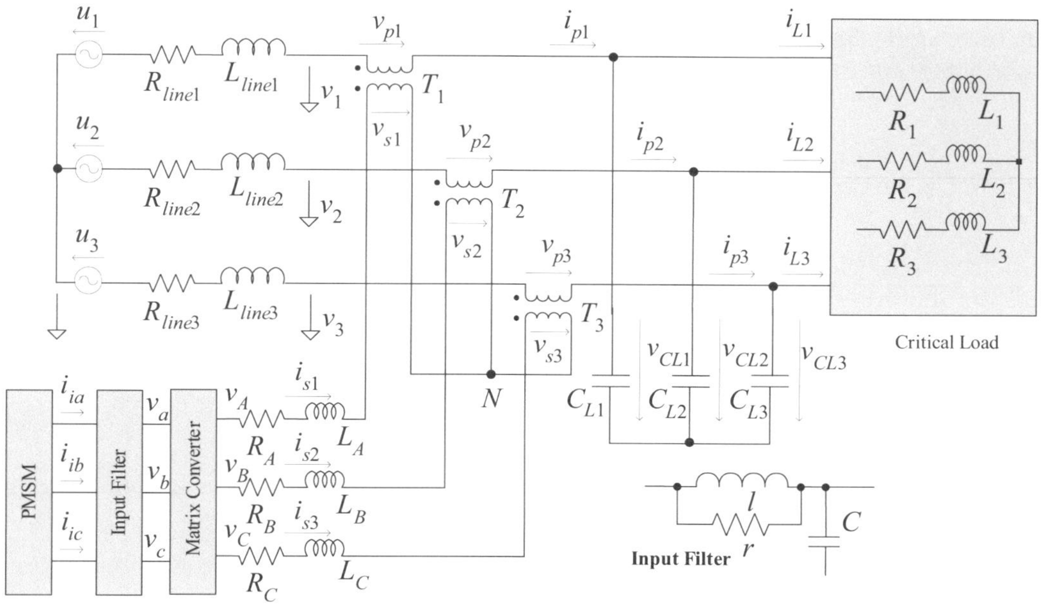

Fig. 2. Schematic diagram of a typical DVR.

2.2 Critical load voltage control using DVRs

To impose the compensating voltage, the needed critical load voltage components in the *dq* frame v_{CLd}, v_{CLq} must be controlled by acting on the matrix converter output current component references i_{sdref}, i_{sqref}, using PI controllers [11], [12].

Gains k_p and k_i are respectively proportional and integral gains, which can be calculated optimizing the resulting closed loop 2nd order system response. The reference values i_{sdref}, i_{sqref} will then be transformed into the αβ frame to establish the references for a predictive current controller for the matrix converter.

Applying the *Kirchhoff* laws to the critical load (Fig.2) and doing some mathematical manipulations, the dynamic equations of the ac voltages $v_{CL1}(t)$, $v_{CL2}(t)$ e $v_{CL3}(t)$, are defined as functions of the circuit parameters and ac currents $i_{p1}(t)$, $i_{p2}(t)$, $i_{p3}(t)$ and $i_{L1}(t)$, $i_{L2}(t)$, $i_{L3}(t)$. The resultant state-space systems model is written in (1). The $i_{p1}=N_2/N_1 i_{s1}$, $i_{p2}=N_2/N_1 i_{s2}$ and $i_{p3}=N_2/N_1 i_{s3}$ is the transformer turns ratio. In this model, the control variables are the output currents of the matrix converter, i_{s1}, i_{s2} and i_{s3}. The currents in the critical load, i_{L1}, i_{L2} and i_{L3} are disturbances for the design of controllers.

$$\begin{bmatrix} \frac{dv_{CL1}}{dt} \\ \frac{dv_{CL2}}{dt} \\ \frac{dv_{CL3}}{dt} \end{bmatrix} = \begin{bmatrix} \frac{1}{C_L}\frac{N_2}{N_1} & 0 & 0 \\ 0 & \frac{1}{C_L}\frac{N_2}{N_1} & 0 \\ 0 & 0 & \frac{1}{C_L}\frac{N_2}{N_1} \end{bmatrix} \begin{bmatrix} i_{s1} \\ i_{s2} \\ i_{s3} \end{bmatrix} - \begin{bmatrix} \frac{1}{C_L} & 0 & 0 \\ 0 & \frac{1}{C_L} & 0 \\ 0 & 0 & \frac{1}{C_L} \end{bmatrix} \begin{bmatrix} i_{L1} \\ i_{L2} \\ i_{L3} \end{bmatrix} \qquad (1)$$

To design the critical load voltage, the system model (1) can advantageously be represented in αβ coordinates. The relationship between the variables X_{123} represented in system coordinates and in αβ coordinates, $X_{\alpha\beta}$, is

$$X_{123} = [C] X_{\alpha\beta} \tag{2}$$

Where C is the *Concordia* transformation (3).

$$C = \sqrt{\frac{2}{3}} \begin{bmatrix} 1 & 0 & \frac{\sqrt{2}}{2} \\ -\frac{1}{2} & \frac{\sqrt{3}}{2} & \frac{\sqrt{2}}{2} \\ -\frac{1}{2} & -\frac{\sqrt{3}}{2} & \frac{\sqrt{2}}{2} \end{bmatrix} \tag{3}$$

Applying (2) and (3) to the model (1), the critical load voltage model (4) in αβ coordinates is obtained (without homopolar component).

$$\begin{bmatrix} \frac{dv_{CL\alpha}}{dt} \\ \frac{dv_{CL\beta}}{dt} \end{bmatrix} = \begin{bmatrix} \frac{1}{C_L}\frac{N_2}{N_1} & 0 \\ 0 & \frac{1}{C_L}\frac{N_2}{N_1} \end{bmatrix} \begin{bmatrix} i_{s\alpha} \\ i_{s\beta} \end{bmatrix} - \begin{bmatrix} \frac{1}{C_L} & 0 \\ 0 & \frac{1}{C_L} \end{bmatrix} \begin{bmatrix} i_{L\alpha} \\ i_{L\beta} \end{bmatrix} \tag{4}$$

The relationship between the variables $X_{\alpha\beta}$ in represented in αβ coordinates and in dq coordinates, X_{dq}, is given by the *Park* transformation:

$$X_{\alpha\beta} = D X_{dq} \tag{5}$$

$$D = \begin{bmatrix} \cos\theta & -\sin\theta \\ \sin\theta & \cos\theta \end{bmatrix} \tag{6}$$

The argument $\theta = \omega t$ is the angular phase of the electrical network ac line voltage. Applying the *Park* transformation (5) to the critical load voltage model (4), in dq coordinates is obtained.

$$\begin{bmatrix} \frac{dv_{CLd}}{dt} \\ \frac{dv_{CLq}}{dt} \end{bmatrix} = \begin{bmatrix} 0 & \omega \\ -\omega & 0 \end{bmatrix} \begin{bmatrix} v_{CLd} \\ v_{CLq} \end{bmatrix} + \begin{bmatrix} \frac{1}{C_L}\frac{N_2}{N_1} & 0 \\ 0 & \frac{1}{C_L}\frac{N_2}{N_1} \end{bmatrix} \begin{bmatrix} i_{sd} \\ i_{sq} \end{bmatrix} - \begin{bmatrix} \frac{1}{C_L} & 0 \\ 0 & \frac{1}{C_L} \end{bmatrix} \begin{bmatrix} i_{Ld} \\ i_{Lq} \end{bmatrix} \tag{7}$$

The mathematical model needed to control the critical load voltage at the load shunt capacitors $C_{L1}=C_{L2}=C_{L3}=C_L$ (Fig. 2), which serve as a load filter and power factor compensation, can be written in the dq frame as:

$$\frac{dv_{CLd}}{dt} = \frac{1}{C_L}\left(C_L \omega v_{CLq} + \frac{N_2}{N_1} i_{sd} - i_{Ld} \right) \tag{8}$$

$$\frac{dv_{cq}}{dt}=\frac{1}{C_L}\left(-C_L\omega v_{CLd}+\frac{N_2}{N_1}i_{sq}-i_{Lq}\right) \tag{9}$$

Where N_2/N_1 is the transformer turns ratio and i_{sd} and i_{sq} are direct and quadrature secondary current components in the transformer. This voltage dynamics is cross-coupled, but can be decoupled introducing auxiliary variables h_d, h_q given by:

$$h_d=C_L\omega v_{CLq}+\frac{N_2}{N_1}i_{sd} \tag{10}$$

$$h_q=-C_L\omega v_{CLd}+\frac{N_2}{N_1}i_{sq} \tag{11}$$

Substituting (10) and (11) in (8), (9), two isolated first order systems are obtained.

$$\frac{dv_{CLd}}{dt}=\frac{1}{C_L}\left(h_d-i_{Ld}\right) \tag{12}$$

$$\frac{dv_{CLq}}{dt}=\frac{1}{C_L}\left(h_q-i_{Lq}\right) \tag{13}$$

Then, the needed critical load voltage components v_{CLd}, v_{CLq} can be controlled by acting on the matrix converter output current component references i_{sdref}, i_{sqref},, using PI controllers (Fig. 3).

$$h_{dref}=\left(k_{pd}+\frac{k_{id}}{s}\right)e_{vCLd}\Leftrightarrow h_{dref}=\left(k_{pd}+\frac{k_{id}}{s}\right)\left(v_{CLdref}-v_{CLd}\right) \tag{14}$$

$$h_{qref}=\left(k_{pq}+\frac{k_{iq}}{s}\right)e_{vCLq}\Leftrightarrow h_{qref}=\left(k_{pq}+\frac{k_{iq}}{s}\right)\left(v_{CLqref}-v_{CLq}\right) \tag{15}$$

Substituting $i_{sd}=i_{sdref}$ and $h_d=h_{dref}$ in (10) and $i_{sq}=i_{sqref}$ and $h_q=h_{qref}$ in (11), the references i_{sdref}, i_{sqref}, are obtained.

$$h_d=C_L\omega v_{CLq}+\frac{N_2}{N_1}i_{sd}\Leftrightarrow i_{sd}=\frac{N_1}{N_2}h_d-C_L\frac{N_1}{N_2}\omega v_{CLq}\xrightarrow[\substack{i_{sd}=i_{sdref}\\ h_d=h_{dref}}]{}i_{sdref}=\frac{N_1}{N_2}h_{dref}-C_L\frac{N_1}{N_2}\omega v_{CLq} \tag{16}$$

$$h_q=-C_L\omega v_{CLd}+\frac{N_2}{N_1}i_{sq}\Leftrightarrow i_{sq}=\frac{N_1}{N_2}h_q+C_L\frac{N_1}{N_2}\omega v_{CLd}\xrightarrow[\substack{i_{sq}=i_{sqref}\\ h_q=h_{qref}}]{}i_{sqref}=\frac{N_1}{N_2}h_{qref}+C_L\frac{N_1}{N_2}\omega v_{CLd} \tag{17}$$

Substituting (14) in (16) and (15) in (17), the matrix converter output current component references i_{sdref}, i_{sqref}, are given by (18) and (19).

$$i_{sdref}=\frac{N_1}{N_2}k_p\left(v_{CLdref}-v_{CLd}\right)+\frac{N_1}{N_2}k_i\int\left(v_{CLdref}-v_{CLd}\right)dt-\frac{N_1}{N_2}C_L\omega v_{CLq} \tag{18}$$

$$i_{sqref} = \frac{N_1}{N_2} k_p \left(v_{CLqref} - v_{CLq} \right) + \frac{N_1}{N_2} k_i \int \left(v_{CLqref} - v_{CLq} \right) dt + \frac{N_1}{N_2} C_L \omega v_{CLd} \tag{19}$$

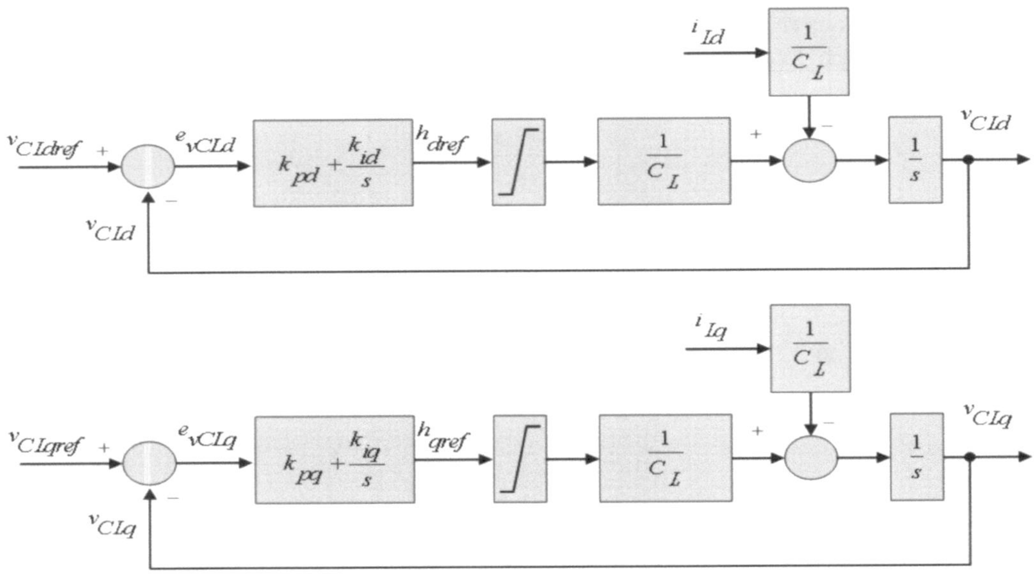

Fig. 3. Block diagram of the voltage components v_{CLd} and v_{CLq}.

Gains k_{pdq} (22) and k_{idq} (23) are respectively proportional and integral gains, which can be calculated minimizing the ITAE criterion in the resulting closed loop 2nd order system (20,21) , provided the zeros $-k_{idq}/k_{pdq}$ are far from the poles.

$$v_{CLd} = \frac{\frac{k_{pd}}{C_L} s + \frac{k_{id}}{C_L}}{s^2 + \frac{k_{pd}}{C_L} s + \frac{k_{id}}{C_L}} v_{CLdref} - \frac{\frac{1}{C_L} s}{s^2 + \frac{k_{pd}}{C_L} s + \frac{k_{id}}{C_L}} i_{Ld} \tag{20}$$

$$v_{CLq} = \frac{\frac{k_{pq}}{C_L} s + \frac{k_{iq}}{C_L}}{s^2 + \frac{k_{pq}}{C_L} s + \frac{k_{iq}}{C_L}} v_{CLqref} - \frac{\frac{1}{C_L} s}{s^2 + \frac{k_{pq}}{C_L} s + \frac{k_{iq}}{C_L}} i_{Lq} \tag{21}$$

$$k_{pd} = k_{pq} = 2\xi \sqrt{C_L k_{idq}} \tag{22}$$

$$k_{id} = k_{iq} = C_L \omega_n^2 \tag{23}$$

The reference values i_{sdref}, i_{sqref} will then be transformed into the αβ frame to establish the references for a predictive current controller for the matrix converter.

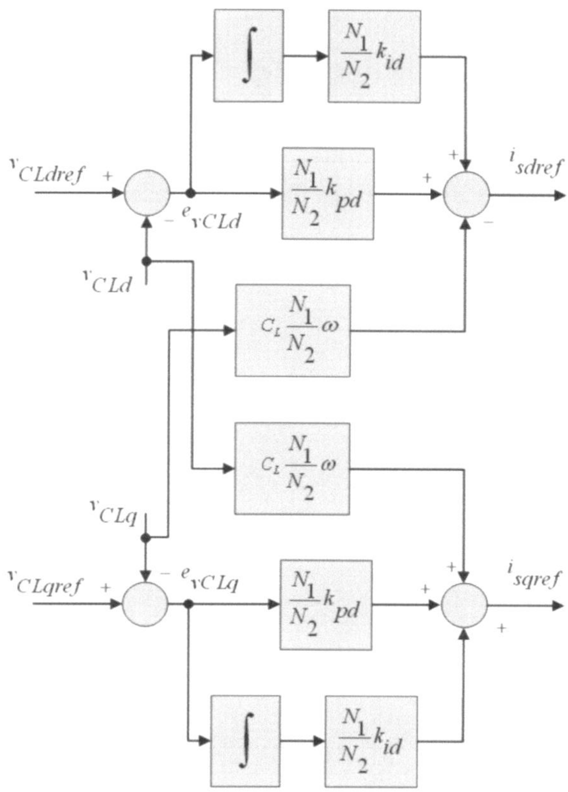

Fig. 4. Block diagram of the reference values components i_{sdref} and i_{sqref}.

3. Predictive control of matrix converters for DVR operation

3.1 Matrix converter

Matrix converters are based upon an association of nine bi-directional switches with turn-off capability, which allow the connection of each one of the three output phases to any one of the three input phases connected to a PMSM through a *rlC* filter (Fig. 5).

A nine–element matrix, with elements S_{ij} representing the state of each bi-directional switch, is used to represent the matrix output voltages (v_A, v_B, v_C) as functions of the input voltages (v_a, v_b, v_c) (24).

$$\begin{bmatrix} v_A \\ v_B \\ v_C \end{bmatrix} = \begin{bmatrix} S_{11} & S_{12} & S_{13} \\ S_{21} & S_{22} & S_{23} \\ S_{31} & S_{32} & S_{33} \end{bmatrix} \begin{bmatrix} v_a \\ v_b \\ v_c \end{bmatrix} \tag{24}$$

The line to line output voltages (v_{AB}, v_{BC}, v_{CA}) are functions of the S_{ij} and of the input line to line voltages (v_{ab}, v_{bc}, v_{ca}):

$$\begin{bmatrix} v_{AB} \\ v_{BC} \\ v_{CA} \end{bmatrix} = \begin{bmatrix} \frac{2}{3}(S_{11}-S_{21})+\frac{1}{3}(S_{13}-S_{23}) & \frac{1}{3}(S_{11}-S_{21})+\frac{2}{3}(S_{12}-S_{22}) & \frac{1}{3}(S_{12}-S_{22})+\frac{2}{3}(S_{13}-S_{23}) \\ \frac{2}{3}(S_{21}-S_{31})+\frac{1}{3}(S_{23}-S_{33}) & \frac{1}{3}(S_{21}-S_{31})+\frac{2}{3}(S_{22}-S_{32}) & \frac{1}{3}(S_{22}-S_{32})+\frac{2}{3}(S_{23}-S_{33}) \\ \frac{2}{3}(S_{31}-S_{11})+\frac{1}{3}(S_{33}-S_{13}) & \frac{1}{3}(S_{31}-S_{11})+\frac{2}{3}(S_{32}-S_{12}) & \frac{1}{3}(S_{32}-S_{12})+\frac{2}{3}(S_{33}-S_{13}) \end{bmatrix} \begin{bmatrix} v_{ab} \\ v_{bc} \\ v_{ca} \end{bmatrix} \quad (25)$$

Each S_{ij} element of the 3×3 matrix represents the state of each bi-directional switch (if switch S_{ij} is off then S_{ij}=0, else S_{ij}=1).

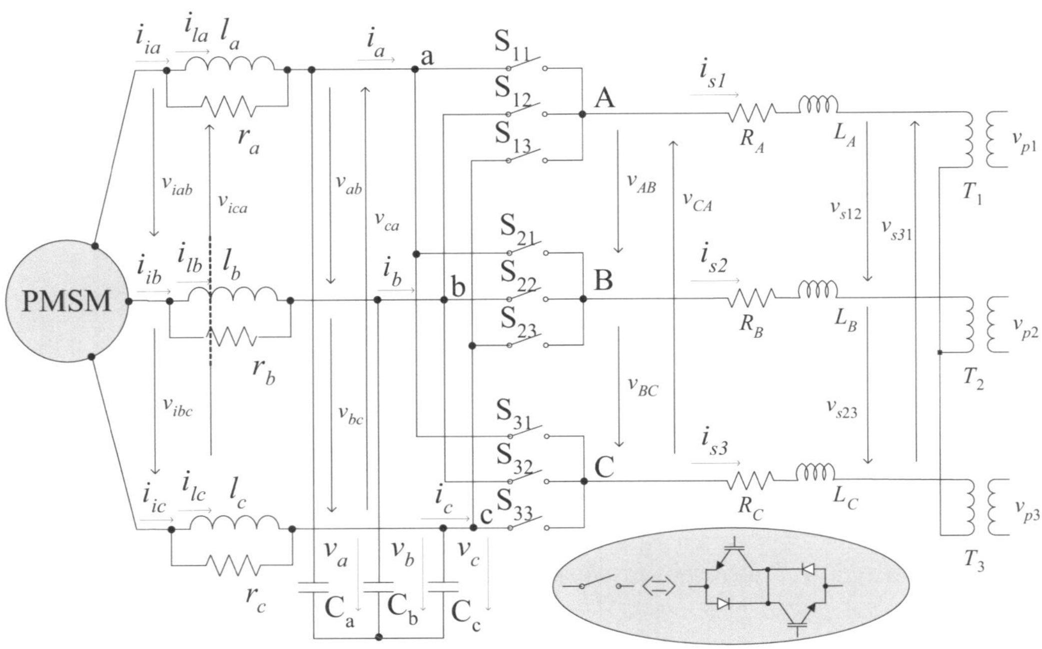

Fig. 5. Matrix converter topology.

The 3-phase matrix converter presents 27 switching combinations, or vectors [10], since for all $k \in \{1,2,3\}$ $\sum_{j=1}^{3} S_{kj} = 1$.

The input phase currents (i_a, i_b, i_c) can be related to the output phase currents (i_{s1}, i_{s2}, i_{s3}) by:

$$\begin{bmatrix} i_a \\ i_b \\ i_c \end{bmatrix} = \begin{bmatrix} S_{11} & S_{21} & S_{31} \\ S_{12} & S_{22} & S_{32} \\ S_{13} & S_{23} & S_{33} \end{bmatrix} \begin{bmatrix} i_{s1} \\ i_{s2} \\ i_{s3} \end{bmatrix} \quad (26)$$

The 27 switching combinations of the nine bi-directional switches S_{ij} (Table 1), can be used as output voltage and/or input current vectors given as functions of each S_{ij} state, meaning that the control of the matrix output voltages and matrix input *rlC* filter currents (Fig. 5) is not independent.

N.º	S_{11}	S_{12}	S_{13}	S_{21}	S_{22}	S_{23}	S_{31}	S_{32}	S_{33}	$v_A(t)$	$v_B(t)$	$v_C(t)$
1	1	0	0	0	1	0	0	0	1	$v_a(t)$	$v_b(t)$	$v_c(t)$
2	0	1	0	0	0	1	1	0	0	$v_b(t)$	$v_c(t)$	$v_a(t)$
3	0	0	1	1	0	0	0	1	0	$v_c(t)$	$v_a(t)$	$v_b(t)$
4	1	0	0	0	0	1	0	1	0	$v_a(t)$	$v_c(t)$	$v_b(t)$
5	0	1	0	1	0	0	0	0	1	$v_b(t)$	$v_a(t)$	$v_c(t)$
6	0	0	1	0	1	0	1	0	0	$v_c(t)$	$v_b(t)$	$v_a(t)$
7	1	0	0	0	1	0	0	1	0	$v_a(t)$	$v_b(t)$	$v_b(t)$
8	0	1	0	1	0	0	1	0	0	$v_b(t)$	$v_a(t)$	$v_a(t)$
9	0	1	0	0	0	1	0	0	1	$v_b(t)$	$v_c(t)$	$v_c(t)$
10	0	0	1	0	1	0	0	1	0	$v_c(t)$	$v_b(t)$	$v_b(t)$
11	0	0	1	1	0	0	1	0	0	$v_c(t)$	$v_a(t)$	$v_a(t)$
12	1	0	0	0	0	1	0	0	1	$v_a(t)$	$v_c(t)$	$v_c(t)$
13	0	1	0	1	0	0	0	1	0	$v_b(t)$	$v_a(t)$	$v_b(t)$
14	1	0	0	0	1	0	1	0	0	$v_a(t)$	$v_b(t)$	$v_a(t)$
15	0	0	1	0	1	0	0	0	1	$v_c(t)$	$v_b(t)$	$v_c(t)$
16	0	1	0	0	0	1	0	1	0	$v_b(t)$	$v_c(t)$	$v_b(t)$
17	1	0	0	0	0	1	1	0	0	$v_a(t)$	$v_c(t)$	$v_a(t)$
18	0	0	1	1	0	0	0	0	1	$v_c(t)$	$v_a(t)$	$v_c(t)$
19	0	1	0	0	1	0	1	0	0	$v_b(t)$	$v_b(t)$	$v_a(t)$
20	1	0	0	1	0	0	0	1	0	$v_a(t)$	$v_a(t)$	$v_b(t)$
21	0	0	1	0	0	1	0	1	0	$v_c(t)$	$v_c(t)$	$v_b(t)$
22	0	1	0	0	1	0	0	0	1	$v_b(t)$	$v_b(t)$	$v_c(t)$
23	1	0	0	1	0	0	0	0	1	$v_a(t)$	$v_a(t)$	$v_c(t)$
24	0	0	1	0	0	1	1	0	0	$v_c(t)$	$v_c(t)$	$v_a(t)$
25	1	0	0	1	0	0	1	0	0	$v_a(t)$	$v_a(t)$	$v_a(t)$
26	0	1	0	0	1	0	0	1	0	$v_b(t)$	$v_b(t)$	$v_b(t)$
27	0	0	1	0	0	1	0	0	1	$v_c(t)$	$v_c(t)$	$v_c(t)$

Table 1. Voltages vectors generated by the matrix converter

3.2 Matrix converter input and output currents dynamic model

Supposing a balanced three-phase inductive load, the first time derivative of the matrix converter output currents (i_{s1}, i_{s2}, i_{s3}) in the αβ frame (*Concordia* transformation) can be defined by (27,28), where $R_A=R_B=R_C=R$ are the parasitic resistances and $L_A=L_B=L_C=L$ are the output inductors of the matrix converter (Fig. 5), and $v_{o\alpha\beta}$ and $v_{s\alpha\beta}$ represent, respectively the line to line output voltage of the matrix converter and the $v_{p\alpha\beta}$ voltages reflected in the transformer secondary.

$$\frac{di_{s\alpha}}{dt} = -\frac{R}{L}i_{s\alpha} + \frac{1}{2L}v_{o\alpha} + \frac{\sqrt{3}}{6L}v_{o\beta} - \frac{1}{2L}v_{s\alpha} - \frac{\sqrt{3}}{6L}v_{s\beta} \tag{27}$$

$$\frac{di_{s\beta}}{dt} = -\frac{R}{L}i_{s\beta} - \frac{\sqrt{3}}{6L}v_{o\alpha} + \frac{1}{2L}v_{o\beta} + \frac{\sqrt{3}}{6L}v_{s\alpha} - \frac{1}{2L}v_{s\beta} \tag{28}$$

This dynamic equation will be used to predict the matrix converter output currents.

Assuming a 2nd order matrix converter input filter (Fig. 5) with resistances $r_a=r_b=r_c=r$, indutances $l_a=l_b=l_c=l$ and capacitors $C_a=C_b=C_c=C_Y$, the dynamics of each inductor current ($i_{l\alpha\beta}$), the dynamics of the capacitor line to line voltages ($v_{c\alpha\beta}$) and the input line to line filter voltage ($v_{ic\alpha\beta}$), in the αβ frame, is given by:

$$\frac{di_{l\alpha}}{dt}=\frac{1}{l}v_{ic\alpha}-\frac{1}{2l}v_{c\alpha}-\frac{\sqrt{3}}{6l}v_{c\beta} \tag{29}$$

$$\frac{di_{l\beta}}{dt}=\frac{1}{l}v_{ic\beta}+\frac{\sqrt{3}}{6l}v_{c\alpha}-\frac{1}{2l}v_{c\beta} \tag{30}$$

$$\frac{dv_{c\alpha}}{dt}=-\frac{3}{2C_Y}i_\alpha+\frac{\sqrt{3}}{2C_Y}i_\beta+\frac{3}{2C_Y}i_{l\alpha}-\frac{\sqrt{3}}{2C_Y}i_{l\beta}-\frac{1}{C_Y}v_{c\alpha}+\frac{3}{2C_Y}v_{ic\alpha}-\frac{\sqrt{3}}{2C_Y}v_{ic\beta} \tag{31}$$

$$\frac{dv_{c\beta}}{dt}=-\frac{\sqrt{3}}{2C_Y}i_\alpha-\frac{3}{2C_Y}i_\beta+\frac{\sqrt{3}}{2C_Y}i_{l\alpha}+\frac{3}{2C_Y}i_{l\beta}-\frac{1}{C_Y}v_{c\beta}+\frac{\sqrt{3}}{2C_Y}v_{ic\alpha}+\frac{3}{2C_Y}v_{ic\beta} \tag{32}$$

$$\frac{di_{l\alpha}}{dt}=\frac{r}{l}i_{i\alpha}-\frac{r}{l}i_{l\alpha} \tag{33}$$

$$\frac{di_{l\beta}}{dt}=\frac{r}{l}i_{i\beta}-\frac{r}{l}i_{l\beta} \tag{34}$$

These dynamic equations will be used to predict the input filter currents $i_{i\alpha}$, $i_{i\beta}$.

3.3 Discrete-time prediction of matrix input and output currents

A first-order difference equation (35,36), with a sampling time T_s, equivalent to the load current differential equation (27,28), can be used to predict the expected values (*Euler-Forward* method) of $i_{s\alpha}$, $i_{s\beta}$ at time instant t_{s+1}, given the values of $i_{s\alpha}$, $i_{s\beta}$, at the tsth sampling instant.

$$i_{s\alpha}(t_{s+1})=\left(1-\frac{RT_s}{L}\right)i_{s\alpha}(t_s)+\frac{T_s}{2L}v_{o\alpha}(t_s)+\frac{\sqrt{3}T_s}{6L}v_{o\beta}(t_s)-\frac{T_s}{2L}v_{s\alpha}(t_s)-\frac{\sqrt{3}T_s}{6L}v_{s\beta}(t_s) \tag{35}$$

$$i_{s\beta}(t_{s+1})=\left(1-\frac{RT_s}{L}\right)i_{s\beta}(t_s)-\frac{\sqrt{3}T_s}{6L}v_{o\alpha}(t_s)+\frac{T_s}{2L}v_{o\beta}(t_s)+\frac{\sqrt{3}T_s}{6L}v_{s\alpha}(t_s)-\frac{T_s}{2L}v_{s\beta}(t_s) \tag{36}$$

The voltages vectors $v_{o\alpha\beta}$ (37,38) can be generated by the matrix converter, and $v_{c\alpha\beta}$ represent the line to line input voltage in the αβ frame (*Concordia* transformation).

$$v_{o\alpha}(t_{s+1})=H_{v\alpha\alpha}(t_{s+1})v_{c\alpha}(t_s)+H_{v\alpha\beta}(t_{s+1})v_{c\beta}(t_s) \tag{37}$$

$$v_{o\beta}(t_{s+1})=H_{v\beta\alpha}(t_{s+1})v_{c\alpha}(t_s)+H_{v\beta\beta}(t_{s+1})v_{c\beta}(t_s) \tag{38}$$

Where $H_{v\alpha\alpha}$, $H_{v\alpha\beta}$, $H_{v\beta\alpha}$ and $H_{v\beta\beta}$ is given by (39).

$$\begin{cases} H_{v\alpha\alpha} = \frac{1}{2}(S_{11} - S_{12} - S_{21} + S_{22}) \\ H_{v\alpha\beta} = \frac{\sqrt{3}}{6}(S_{11} + S_{12} - 2S_{13} - S_{21} - S_{22} + 2S_{23}) \\ H_{v\beta\alpha} = \frac{\sqrt{3}}{6}(S_{11} - S_{12} + S_{21} - S_{22} - 2S_{31} + 2S_{32}) \\ H_{v\beta\beta} = \frac{1}{6}(S_{11} + S_{12} - 2S_{13} + S_{21} + S_{22} - 2S_{23} - 2S_{31} - 2S_{32} + 4S_{33}) \end{cases} \quad (39)$$

Applying *Euler-Backward* method to the model (29,30) and (31,32), the expected values of the $il_{\alpha\beta}(t_{s+1})$ are obtained. Applying *Euler-Backward* method to the model (33,34) and replacing $il_{\alpha\beta}(t_{s+1})$, the expected values of the $i_{i\alpha\beta}(t_{s+1})$ are obtained. The discrete–time difference equations (40,41) (sampling time T_s) of the matrix input filter current dynamic equations, can be used to predict the expected values of the input filter currents at the t_{s+1} sampling instant.

$$\begin{aligned} i_{i\alpha}(t_{s+1}) = {} & \frac{C_Y lr}{rT_s^2 + lT_s r + C_Y lr} i_{l\alpha}(t_s) + \frac{T_s(l + T_s r)}{rT_s^2 + lT_s r + C_Y lr} i_{\alpha}(t_{s+1}) + \frac{C_Y(l + T_s r)}{2rT_s^2 + 2lT_s r + 2C_Y lr} v_{ic\alpha}(t_{s+1}) + \\ & + \frac{\sqrt{3}C_Y(l + T_s r)}{6rT_s^2 + 6lT_s r + 6C_Y lr} v_{ic\beta}(t_{s+1}) - \frac{C_Y(l + T_s r)}{2rT_s^2 + 2lT_s r + 2C_Y lr} v_{c\alpha}(t_s) - \frac{\sqrt{3}C_Y(l + T_s r)}{6rT_s^2 + 6lT_s r + 6C_Y lr} v_{c\beta}(t_s) \end{aligned} \quad (40)$$

$$\begin{aligned} i_{i\beta}(t_{s+1}) = {} & \frac{C_Y lr}{rT_s^2 + lT_s r + C_Y lr} i_{l\beta}(t_s) + \frac{T_s(l + T_s r)}{rT_s^2 + lT_s r + C_Y lr} i_{\beta}(t_{s+1}) - \frac{\sqrt{3}C_Y(l + T_s r)}{6rT_s^2 + 6lT_s r + 6C_Y lr} v_{ic\alpha}(t_{s+1}) + \\ & + \frac{C_Y(l + T_s r)}{2rT_s^2 + 2lT_s r + 2C_Y lr} v_{ic\beta}(t_{s+1}) + \frac{\sqrt{3}C_Y(l + T_s r)}{6rT_s^2 + 6lT_s r + 6C_Y lr} v_{c\alpha}(t_s) - \frac{C_Y(l + T_s r)}{2rT_s^2 + 2lT_s r + 2C_Y lr} v_{c\beta}(t_s) \end{aligned} \quad (41)$$

Where, considering (40,41), the matrix input currents $i_a(t_{s+1})$, $i_b(t_{s+1})$ and $i_c(t_{s+1})$, at the t_{s+1} sampling instant are established by equations (42), (43) and (44).

$$i_a(t_{s+1}) = S_{11}(t_{s+1})i_{s1}(t_s) + S_{21}(t_{s+1})i_{s2}(t_s) + S_{31}(t_{s+1})i_{s3}(t_s) \quad (42)$$

$$i_b(t_{s+1}) = S_{12}(t_{s+1})i_{s1}(t_s) + S_{22}(t_{s+1})i_{s2}(t_s) + S_{32}(t_{s+1})i_{s3}(t_s) \quad (43)$$

$$i_c(t_{s+1}) = S_{13}(t_{s+1})i_{s1}(t_s) + S_{23}(t_{s+1})i_{s2}(t_s) + S_{33}(t_{s+1})i_{s3}(t_s) \quad (44)$$

Applying (2) and (3) to equations (42), (43) and (44), the input currents at the t_{s+1} sampling instant, in αβ coordinates is obtained (45,46).

$$i_\alpha(t_{s+1}) = H_{i\alpha\alpha}(t_{s+1})i_{s\alpha}(t_s) + H_{i\alpha\beta}(t_{s+1})i_{s\beta}(t_s) \quad (45)$$

$$i_\beta(t_{s+1}) = H_{i\beta\alpha}(t_{s+1})i_{s\alpha}(t_s) + H_{i\beta\beta}(t_{s+1})i_{s\beta}(t_s) \quad (46)$$

Where $H_{i\alpha\alpha}$, $H_{i\alpha\beta}$, $H_{i\beta\alpha}$ and $H_{i\beta\beta}$ is given by (47).

$$\begin{aligned}
H_{i\alpha\alpha} &= \frac{2}{3}S_{11} - \frac{1}{3}S_{12} - \frac{1}{3}S_{13} - \frac{1}{3}S_{21} + \frac{1}{6}S_{22} + \frac{1}{6}S_{23} - \frac{1}{3}S_{31} + \frac{1}{6}S_{32} + \frac{1}{6}S_{33} \\
H_{i\alpha\beta} &= \frac{2\sqrt{3}}{6}S_{21} - \frac{\sqrt{3}}{6}S_{22} - \frac{\sqrt{3}}{6}S_{23} - \frac{2\sqrt{3}}{6}S_{31} + \frac{\sqrt{3}}{6}S_{32} + \frac{\sqrt{3}}{6}S_{33} \\
H_{i\beta\alpha} &= \frac{2\sqrt{3}}{3}S_{12} - \frac{2\sqrt{3}}{6}S_{13} - \frac{\sqrt{3}}{6}S_{22} + \frac{\sqrt{3}}{6}S_{23} - \frac{\sqrt{3}}{6}S_{32} + \frac{\sqrt{3}}{6}S_{33} \\
H_{i\beta\beta} &= \frac{1}{2}S_{22} - \frac{1}{2}S_{23} - \frac{1}{2}S_{32} + \frac{1}{2}S_{33}
\end{aligned} \tag{47}$$

3.4 Quadratic cost functional for the matrix output current errors and matrix input power factor error

The αβ errors $e_{s\alpha}$ (48) and $e_{s\beta}$ (49) of the matrix output currents $i_{s\alpha}$ and $i_{s\beta}$ are defined as differences between the current references $i_{s\alpha ref}$ and $i_{s\beta ref}$ and the actual output currents $i_{s\alpha}$ and $i_{s\beta}$.

$$e_{s\alpha}(t_{s+1}) = i_{s\alpha ref}(t_s) - i_{s\alpha}(t_{s+1}) \tag{48}$$

$$e_{s\beta}(t_{s+1}) = i_{s\beta ref}(t_s) - i_{s\beta}(t_{s+1}) \tag{49}$$

For the matrix input filter current errors, a near unity input power factor for the matrix converter is assumed. Then, the reactive power reference Q_{ref} should be zero. Therefore, the reactive power error e_Q is (50).

$$e_Q(t_{s+1}) = Q_{ref} - \left[-\frac{1}{2}v_{ic\alpha}(t_{s+1})i_{i\beta}(t_{s+1}) - \frac{\sqrt{3}}{6}v_{ic\beta}(t_{s+1})i_{i\beta}(t_{s+1}) - \frac{\sqrt{3}}{6}v_{ic\alpha}(t_{s+1})i_{i\alpha}(t_{s+1}) + \frac{1}{2}v_{ic\beta}(t_{s+1})i_{i\alpha}(t_{s+1})\right] \tag{50}$$

The input filter voltages v_{iab}, v_{ibc} and v_{ica} at the t_{s+1} sampling instant, in αβ coordinates is obtained (51,52).

$$v_{ic\alpha}(t_{s+1}) = v_{ic\alpha}(t_s)\cos(2\pi f T_s) - v_{ic\beta}(t_s)\sin(2\pi f T_s) \tag{51}$$

$$v_{ic\beta}(t_{s+1}) = v_{ic\beta}(t_s)\cos(2\pi f T_s) + v_{ic\alpha}(t_s)\sin(2\pi f T_s) \tag{52}$$

The cost functional to be chosen must penalize larger errors, while forgiving very small errors. Thus, the most suitable cost evaluator is a quadratic cost functional F (53), which computes the norm of the weighted matrix output current errors and input power factor error.

$$F = \sqrt{e_{s\alpha}^2(t_{s+1})k_\alpha + e_{s\beta}^2(t_{s+1})k_\beta + e_Q^2(t_{s+1})k_Q} \tag{53}$$

In the cost functional (53), k_α, k_β, k_Q are weighting constants, degrees of freedom of the predictive optimal controller. In Fig. 6 it is shown the predictive optimal matrix converter control algorithm.

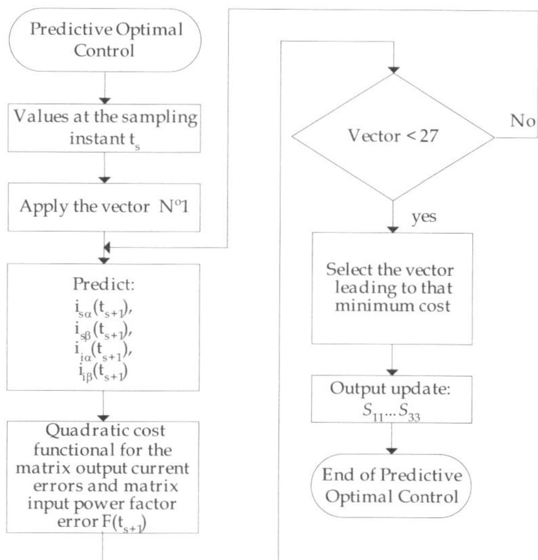

Fig. 6. Predictive Optimal Matrix Converter Control Algorithm.

3.5 Predictive optimal matrix converter control

The predictive optimal matrix controller uses the discrete-time model (35, 36, 40, 41) to predict, for the next sampling instant, the values of the input and output currents and the errors for all the 27 possible vectors of the matrix converter. An optimal algorithm (Fig. 7) minimizes input and output current errors using the minimum value of the weighted cost functional (53) evaluated for all 27 vectors, to select the vector leading to that minimum cost.

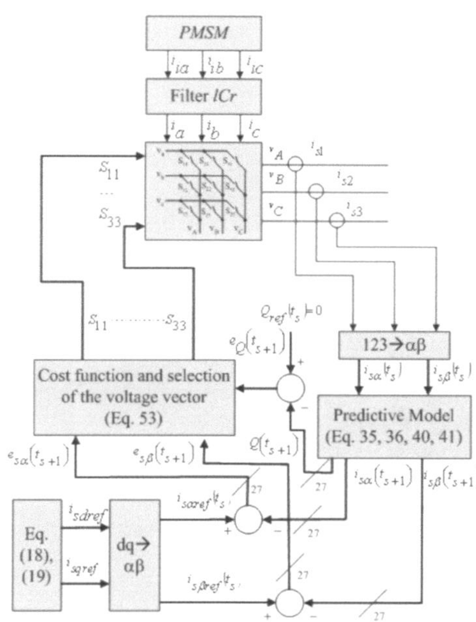

Fig. 7. Block diagram of the predictive optimal matrix converter control.

4. The experimental kinetic energy storage system

4.1 The concept of flywheel

An experimental kinetic energy storage (Fig. 8, 9) was designed and built to have a moment of inertia of 4.2kgm^2 and a maximum rotating speed of 2500 rpm.
A PMSM (2.9kW, 3000rpm) was selected as the electromechanical energy transfer device. A steel seamless tube (fig. 8) was selected as a rotating mass, given its availability and ease of assembly. This part of the device is characterized by parameters which depend on material (in this case steel), and parameters and dimensions of the barrel (Fig. 8).
The volume of the steel barrel is:

$$V = 2\pi\left(r_2^2 - r_1^2\right)h = 0.0062m^3 \tag{54}$$

Where r_1, r_2 are respectively the internal and external radius of the barrel. Its mass is:

$$m = \rho V = 70.0kg \tag{55}$$

Where $\rho=7.8\times10^3$[kg/m^3] is the steel density.
The moment of inertia is given as:

$$J = \frac{1}{2}m\left(r_2^2 + r_1^2\right) = 4.2kgm^2 \tag{56}$$

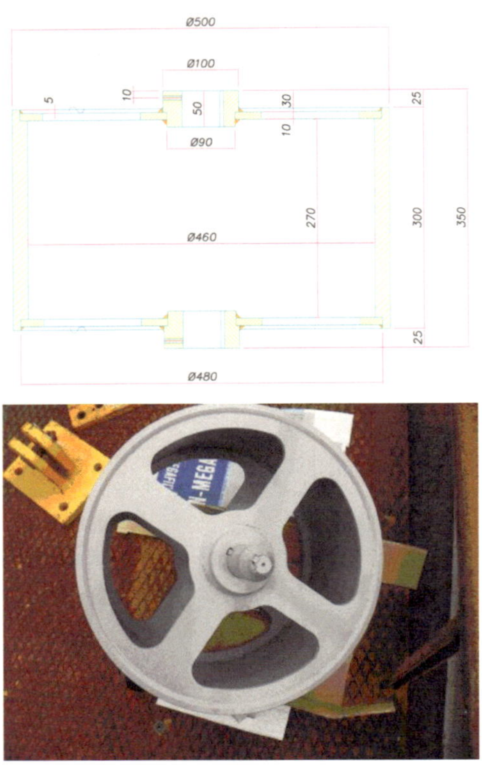

Fig. 8. Dimension and view of the steel barrel.

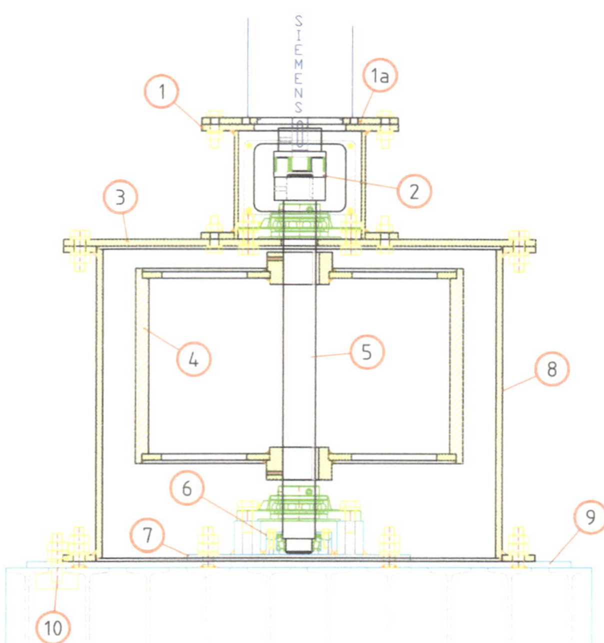

Fig. 9. Flywheel energy storage prototype.

Using r_1=250mm, r_2=240mm, m=70kg, and ω_{max}=2500×2π/60[rad/s] as the maximum angular speed of the barrel, the maximum energy stored in the FES is:

$$E_{k\max} = \frac{1}{2} J\omega_{\max}^2 = 144kJ \tag{57}$$

This energy cannot be completely extracted from the FES, as the energy conversion becomes inefficient when the angular speed drops below a certain value (roughly 50% of the rated speed) [11]. Supposing this limit equals 50% of the maximum speed, the amount of useful energy is 75% of the total energy and, in this case, 0.75×144[kJ]=108 [kJ]. Given this energy, Fig. 10 relates the maximum power available from the flywheel for a given time. For example, if it is necessary to have power available during 0.5s, then nearly 200kW can be supplied.

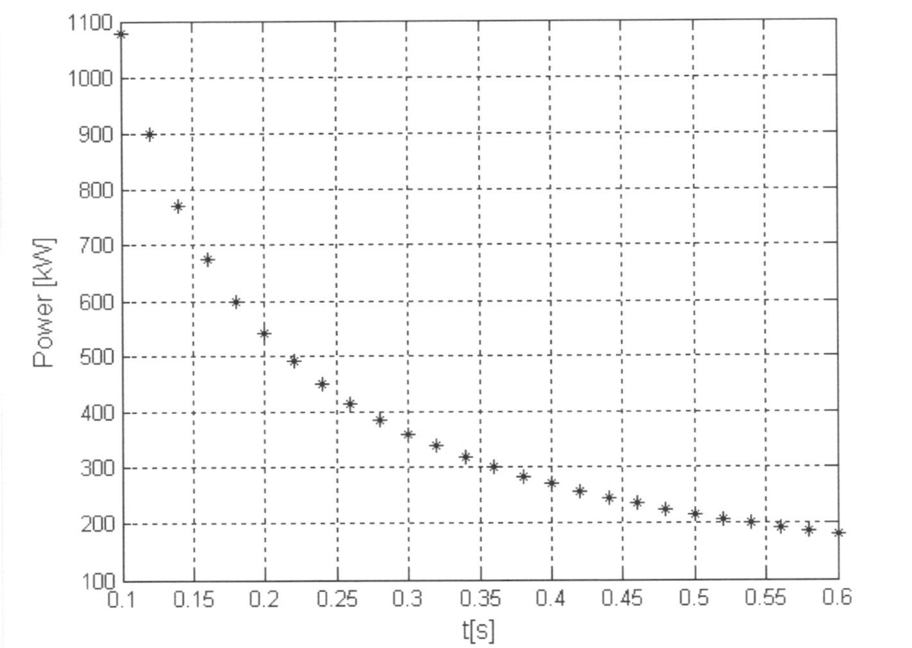

Fig. 10. Output power of the flywheel energy storage.

4.2 Flywheel dynamics

The Flywheel dynamics can be given as:

$$T = J\frac{d\omega}{dt} + K_D\omega + K_C \tag{58}$$

Where J is the combined inertia of rotor PMSM and Flywheel, K_D is the friction coefficient, K_C is the Coulomb friction and T is the resultant torque. To obtain parameters K_D and K_C an experimental deceleration (from 1500rpm to 0rpm) test was made (Fig. 11).

From Fig. 11, the flywheel parameters K_D=0.01Nms and K_C=1.04Nms can be obtained.

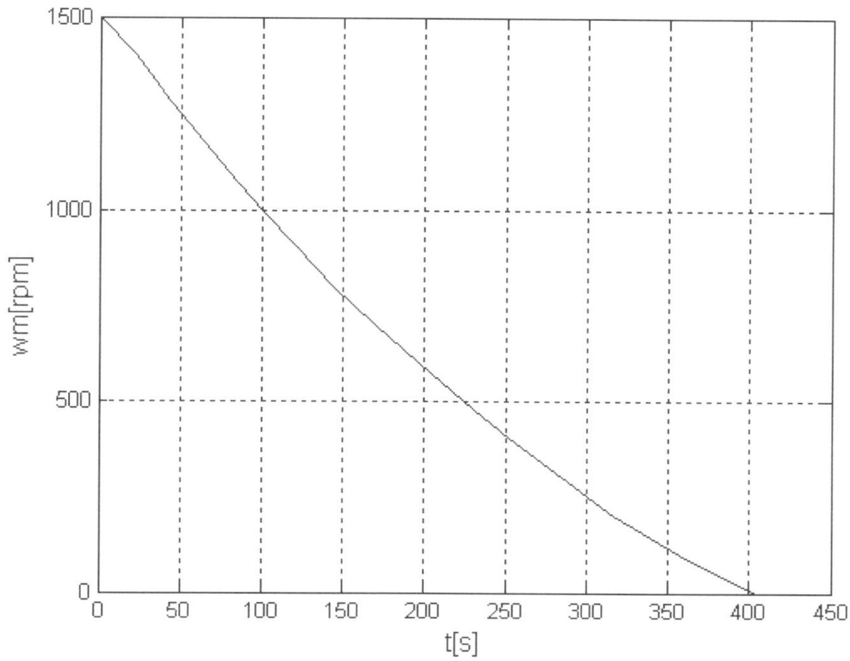

Fig. 11. Experimental results of flywheel deceleration.

In the project, design, construction and assembly in the laboratory were considered the following factors:

- Equilibrium of the moving parts for maximum angular speed of 2500rpm;
- Possibility of connecting several types of electric machines (Fig. 12);
- The vertical assembly was selected (Fig. 9);
- For security reasons, the flywheel is enclosed in another steel tube (Fig. 9);
- The wheel has a modular structure allowing easy assembly, disassembly and transportation;
- The flywheel total weight is about 350kg.

4.3 Electric drive

The Permanent Magnet Synchronous Motor (PMSM), manufacturer *Siemens*, model 1FK6063 – 6AF71 – 1AA0 (Fig. 12), has the following parameters:

- Rated speed: 3000rpm;
- Machine pole-pair number: 3
- Rated torque: 6Nm;
- Rated current: 4.7A
- Inertia of rotor: 16.1×10^{-4}kgm^2
- Torque constant: 1.39Nm/A
- Stator resistance: 0.83Ω
- Rotating field inductance: 6.5mH
- Incremental encoder with 2048ppr

Fig. 12. View of the PMSM machine.

4.4 Matrix converter

The AC-AC Matrix Converter, supplying the machine, was built using the Eupec ECONOMAC matrix module (Fig. 13.a). As an example, Fig. 13.b shows one of the IGBT's drivers. The matrix converter input filter (Fig. 5) has the following parameters: r_a=r_b=r_c=25Ω; l_a=l_b=l_c=6.5mH and C_a=C_b=C_c= 3×6.8μF. The matrix output inductor (Fig. 2) has the following parameters: R_A=R_B=R_C=0.1Ω and L_A=L_B=L_C=10mH [13].
The critical load and filter (Fig. 2) has parameters are: R_1=R_2=R_3=100Ω, L_1=L_2=L_3=10mH and C_{L1}=C_{L2}=C_{L3}=5μF.

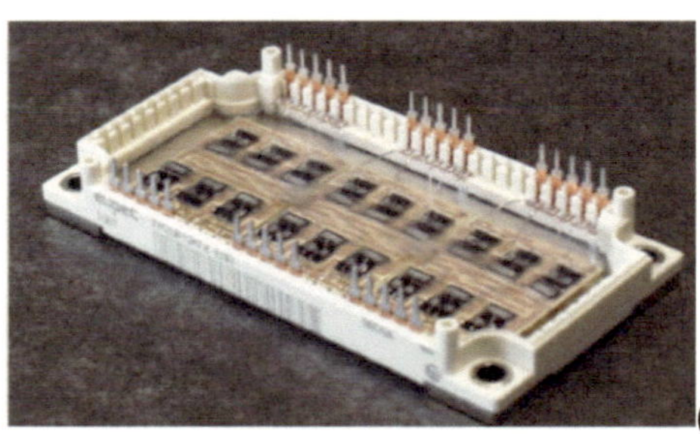

(a) (b)

Fig. 13. a) The Eupec ECONOMAC matrix module; b) View of the gate driver printed circuit board.

5. DVR simulations results

The matrix converter with input filter, PMSM, voltage source, series transformers, critical load and the real-time predictive optimal controller were simulated in the *Matlab/Simulink* environment to evaluate the performance of the proposed predictive optimal matrix converter based DVR.

The DVR with the matrix predictive controller is applied to compensate critical load voltage sags and swells. Two different tests were considered. In test 1, mitigation of balanced and unbalanced voltage sags is evaluated. In test 2 the performance of the matrix DVR is demonstrated for balanced and unbalanced voltage swells. The total harmonic distortion (THD) of the critical load voltage is also evaluated.

5.1 DVR response to voltage sags

In the first test for balanced sags, it is assumed that there is a 40% three-phase voltage sag in the supply voltage, initiated at 0.6s and lasting 0.08s. For unbalanced sags, 20% (phase 1) and 40% (phase 2) voltage dip is considered.

Fig. 14 (balanced sag v_{s123}) and Fig. 15 (unbalanced sag v_{s123}) show the result of the voltage sag compensation using the predictive optimal matrix converter control. The serial injected voltage components (v_{p123}) compensate the critical load voltage (v_{CL123}) without showing delays, voltage undershoots or overshoots. This illustrates the fast response of the predictive optimal controllers and the enhanced controllability of the matrix converter since their input-output interdependency does not disturb the critical load voltages, which are maintained balanced and at their nominal value (400V).

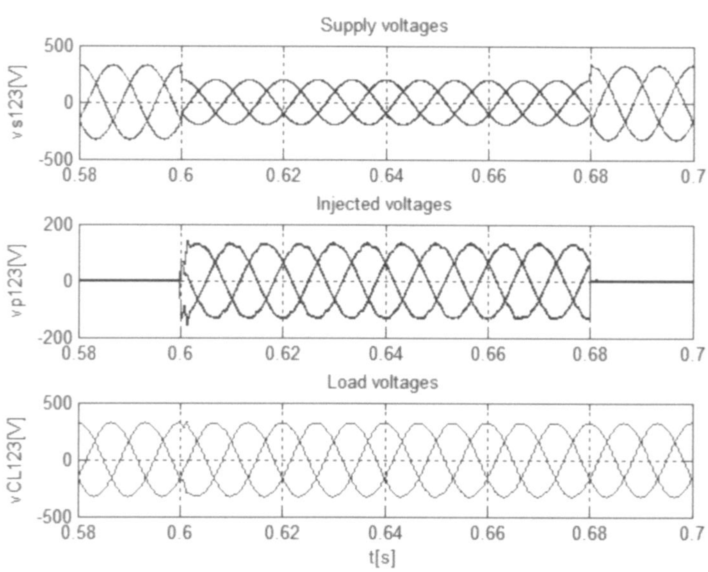

Fig. 14. Simulation result of DVR response to balanced voltage sag: Supply voltages (v_1,v_2,v_3), Injected voltage (v_{p1},v_{p2},v_{p3}) and Load voltage (v_{CL1}, v_{CL2}, v_{CL3}).

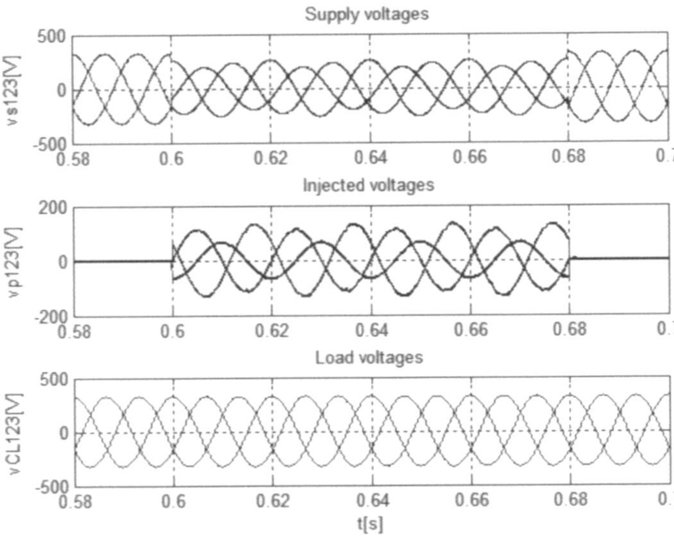

Fig. 15. Simulation result of DVR response to unbalanced voltage sag: Supply voltages (v_1,v_2,v_3), Injected voltage (v_{p1},v_{p2},v_{p3}) and Load voltage (v_{CL1}, v_{CL2}, v_{CL3}).

5.2 DVR response to voltage swells

In the second test, the DVR performance for a voltage swell condition is investigated. A balanced voltage swell with 40% three-phase voltage increase, which starts at 0.6s and ends at 0.68s, is considered. For unbalanced swells, 50% (phase 1) and 20% (phase 2) voltage swell is considered. The performance of DVR is illustrated in Fig. 16 (balanced swell) and Fig. 17 (unbalanced swell).

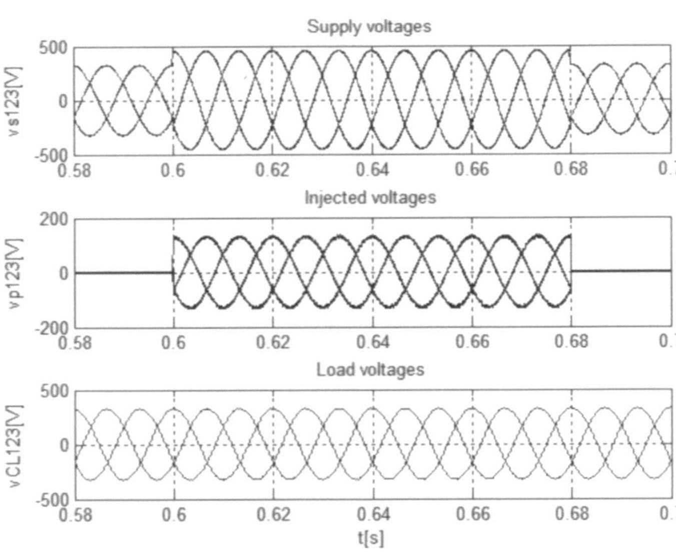

Fig. 16. Simulation result of DVR response to balanced voltage swell: Supply voltages (v_1, v_2, v_3), Injected voltage (v_{p1},v_{p2},v_{p3}) and Load voltage (v_{CL1}, v_{CL2}, v_{CL3}).

Again, the DVR injected voltage components (v_{p123}) compensate the critical load voltage (v_{CL123}) without showing any delays, voltage undershoots or overshoots. The DVR is able to correct the voltage swells showing response times far lower than the voltage supply period.

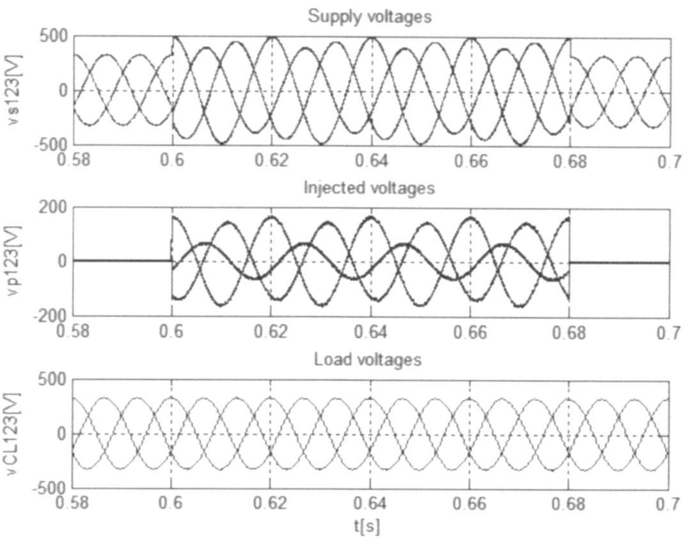

Fig. 17. Simulation result of DVR response to unbalanced voltage swell: Supply voltages (v_1, v_2, v_3), Injected voltage (v_{p1},v_{p2}, v_{p3}) and load voltage (v_{CL1}, v_{CL2}, v_{CL3}).

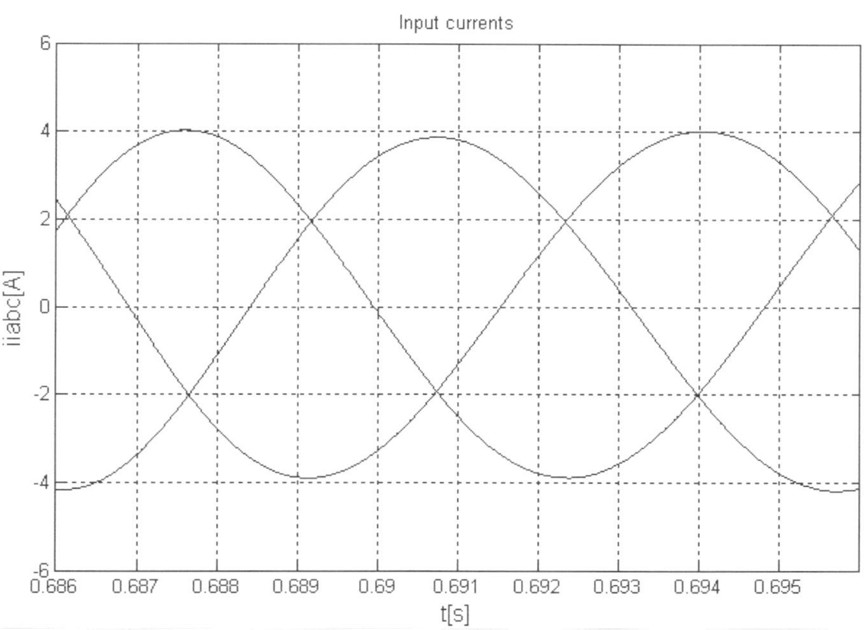

Fig. 18. Matrix input currents (i_{ia}, i_{ib}, i_{ic}).

Fig. 18 present the matrix input currents (i_{ia}, i_{ib}, i_{ic}). Simulations show near sinusoidal input currents.

5.3 Critical load voltage THD

Fig. 19 presents the frequency spectrum of the critical load voltage. Besides the tracking capabilities of the predictive optimal control method, it presents a very useful characteristic, the very low harmonic content of critical load voltage. The spectrum shows the fundamental frequency at 50Hz (100%) and some very low amplitude (<1%) harmonics. The critical load voltage THD is less than 1%.

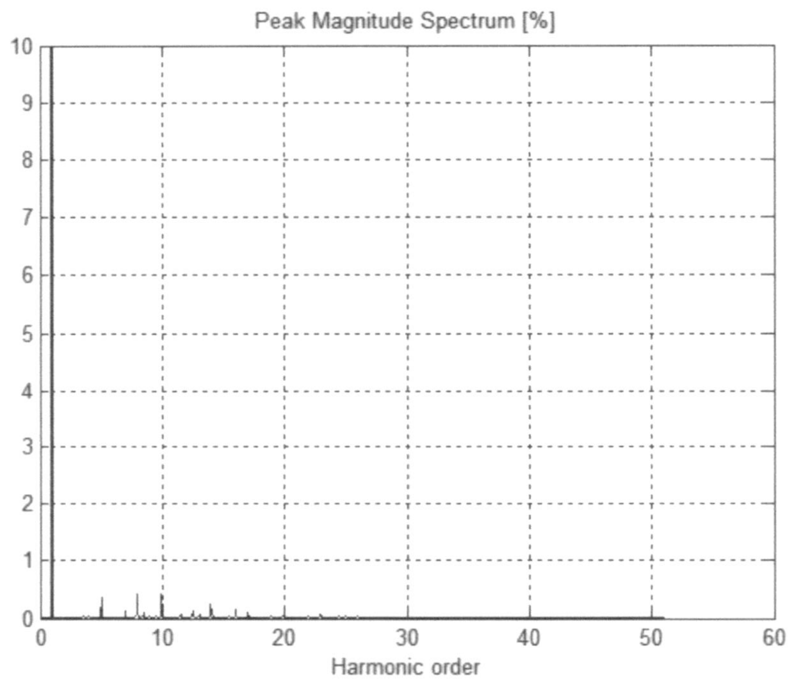

Fig. 19. Load voltage spectrum (v_{CL1}).

6. Conclusions

This Chapter has shown that predictive optimal control is a powerful control technique that will open new possibilities in the control of AC-AC matrix and other advanced power converters, as well as in the control of matrix driven permanent magnet synchronous motors. The predictive optimal control allows the control designer to overcome the matrix converter input/output coupling, easing the use of all the available converter vectors.

The presented predictive optimal control does not need any modulator. However, it achieves a very effective control of input and output currents and rotor speed. Therefore, the proposed technique is a good alternative for established control methods like hysteresis current control, sliding mode control and PWM control.

A flywheel energy storage system, including a Permanent Magnet Synchronous Machine driven by the predictive optimal controlled matrix converter was used as a DVR to mitigate voltage sags, swells and THD. It was pointed out that the proposed predictive optimal matrix converter control presents very good performance when used in dynamic voltage restorers to mitigate voltage sags, swells and THD at critical loads, without showing delays, voltage undershoots or overshoots.

7. References

[1] R. Hebner, J. Beno, A. Walls (2006). "Flywheel batteries come around again", IEEE Spectrum, April 2002, pp. 46-51.

[2] T. Siostrzonek, S. Piróg (2006). "*The Flywheel Energy Storage with Brushless DC Motor - the Practical Results*", EPE-PEMC 2006 - September 2006, Portoroz, Slovenia.

[3] B. Wang, G. Venkataramanan (2007). "*Dynamic Voltage Restorer Utilizing a Matrix Converter and Flywheel Energy Storage*", Industry Applications Conference, 2007. 42nd IAS Annual Meeting. Conference Record of the 2007 IEEE, Volume , Issue , 23-27 Sept. 2007 Page(s):208–215.

[4] J. Silva, B. Borges,. Anunciada (1991). "*Improving Control Strategies for HF Resonant Link Converters: the Current Mode Predictive Modulator*", Power Electronics Specialists Conference, IEEE PESC'91, ISSN 0275-9306, pp 268-274, USA, June 1991.

[5] J. Rodriguez, J. Pontt, C. Silva, P. Cortés, and S. Rees (2004). "*Predictive Current Control of a Voltage Source Inverter*", 35th Annual IEEE Power Electronics Specialists Conference. Aachen, Germany, June 2004.

[6] J. Barros, J. Silva, S. Pinto (2005). "*Real - Time Based Optimal Controller For Four Multilevel Converters*", Proc. IEEE - CEE'05 - International Conference on Electrical Engineering - CEE05, CD-ROM paper 3w55, October 2005, Coimbra, Portugal.

[7] E. Wiechmann, A. Garcia, L. Salazar and J. Rodriguez (1997). "*High-Performance Direct-Frequency Converters Controlled by Predictive-Current Loop*", IEEE Transactions on Power Electronics, Vol. 12, No. 3, May 1997.

[8] P. Gambôa, S. Pinto, J. Silva, E. Margato (2007). "*Predictive Optimal Control of Unity-Power-Factor Matrix Converters used in Field Oriented Controlled Induction Motor Drives*", International Symposium on Industrial Electronics, Vigo, Spain, 4-7 June 2007.

[9] P. Gambôa, S. Pinto, J. Silva, E. Margato (2008). "*A Flywheel Energy Storage System with Matrix Converter Controlled Permanent Magnet Synchronous Motor*", ICEM08-XVIII International Conference on Electrical Machines, Vilamoura, Portugal, 6-9 September 2008.

[10] L. Huber, D. Borojevic, N. Burany, "*Analysis, Design and Implementation of the Space-Vector Modulator for Forced-Commutated Cycloconverters*" IEE Proceedings-B, Vol. 139, no 2, March 1992, pp. 103-113.

[11] P. Gambôa, S. Pinto, J. Silva, E. Margato (2009). "*Predictive Optimal Matrix Converter Control for a Dynamic Voltage Restorer with Flywheel Energy Storage*", The 35th Annual

Conference of the IEEE Industrial Electronics Society, Porto, Portugal, 3-5 November 2009.

[12] Barros, J.D.; Silva, J.F.; "*Multilevel Optimal Predictive Dynamic Voltage Restorer*", Industrial Electronics, IEEE Transactions on, Aug. 2010, Volume: 57 Issue:8, pages: 2747 - 2760, ISSN: 0278-0046.

[13] S. F. Pinto, J. F. Silva, "*Input Filter Design of a Mains Connected Matrix Converter*", Session on Power Electronics Mitigation Methods for Power Quality, IEEE-Power Engineering Society International Conference on Harmonics and Quality of Power, October 2006, Cascais, Portugal.

19

Unified Power Flow Controllers Without Energy Storage: Designing Power Controllers for the Matrix Converter Solution

Joaquim Monteiro[1], J. Fernando Silva[2], Sónia Pinto[2] and João Palma[3]
[1]*Cie3 and ISEL – Polytechnic Institute of Lisbon,*
[2]*Cie3 and IST – Technical University of Lisbon,*
[3]*SIC – National Laboratory for Civil Engineering,*
Portugal

1. Introduction

In the last years the growing economic, environmental and social concerns have increased the difficulty to use fossil fuels, as well as to obtain new licenses to build either transmission lines (right-of-way) or high power facilities. This led to the continuous growth of decentralized electricity generation (using renewable energy resources) (Hingorani, 2000). This scenario has introduced new problems and technical challenges to power systems researchers and electricity markets participants. One of the main consequences of these changes has been the substantial increase of power transfer within transmission networks, approaching their rated capacity and requiring adequate control capability to supply the continuously growing demand of electric power.

To solve these issues Flexible AC Transmission Systems (FACTS) became a well known power electronics based solution to control power flow in transmission lines. These systems are switching controlled converters that operate in real time increasing the transmission lines power flow capacity up to their thermal limits. Currently, Unified Power Flow Controllers (UPFC) are the most versatile and complex FACTS enabling accurate and reliable control of both active and reactive power flow over networks, through load sharing between alternative line paths (Song et al , 1999).

The original UPFC concept was proposed by L. Gyugyi (Gyugyi, 1992), and consisted of the combination of a Static Synchronous Compensator (STATCOM) and a Static Synchronous Series Compensator (SSSC) connected by a common DC link, using large high-voltage DC storage capacitors. The AC converters sides of these compensators are connected to a transmission line, through coupling transformers, in shunt and series connection with the line. This arrangement operates as an ideal reversible AC-AC switching power converter allowing shunt and series compensation and bidirectional power flow, between the AC terminals of the two converters.

The DC capacitor bank used in the UPFC topology to link the two back-to-back converters increases the UPFC weight, cost, occupied area and introduces additional losses. Replacing the double three-phase inverter by one three phase matrix converter the DC link capacitors are eliminated, reducing costs, size, maintenance, and increasing reliability and lifetime. The

AC-AC matrix converter, also known as all silicon converter, processes the energy directly without large energy storage needs, allows bi-directional power flow, while it guarantees input and output sinusoidal voltages and currents with variable amplitude and frequency and adjustable power factor. On the other way, the matrix converter control is more complex than the control of the back-to-back converter.

Over the years the interesting properties presented by the matrix converter pushed the design of their controllers, so that matrix converters are being used quite successfully in many industrial applications, such as in power sources for electrical drives with variable speed (Matsuo et al, 1996), in applications related to power quality enhancement in the electrical grid (Galkin et al, 2001), in renewable power supply systems (Nikkhjoei et al, 2005) and also in the compensation of harmonics in power network as dynamic voltage restorers (DVR) (Wang et al, 2009).

In general, the conventional control methods of UPFCs are based on power systems linearized models, valid around an operating point. Usually, these linearized models do not guarantee robustness and insensitivity to the parameters and may give rise to poor dynamic response and/or undesired instability, since most of these controllers do not have the capacity to adapt to nonlinearities or continuously changing dynamics of the power system (Monteiro et al, 2005), (Liu et al, 2007). In addition, many of the control strategies used in the UPFC are based in proportional integral controllers obtained from its dynamic model in *dq* coordinates to improve performance and reduce the interaction between the control of active and reactive power (Round et al, 1996).

In this chapter, the use of a UPFC without energy storage, based on a matrix converter topology, is proposed to control the active and reactive power flow in the transmission line (section 2). Decoupled controllers (Verveckken et al, 2007) using the inverse dynamics linearization approach are proposed for active and reactive power control. These controllers allow the elimination of the cross-coupling effect between active and reactive power controllers and fast response (section 3). The designed controllers are implemented using digital signal processing (DSP) hardware together with a matrix converter prototype and laboratory equipment to emulate the power network (section 4). The dynamic and steady-state performance of the proposed power control methods are evaluated both by simulation and by experimental results (section 5). Finally, conclusions are listed regarding the behaviour of the overall matrix converter based UPFC when operated with the proposed active and reactive power controllers (section 6).

2. Modelling of UPFC power system

A simplified power transmission network using the proposed matrix converter based UPFC is presented in Fig. 1. In this scheme V_S and V_R are, respectively, the sending-end and receiving-end sinusoidal voltages of the G_S and G_R generators feeding load Z_L consisting of a resistance R_L and an inductance L_L. The matrix converter is connected to transmission line 2, represented as a series inductance and resistance (L_{L2}, R_{L2}), through coupling transformers, T_1 in the shunt side and T_2 in the series side.

A detailed diagram of the UPFC system showing the connection of the matrix converter to the transmission line, in Fig. 2, includes three-phase shunt input transformer (T_a, T_b, T_c), three-phase series output transformer (T_A, T_B, T_C) and a three-phase matrix converter. In this diagram, the three-phase *LCr* input low pass filter is required to re-establish a voltage–source boundary to the matrix converter, also enabling smooth input currents.

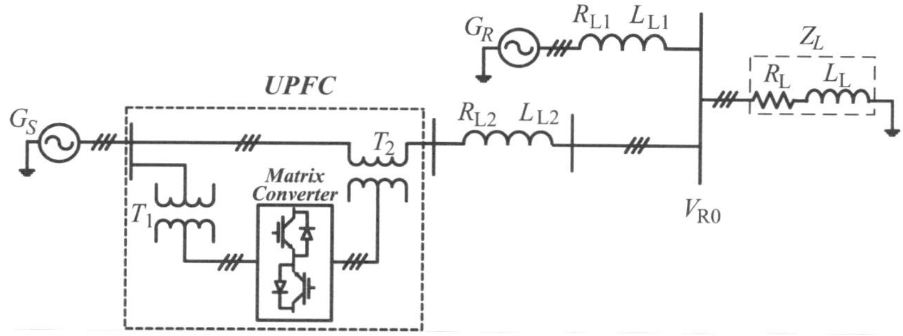

Fig. 1. A transmission network with a matrix converter UPFC.

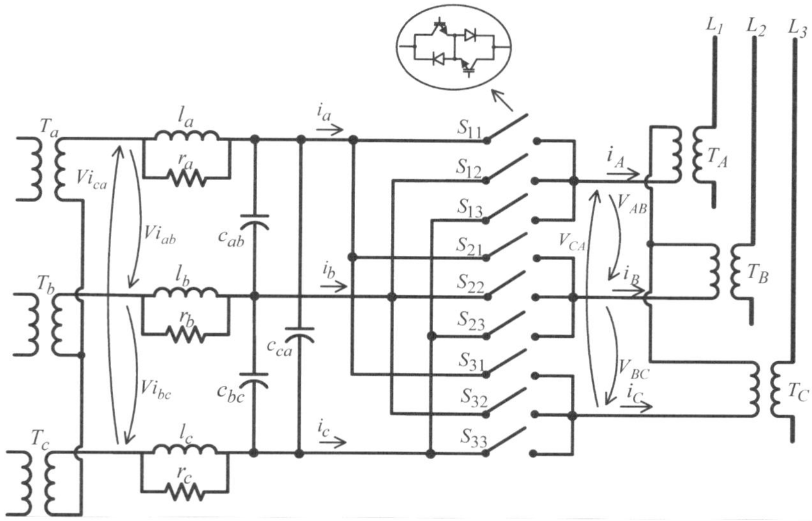

Fig. 2. Detailed matrix converter based UPFC.

The next subsections will detail the matrix converter and the UPFC dynamic model.

2.1 Matrix converter model

The matrix converter UPFC system (Fig. 2) modelling assumes ideal voltage sources and ideal shunt and series transformers. Considering also ideal power semiconductors, each matrix converter bi-directional switch S_{kj} ($k, j \in \{1,2,3\}$) can only have two possible states: "$S_{kj}=1$" if the switch is closed or "$S_{kj}=0$" if the switch is open. The nine switches of the matrix converter can be represented as a 3x3 matrix (1):

$$S = \begin{bmatrix} S_{11} & S_{12} & S_{13} \\ S_{21} & S_{22} & S_{23} \\ S_{31} & S_{32} & S_{33} \end{bmatrix} \tag{1}$$

The matrix converter topological constraints imply that $\sum_{j=1}^{3} S_{kj} = 1$, for all $k \in \{1,2,3\}$.

According to (1), the relationship between load and input voltages can be expressed as:

$$\begin{bmatrix} v_A & v_B & v_C \end{bmatrix}^T = S \begin{bmatrix} v_a & v_b & v_c \end{bmatrix}^T \tag{2}$$

Using the transpose of matrix S, the input phase currents can be related to the output phase currents, by (3).

$$\begin{bmatrix} i_a & i_b & i_c \end{bmatrix}^T = S^{\,T} \begin{bmatrix} i_A & i_B & i_C \end{bmatrix}^T \tag{3}$$

From the 27 possible switching patterns (3^3), time variant vectors can be obtained (Pinto et al, 2001), representing the output voltages and input currents in $\alpha\beta$ coordinates.
The command of the matrix converter switches can be accomplished using a Venturini based high frequency PWM modulator (Alesina et al, 1981) , (Wheeler, 2002) (4).

$$\begin{aligned} m_{kj}(t) = 1/3 + (1/3)\Big\{ & \left(v_k v_j\right)/V_i^2 + \\ & + \left(4/3\sqrt{3}\right) q\, sen\left(\omega_i t - \beta_k\right) sen\left(3\omega_i t\right)\Big\} \end{aligned} \tag{4}$$

$$k = a, b, c;\; j = A, B, C;\; q = V_o/V_i;\; \beta_k = 0, 2\pi/3, 4\pi/3$$

This PWM method yields near sinusoidal output voltages with amplitude defined by an active power controller and phase defined by a reactive power controller, as well as almost sinusoidal input currents with near unity input power factor, if needed.

2.2 UPFC dynamic model

The scheme presented in Fig. 3 shows the simplified three-phase equivalent circuit of matrix UPFC transmission system model. For dynamic system modelling, the power sources and the coupling transformers are all considered ideal, including the matrix converter represented as a controllable voltage source, with amplitude V_C and phase ρ.
In this circuit L_2 and R_2 are, respectively, the Thévenin equivalent inductance and resistance calculated by: $L_2 = L_{L2} + L_{L1} // L_L$ and $R_2 = R_{L2} + R_{L1} // R_L$. Besides, V_{R0} is the voltage at the load bus.

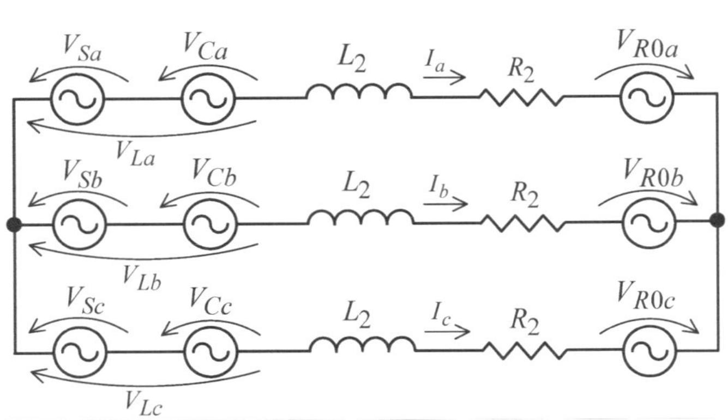

Fig. 3. Three phase equivalent circuit of matrix UPFC and transmission line.

Considering a symmetrical and balanced three phase system and applying Kirchhoff laws to the three phase equivalent circuit in Fig. 3, the dynamic equations of AC line currents are obtained in *dq* coordinates as follows:

$$\frac{dI_d}{dt} = \omega I_q - \frac{R_2}{L_2} I_d + \frac{1}{L_2}\left(V_{Ld} - V_{R0d}\right) \tag{5}$$

$$\frac{dI_q}{dt} = -\omega I_d - \frac{R_2}{L_2} I_q + \frac{1}{L_2}\left(V_{Lq} - V_{R0q}\right) \tag{6}$$

In the last equations, V_{Ld} and V_{Lq} are introduced for notation simplicity and are given by: $V_{Ld}=V_{Sd}+V_{Cd}$ and $V_{Lq}=V_{Sq}+V_{Cq}$.
Applying Laplace transform to system transfer functions (5) and (6), equation (7) is obtained:

$$\left(s + \frac{R_2}{L_2}\right)\begin{bmatrix} I_d \\ I_q \end{bmatrix} = \omega \begin{bmatrix} I_q \\ -I_d \end{bmatrix} + \frac{1}{L_2}\begin{bmatrix} V_{Ld} - V_{R0d} \\ V_{Lq} - V_{R0q} \end{bmatrix} \tag{7}$$

Solving (7) in order to the line currents I_d I_q, equation (8) may be obtained as a function of voltages V_{Ld}, V_{Lq} and V_{R0d}, V_{R0q} at the receiving end.

$$\begin{bmatrix} I_d \\ I_q \end{bmatrix} = \frac{\frac{1}{L_2}\begin{bmatrix} s + \frac{R_2}{L_2} & \omega \\ -\omega & s + \frac{R_2}{L_2} \end{bmatrix}}{\left(s + \frac{R_2}{L_2}\right)^2 + \omega^2}\begin{bmatrix} V_{Ld} - V_{R0d} \\ V_{Lq} - V_{R0q} \end{bmatrix} \tag{8}$$

Active and reactive powers of sending end generator are given in *dq* coordinates by (9):

$$\begin{bmatrix} P \\ Q \end{bmatrix} = \begin{bmatrix} V_{Sd} & V_{Sq} \\ V_{Sq} & -V_{Sd} \end{bmatrix}\begin{bmatrix} I_d \\ I_q \end{bmatrix} \tag{9}$$

The active and reactive power controllers will be designed based on the previous equations.

3. Designing active and reactive power controllers

In this chapter, new linear controllers will be derived in *dq* coordinates to guarantee no cross-coupling between active and reactive power controllers and fast response, using inverse dynamics linearization. The synthesis of these controllers is also based on a modified Venturini high frequency modulator.

3.1 Matrix converter UPFC controllers design by inverse dynamics linearization

The dynamic equations of UPFC model (8) show that there is no dynamics related to the power sources voltages, which are considered ideal. So in *dq* Laplace domain the power sources voltages are constant. Assuming both V_{R0d} and V_{Sd} as constant and a rotating reference frame synchronized to the V_S source so that $V_{Sq}=0$, active and reactive power P and Q will be obtained by (10) and (11).

$$P = V_{Sd} I_d \tag{10}$$

$$Q = -V_{Sd} I_q \tag{11}$$

The synthesis of active and reactive power controllers is obtained substituting the previously calculated currents in dq coordinates (8) on (10) and (11). Active and reactive powers are obtained as a function of transmission line parameters, load bus and sources voltages.

$$\begin{aligned} P &= V_{Sd} \frac{(sL_2 + R_2)(V_{Sd} - V_{R0d}) - \omega L_2 V_{R0q}}{(sL_2 + R_2)^2 + (\omega L_2)^2} \\ &+ V_{Sd} \frac{(sL_2 + R_2)V_{Cd} + \omega L_2 V_{Cq}}{(sL_2 + R_2)^2 + (\omega L_2)^2} \end{aligned} \tag{12}$$

$$\begin{aligned} Q &= V_{Sd} \frac{(sL_2 + R_2)V_{R0q} + \omega L_2 (V_{Sd} - V_{R0d})}{(sL_2 + R_2)^2 + (\omega L_2)^2} \\ &+ V_{Sd} \frac{\omega L_2 V_{Cd} - (sL_2 + R_2)V_{Cq}}{(sL_2 + R_2)^2 + (\omega L_2)^2} \end{aligned} \tag{13}$$

Both active and reactive powers, obtained respectively by equations (12) and (13), consist of an uncontrollable constant part (P_i, Q_i) based on sending end power source voltages and line impedance, and a controllable dynamic part (ΔP, ΔQ) determined by the matrix converter voltages. These relationships are presented in (14) and (15).

$$P = P_i + \Delta P \tag{14}$$

$$Q = Q_i + \Delta Q \tag{15}$$

From (12) and (13), the controllable part of steady-state active and reactive power (ΔP, ΔQ) can be obtained expressed as a function of matrix converter voltages in dq coordinates according to (16).

$$\begin{bmatrix} \Delta P \\ \Delta Q \end{bmatrix} = \frac{V_{Sd}}{(sL_2 + R_2)^2 + (\omega L_2)^2} \begin{bmatrix} sL_2 + R_2 & \omega L_2 \\ \omega L_2 & -(sL_2 + R_2) \end{bmatrix} \cdot \begin{bmatrix} V_{Cd} \\ V_{Cq} \end{bmatrix} \tag{16}$$

The controllable part of the active and reactive power flow components (16) may be written as in (17), introducing a matrix G_C for notation simplicity.

$$\begin{bmatrix} \Delta P \\ \Delta Q \end{bmatrix} = -\frac{V_{Sd}}{\det[G_C]} G_C \begin{bmatrix} V_{Cd} \\ V_{Cq} \end{bmatrix} \tag{17}$$

In (17) matrix G_C depends on the transmission line parameters (18) and its determinant is $\det\left[G_C\right]=-\left(sL_2+R_2\right)^2-\left(\omega L_2\right)^2$.

$$G_C=\begin{bmatrix} sL_2+R_2 & \omega L_2 \\ \omega L_2 & -\left(sL_2+R_2\right) \end{bmatrix} \tag{18}$$

The next section will present the controllers design based on the inverse dynamics linearization of the power system model.

3.2 Inverse dynamic linearization of the power system model

The proposed power controllers design use the inverse model of the power system to linearize and decouple active and reactive power control, calculating the control signals V_{Cd} and V_{Cq} as a function of the active and reactive power flow components ΔP, ΔQ. Knowing that $G_C=-\left(\det\left[G_C\right]\right).G_C^{-1}$, equation (19) is obtained.

$$\begin{bmatrix} V_{Cd} \\ V_{Cq} \end{bmatrix}=\frac{1}{V_{Sd}}G_C\begin{bmatrix} \Delta P \\ \Delta Q \end{bmatrix} \tag{19}$$

Considering a feedback loop controller topology for the power components, using the inverse system model (19) to design the controllers, and adding a linear integral controller to obtain zero static error, two decoupled equivalent systems with a first order system behavior with time constant T_P (being $[\Delta P\ \Delta Q]^T=1/(sT_P+1)[\Delta P_{ref}\ \Delta Q_{ref}]^T$ can be obtained using (20).

$$\begin{bmatrix} \Delta P \\ \Delta Q \end{bmatrix}=\frac{1}{sT_p}\begin{bmatrix} \Delta P_{ref}-\Delta P \\ \Delta Q_{ref}-\Delta Q \end{bmatrix} \tag{20}$$

Multiplying both members of equation (20) by G_C/V_{Sd} control variables V_{Cd} and V_{Cq} are then obtained in (21) as independent functions of the active and reactive power errors, respectively.

$$\begin{bmatrix} V_{Cd} \\ V_{Cq} \end{bmatrix}=\frac{G_C}{V_{Sd}}\frac{1}{sT_p}\begin{bmatrix} \Delta P_{ref}-\Delta P \\ \Delta Q_{ref}-\Delta Q \end{bmatrix} \tag{21}$$

Based on the previous equations, the block diagram of Fig. 4 is obtained, representing the closed loop control system with decoupled active and reactive powers.

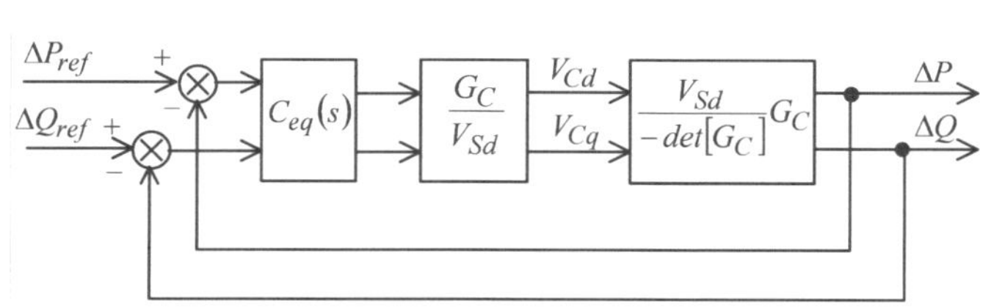

Fig. 4. Block diagram in closed loop control with active and reactive power decoupling.

Substituting in (21) the matrix G_C, an overall control system with decoupled controllers for active and reactive power is obtained (22).

$$\begin{bmatrix} V_{Cd} \\ V_{Cq} \end{bmatrix} = \frac{1}{V_{Sd}} \begin{bmatrix} \frac{sL_2 + R_2}{sT_p} & \frac{\omega L_2}{sT_p} \\ \frac{\omega L_2}{sT_p} & -\frac{sL_2 + R_2}{sT_p} \end{bmatrix} \begin{bmatrix} \Delta P_{ref} - \Delta P \\ \Delta Q_{ref} - \Delta Q \end{bmatrix} \tag{22}$$

The overall block diagram of the active and reactive power closed loop controllers, shown in Fig. 5, uses proportional integral controllers and integrators to generate the control variables (V_{Cd}, V_{Cq}) applied to the matrix converter modulator.

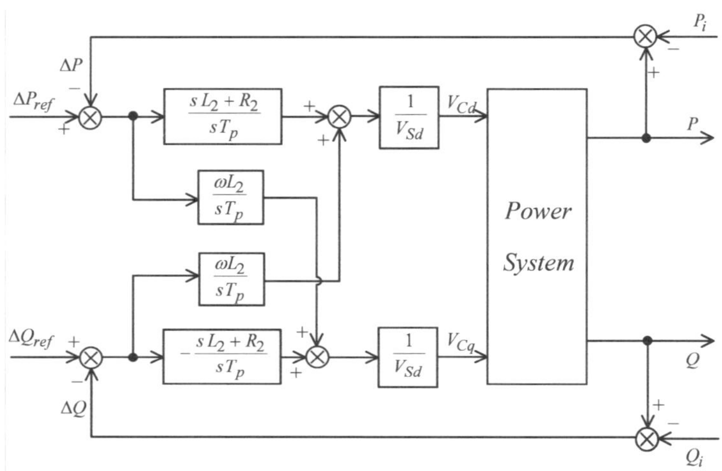

Fig. 5. The overall block diagram in closed loop control with controllers and power system.

This global diagram has also a power system block representing the sending end and receiving end voltage sources, the transmission lines, a three phase load and a three phase matrix converter connected to the transmission line through series and parallel power transformers.

4. Implementation of the power controllers

The implementation scheme of the active and reactive power controllers is shown on Fig. 6. This diagram presents the voltage sources (V_S and V_R) the transmission lines with resistance and inductance, an inductive three phase load and a three phase matrix converter connected to the line through power transformers (T_1, T_2). A digital signal processing (DSP) was used to implement the designed linear controllers for matrix vector selection at successive time steps.
To achieve safe commutation between matrix converter bidirectional switches, the four-step output current strategy was used (Huber et al, 1992). This commutation process was implemented in a field programmable gate array (FPGA) included in Fig. 6.
As shown in the previous diagram, the control of the instantaneous active and reactive power componets requires the measurement of G_S voltages, input currents and output currents, allowing the compensation of the active and reactive power errors, in real time.

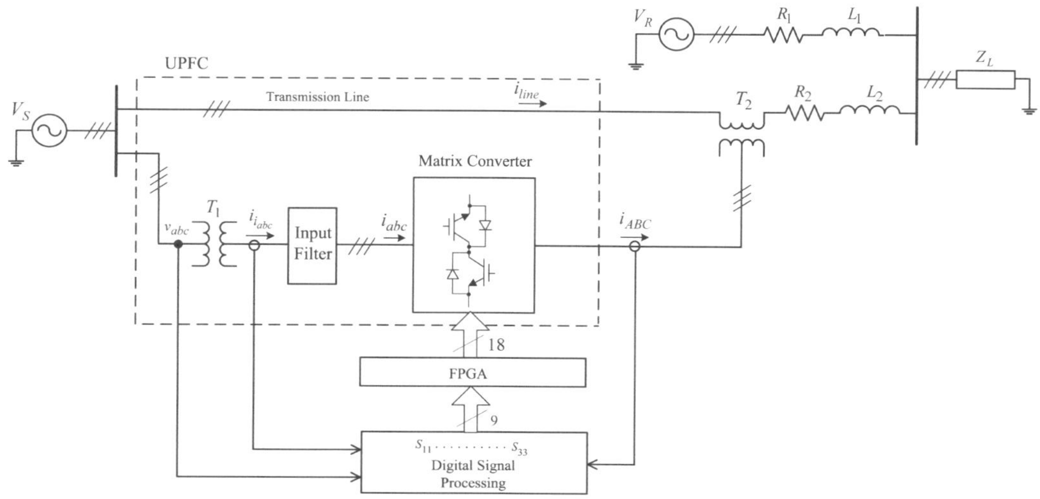

Fig. 6. Control scheme for the proposed method of a matrix converter operating as UPFC.

Based on this approach, at each time step the control system chooses the most suitable matrix vector upon the available set of discrete values corresponding to the set of acceptable switching states of all power semiconductors of the matrix converter.

5. Simulation and experimental results

In this section the dynamic and steady-state performance of the proposed active and reactive power control method is evaluated and discussed using detailed simulations and experimental implementation. A detailed simulation model representing matrix converter, series and shunt transformers, sources, transmission lines and additional blocks to simulate the control system is designed using MATLAB/Simulink and SimPowerSystems, being the matrix converter semiconductors represented as ideal switches.

Simulations are experimentally validated using a 3 kW prototype matrix converter, built with integrated power semiconductor modules, each one with six 1200V, 25A insulted gate bipolar transistors (IGBT) with anti-parallel diode in a common collector arrangement. The IGBT gates are driven by optically isolated drives (TLP250). The matrix converter input second order filter uses $l = 4.2\,mH$, $C = 6.6\,\mu F$, $r = 25\,\Omega$.

This prototype was connected to the laboratory low voltage network operating as UPFC (Fig. 6) with an input phase-to-phase matrix converter voltage equal to 120 V_{RMS}. Load power is 1.5 kW and the transmission lines 1 and 2 are emulated by inductances $L_1 = 12\,mH$, $L_2 = 15\,mH$ and series resistances $R_1 = R_2 = 0.2\,\Omega$, respectively, for lines 1 and 2. Current and voltage sensors were used respectively to measure matrix input and output currents (LEM LA25NP) and to measure the power network phase to phase voltages (LEM LV 25-P). Designed controllers were implemented using a DSP PowerPC board (dSPACE) and a FPGA from Xillinx for driving the commutation process.

Experimental and simulation results for the active and reactive power UPFC controllers using the proposed control method based in the linearized inverse dynamics were sized using time constant T_p=0.1ms, and T_s approximately equal to 23 μs.

Fig. 7 shows simulation (a) and experimental (b) results for the proposed controllers considering active and reactive power step response ($\Delta P_{ref} = +0.4\ pu$, $\Delta Q_{ref} = +0.2\ pu$) and initial reference values: P_{ref} =0.4 *pu*, Q_{ref} =0.2 *pu*. Results show decoupled active and reactive powers, fast response and a small ripple in steady state.

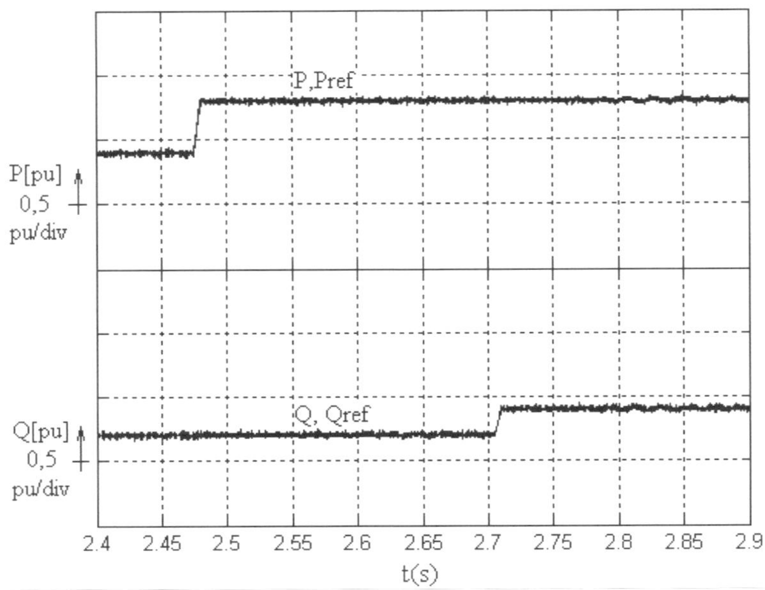

a)

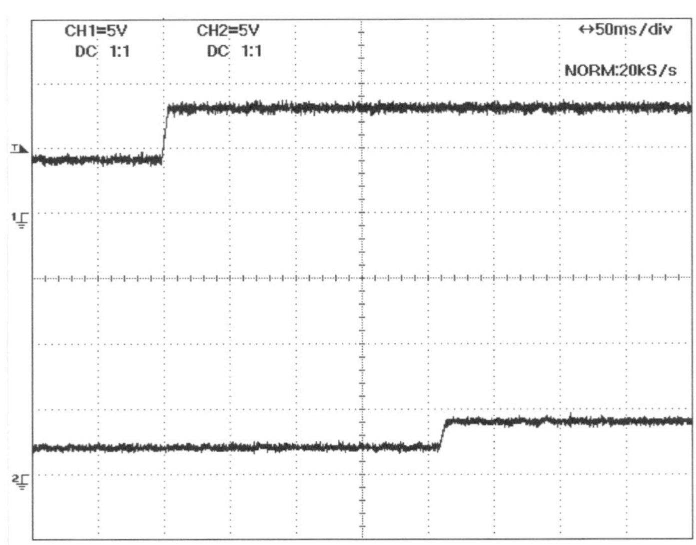

b)

Fig. 7. Active and reactive power response for a P and Q step when $\Delta P_{ref} = +0.4\ pu$ and $\Delta Q_{ref} = +0.2\ pu$. a) Simulation results. b) Experimental results.

Simulation and experimental results of Fig 8 show steady-state line and matrix converter input currents using the controllers proposed for P_{ref} =0.4 *pu*, Q_{ref} =0.2 *pu*. The results show that currents are almost sinusoidal with small ripple content.

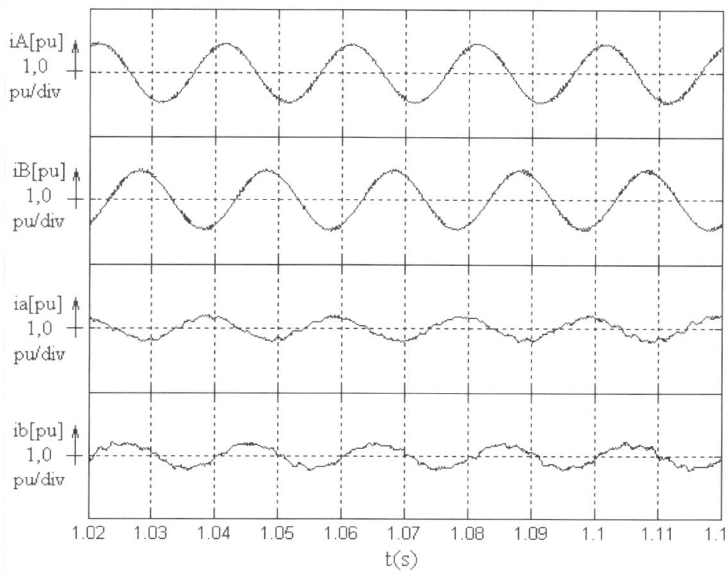

a)

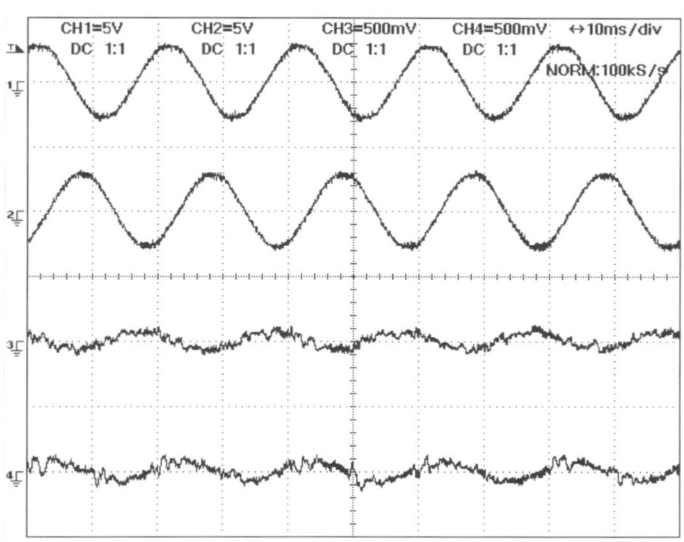

b)

Fig. 8. Line currents (i_A, i_B) and input matrix converter currents (i_b, i_c) for $P_{ref} = 0.4\ pu$ and $Q_{ref} = 0.2\ pu$. a) Simulation results. b) Experimental results.

The experimental power spectral density of transmission line and matrix converter current for the proposed controllers, in Fig. 9, shows that the main harmonics are nearly 33 dB below the 50Hz fundamental for the line current, and 20 dB below the 50Hz fundamental for the matrix converter current.

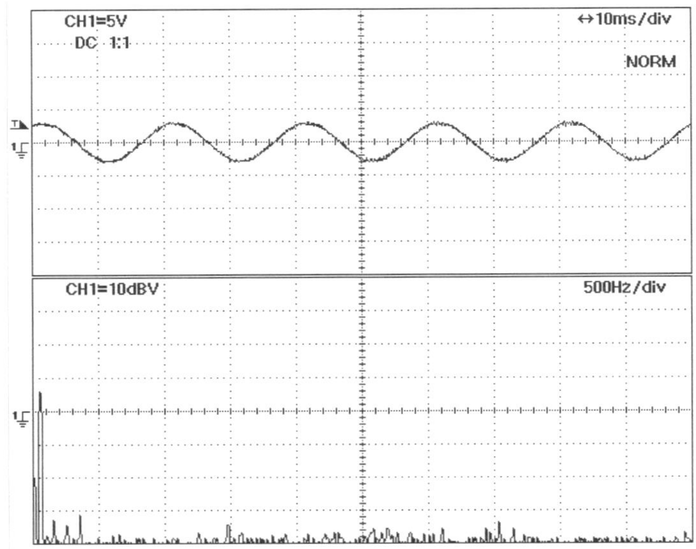

a)

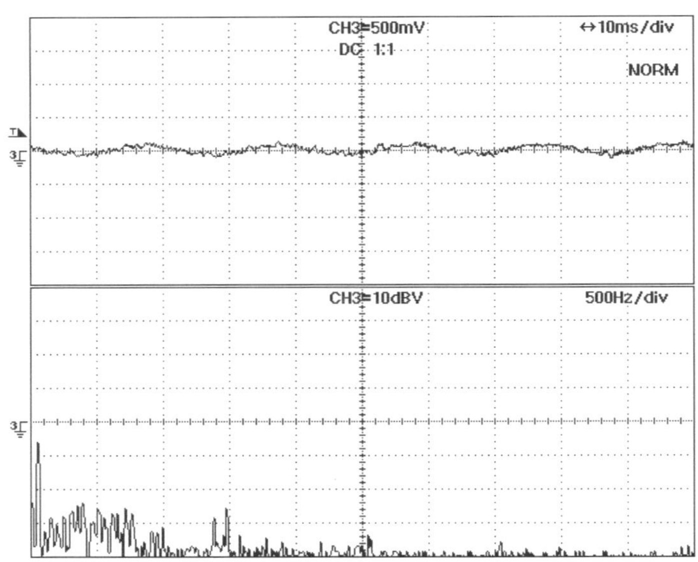

b)

Fig. 9. a) Power spectral density of line current (i_A). b) Power spectral density of input matrix current (i_b).

6. Conclusions

In this chapter, after a brief discussion about the use of static power converters for power conditioning and quality improvement within electric power networks, an application was proposed for a three-phase matrix converter based UPFC without energy storage, exploiting the ease of interaction of this converter with the electrical grid, and their benefits in replacing the classical topologies with associations of back-to-back converters. For this new UPFC topology, dynamic models of matrix converter based UPFC in *dq* coordinates were obtained. Despite its control complexity, matrix converters have advantages (weight, size and maintenance) and present good performance as power flow controllers in the electrical power grid in comparison to back-to-back converter topologies.

The solution here proposed uses inverse dynamics decoupled linear controllers based in control algorithms applied to a multivariable system, and have the ability to compensate the cross-terms existing in the control loop.

The controller design was validated using simulation and experiments. Results presented in this chapter show that active and reactive components of power flow can be advantageously controlled using the proposed matrix converter based UPFC.

The design of the inverse dynamics decoupled linear controllers is dependent on the transmission line parameters, the controllers being sensitive to the system parameters. Nevertheless, simulation and experimental results show that the active and reactive power controllers, as designed, exhibit almost no cross-coupling between active and reactive power and guarantee fast response, as well as very little sensitivity to system parameters.

7. References

Alesina A. and Venturini, M. G. B. "Solid-state power conversion: A Fourier analysis approach to generalized transformer synthesis" IEEE Trans. Circuits Systems, vol. CAS-28, pp. 319–330, Apr. 1981.

Galkin, I.; Ribickis, L.; "Control of Reactive Power by means of Matrix Converters"; Proc. EPE'01 Conf., Graz, Lausanne, Aug. 2001.

Gyugyi, L., "Unified Power Flow Control Concept for Flexible AC Transmission Systems" IEE PROCEEDINGS-C, vol. 139, No 4, Jul. 1992.

Hingorani N.; Gyugyi, L.; Understanding FACTS – Concepts and Technology of Flexible AC Transmission Systems, IEEE Press, A John Wiley &Sons Inc., Publication, 2000.

Huber L., Borojevic, D. and Burany N., "Analysis design and implementation of the space-vector modulator for forced-commutated cycloconvertors", IEE Proceedings-B Vol. 139 No.2, March 1992, pp 103-113.

Liu, L.; Zhu, P.; Kang, Y.; Chen, J.; "Power-Flow Control Performance Analysis of a Unified Power-Flow Controller in a Novel Control Scheme", IEEE Transactions on Power Delivery, vol. 22, No 3, pp. 1613-1619, July 2007.

Matsuo, T.; Bernet, S.; Colby, R. S.; Lipo, T.; "Modelling and Simulation of Matrix Converter/Induction Motor Drive"; Proc. ELECTRIMACS'96 Conf., pp. 1-10, Saint-Nazaire, France, Sep 1996.

Monteiro, J.; Silva, J.: Pinto, S.; Palma, J.; "Unified Power Flow Controllers without DC Bus: Designing Controllers for the Matrix Converter Solution", Proc. of the International Conference on Electrical Engineering (CEE'05), Coimbra, Portugal 2005.

Nikkhajoei H., Iravani, M. "A Matrix Converter Based Micro-Turbine Distributed Generation System", IEEE Trans. On Power Delivery, Vol. 20, No. 3, July 2005.

Pinto, S.; Silva, J.; "Sliding Mode Direct Control of Matrix Converters", IET Electric Power Applications, 1 (3), pp. 439-448, 2007.

Round, S. D.; Yu Q.; Norum, L. E.; Undeland, T. M.; "Performance of a unified power flow controller using a d-q control system". Conference Publication Ac and Dc Power Transmission, 1(423), 1996.

Song, Y.H. and Johns, A.T.; "Flexible AC Transmission Systems", IEE Power and Energy series, UK, 1999.

Verveckken, J.; Silva, F.; Driesen, J. "Design of Inverse Controller with Cross-Coupling Suppression for UPFC Series Converter" EUROCON 2007, The International Conference on "Computer as Tool" Warsaw, September 9-12.

Wang, B., Venkataramanan, G. "Dynamic Voltage Restorer Utilizing a Matrix Converter and Flywheel Energy Storage", IEEE Tran. on Industry Applications, Vol. 45, No 1, pp. 222- 231, Jan/Feb 2009.

Wheeler, P.; Rodriguez, J.; Clare, J.; Empringham, L.; Weinstein, A., "Matrix Converters: A Technology Review," IEEE Trans. Industrial Electronics, vol. 49, No 2, pp. 276–288, Apr. 2002.

20

The Benefits of Device Level Short Term Energy Storage in Ocean Wave Energy Converters

D. O'Sullivan, D. Murray, J. Hayes, M. G. Egan and A. W. Lewis
University College Cork,
Ireland

1. Introduction

This chapter presents an outline of the requirements for, and the benefits of, short term energy storage at the level of individual wave energy devices, in the field of ocean wave energy conversion. A general background introduction to ocean renewable energy from the perspective of industry growth and incentives, as well as an overview of the different technology types is provided. The unique and challenging features of the short term variability of wave energy is presented and its implications for equipment and grid connectivity are outlined. Short term energy storage is considered as a possible element in the amelioration of this fluctuating output. A case study of a supercapacitor based storage system is presented for an oscillating water column type wave energy device. The issue of supercapacitor lifetime is then addressed in a comprehensive manner in conjunction with results from a lifetime testing rig. Finally, some of the ancillary benefits associated with such a short term energy storage system are briefly described.

2. Ocean renewable energy

Renewable energy technology is steadily gaining importance in the world energy market, due to the limited nature and unstable costs of fossil fuel supplies, national requirements for security of supply, as well as political pressure towards the reduction of carbon emissions. In the European context, wind energy is leading the way in terms of installed capacity, with over 84 GW of cumulative installed capacity in the EU by the end of 2010, representing almost 10% of total installed power capacity (European Wind Energy Association, 2011). The vast majority of this installed capacity is located onshore with a mere 2.9 GW located in the near or off-shore environment. However, offshore installations had a record-breaking year in 2010 with 883 MW of new installed capacity, reflecting an underlying trend of a gradual movement towards the offshore environment. Interestingly solar PV installations in the EU represented the largest single block of renewable energy sources installed capacity in 2010 with 12 GW installed, although the total installed capacity in solar PV still lags behind that of wind with 25 GW of total installed capacity within the EU (European Wind Energy Association, 2011).

The next wave of renewable energy development is anticipated to be offshore renewable energy, which mainly comprises offshore wind, ocean wave and ocean tidal flow technologies. As an industry, ocean energy is still in its relative infancy, although there has been a rapid acceleration in recent years in research and development funding, infrastructure creation, foreshore license policy streamlining, and policy development.

2.1 Roadmaps and targets

The European Union Ocean Energy Association (EU-OEA) has created a 2010-2050 roadmap for the development of the ocean energy industry in Europe, which aims to enable the industry to reach 3.6 GW of installed capacity by 2020 and close to 188 GW by 2050. This roadmap trend is plotted in conjunction with current and projected trends in onshore and offshore wind energy development in fig. 1.

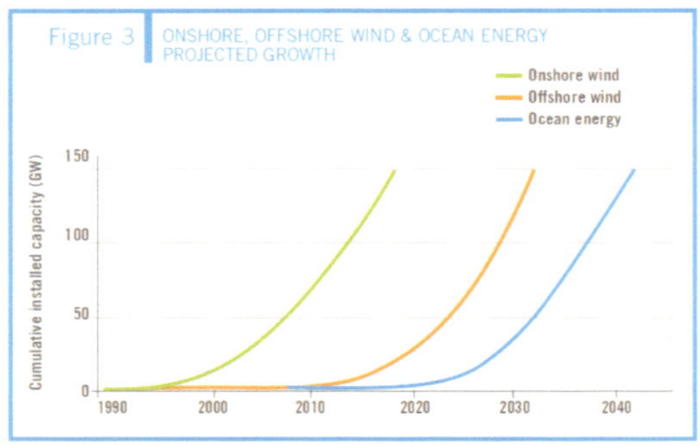

Fig. 1. Ocean Energy Roadmap and Trends in Wind Energy

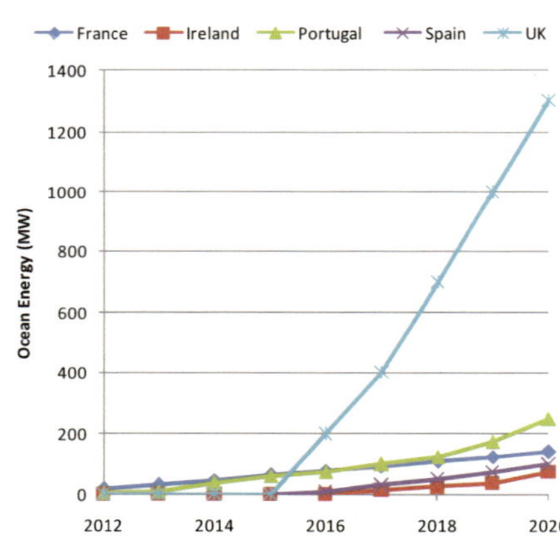

Fig. 2. Ocean Energy Deployment Scenarios by Country

A further incentive to the ocean energy industry and to national and regional funding bodies has been the targets or scenarios for the deployment of ocean energy technology produced by the EU, national governments, as well as industry associations such as the EU-OEA. These range from aspirational roadmaps such as that represented in fig. 1, to legally binding targets such as EU Directive 2009/28/EC. This directive mandates a percentage target for share of energy from renewable sources for each EU member state country in gross final consumption of energy by 2020. These targets are legally binding and can be met by the individual member state across electricity, heat and transport sectors in whatever proportion they see fit, once the overall target is complied with. Each member state is also required to produce a National Renewable Energy Action Plan (NREAP) detailing how they intend to meet their targets. The NREAP generally results in a set of scenarios for different levels of deployment of various renewable energy technologies, consistent with the natural resources of the specific member state. In particular, some member states have specifically identified the deployment of various levels of ocean energy as part of their renewable electricity contribution. Unlike the overall renewable targets outlined in the Directive, the specific mix of contributions contained in the NREAPs is not legally binding. It will nonetheless act as a driver of policy, technology development and investment. A summary of some of the more important ocean energy deployment scenarios as outlined in the individual member state NREAPs (e.g. (Department of Communications, Energy and Natural Resources, Irish Government, 2010)) is portrayed in fig. 2. These scenarios include ocean wave and tidal current plant (Beurskens & Hekkenberg, 2011).

2.2 Electricity grid developments

In parallel with the policy developments and incentives outlined in the previous section, electricity network operators have been active in working to facilitate the large scale integration of renewable energies into transmission and distribution networks. This has been focussed primarily on the wind sector, however, the reinforcements and upgrades to the network to facilitate wind energy will in many cases indirectly facilitate the development of ocean energy farms also. For instance, in Ireland the greatest wind resource is located along the western seaboard. This is also the location of the majority of the ocean energy resource in the form of wave energy. The anticipated creation of a North Sea offshore grid, as well as feasibility studies such as the ISLES project (Scottish Govt., Northern Ireland Executive & Govt. of Ireland, 2011) will further prepare the ground for large scale interconnection of offshore wind, wave and tidal resources. A possible projection of the 2050 electricity grid in Europe (European Climate Foundation, 2010) in the scenario of 80% renewables penetration is illustrated in fig. 3.

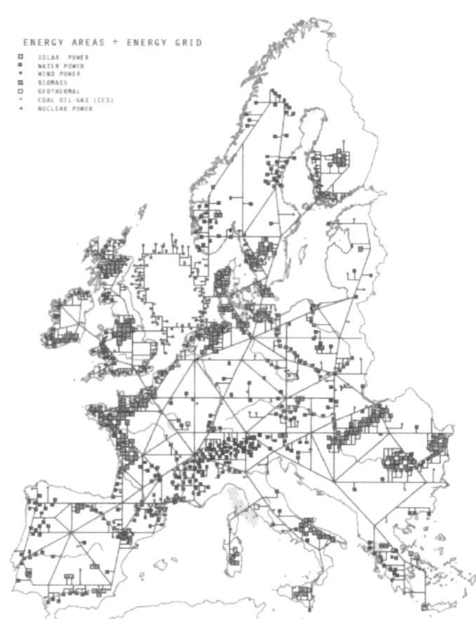

Fig. 3. Projected European Electricity Grid

2.3 Job creation

As well as being a vehicle for satisfying renewable energy targets, the development of the marine renewable sector is seen as a potential catalyst for economic growth and job creation.

Ocean energy is well positioned to contribute to regional development in Europe, especially in remote and coastal areas. This is of particular value in locations where the redeployment of resources and infrastructure previously associated with more traditional marine sectors such as fisheries can revitalise economically depressed communities. The manufacturing, transportation, installation, operation and maintenance of ocean energy facilities will generate revenue and employment. Studies suggest that ocean energy has a significant potential for positive economic impact and job creation(Department of Communications, Energy and Natural Resources, Irish Government, 2010). Parallels can also be drawn with the growth of the wind industry. Export of clean technology now accounts for €7.1 billion annually in Denmark, while in Germany export of wind technology alone is worth over €5.1 billion. Based on the projections for installed capacity in the EU-OEA report, it is anticipated that by 2020 the ocean energy sector will generate over 26,000 direct and 13,000 indirect jobs, increasing to over 300,000 direct and over 150,000 indirect jobs by 2050 assuming the targeted 188 GW is installed (Department of Communications, Energy and Natural Resources, Irish Government, 2010).

2.4 Ocean energy technology

The term ocean energy can encompass a wide range of technologies including ocean wave, tidal current, ocean thermal energy conversion, and ocean salinity gradient. Practically speaking, only tidal current and ocean wave energy are currently anywhere close to commercial operation.

There are many different methods for wave and tidal current energy conversion. The majority of devices, however, follow an approximately similar general outline in terms of energy conversion and capture. This section looks at the various stages in the energy conversion process and discusses the different methodologies used within the main converter technologies.

The energy conversion process can be broken down as follows:-

Primary Energy Capture: This is the means through which the device interacts with the energy source, transferring energy from the waves or tidal currents to a medium which can be captured by a 'prime mover'.

Prime Mover: This is a component which can convert the energy captured at the primary energy capture stage to a more useful form of energy, usually mechanical energy, which can be connected to a generator. In some devices, such as tidal turbines, the primary energy capture and prime mover functions are embodied in the same component. In such a case, this component will be referred to as the primary energy capture component as this more completely describes its functionality.

Generator: The generator converts the mechanical energy of the prime mover into electrical energy and can also act as one of the main control elements in the system.

Storage: Energy storage is used to smooth the time variation of the output electrical power, thus enhancing the power quality of the device.

Control: Control systems are required to optimise, coordinate and control the operating points of some, or all, of the power take-off components and also to protect the device in undesirable operating conditions.

In an attempt to group and classify ocean energy devices, the primary energy capture technique is typically used as a demarcation between device classes. Often, the same or similar prime movers and generators are employed in very different devices, and so it is reasonable to classify devices according to the dynamics of the primary energy capture method as mentioned previously, in some tidal current devices the primary energy capture component can also be considered as a prime mover. The following device classifications are representative of the majority of ocean energy devices (O' Sullivan et al., 2010).

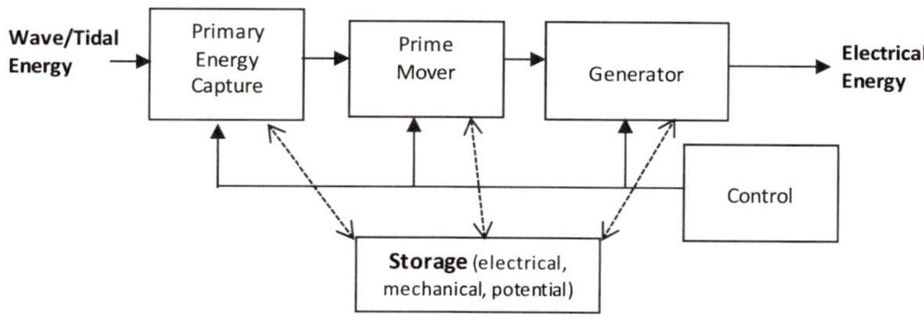

Fig. 4. Typical Ocean Energy Conversion Process

Wave Energy	Tidal Current Energy
Oscillating Water Column (OWC)	Tidal Turbine
Attenuator	Oscillating Hydrofoil
Point Absorber	Tidal Sail Device
Submerged Pressure Differential	Venturi Effect Device
Oscillating Wave Surge Converter	
Overtopping Device	

Table 1. Major Device Classifications

The focus in this article is on ocean wave energy so a brief description of each of the primary power capture processes is provided for the wave energy technologies. Most of these technologies are described in more detail in other technology overview publications (2008, Khan & Bhuyan, 2009).

2.4.1 Oscillating water column

The Oscillating Water Column (OWC) device (Evans, 1978, Falcão, 2002) converts wave motion into pneumatic energy within an enclosed air chamber through the action of external wave pressure fluctuations on a column of water tuned to resonate with the dominant wave frequency. The air is then passed through a turbine which is connected to a generator. The air turbine is typically a Wells turbine (Raghunathan, 1995) or impulse turbine (Setoguchi & Takao, 2006) both of which have the ability to convert bidirectional airflow into a unidirectional torque. An illustration of a typical system is shown in fig. 5:

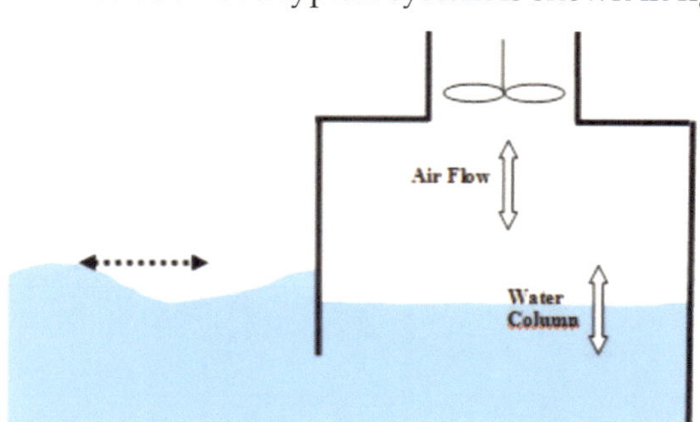

Fig. 5. OWC Illustration

An example of an OWC device is the Ocean Energy Buoy (O'Sullivan et al. 2011).

2.4.2 Attenuator

Attenuators (Henderson, 2006) are floating devices aligned to the incident wave direction. Passing waves cause movements along the length of the device. Energy is extracted from this motion.

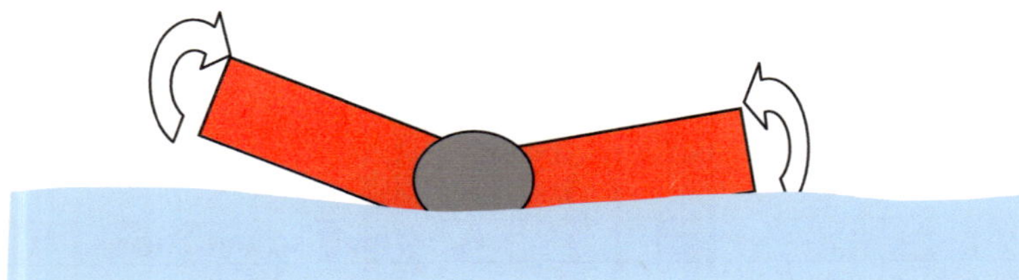

Fig. 6. Attenuator Illustration

These types of devices are typically long multi-segment structures. The device motion follows the motion of the waves. Each segment, or pontoon, follows oncoming waves from crest to trough. The floating pontoons are usually located either side of some form of power converting module. Passing waves create a relative motion between each pontoon. This relative motion can then be converted to mechanical power in the power module, through either a hydraulic circuit (most common) or some form of mechanical gear train. An example of an attenuator is the Pelamis device (Henderson, 2006).

2.4.3 Point absorber

Point absorber devices (Ricci et al., 2009) are generally axi-symmetric about a vertical axis. They are small in comparison to the incident wave length. Point absorber devices usually consist of two main components - a displacer which is a buoyant body which moves with wave motion, and a stationary or slow moving reactor. Energy can be extracted through the relative motion between the displacer and the reactor. This can be accomplished using electromechanical or hydraulic energy converters. The hydraulic converters usually involve hydraulic rams, rectifying valves, gas accumulators and hydraulic motors. An illustration of a typical point absorber is given in fig. 7.

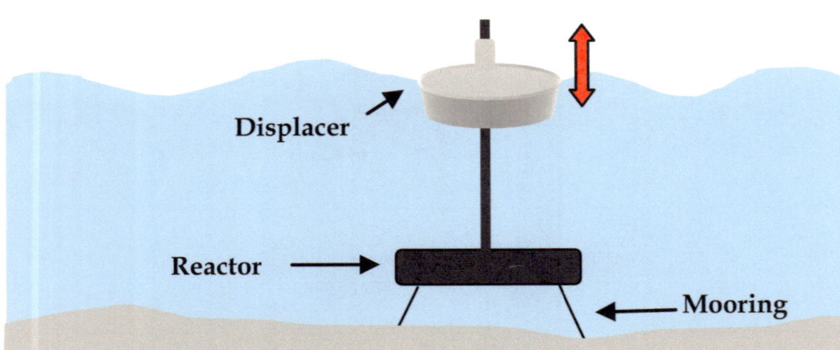

Fig. 7. Point Absorber Illustration

2.4.4 Submerged pressure differential

This type of device can be considered to be a fully submerged point absorber (Polinder et al., 2004). The PTO for the device consists of two main components, a reactor and a displacer. Passing waves cause the sea surface elevation above the device to rise and fall. A pressure differential is created above the device as waves pass. This causes an air chamber within the displacer to decompress and compress, thus causing the displacer to rise and fall. The reactor is typically secured to the sea bed. Power can be extracted from the relative motion between the displacer and reactor, by using a hydraulic or electromechanical system connected between the displacer and reactor.

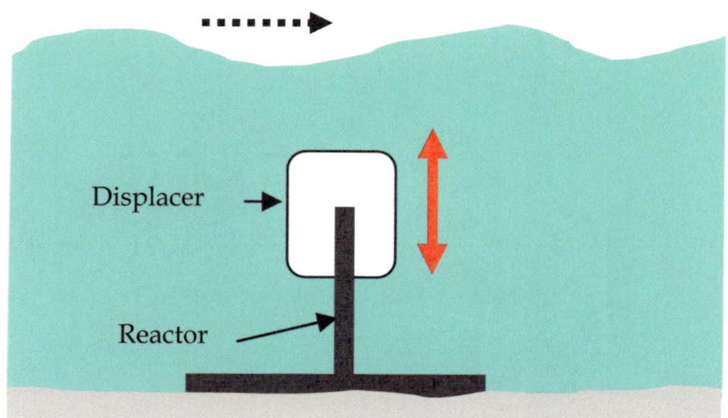

Fig. 8. Submerged Pressure Differential Device Principle

2.4.5 Oscillating Wave Surge Converter

The Oscillating Wave Surge Converter (OWSC) (Chaplin et al., 2009) extracts the energy caused by wave surges and the movement of water particles with them. At the sea bed, on or near the shore, the water particle motion becomes a back and forth motion. It is from this oscillating surge motion that the OWSC extracts energy. The devices can be secured to the sea bed, on or near the shore. They consist of a surge displacer which can be hinged at the top or bottom. Energy is typically extracted using hydraulic converters secured to a reactor.

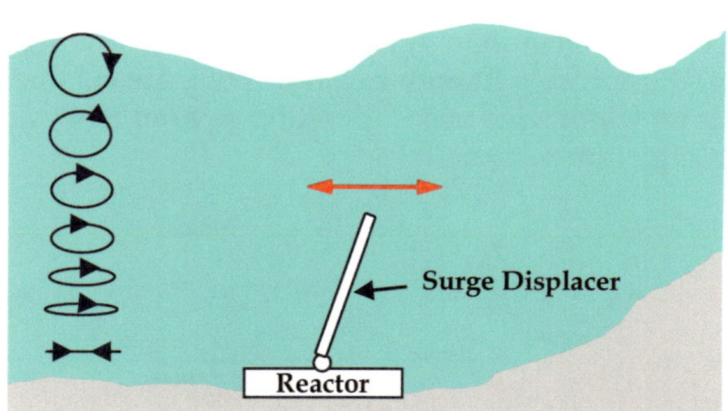

Fig. 9. OWSC Schematic

It is also common to place the device on the shoreline and hinge the displacer above the water. Incoming waves first impact on the displacer and are then captured within the device to form a water column. This water column then empties, moving the displacer in the opposite direction, and the water is returned to the sea. It is also possible to use the surge action of the waves to trap and compress air within a pneumatic chamber (Kemp, 2011). In this case, the OWSC is usually semi-submerged, to allow for the trapping of air at the surface of the wave troughs.

2.4.6 Overtopping devices

Overtopping devices (Jasinski et al., 2007) extract energy from the sea by allowing waves to impinge on a structure such that they force water up over that structure thus raising its potential energy. The water can then be stored in some form of a reservoir. The potential energy of the water is converted to kinetic energy using a conventional hydro turbine. After exiting the turbine, the water is then returned to the sea.

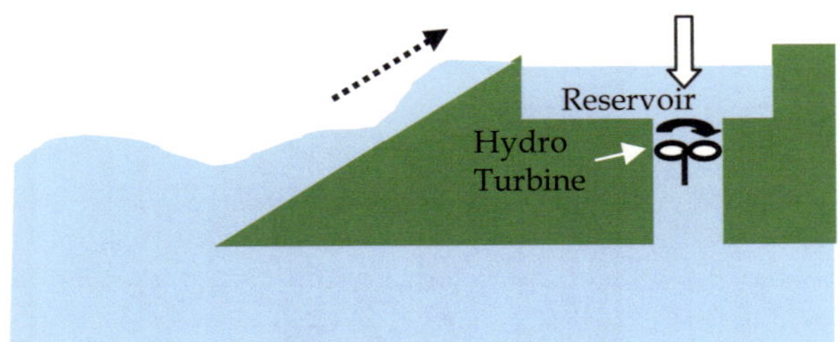

Fig. 10. Overtopping Device Illustration

These devices are fundamentally low-head hydro power plants, except the source of water is from the sea rather than rivers or lakes. They tend to be typically much larger than other devices as significant volumes of water capture are necessary. These devices have one clear advantage over other wave energy devices - the inclusion of a reservoir allows for inherent energy storage. This can be used to produce a more consistent level of power supplied to an electrical grid.

2.5 Wave energy variability

Time variability in wave energy occurs over both long and short time horizons. Long term variability follows somewhat similar patterns to those seen in wind energy in that wave action is generally much lower in summer than in winter, as illustrated in fig. 11 for the Irish wave atlas (Marine Institute/ Sustainable Energy Ireland, 2005). This is due to the fact that wave energy largely depends on wind: wind speed, duration of wind blow, and fetch area typically define the amount of energy transferred.

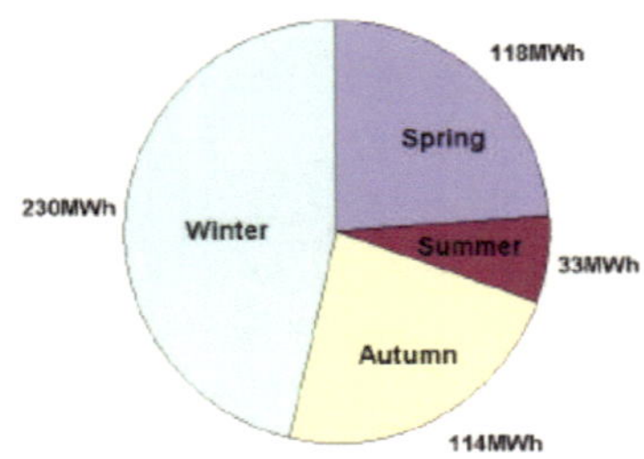

Fig. 11. Seasonality of Technical Wave Energy Resource off Ireland in MWh/m width of wave front (2005)

Despite this wind dependency, the fluctuations seen in wave energy are quite different. The waves in effect 'integrate' the wind energy, smoothing out some of the more rapid fluctuations seen in the wind. Wave energy thus builds up more slowly over large areas of water, and in deep water tends to lose energy only very slowly. Thus over time periods of days, wave energy is more predictable and more persistent than wind energy. This represents a considerable advantage in the integration of wave energy into the electricity grid, as despatch planning becomes significantly easier.
It is at time periods of seconds that significant divergence takes place between wind and wave energy in power time variability or fluctuation. Wind speed, and hence power, fluctuations occur as divergence around a mean value, however, wave power, which is a function of wave elevation returns to zero twice in every wave period. This is illustrated in fig. 12 over a 40 s time window for normalised traces of wind speed and wave elevation.

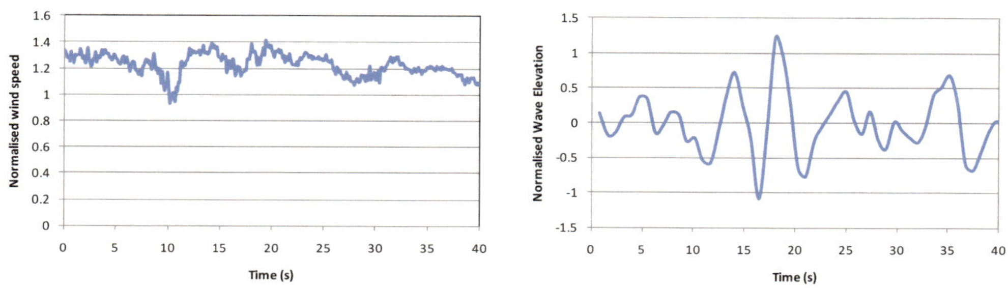

Fig. 12. Comparison of wind speed and wave elevation short term variability

It should be clear from the previous sections that the widely divergent operation modes of the different wave energy converters will result in differing output power responses to the same wave input. Evidence of this is illustrated in fig. 13, where the generated electrical output power from an attenuator wave energy converter with significant hydraulic accumulator storage (Henderson, 2006) is contrasted with the electrical power output from a floating OWC device (O'Sullivan et al., 2011) with an impulse turbine and limited inertial energy storage. Both devices have a mean power output close to 200 kW. However, the peak power output of the OWC is close to 2 MW whereas the peak power output of the hydraulic attenuator device is around 300 kW.
In the case of the attenuator device, the inherent operation of the converter and its own internal energy storage in the hydraulic accumulators act to significantly filter the fluctuations in the wave power incident on the device. In the case of the OWC device, the only inherent energy storage within the conversion chain is some inertial storage in the rotating mass of the turbine and generator. Variable speed control of the turbine allows for this rotating mass to be used to absorb some of the power fluctuations (Justino & Falcao, 1999). It is clear however, from fig. 13 that its impact is not as significant as that of the hydraulic accumulators in the attenuator device.
Apart from the inherent operation of the converters themselves, the control of wave energy converters can also influence the extent of the power fluctuations seen in the output power. Clearly control action in conjunction with energy storage will have an impact on power fluctuation, however some control schemes whose objective is to enhance power output can inadvertently result in significantly larger power fluctuation. The control scheme known as latching, for instance, forces a reduction in the duty cycle of the power take-off period in order to maximise the output power. This has the effect of producing shorter, higher amplitude pulses of output power, in effect increasing the power fluctuations.

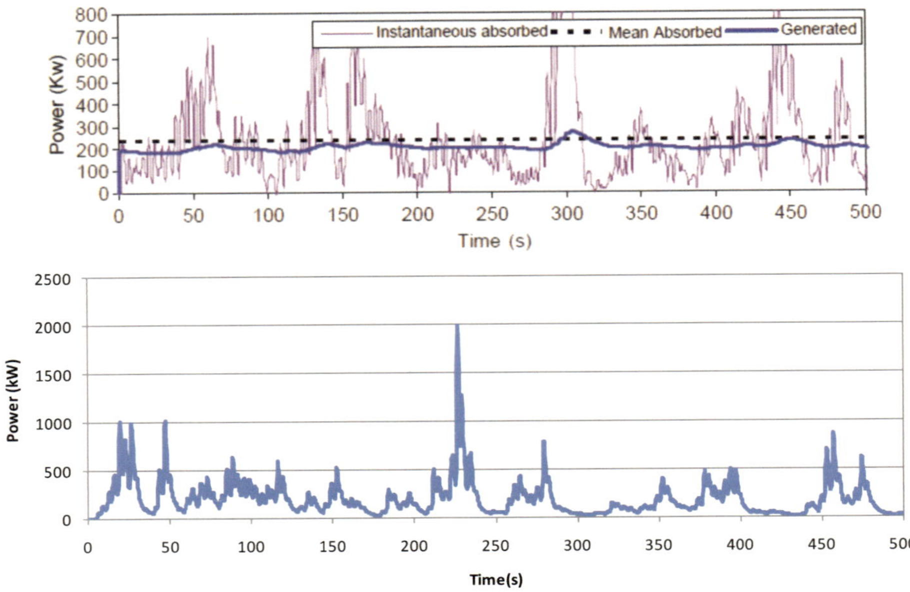

Fig. 13. Output electrical power time series from (top) attenuator device with significant hydraulic accumulator storage (bottom) OWC device with impulse turbine.

2.6 Impact of power fluctuations

The impact of large power fluctuations in a grid connected wave energy converter device or array is generally detrimental. Four main areas of concern can be readily identified:

- Equipment rating
- Equipment lifetime
- System losses
- Power quality

2.6.1 Equipment rating

The issue of equipment rating is related to the peak-to-average ratio of nominal power output (i.e. apart from fault conditions or transient overloads). A 1:1 ratio is optimal, as the usage of the equipment is optimised. This results in enhanced performance since the equipment is being operated at its design point and best cost since needless over-rating is not being purchased. It is evident that the peak to average power output ratio of some wave energy converters is significantly higher than 1:1 as already illustrated in fig. 13. This is a particular issue for devices that utilise power electronics converters as part of the generator and grid power control. In reality, the vast majority of renewable power generation equipment requires the use of power electronic converters in order to optimise the power control. Power electronics converters do not have a long thermal time constant. It is typically of the order of hundreds of milliseconds, so in effect, for wave energy devices, the power converters must be rated for the peak power output. There is some flexibility in the rating of other equipment such as machines, cables and transformers. These will typically have thermal time constants of the order of minutes, and so can be operated transiently at higher peak powers than their mean rating. However, without some means of mitigation such as inherent or added short term energy storage, or deliberate power release, rating of all of the electrical equipment can be several times higher than the mean power output of the device.

2.6.2 Equipment lifetime

One of the main factors in shortening the lifetime of electrical equipment is the extent and frequency of the thermal cycling that takes place within the equipment. This has a particular impact on the power electronic converters. The transistor modules in these converters are inter-connected through wire bond technology. Differing thermal coefficients stresses the interface between the wire bond and the silicon, and this eventually leads to transistor failure. Hence, component lifetime is directly related to the number and depth of the thermal cycles endured by the equipment. Clearly, fluctuating power in the system results in a fluctuating thermal profile which in turn leads to degradation of system lifetime and reliability. Once again, the power electronic components are the most susceptible due to their very low thermal time constant.

2.6.3 System losses

Large power fluctuations result in increased power losses in system equipment when compared to a system with the same mean power output and the same equipment, but with no power fluctuations. This is mainly due to the fact that resistive power losses are proportional to the square of the current. Hence, a system with fluctuating power has an additional conductive power loss component ΔP_{loss} where

$$\Delta P_{loss} = R_{sys}\left[\frac{1}{T}\int_{0}^{T} I_{rms}{}^{2}(t)\,dt - I^{2}{}_{av}\right] \tag{1}$$

R_{sys} represents the total equivalent resistance in the system incorporating all loss mechanisms, T is the time window in consideration, I_{av} is the mean rms current over the time T and $I_{rms}(t)$ is the quasi-static approximation of the time varying rms current in the system.

2.6.4 Power quality

Power quality issues arise due to the interaction of fluctuating current with the impedance of the electrical network. This results in voltage fluctuations in the network that are proportional to the current fluctuation levels and also to the short circuit impedance of the network. Weaker networks have higher impedance, and thus the voltage fluctuations will be more evident.

In a case study assessing the impact of the integration of a small wave farm at the national wave energy test site in Belmullet, off the north west coast of Ireland, the impact on the local network voltage of varying levels of power fluctuation was examined (Santos et al., 2011). A 3% limit is applied to the voltage change magnitude, and the maximum allowed power fluctuation amplitude is plotted. The power output is assumed to consist of a mean power level added to the sum of three sinusoidal terms representing the dominant wave periods within the wave spectrum. The resultant fluctuation amplitude is defined as the ratio of the peak power to the mean power, i.e. 100% fluctuation implies that the output power increases to twice the mean power and drops to zero over the course of several wave periods.

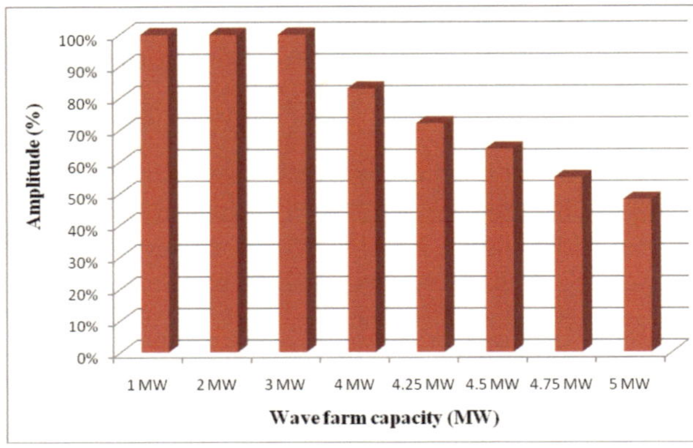

Fig. 14. Maximum allowed power fluctuation amplitude

The results are graphed in fig. 14. It is evident that above a certain power level, the power fluctuation amplitude must be reduced in order to avoid breaching the 3% voltage fluctuation limit.

3. Short term energy storage options

Energy storage can be a very useful feature in ocean energy applications. Due to the highly varying nature of the resource, particularly the wave resource, designing a device that can deliver a relatively constant electrical power output at an optimum efficiency is an onerous task. Large scale electrical storage would be an ideal scenario as devices could store the varying power produced, and supply it to the electricity grid at a constant rate when required. This would not only improve the efficiency of the device but it would also enable grid code requirements to be met with greater ease. The injection of a rapidly varying power output into a weak electricity network can result in significant voltage deviation that may be in danger of breaching grid code requirements, as discussed in the previous section.

However, although the technology for large scale electrical storage currently exists it is extremely expensive and its use would render most ocean energy projects uneconomical. Despite this, developers continue to investigate other methods for some form of energy storage for their devices.

There are a number of wave energy devices that *inherently* contain energy storage methods i.e. energy storage forms part of their fundamental operation mode, as opposed to being explicitly added to the device. The most obvious is the overtopping device - this contains a reservoir which is essentially a large storage tank for potential energy. The reservoir is often an integral part of these devices, so it does not cost an energy loss to include this storage method. Also, devices containing a gas accumulator within a hydraulic circuit are inherently capable of storing energy, although, generally only relatively small amounts of energy can be stored within accumulators. Furthermore, energy is released over a relatively short periods of time. This factor means that accumulators are not good for long term energy storage, but can be used over the short term to reduce power fluctuations in the hydraulic circuit.

It is also worth noting that rotating turbines in both tidal and wave devices can contain significant mechanical inertia which is effectively a form of energy storage. The energy of a rotating turbine with inertia J between two speed limits, ω_1 and ω_2 is:

$$W_{turbine} = \frac{1}{2} J\left(\omega_1^2 - \omega_2^2\right) \tag{2}$$

To utilise this inherent energy storage device, a variable speed control scheme is needed which accounts for the power flow and speed variation of the turbine. The basic governing equation is:

$$P_{mech} = T_{mech}\omega = P_{gen} + P_{loss} + \frac{d}{dt}\left(\frac{1}{2} J\omega^2\right) \tag{3}$$

where P_{mech} is the mechanical power applied by the turbine, T_{mech} and ω are the corresponding mechanical torque and speed, P_{gen} is the electrical power output through the generator, P_{loss} represents the power losses in the rotating system, and J is its combined inertia. Depending on the level of inertia available, the electrical power output can be maintained relatively constant or at least with reduced power fluctuation through control of the system speed.

With so many WECs in development, it is recognised that any implemented variable speed strategy is unique to each device and its location. Factors to be considered when devising a control strategy are discussed in (Justino & Falcao, 1999) and consist of

i. remaining within speed limits
ii. efficient performance
iii. power quality to the grid
iv. a realistic control procedure where measurable quantities are used such as pressure and speed

Utilising the turbine as an effective flywheel, or utilising a separate flywheel where the speed variation will not directly affect the power take-off, is a proven, robust method of energy storage.

Devices without inherent energy storage are reliant on conventional added energy storage techniques. These include compressed air storage, hydrogen storage, supercapacitors, batteries (including flow batteries and fuel cells) and flywheels. These options all have their own advantages and limitations (Santos et al., 2011).

3.1 Lifetime requirements

Maintenance intervals in offshore wave energy devices should be long and not limited by a prototype energy storage system. The difficulty in carrying out on-board maintenance on an offshore WEC is highlighted in (O'Sullivan & Lewis, 2008), where docking issues and working in an unstable environment are key concerns and results in severe costs. A typical desired interval for non-routine, disruptive maintenance in an offshore plant is five years, giving the minimum desired lifetime of any employed energy storage element.

An average wave period of 10 s is typical for most full scale WECs. Due to the unidirectional turbine torque from the bidirectional airflow in OWCs, the average input pneumatic power period is half this value. This calculates the total number of wave power cycles on an offshore wave energy converter over a five year maintenance period, taking account of the expected operational time and availability, to be around 21 million. This poses serious lifetime issues for any energy storage equipment that is likely to be cycled at every wave cycle.

3.2 Electrical storage

Electrical storage technologies include batteries, supercapacitors and Superconducting Magnetic Energy Storage (SMES).

3.2.1 SMES

SMES is currently costly and consists of many essential parts, including a cryogenically cooled refrigerator, that increase breakdown vulnerability in the harsh offshore wave climate, as well as increasing the necessary available space and mechanical support. For these reasons SMES has not yet been considered for offshore ocean energy applications.

3.2.2 Batteries

Batteries are high energy density electrical storage devices that have undergone significant development in recent times. With the increased research into electric vehicles, suitable rechargeable batteries are being developed. Currently lithium ion batteries are the chosen technology installed in new electric vehicles as their improved performance over NiMH batteries are now being realised as production costs decrease. Some lithium ion batteries for electric and hybrid electric vehicles have energy densities as high as 140 Wh/kg and power densities of up to 745 W/kg (Amjad et al., 2009). Their cycle durability at present is in the range of several thousand. These small power densities and cycle lifetimes prevent lithium ion batteries from making contributions of power smoothing over time periods near those of the ocean waves.

3.2.3 Supercapacitors

Supercapacitors (SCs) are also known variously as electric double layer capacitors, Ultracapacitors, and electrochemical double layer capacitors (EDLC). They utilise high surface area porous carbon based electrodes, and have capacitances ranging from a few farads up to 5,000 farads. Due to the very small charge separation distance in the 'double layer', voltage ratings are low; close to 2.7 V. To achieve higher voltages, strings of series-connected supercapacitors are created. Usually voltage balancing circuits are added, as due to the manufacturing process relatively large tolerance values of capacitance exist between individual SCs.

SCs are governed by the same equations as conventional capacitors. While SCs cannot compete with batteries in terms of energy density, their much longer cycle life, power density, operational temperature range, and ability to fully discharge make them an energy storage option that must be considered in many applications. A typical supercapacitor has an energy density of over 5 Wh/kg, a power density of over 6,000 W/kg, and a rated lifetime of 1 million cycles. Coupled with this, SCs have charge/discharge efficiencies ranging from 0.85 to 0.98 (Douglas & Pillay, 2005).

SCs have a demonstrated robustness. Applications with photovoltaics were shown in (Glavin et al., 2008), (Weeren et al., 2006) where the supercapacitors complemented battery storage and improved system performance and battery lifetime. The ability of SCs to operate at sea for long periods of time was shown in (Weeren et al., 2006). SCs have also been used in wind turbine pitch systems, hybrid vehicles, trains, buses, and lift trucks. The time constant of SCs is typically around one second, and their small energy density but large power density suggest they are ideal short term energy storage options, especially for ocean energy applications if their lifetimes can be shown to be compatible with the required service life of such equipment in an offshore wave energy converter.

4. Case study

Power smoothing in a full-scale offshore Oscillating Water Column (OWC) Wave Energy Converter (WEC) was investigated by integrating supercapacitors (SCs) with the inertia of a Wells turbine controlled at variable speed. In effect, this case study examines an integrated approach for short term energy storage combining inertial and electrical methods.
An experimentally derived Simulink model represents the full-scale 500 kW WEC system, and available sea-state data is utilised to obtain the full-scale power flows and system speed response. From this, a SC system is sized and integrated into the Simulink model.
In an effort to help validate SC cycle life and robustness, lifetime testing was also carried out. Test setups were built to establish the SC lifetimes under standard and application test conditions as at the time of writing results of documented SC cycle lifetime testing did not approach the 1 million cycle lifetime figure often quoted in datasheets.

4.1 Case study description

4.1.1 The system model and speed control

The Oscillating Water Column (OWC) WEC considered employs a Wells turbine without actuating valves. To predict realistic power take-off (PTO) data from the device at variable speed, the Simulink model created in (Cashman et al., 2009) is used. This was based on experimental data from a quarter-scale prototype operated offshore in an Atlantic test site. The inputs to the model are pneumatic power and turbine speed, and the output is turbine torque and takes account of the effect variable speed has on the pneumatic power production. The model uses non-dimensional quantities to allow scaling to full size.
Variable speed strategies developed in (Duquette et al., 2009, Falcão, 2002, Justino & Falcao, 1999) were examined and compared using the turbine model described above with sea state data. The strategy that produced optimum performance was developed in (Falcão, 2002) where generator torque is evaluated from a measure of turbine speed.
This control scheme consisted of two parts. The first part was developed by measuring the average mechanical power produced at a fixed machine speed. The fixed machine speed maximising the average power for each of the 13 sea states was found and these speeds and powers were plotted. Results from two mid-power examined sea states are shown in fig. 15.
Using the curve fitting tool in Matlab with these maximum average powers and corresponding fixed speeds for each sea state, the power coefficients in (4) were derived (producing an R-square value of 0.9996). This curve is also shown in fig. 15.

$$P_{gen} = 0.0005307\,\omega^{3.797} \tag{4}$$

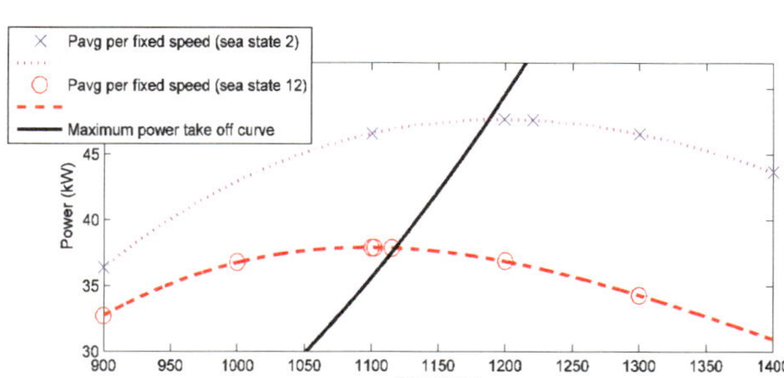

Fig. 15. Average mechanical power versus fixed speed for two of the thirteen case study sea states and maximum power curve

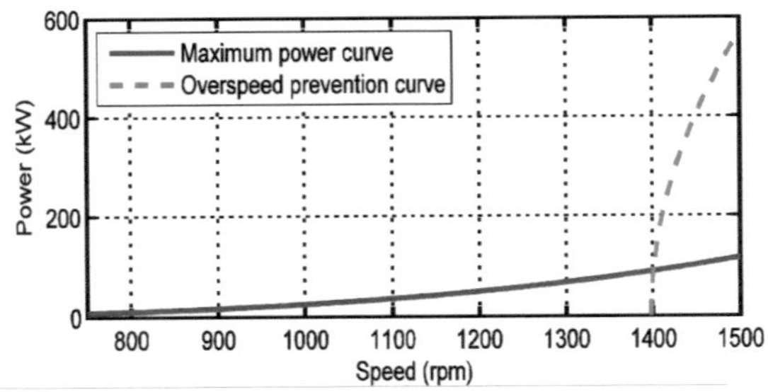

Fig. 16. Speed control power curve of generator power drawn versus speed

The second part of the developed control scheme limits the generator power as shown in (5) and ensures that the turbine does not over-speed to avoid mechanical stress and possible failure.

$$P_{gen} = \left[P_{\max}^2 - J\left|\frac{dP_{gen}}{dt}\right|\left(\omega_{\max}^2 - \omega^2\right)\right]^{\frac{1}{2}} \tag{5}$$

where $J\,|\,dP_{gen}/dt\,| = 100$ MW s^{-2} kg m^2 as in (Falcão, 2002), and turbine inertia was set at 595 kg m^2 (in line with other full scale OWC Wells turbines (Falcão, 2002)).

The control algorithm sets the generator power to the maximum value evaluated from (4) and (5) according to the turbine speed as shown in fig. 16. Simulated plots of input pneumatic power, electrical power and speed are shown in fig. 17 (a), (b) and (c). Chattering of the generator power occurs around the speed where (5) comes into effect. To prevent this chattering, a switched controller is used where the local maximum generator torque achieved is maintained until the speed drops by a predetermined level (this hysteresis value was set at 80 rpm). The resulting power profile and turbine speed are shown in fig. 17 (d) and (e).

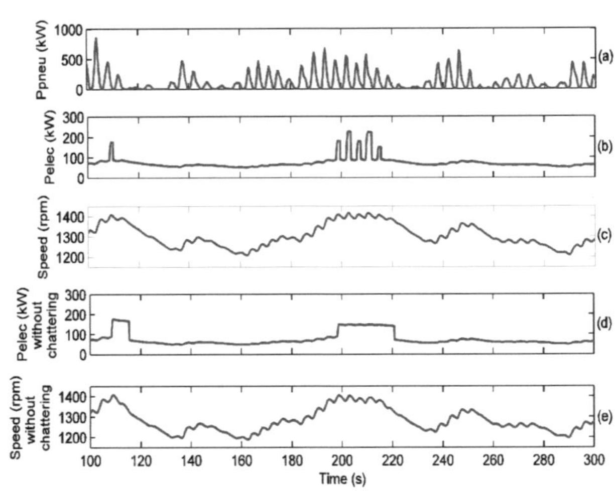

Fig. 17. Generator power and speed with and without the switched controller to prevent chattering for a given input pneumatic power.

4.1.2 Power smoothing with supercapacitors

As shown in fig. 17, the generator electrical power contains large peaks that occur only occasionally. It is proposed to further smooth this power with SCs connected to the dc-bus of the back-to-back power electronics frequency converter which couples the generator to the grid.

The number of generator power peaks for each sea state were measured, and multiplied out by occurrence data values to evaluate the total number of peaks over the five-year WEC maintenance interval. It was assumed that the WEC would not be operational in very low or high energy sea states. Therefore, the device would be operational over 70% of the time with approximately 990,000 peaks of electrical power to be smoothed. This number of peaks is within the specified lifetime of many SC modules.

The discharge strategy attempts to maintain the SCs at their lowest operational voltage (half rated voltage) to make the SC energy capacity available for absorbing power peaks. Once the generator power exceeds a predetermined value (dependent on the sea state), the SCs prevent any excess power flowing to the grid and absorb the difference. Once the input power drops below this value, the SCs maintain this power to the grid until their minimum voltage is achieved. A voltage hysteresis band prevents discharge of the SCs until the band is exceeded toto prevent rapid charge and discharge cycles occurring.

The SCs are sized for the maximum energy sea state of the WEC which produces 152 kW on average. Sizing was based on multiples of the BMOD0063 P125 63 F 125 V module from Maxwell Technologies (utilising SCs of the same technology as the SCs under test). Five parallel strings of two modules in series satisfied all ratings and limited the grid power to 185 kW.

4.1.3 Supercapacitor lifetime testing

While SC lifetime has been tested before, it has typically been accelerated testing, where elevated voltages and temperatures were used. Based on changes in lifetime at small deviations of voltage and temperature at elevated values, typical lifetimes at normal conditions were determined from extrapolations (El Brouji et al., 2008, Lajnef et al., 2007, Paul et al., 2009). Maxwell Technologies provide some results from their lifetime testing but only up to 150,000 cycles and then extrapolate to one million (Maxwell, 2011). Also, this testing procedure provided 15 seconds of rest between every cycle.

Two different types of lifetime testing are carried out - the first is standard lifetime testing at rated current levels, and the second is application testing with the type of power profile expected in a wave energy converter device.

Thirty BCAP0005 P270 cells have been characterised. Each SC is charged at the rated current of 1.6 A to the rated voltage of 2.7 V, undergoes a five second rest period (approximately five time constants) and the voltage and time are measured. The SC is discharged at rated current to half rated voltage (1.35 V), and another five second rest period takes place before measuring the final voltage. Plots of this characterisation profile are shown in fig. 18. From this the capacitance, C and the equivalent series resistance (ESR), R are evaluated according to (6) and (7).

$$C = \frac{I_{rated} t_d}{V_{start} - V_{finish}} \tag{6}$$

$$R = \frac{V_{finish} - V_{rated}/2}{I_{rated}} \tag{7}$$

A SC with close to average specifications was chosen for testing. The test setup, shown in fig. 19, consists of a power supply to charge the SC, an electronic load, a high precision voltmeter, and a thermocouple monitor taking temperature readings of the top, body and leg of the SC, as well as the ambient temperature. These devices are operated using GPIB hardware under the control of a Matlab file. The testing is carried out at ambient temperature with continuous rated current.

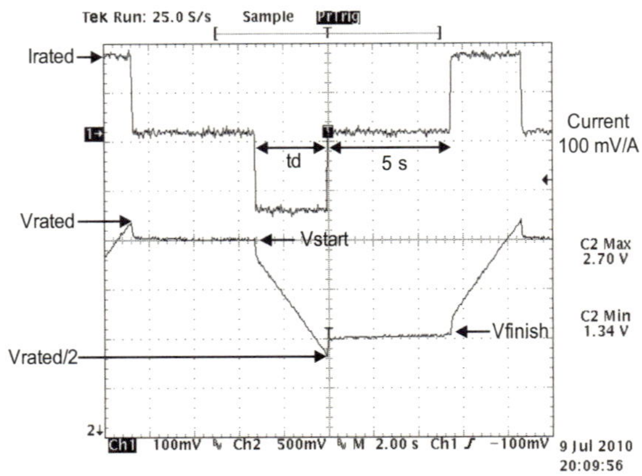

Fig. 18. SC current and voltage during characterisation

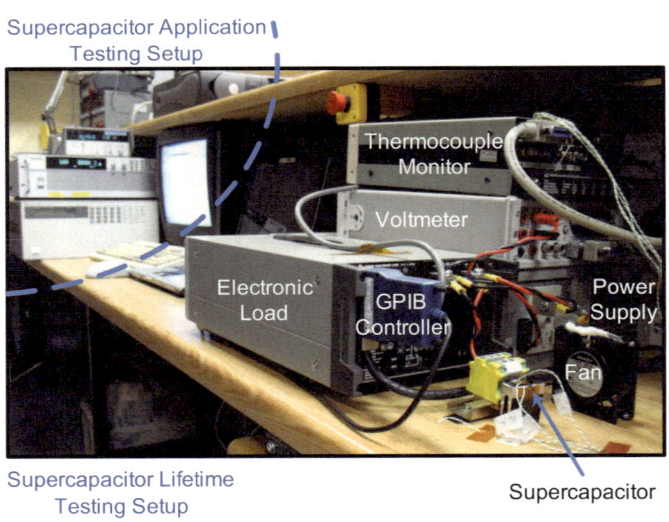

Fig. 19. SC lifecycle test setup

Constant current cycling between rated and half rated voltage is carried out continuously during the day and the apparatus is shut down at night. There is no rest time between charge and discharge cycles except for during characterisation. Characterisation tests occur every 100 cycles and are performed over five consecutive cycles, from which average values are obtained giving more accurate readings. The BCAP0005 P270 SC has a specified lifetime of 500,000, where end of life is specified as a 30% reduction in capacitance, or a 100% increase in ESR. The degradation is shown in the results section.

From the modelling work, the full scale SC power profile is obtained. Using Froude scaling (Wavenet, 2003), these powers are scaled down to values relevant to the BCAP0005 P270 SC under test. A scale factor of 21.135 was chosen and the resultant scaled values compared to the tested SC ratings are shown in table 2. As the resultant usable energy of the SC is lower than the scaled value, the maximum voltage limit is expected to be reached during the application testing.

The model SC power profile is developed from the most occurring sea state before voltage limits are encountered, with the grid power limited to 150 kW. This sea state contains over 30 minutes of data and produces 10 power peaks; close to the average power peak rate over yearly operation. The application SC test utilises similar equipment as outlined in the lifecycle testing. Due to Froude scaling the applied power profile lasts 395 seconds. This is looped three times before characterisation tests are carried out. Again, the process is continued for the day and the apparatus is switched off at night.

	SC modules scaled	Tested SC
Continuous power (W)	4.32	4.32
1 sec power (W)	10.17	9.18
Usable energy (J)	18.5	13.7

Table 2. SC modules scaled to values relevant to tested SC.

4.2 Case study results

The modelling work produced the following results for the most common sea-state: a peak-to-average mechanical power on the turbine of 6.8, a peak-to-average generator power of 4.6, and a peak-to-average grid power of 2.3. A further level of power smoothing was indicated by measuring the standard deviations of the different powers in the system. These results were: 1 pu for mechanical power, 0.43 pu for electrical power, and 0.33 pu for grid power.

Over 750,000 cycles have been tested on the SC under standard test at constant room temperature. The degradation of capacitance and ESR are seen in fig. 20 and fig. 21 respectively. The SC itself is rated for 500,000 cycles, and to date the authors have not found this validated in another source. Also, all initial values are within manufacturer's specifications. Fig. 20 and fig. 21 validate SC performance discussed in where there is an exponential decrease of capacitance initially before capacitance degradation becomes more linear. It is expected that near end of life an exponential fall off of capacitance will occur.

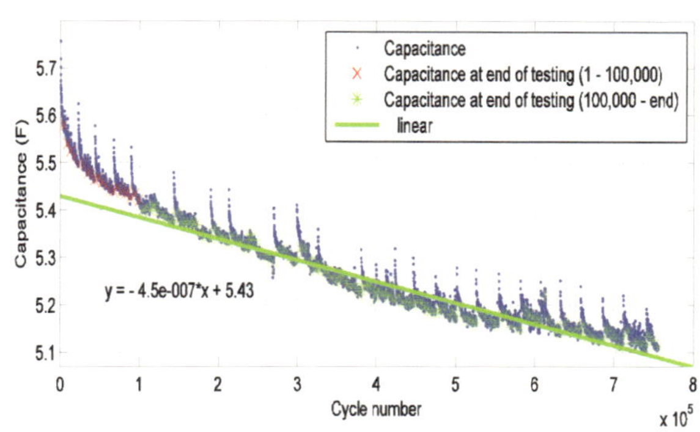

Fig. 20. Capacitance versus cycle number during cycle lifetime testing

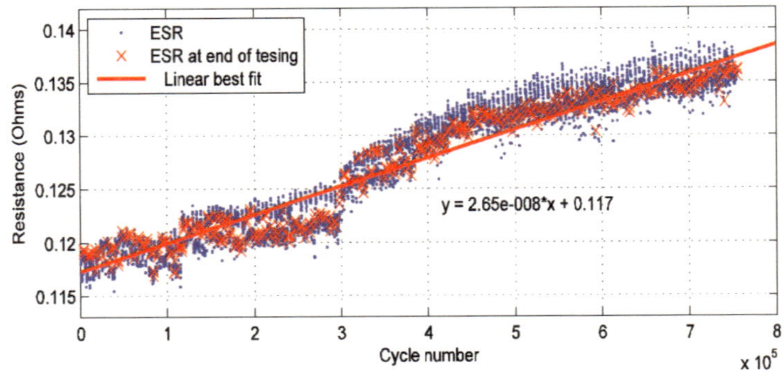

Fig. 21. ESR versus cycle number during cycle lifetime testing

Application lifetime testing has achieved over 85,000 cycles to date. This corresponds to almost six months operation at full scale. The degradation of capacitance and ESR are seen in fig. 22 and fig. 23 respectively. If these trends continue, capacitance will reach end of life first after just over one million cycles, corresponding to over five years operation.

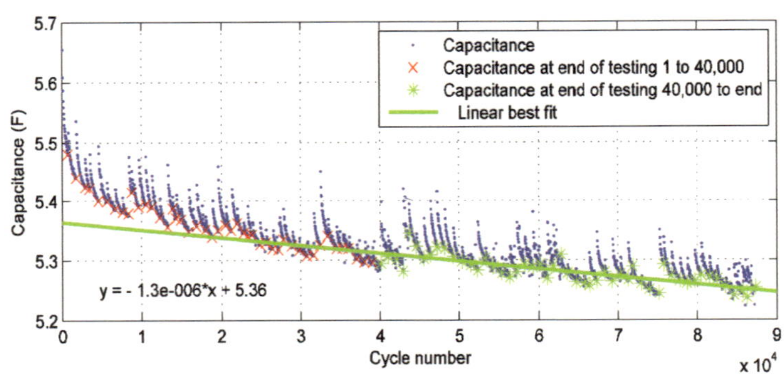

Fig. 22. Capacitance versus cycle number during application testing

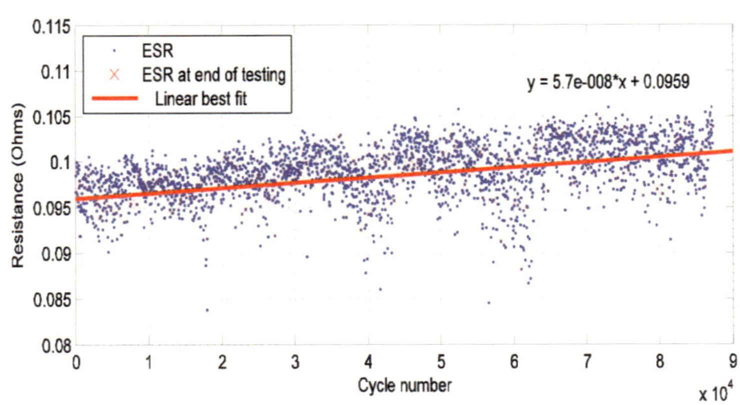

Fig. 23. ESR versus cycle number during application testing

4.3 Ancillary benefits

In a separate study (Murray et al., 2009) other ancillary uses of SCs were investigated. A similar Matlab model was created representing an OWC utilising a Wells turbine. The Wells turbine is a non self-starting device. It was found that supercapacitors had suitable energy and power capabilities to quickly speed up the turbine from rest. The study employed two Maxwell modules to speed up the 135 kg m^2 turbine to 1,000 rpm in 10 to 17 seconds. This feature can be of value where significant import power is required for start-up of offshore equipment, as import power rating can be quite costly.

Low voltage ride through is a problem when a nearby grid fault causes a reduction in the grid voltage at the generator grid connection. This limits the power that can be extracted from the device. If there is large input power, and the powers are not controlled, the power imbalance leads to an unregulated increase in the turbine speed, or dc bus voltage of the back-to-back converter. The study demonstrated that the SC bank could prevent turbine over-speed, maintain dc-bus voltage ratings, and satisfy the grid requirements.

5. Conclusion

This chapter has considered the use and value of short term energy storage in the field of ocean wave energy converters. A background to the area of ocean renewable energy has been outlined, emphasising the development of the industry, its potential benefits to society, and an overview of the different device technologies. Some detail has been given on the main device categories relevant to ocean wave energy. The particular challenges of short term power variability in ocean wave energy technology have been illustrated, and the disadvantageous effect this can have on power quality, power performance, and equipment rating demonstrated. Options for amelioration through short term energy storage have been explored, and a judicious combination of mechanical and electrical energy storage in an OWC device has been selected as an appropriate case study. The lifetime issues, particularly associated with supercapacitors have been examined, and tested as an integral part of the case study. Standard life cycle testing has been performed along with application testing under the charge-discharge profiles likely to be seen in the study device. The findings of this study indicate the following:

- A combination of mechanical and electrical energy storage can reduce power fluctuations to the grid from a single OWC, reducing peak-to-average ratios almost threefold.
- Combining mechanical and electrical storage in an intelligent algorithm enables the supercapacitor usage to be extended to a sensible service life of 5 years.
- Application based life cycle testing indicates that the supercapacitors will have a cycle lifetime of one million cycles, enabling the 5 year target to be met.

Finally, some ancillary benefits of the presence of short term energy storage in the OWC system have been described.

6. References

Amjad, S., Neelakrishnan, S. & Rudramoorthy, R. (2009). Review of design considerations and technological challenges for successful development and deployment of plug-in hybrid electric vehicles. *Renewable and Sustainable Energy Reviews, 14* (3). 1104-1110.

Energy Research Centre of the Netherlands. (2011). In:*Renewable Energy Projections as Published in the National Renewable Energy Action Plans of the European Member States*, Available from: <http://www.ecn.nl/docs/library/report/2010/e10069.pdf>

Cashman, D.P., O'Sullivan, D.L., Egan, M.G. & Hayes, J.G. (2009). Modelling and analysis of an offshore oscillating water column wave energy converter. *Proceedings of 8th European Wave and Tidal Energy Conference*, (Uppsala, Sweden, 2009), pp. 924-933.

Chaplin, R.V., Rahmati, M.T., Gunura, K., Ma, X. & Aggidis, G.A. (2009). Control systems for Wraspa. *Proceedings of International Conference on Clean Electrical Power*, (2009), 93-97.

Department of Communications, Energy and Natural Resources, Irish Government. (2010). In:*DRAFT National Renewable Energy Action Plan (NREAP)*, Available from: <http://www.dcenr.gov.ie/NR/rdonlyres/0E9749D9-BB72-49D6-B5BC-DC4EE41A6302/0/DraftNREAPv17June2010forwebsite.pdf>

Douglas, H. & Pillay, P. (2005). Sizing ultracapacitors for hybrid electric vehicles. *Proceedings of 31st Annual Conference of IEEE Industrial Electronics Society, 2005.*

Douglas Westwood Ltd. (2008). In:*The World Wave & Tidal Market Report 2009-2013.*

Duquette, J., O'Sullivan, D., Ceballos, S. & Alcorn, R. (2009). Design and Construction of an Experimental Wave Energy Device Emulator Test Rig. *Proceedings of European Wave and Tidal Energy Conference*, (Uppsala, Sweden, 2009).

El Brouji, H., Briat, O., Vinassa, J.M., Bertrand, N. & Woirgard, E. (2008). Comparison between changes of ultracapacitors model parameters during calendar life and power cycling ageing tests. *Microelectronics Reliability, 48* (8-9). 1473-1478.

European Climate Foundation. (2010). In:*Roadmap 2050: A Practical Guide to a Prosperous, Low-Carbon Europe*, Available from: <http://www.roadmap2050.eu/attachments/files/Volume3_FullBook.pdf>

European Ocean Energy Association. (2010). In:*Oceans of Energy: European Ocean Energy Roadmap 2010-2050*, Available from: < http://www.eu-oea.com/euoea/files/ccLibraryFiles/Filename/000000000827/ocean%20vectorise%2010%20mai%202010.pdf>

European Wind Energy Association. (2011). In:*Wind in Power: 2010 European Statistics*, Available from: <http://www.ewea.org/fileadmin/ewea_documents/documents/statistics/EWEA_Annual_Statistics_2010.pdf>

Evans, D.V. (1978). The Oscillating Water Column Wave-energy Device. *IMA J Appl Math, 22* (4). 423-433.

Falcão, A.F.d.O. (2002). Control of an oscillating-water-column wave power plant for maximum energy production. *Applied Ocean Research, 24* (2). 73-82.

Glavin, M.E., Chan, P.K.W., Armstrong, S. & Hurley, W.G. (2008). A stand-alone photovoltaic supercapacitor battery hybrid energy storage system. *Proceedings of Power Electronics and Motion Control Conference*, (2008), 1688-1695.

Henderson, R. (2006). Design, simulation, and testing of a novel hydraulic power take-off system for the Pelamis wave energy converter. *Renewable Energy, 31* (2). 271-283.

Jasinski, M., Knapp, W., Faust, M. & Fris-Madsen, E. (2007). The Power Takeoff System of the Multi-MW Wave Dragon Wave Energy Converter. *Proceedings of European Wave and Tidal Energy Conference*, (2007).

Justino, P.A.P. & Falcao, A.F.d.O. (1999). Rotational Speed Control of an OWC Wave Power Plant. *Journal of Offshore Mechanics and Arctic Engineering*, *121* (2). 65-70.

Kemp, J. (2011). In: *The Owel Wave Energy Converter*, Mar 2011, Available from: <http://www.owel.co.uk/owel.htm, (Link to News and Press Releases)>

Khan, J. & Bhuyan, G. (2009). In: *Ocean Energy: Global Technology Development Status*, Mar 2011, Available from: < www.iea-oceans.org>

Lajnef, W., Vinassa, J.M., Briat, O., El Brouji, H. & Woirgard, E. (2007). Monitoring fading rate of ultracapacitors using online characterization during power cycling. *Microelectronics Reliability*, *47* (9-11). 1751-1755.

Marine Institute/ Sustainable Energy Ireland. (2005). In:*Accessible Wave Energy Resource Atlas : Ireland : 2005*, Available from: <http://www.seai.ie/Renewables/Ocean_Energy/Ireland%E2%80%99s_Wave_Energy_Resource/Wave_Energy_Resource_Atlas_Ireland_2005.pdf>

Maxwell. (2011). In: *Maxwell Technologies*, Mar 2011, Available from: www.maxwell.com

Murray, D. B., Egan, M. G., Hayes, J. G., O'Sullivan, D. L. (2009). Applications of supercapacitor energy storage for a wave energy converter system, Proceedings of 8th European Wave and Tidal Energy Conference (2009), pp. 786-795

O'Sullivan, D., Griffiths, J., Egan, M.G. & Lewis, A.W. (2011). Development of an electrical power take off system for a sea-test scaled offshore wave energy device. *Renewable Energy*, *36* (4). 1236-1244.

O'Sullivan, D. & Lewis, A.W. (2008). Generator Selection for Offshore Oscillating Water Column Wave Energy Converters. *Proceedings of European Power Electronics and Motion Control (EPE-PEMC) Conference*, (2008).

O' Sullivan, D., Mollaghan, D., Blavette, A. & Alcorn, R. (2010). In: *Dynamic characteristics of wave and tidal energy converters & a recommended structure for development of a generic model for grid connection*, Mar 2011, Available from: <www.iea-oceans.org.>

Paul, K., Christian, M., Pascal, V., Guy, C., Gerard, R. & Younes, Z. (2009). Constant power cycling for accelerated ageing of supercapacitors. *Proceedings of 13th European Conference on Power Electronics and Applications* (2009), 1-10.

Polinder, H., Damen, M.E.C. & Gardner, F. (2004). Linear PM Generator system for wave energy conversion in the AWS. *Energy Conversion, IEEE Transactions on*, *19* (3). 583-589.

Raghunathan, S. (1995). The Wells air turbine for wave energy conversion. *Progress in Aerospace Sciences*, *31* (4). 335-386.

Ricci, P., Lopez, J., Santos, M., Villate, J.L., Ruiz-Minguela, P., Salcedo, F. & O.Falcão, A.F.d. (2009). Control Strategies for a simple Point-Absorber Connected to a Hydraulic Power Take-off. *Proceedings of European Wave and Tidal Energy Conference*, (2009).

Santos, M., Salcedo, F., Haim, D.B., Mendia, J.L., Ricci, P., Villate, J.L., Khan, J., Leon, D., Arabi, S., Moshref, A., Bhuyan, G., Blavette, A., O' Sullivan, D. & Alcorn, R. (2011). In: *Integrating Wave and Tidal Current Power: Case Studies through Modelling and Simulation*, Mar 2011, Available from: www.iea-oceans.org

Scottish Govt., Northern Ireland Executive & Govt. of Ireland. (2011). In:*Irish-Scottish Links on Energy Study (ISLES)*, Available from: <http://www.islesproject.eu/>

Setoguchi, T. & Takao, M. (2006). Current status of self rectifying air turbines for wave energy conversion. *Energy Conversion and Management*, *47* (15-16). 2382-2396.

Wavenet. (2003). In:*Results from the work of the European thematic network on wave energy*, Available from: <http://www.wave-energy.net/Library/WaveNet%20Full%20Report%2811.1%29.pdf >

Weeren, D.V., Joosten, H., Scrivens, R. & Schnewly, A. (2006). In: *Ultracapacitors double operational life of wave measurement buoys*, Mar 2011, Available from: <www.maxwell.com>

21

A New On-Board Energy Storage System for the Rolling Stock

Masao Yano
Toyo University,
Japan

1. Introduction

The increasing environmental problems, such as running out of fossil fuels, global warming, and pollution give a major impetus to the development of the on-board energy storage system for the rolling stock, using such as rechargeable batteries and electric double layer capacitors (EDLCs). In addition to the energy saving, an on-board energy storage system enables to realize a contact-wireless type railcar, by which the landscape is improved and the maintenance cost of overhead contact wires is reduced (Ogiwara, 2010).

Fig. 1. Typical commuter electric train in Japan

In case of the system with the catenary wire such as shown in Figure 1, the electrical regenerative braking has reduced total energy consumption in electric railway systems. However, if the energy is not absorbed by another train, the catenary voltage rises and regenerative failure is occurred under DC power feeding system. The train must be decelerated only with a mechanical brake. Therefore on-board energy storage system is a promising tool to prevent regenerative energy failure for rolling stock. As EDLCs have superiorities such as long lifetime and quick charge/discharge capability compared to rechargeable batteries, some systems using EDLC were designed and tested (Sekiyama et al., 2007), (Baklan, 2009), (Drabek & Streit, 2009).
Energy storage devices and their charge/discharge converters are shunt connected to main DC power source in the typical configuration as shown in Figure 2. In this configuration, the regeneration power is absorbed with EDLCs and braking is realized without other energy consuming trains. Figure 3 shows typical traction effort and power vs. speed characteristics in conventional traction systems using AC motors.

Costs and weights of energy storage devices are considerably large. If we fully use the ability of energy storage devices, the increased power and the traction effort can be realized both at powering and breaking. Prof. Sone et al. proposed two methods to realize increased power at high speed region (Sone et al., 2005). In this chapter, these two methods were investigated and compared in detail, and the increased voltage method was adopted preferably. Then the feasibility of the system was verified with simulations and experiments.

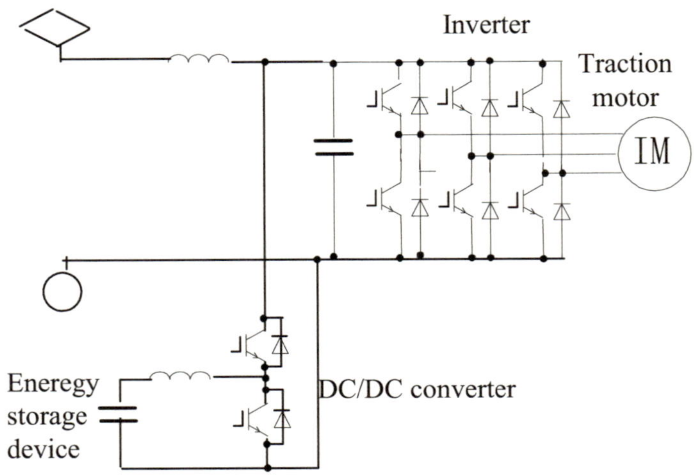

Fig. 2. Shunt connected energy storage device for DC electric rolling stock

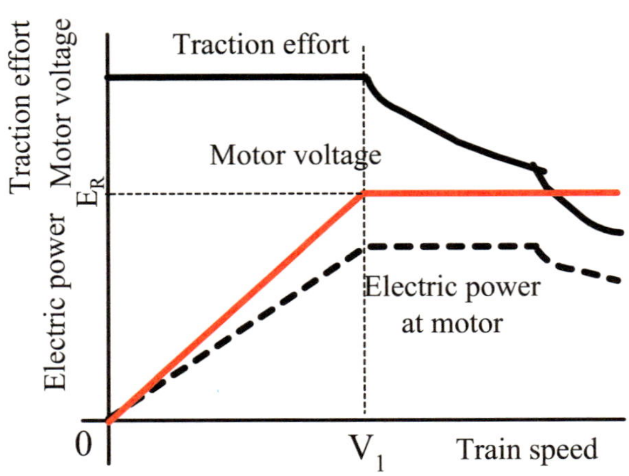

Fig. 3. Traction effort, power and voltage vs. speed characteristics in conventional systems

2. Increased power method at high speed region

Two methods were proposed to realize increased power at high speed region. One is the increased current method, in which the rated motor current is doubled and the motor rated voltage E_m is reduced to half at speed V_1 for example, and the motor voltage is increased to the maximum voltage E_R determined by the catenary voltage at the speed region over V_1 as shown in Figure 4 (Abe et al. , 2006). The circuit configuration of this method is the same as that of the conventional method, shown in Figure 2.

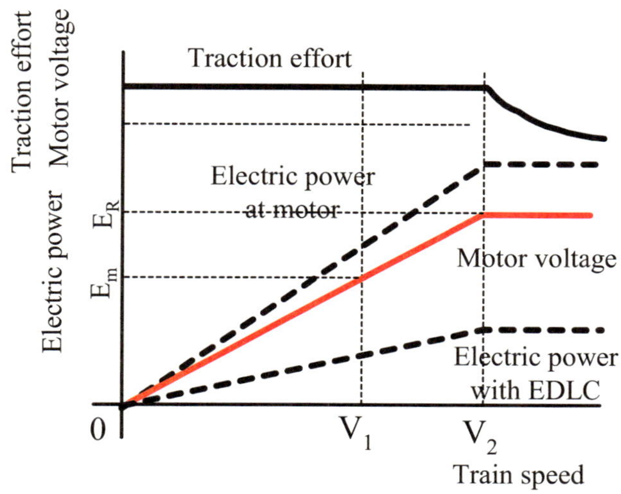

Fig. 4. Traction effort, power and voltage vs. speed characteristics in the system of increased current method

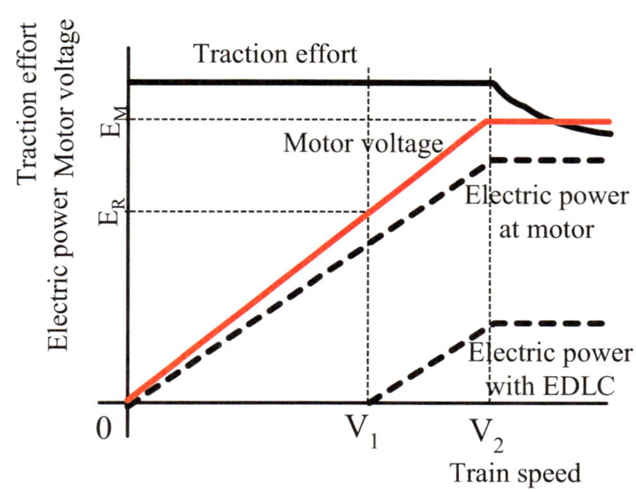

Fig. 5. Traction effort, power and voltage vs. speed characteristics in the system of increased voltage method

The other is the increased voltage method, in which the output voltage of the energy storage devices is added to the voltage generated by the overhead source at the speed region over V_1. The motor voltage is increased, and increased power and traction effort are realized both at powering and breaking as shown in Figure 5. In this system, increased power of the traction motor is realized by increased voltage rather than increased current, and for DC side, increased power is fed from or stored to the energy storage devices, without increased catenary current. In the published papers (Yano et al., 2007), (Sone et al., 2005), a new system of increased voltage method using interface transformers between auxiliary inverters and motor windings was proposed. As we could dispense with these transformers with further studies, we developed the new energy storage systems, using EDLCs connected in series to main motor windings with such a simple circuit configuration as shown in Figure 6 (Yano et al., 2009).

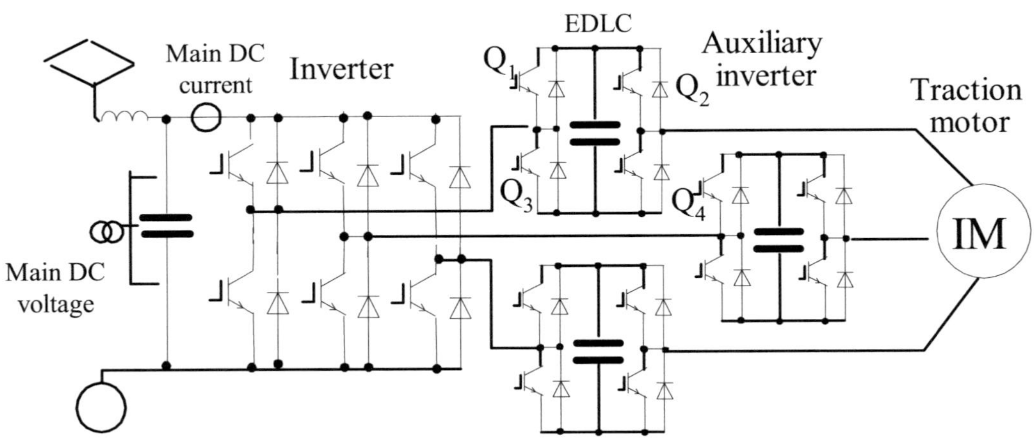

Fig. 6. Circuit configuration of energy storage system connected in series to main motor windings

3. System comparison

In the new systems of increased power with energy storage devices following the operating pattern shown in Figure 4 and Figure 5, the maximum speed V_2 of increased power depends on the output/input power of the DC/DC converter and the auxiliary inverter respectively. Thus the capacities of the DC/DC converter and the auxiliary inverter are selected, considering the desirable maximum speed V_2. For simple estimations, V_2 is selected to be twice V_1, and TABLE 1 shows evaluation results among three methods.

In the increased voltage method, the capacity of the auxiliary inverter is the same as the main inverter, and we can dispense with DC/DC converters which are required in the conventional method and the increased current method. In the current increased method, the motor of high current capacity is required. On the other hand, the power and the voltage of the traction motor are increased to be twice those in conventional systems in the voltage increased method. However the magnetic capacity of the motor (the product value by voltage and time) is the same value as that in the conventional method, and the physical scale of the motor does not increase considerably. For this reason, we prefer the voltage increased method rather than the current increased method.

	Conventional method	Increased current method	Increased voltage method
Traction effort at V_2	Half of rated torque	Rated torque	Rated torque
Traction motor	Rated voltage, rated current	Rated voltage, twice rated current	Twice rated voltage at V_2, rated current
Main inverter	Rated voltage, rated current	Rated voltage, twice rated current	Rated voltage, rated current
Auxiliary inverter	None	None	Rated voltage, rated current
EDLC charge circuit	Rated voltage, rated current	Rated voltage, rated current	None
System elements such as wires	Rated voltage, rated current	Rated voltage, twice rated Current	Twice rated voltage, rated current

Table 1. System comparison

4. Circuit configuration and control

The voltage increased method was adopted. The performance and feasibility are verified by simulation case studies and experiments under several conditions. For experiments, a 3 kVA inverter with the carrier frequency of 7 kHz, which consists of IGBTs, EDLCs, and the digital control system of dSpace's rt1103, was constructed and tested.

4.1 Circuit configuration

The basic configuration of the studied system is shown in Figure 6. In the regular mode, the output voltages of the auxiliary inverters (3 single-phase inverters) are added to the output voltage of the main inverter. In the bypass mode (Q_1 and Q_2:on, Q_3 and Q_4: off), the output voltages of the auxiliary inverters are kept zero, thus the traction motors are fed only with the main inverter.

In the experiments and simulations, an induction motor with ratings of 2.2 kW, 200 V, 9.4 A, 50 Hz and 1430 rpm was used, and 3 banks of 0.93 F EDLC were respectively used as energy storage devices of 3 phases. In each bank, 60 small cell capacitors (2.3 V, 56 F) were connected in series.

4.2 Control system

The basic block diagram of the main circuit and control is shown in Figure 7. A standard indirect coordinate-vector controller was adopted for the main PWM inverter (Yano & Iwahori, 2003), and the voltage command signal of the controller is used for the control of the auxiliary inverter as well. The output signal of the EDLC controller is multiplied with the voltage command signal of the vector controller and the share ratio of the burdens between the main inverter and auxiliary inverters is adjusted. Control circuits of initial charging of EDLC banks and voltage balancing among 3 EDLC banks were developed and implemented as well.

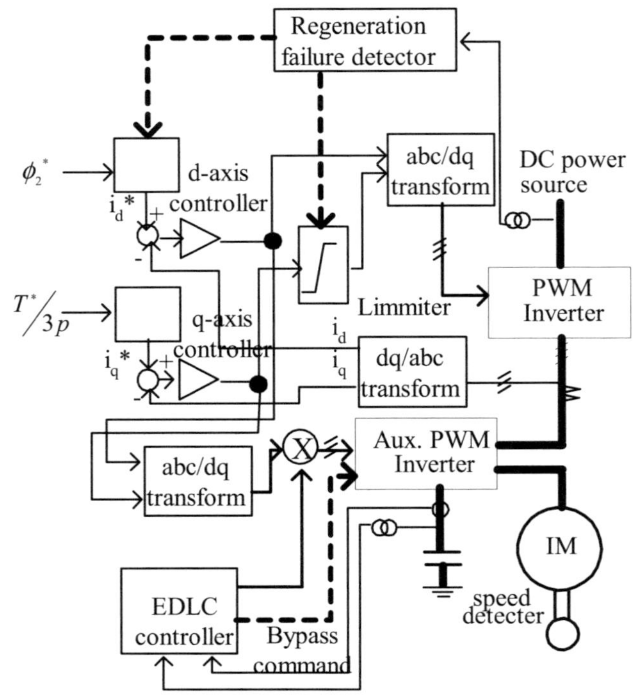

Fig. 7. Control block diagram

5. Simulation case study

Simulations using MATLAB/SIMULINK were carried out to prove the validity of the proposed method and to compare the method with conventional methods. The basic configuration of the studied system and system parameters are shown in Figure 6, and TABLE 2.

Main DC circuit voltage	300 V
Main IGBT inverter	3 kVA, carrier frequency= 7 kHz, operation frequency= 0 - 122 Hz
Auxiliary IGBT inverter	3 sets of 1 kVA, carrier frequency= 7 kHz, operation frequency= 0 - 122 Hz
EDLC	3 sets of 0.93 F, 60 cell capacitors connected in series
Induction motor ratings	2.2 kW, 200 V, 9.4 A, 50 Hz, 1430 rpm, rated d-axis current= 6.62 A, rated q-axis current= 10.5 A
Motor parameters	R_1= 0.881Ω, R_2= 0.658Ω, L_1= 0.815 H, L_2= 0.815 H, M= 0.0786 H
Inertia of the system	0.025 kg-m^2 at simulations, 28.9 kg-m^2 at experiments
Operations at conventional method	0 – 1430 rpm: constant-torque region, 1430 – 3000 rpm: constant-power region

Table 2. System parameters and motor data

5.1 Simulation results with conventional operating pattern

Figure 8 shows simulation results of the operation following the conventional operating pattern. At the speed weakening region over 1430 rpm, the d-axis current (magnetizing component) is reduced and the motor power is kept constant as 2.2 kW. The voltage of EDLC changes from 200 V to 199 V during discharging time with regular mode, and vice versa during charging time. Only the inertia of the motor is considered in simulations in order to reduce simulation time, and voltage changes of EDCLs are rather small.

Approximately 20 % - 40 % of the total power of the motor is shared by the storage devices. Traction effort vs. speed characteristics of this operation are the same as those of conventional trains, in which the regenerative power is absorbed by other trains through catenaries.

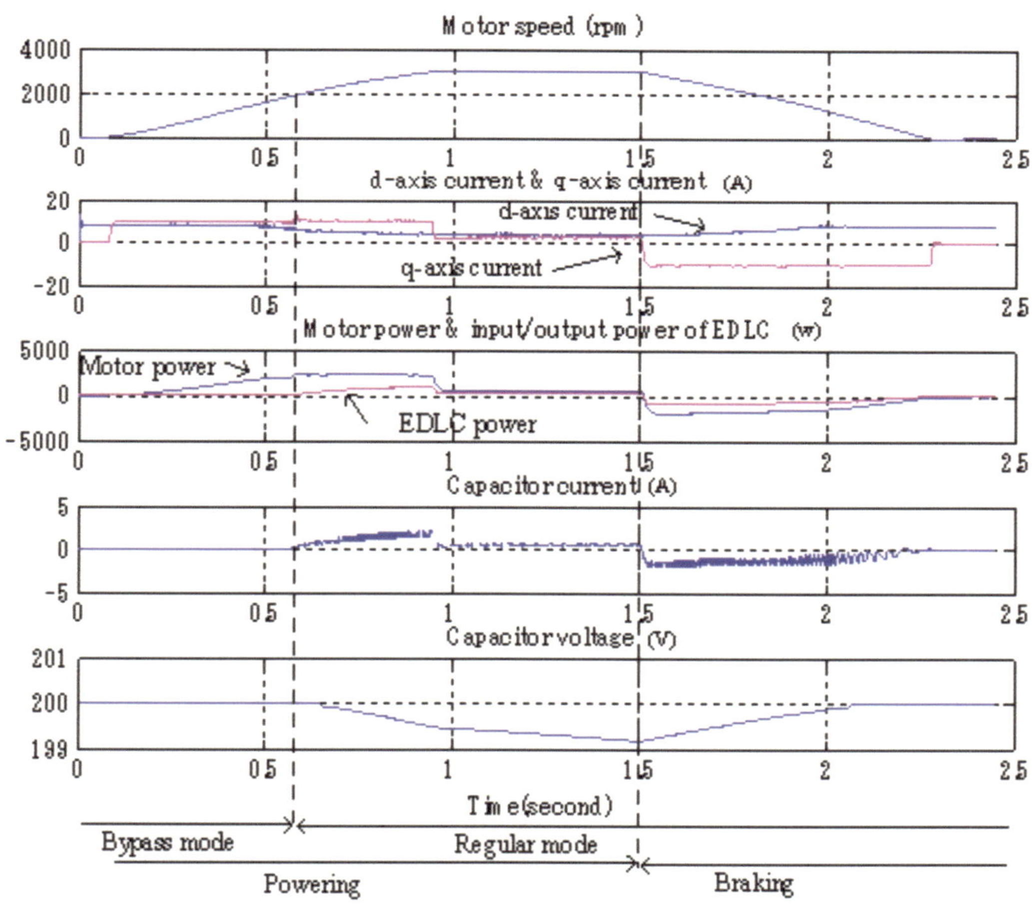

Fig. 8. Simulation wave forms at conventional operating pattern

5.2 Operation of regeneration failure at conventional operating pattern

Figure 9 shows operating results of regeneration failure at conventional operating pattern. Following the detection of regeneration failure due to the absence of other running trains, the regeneration power is only absorbed by on-board EDLCs. Although the regeneration power is reduced considerably, operations are maintained smoothly.

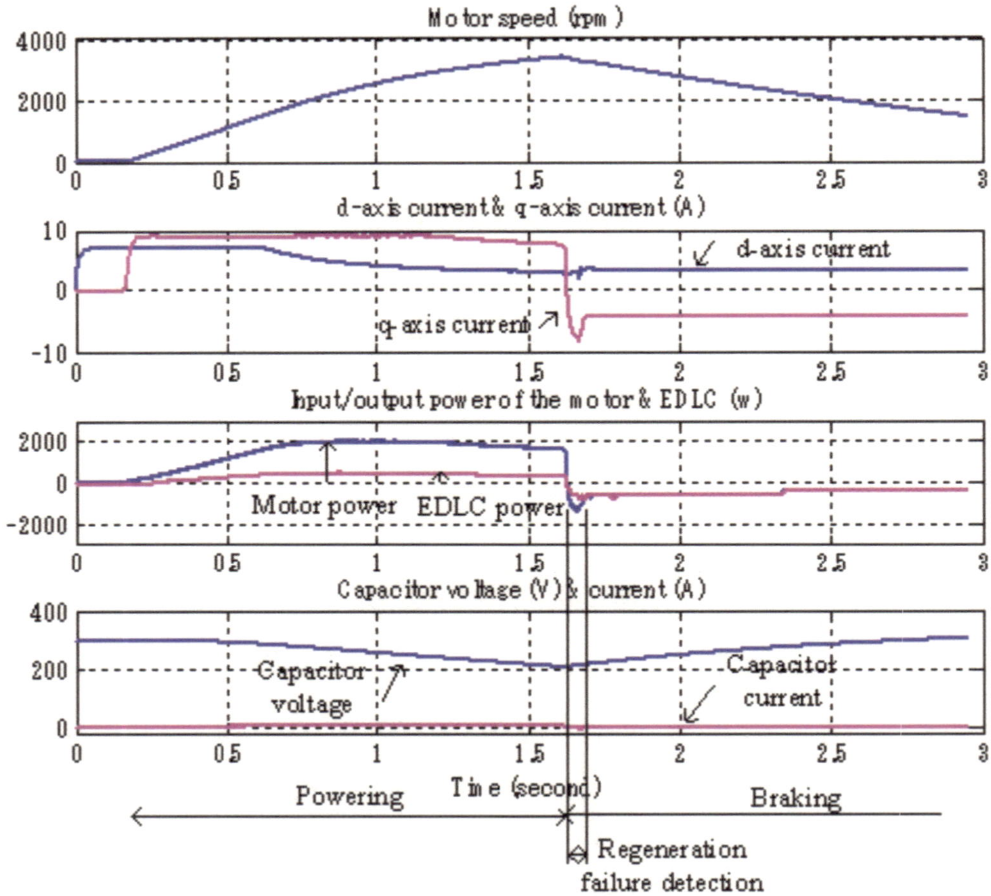

Fig. 9. Simulation wave forms at regeneration failure

5.3 Operation of increased power with energy storage devices

Figure 10 shows simulation results of the operation of increased power with energy storage devices. At all speed regions, the d-axis current (magnetizing component) is kept constant. At the speed region under 1430 rpm, the auxiliary inverter is at bypass mode, and plays no role in energy treatment. At the speed region over 1430 rpm, the auxiliary inverter is at regular mode, and shares the storage and the supply of the electric power, thus maximum motor power increases to 4.4 kW, twice that in conventional traction systems. The capacity of the auxiliary inverter is the same as that of the main inverter.

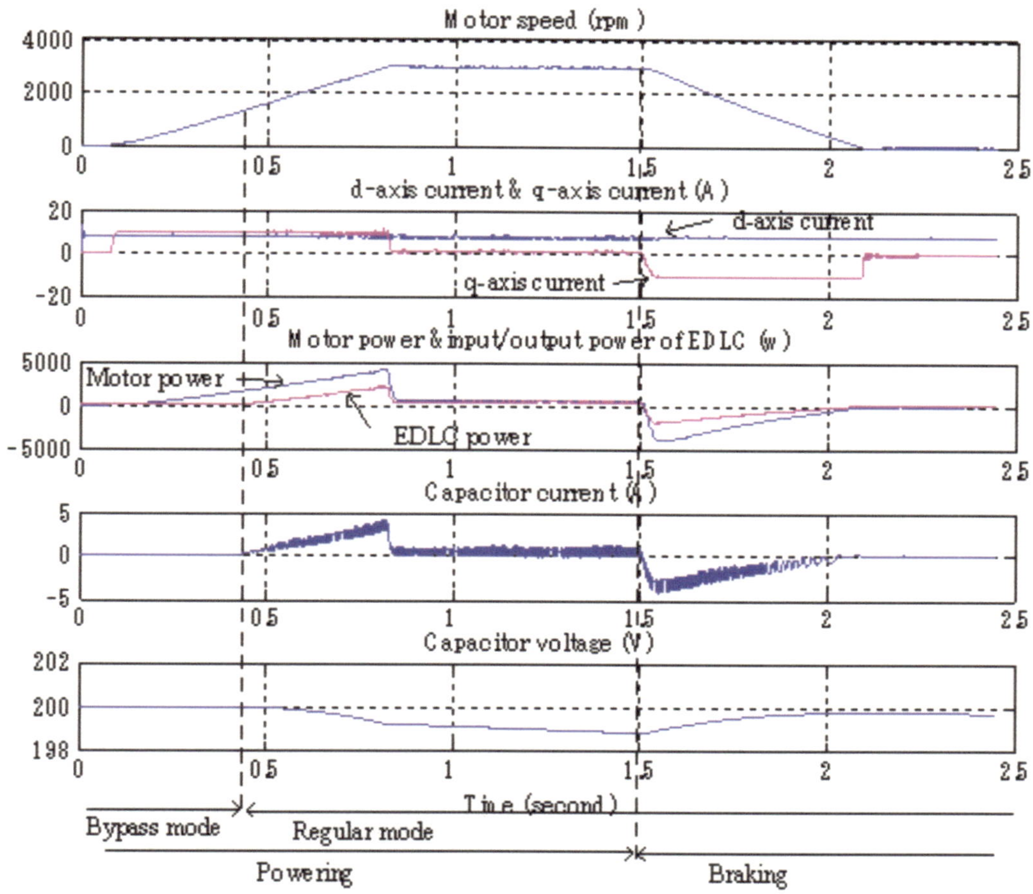

Fig. 10. Simulation wave forms in the new system of increased power with energy storage device

6. Experiments

The mini-model following the voltage increased method was constructed in the circuit configuration shown in Figure 6, and tested.

6.1 Experimental setup

A PWM inverter rated 3 kVA and auxiliary input/output systems for EDLCs, whose parameters are shown in TABLE 2, were designed and constructed. Figure 11 shows the outline of the mini model and the experimental setup. A flywheel shown in Fig. 12 was connected to the main motor to simulate the large inertia of the system. The flywheel, whose diameter and thickness are 50 cm and 30 cm respectively, was made with the iron steel and its inertia was 28.9 kg-m^2. Starting time from standstill to rated speed 1430 rpm is about 295 s. 3 sets of 0.93 F EDLC were respectively used as energy storage devices of 3 phases. In each set, 60 cell capacitors (2.3 V, 56 F) were connected in series.

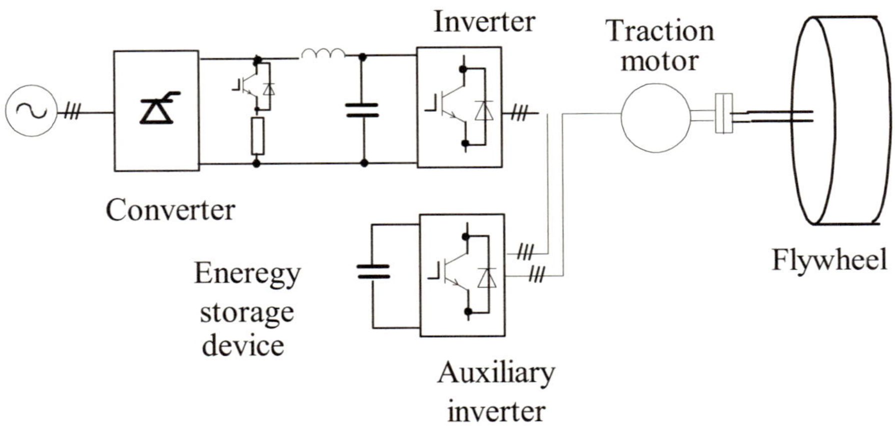

Fig. 11. Mini model and experimental setup

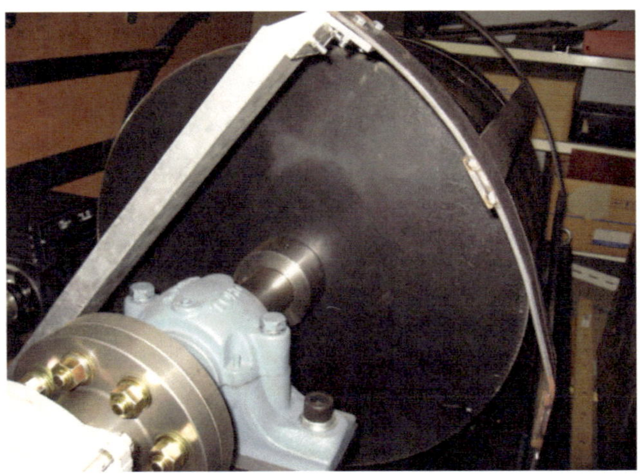

Fig. 12. Fly-wheel

6.2 Starting and low speed operation

Figure 13 shows experimental wave forms in starting from standstill to 40 rpm and braking from that speed to zero as shown as (d). The regular mode took place of the bypass mode about 3 s after starting, but motor currents (c) kept regular sinusoidal wave forms and no transient effects were observed. Torque current (q-axis current) (e) was kept 10.5 A during mode change, and changed to -10.5 A at braking. Magnetizing current (d-axis current) was also observed to be constant during whole operation.

At the mode change, main DC circuit current (b) began to decrease, and discharging current of EDLC (g) began to flow. As EDLCs were divided among each phase, EDLC current showed ripple of twice operating frequency. EDLC voltage (f) decreased due to discharging, but EDLC was not charged significantly during braking, as the regenerative energy was small at low speed operation.

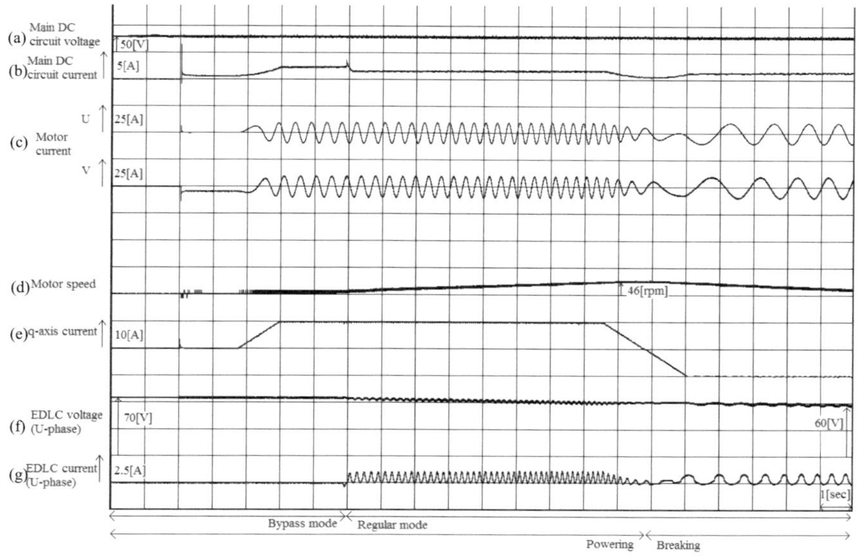

Fig. 13. Experimental wave forms in starting and low speed operation

6.3 Normal speed operation

Fig. 14 shows experimental waveforms at normal speed operation. The bypass mode was changed to the regular mode at the speed of 460 rpm during acceleration with rated torque, and the regenerative braking was added at the speed of 500 rpm. At the mode change, main DC circuit current (b) began to decrease, and the discharging current of EDLC (g) began to flow. EDLC voltage (f) decreased to 47 V from 70 V. Q-axis current of the motor (e) was kept constant, and was shared with the main inverter and the auxiliary inverters. After several seconds, torque command was changed from powering to braking, and EDLC was charged with braking power, thus its voltage increased from 47 V to 85 V. Charge/discharge power was about 200 W and total discharge energy was 3800 J. All operations including starting transients were smooth as expected.

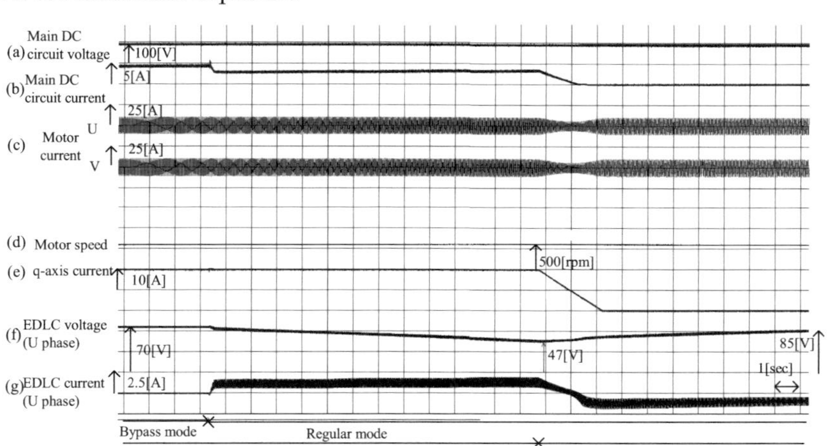

Fig. 14. Experimental wave forms in normal speed operation of the new system

6.4 Initial charging and voltage balancing of EDLC

At the first stage of operations, three banks of EDLC are individually charged in the proper order with the main inverter. In charging of U-phase EDLC, all arms of U-phase auxiliary inverter are off and the inverter works as a rectifier. The other two auxiliary inverters are at bypass mode, thus work as short circuits. Reference signals of output voltages of V-phase and W-phase of the main inverter are kept zero and the charging current is controlled by two U-phase arms of the main inverter. Experimental wave forms in Figure 15 shows that each bank of EDLC was charged with constant current smoothly.

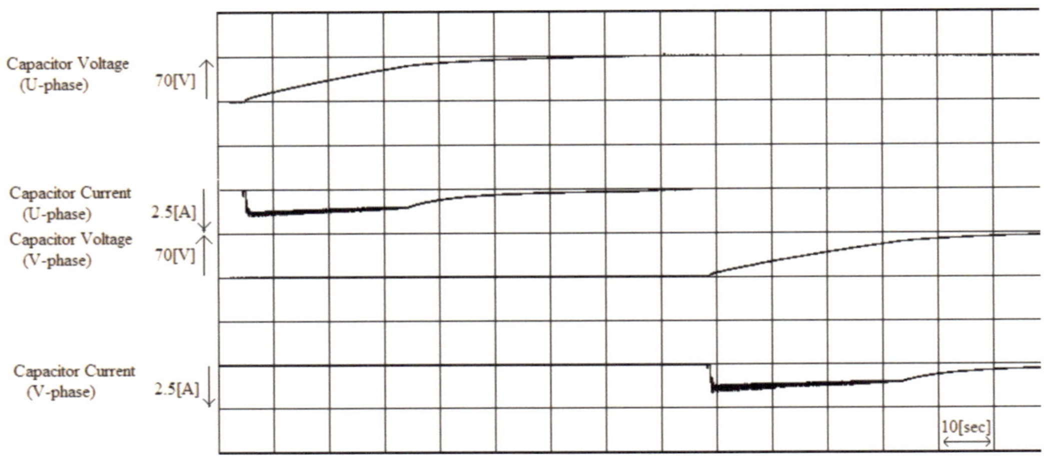

Fig. 15. Experimental results of initial charge operation

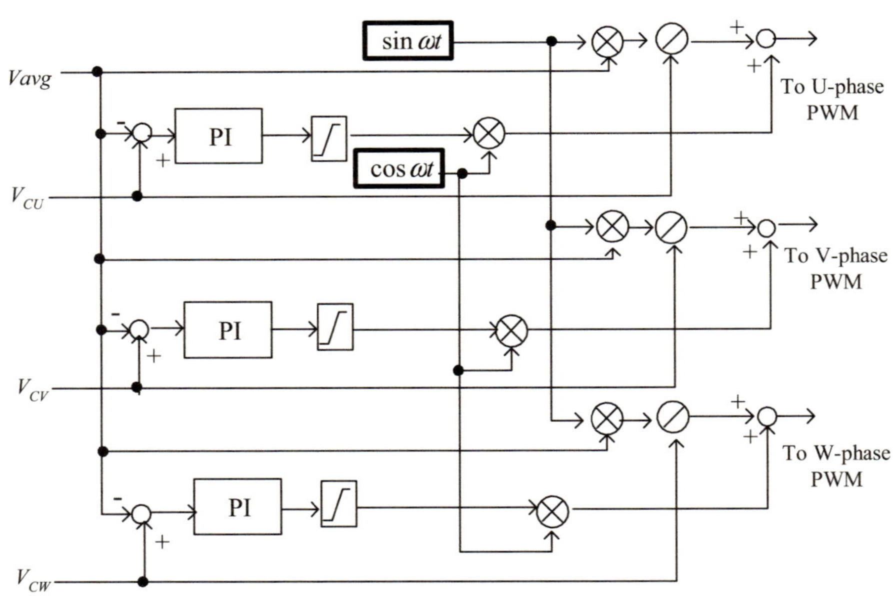

Fig. 16. Block diagram of balancing control

Balancing control of EDLC voltages among three banks takes place both after the initial charging and at the standstill operation. The output voltage of the main inverter is kept to zero and only auxiliary inverters work to equalize three voltages of EDLC banks. Figure 16 shows block diagram of balancing control. Three auxiliary inverters generate three alternative voltage of 50 Hz. Amplitudes of the direct component of the voltage ($\sin \omega t$) are controlled to be the same among three outputs and the quadratic components of the voltage ($\cos \omega t$) are added depending on the difference between EDLC voltage of each phase (V_{CU}, V_{CV}, V_{CW}) and the average value among three banks (V_{avg}). The phase angle of the output voltage for the EDLC bank of higher voltage leads those of other voltages, and energy flows from this bank to other banks, so that EDLC voltages are equalized among three banks. Figure 17 shows experimental results of balancing control. Leading or lagging phase angles were limited to 8°, and currents of charging and discharging were kept within constant values, thus smooth charging and discharging were observed.

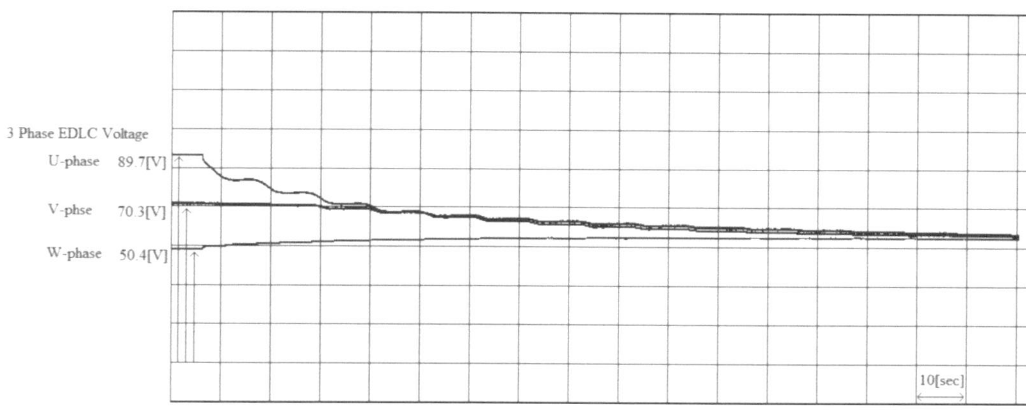

Fig. 17. Experimental results of balancing control

7. Capacity estimation of EDLC

The selection of the capacity of energy storage device depends on many factors such as railway conditions, traction efforts of rolling stock, the setting of desirable maximum power, and so on. Some specifications of on-board energy storage devices were shown in materials (Takahara & Wakasa, 2002), (Technical committee of IEEJ, 2004). Hereafter some estimates are presented referring these materials.

At a standard Japanese commuter 1M1T electric train of 65 t weight including passengers m, 80 % of moving energy is assumed to be charged as electric energy, and the electric energy Q is calculated as follows.

At rated speed of 70 km/h (=19.4 m/s) $v_{\mathbf{rated}}$,

$$Q_{\mathbf{rated}} = 0.8\text{x}0.5\text{x}mv_{\mathbf{rated}}^{\ 2} = 0.8\text{x}0.5\text{x}65\text{x}10^3\text{x}19.4^2 = 10\ [\text{MJ}] = 2.78\ [\text{kWh}]. \quad (1)$$

At top speed of 100 km/h (=27.8 m/s) $v_{\mathbf{top}}$,

$$Q_{\mathbf{top}} = 0.8\text{x}0.5\text{x}mv_{\mathbf{top}}^{\ 2} = 0.8\text{x}0.5\text{x}65\text{x}10^3\text{x}27.8^2 = 20\ [\text{MJ}] = 5.56\ [\text{kWh}]. \quad (2)$$

The energy difference between Q_{top} and Q_{rated} is 10 [MJ]. Figure 18 shows simplified power share of EDLC referring to total motor power. This indicates that rated traction or braking effort can be realized at the top speed, if around a quarter of this energy (2.5 [MJ]), is charged /discharged with on board EDLC. This train is driven with 4 motors of 150 kW, and the catenary voltage is DC 1.5 kV.
At top speed, total traction/braking power is 857 kW(=4x150x100/70) and it corresponds to DC power of 400 A and 2142 V.
Charge energy of 2.5 F EDLC from 1.15 kV to 1.85 kV Q_{cap}, is calculated as follows and meets the requirement.

$$Q_{\mathbf{cap}} = 0.5\text{x}2.5\text{x}\left(\left(1.85\text{x}10^3\right)^2 - \left(1.15\text{x}10^3\right)^2\right) = 2.63\left[\text{MJ}\right] \quad (3)$$

Figure 19 shows the picture of a EDLC case unit of 160 V, 3.7 F and 50 A, and one case unit houses 70 EDLC cells (Meidensya, 2009). Using these case units, 3 sets of 0.85 F, 2.0 kV EDLC can be constructed with 117 case units. 1 set consists of 3 parallel circuits of 13 case units in series. Total estimated weight of EDLC is around 1 t.
The traction effort F can be calculated as follows.

$$F = (1/9.8)\text{x}P_{\text{rated}} / v_{\mathbf{rated}} = (1/9.8)\text{x}600/19.4 = 3160\left[\text{kg} - \text{weight}\right] \quad (4)$$

If the equivalent weight of rotation inertia is assumed to be 9% of total weight m, and the travelling resistance of the train is neglected, the acceleration/deceleration rate $\boldsymbol{\alpha}$ can be calculated as follows.

$$\alpha = 3.6\text{X}9.8F / \left(1.09\text{x}m\text{x}10^3\right) = F/(31\text{x}m) = 1.57\left[\text{km/h/s}\right] \quad (5)$$

$$\alpha = 3.6\text{X}9.8F / \left(1.09\text{x}m\text{x}10^3\right) = F/(31\text{x}m) = 1.57\left[\text{km/h/s}\right] \quad (5)$$

This traction effort and the acceleration/deceleration rate can be realized through all operations. However the friction brake must be used additionally with this installation, when receptive conditions of the catenary are insufficient or lost.

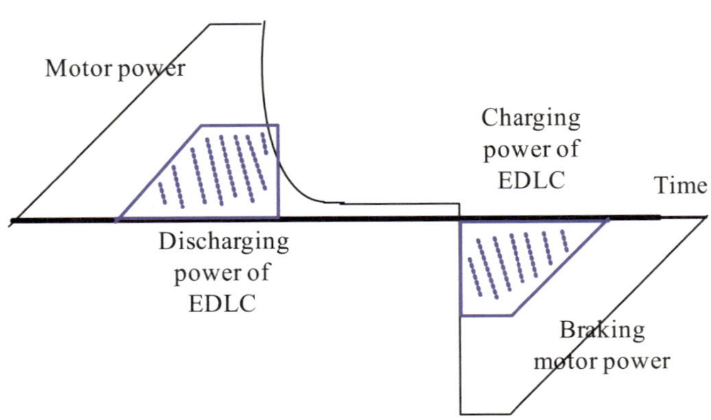

Fig. 18. Total motor power & charging/discharging power of EDLC

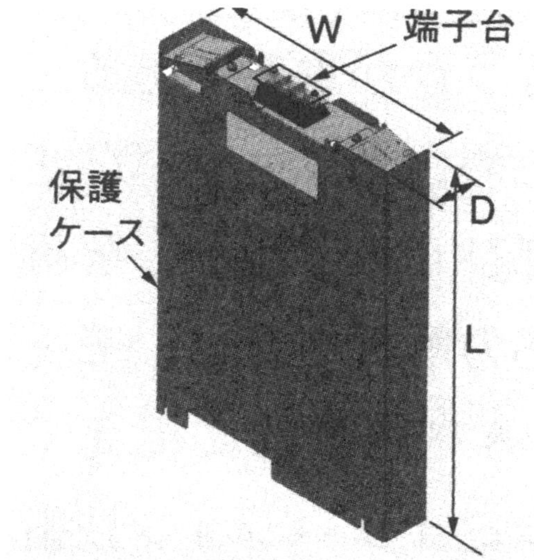

Fig. 19. EDLC case unit-160 V, 3.7 F, 50 A, 8.8 kg, W: 272 mm, L: 387 mm, D: 51 mm

8. Conclusion

A new configuration of an energy storage system, consisting of EDLCs and their input-output auxiliary inverters connected in series to main motor windings was presented, and the superiority of this system was verified by simulation case studies and experiments. This configuration can increase maximum traction and regeneration power substantially without increasing the motor current and the catenary current. Wide assessment, which includes the setting of desirable maximum power, the selection of the capacity of energy storage device, and estimates for cost, weight and size, must be carried out for specified rolling stock. As a result, the adoption of this system to actual electric rolling stock is expected.

9. Acknowledgment

This work has been supported by the Faulty of Engineering, Toyo University, in Japan.

10. References

Abe, T. ; Sone, S. ; Takano, S. et al. (2006). A Study of Improved Regeneration over High-Speed Region by Increased Current Method and Verification by Model Experiment, *IEEJ Annual Conference 2006*, pp. 5-213-214

Baklan, Mark-M. (2009). A Power Electronics View on Rail Transportation Applications, *Proceedings of EPE 2009*, ISBN 978-907-5815-009, Barcelona, Spain, October, 2009

Drabek, P. & Streit, T. (2009). The Energy Storage System for Public Transport Vehicles, *Proceedings of EPE 2009*, ISBN 978-907-5815-009, Barcelona, Spain, October, 2009

Meidensya (2009). MEICAP specification, *Meidenjiho* Vol.1, 2009, p 7

Ogihara, T. (2010). Chapter 6 :Development of contact-wireless type railcar by lithium ion battery, In:*Electric Energy Storage*, InTech, ISBN 978-953-307-058-2

Sekijima, Y.; Inui, M.; Aoyama, I. & Monden, Y. (2007). A trial of regenerated energy storage with an electric double layer capacitor for rolling stock, *IEEJ-IAS Annual Conference 2007*, 1-05-5, pp. I-125-128

Sone, S.; Satoh, T. & Kamiyama, J. (2005). Proposal and Discussion of High-Speed Regenerative Braking - For Realizing Genuine Pure Electric Braking, *The Papers of Technical Meeting on Transportation and Electric Railway, IEEJ*, TER-05-26, pp. 71-74, 2005

Takahara, E. & Wakasa, T. (2002). Electric Double Layer Capacitors for Electric Railway Vehicles, *The Papers of Technical Meeting on Transportation and Electric Railway, IEEJ*, TER-02-33, pp. 27-30, 2002

Technical committee of IEEJ (2004). Power Electronics Applications in Railways, *IEEJ Technical Report*, No. 979, ISSN0919-9195, pp. 64-66

Yano, M. ; Kurihara, M. & Kuramochi, S. (2009). A New On-board Energy Storage System for the Railways Rolling Stock Utilizing the Overvoltage Durability of Traction Motors, *Proceedings of EPE 2009*, ISBN 978-907-5815-009, Barcelona, Spain, October, 2009

Yano, M.; Mizumura, T. & Kuramochi, S. (2007). A New Energy Storage System for Railway Rolling Stock Using Transformers Connected in Series to Motor Windings, *Proceedings of The IEEE International Electric Machines and Drives Conference*, pp. 112-117, Antalya, Turkey, May, 2007

Yano, M. & Iwahori, M. (2003). Transition from Slip-Frequency Control to Vector Control for Induction Motor Drives for Traction Application in Japan, *Proceedings of the IEEE International Conference on Power Electronics and Drive Systems*, pp. 1246-1251, Singapore, November, 2003